T0212673

Lecture Notes in Computer Science 9368

Commenced Publication in 1973
Founding and Former Series Editors:
Gerhard Goos, Juris Hartmanis, and Jan van Leeuwen

Sara Irina Fabrikant · Martin Raubal
Michela Bertolotto · Clare Davies
Scott Freundschuh · Scott Bell (Eds.)

Spatial Information Theory

12th International Conference, COSIT 2015
Santa Fe, NM, USA, October 12–16, 2015
Proceedings

 Springer

Editors

Sara Irina Fabrikant
Department of Geography
University of Zurich
Zurich
Switzerland

Martin Raubal
Institute of Cartography and Geoinformation
ETH Zurich
Zurich
Switzerland

Michela Bertolotto
School of Computer Science and Informatics
University College Dublin
Dublin
Ireland

Clare Davies
Department of Psychology
University of Winchester
Winchester
UK

Scott Freundschuh
Geography and Environmental Studies
The University of New Mexico
Albuquerque
USA

Scott Bell
Department of Geography and Planning
University of Saskatchewan
Saskatoon
Canada

ISSN 0302-9743   ISSN 1611-3349   (electronic)
Lecture Notes in Computer Science
ISBN 978-3-319-23373-4   ISBN 978-3-319-23374-1   (eBook)
DOI 10.1007/978-3-319-23374-1

Library of Congress Control Number: 2015948776

LNCS Sublibrary: SL1 – Theoretical Computer Science and General Issues

Springer Cham Heidelberg New York Dordrecht London
© Springer International Publishing Switzerland 2015

Printed on acid-free paper

Springer International Publishing AG Switzerland is part of Springer Science+Business Media
(www.springer.com)

Preface

Since its inception in Elba (Italy) in 1993 (COSIT was born out of COSIT 0 in Pisa, Italy in 1992), the COSIT biennial conference series (www.cosit.org) has brought together leading researchers from all cognate disciplines reflecting the interdisciplinary breadth of spatial information theory, including (but not limited to) geography, psychology, cognitive science, computer science, information science, and linguistics.

Following the conference on the North Sea coast in Scarborough (UK) in 2013, the 12th COSIT conference returned to the USA for the fifth time. The COSIT 2015 conference was held in Santa Fe, New Mexico, during October 12–16, 2015 in the oldest capital city of the USA, located near the foothills of the Sangre de Cristo Mountains.

We received 52 full papers, which were each thoroughly reviewed by at least three Program Committee members; 22 were selected for presentation at the conference and are included in this volume.

The breadth of the topics in this volume also reflects the breadth of the disciplines involved in fundamental research related to geographic information theory. Excitingly, traditional research topics, such as space-time representations, spatial relations, navigation, (strong) spatial cognition, etc., are still alive and well. Empirical research on how to extract and analyze spatial information from rapidly growing user-generated online multimedia databases, for example, produced in a citizen science context, has clearly emerged as a new and popular research frontier in the field. Meanwhile, "big picture" theories and human behavioral studies have recently yielded fewer contributions (although still represented herein), despite being of great value to this interdisciplinary field.

In addition to the single-track paper session, COSIT 2015 also offered four peer-reviewed workshops and one tutorial before the conference, and a doctoral colloquium after the main conference as in previous years. These events were intended as complementary opportunities to additionally facilitate dialogue across disciplinary boundaries and research expertise. Two keynote speakers, a poster session, as well as social events rounded off the stimulating COSIT 2015 conference activities in the beautiful city of Santa Fe in the U.S. South West, renowned for the natural beauty of its landscape.

Organizing a successful conference is not possible without the commitment, additional effort, and diligent help of many people. We would like to thank the international Program Committee for their timely and thorough reviews and the sponsors and supporters for providing travel support for students and keynote speakers, for supplying materials at the conference, and for supporting social events. Furthermore, the organizers of the workshops, tutorials, and doctoral colloquium contributed an important part of the overall program. We would also like to thank the conference organizing crew for all the hard work in front of and behind the scenes. Our special thanks go to Tumasch Reichenbacher in the Department of Geography at the University of Zurich who efficiently handled proceedings production matters, and Danqing Xiao in the

Department of Geography at the University of New Mexico for setting up and managing conference registration. Finally, we would like to thank the most important people at any conference – those who attended COSIT 2015 to present and discuss their work, and who by so doing demonstrated the continuing strength of spatial information theory as a research field in its own right.

July 2015

Sara Irina Fabrikant
Martin Raubal
Michela Bertolotto
Clare Davies
Scott Freundschuh
Scott Bell

Organization

General Chairs

Scott Freundschuh   University of New Mexico, USA
Scott Bell   University of Saskatchewan, Canada

Steering Committee

Christophe Claramunt   Naval Academy Research Institute, France
Anthony Cohn   University of Leeds, UK
Matt Duckham   RMIT University, Australia
Max Egenhofer   University of Maine, USA
Andrew Frank   Vienna University of Technology, Austria
Christian Freksa   University of Bremen, Germany
Nicholas Giudice   University of Maine, USA
Stephen Hirtle   University of Pittsburgh, USA (Chair)
Werner Kuhn   University of California Santa Barbara, USA
David Mark   State University of New York at Buffalo, USA
Daniel Montello   University of California Santa Barbara, USA
Reinhardt Moratz   University of Maine, USA
John Stell   University of Leeds, UK
Kathleen Stewart   University of Iowa, USA
Thora Tenbrink   Bangor University, UK
Sabine Timpf   Augsburg University, Germany
Stephan Winter   University of Melbourne, Australia
Zena Wood   University of Exeter, UK
Michael Worboys   University of Greenwich, UK

Program Chairs

Sara Irina Fabrikant   University of Zurich, Switzerland
Martin Raubal   ETH Zurich, Switzerland

Program Committee

Thomas Barkowsky   University of Bremen, Germany
John Bateman   University of Bremen, Germany
Kate Beard   University of Maine, USA
Itzhak Benenson   Tel Aviv University, Israel
Brandon Bennett   University of Leeds, UK
Sven Bertel   Bauhaus-University Weimar, Germany

James Pustejovsky	Brandeis University, USA
Jochen Renz	The Australian National University, Australia
Kai-Florian Richter	University of Zurich, Switzerland
Andrea Rodriguez	University of Concepción, Chile
Victor Schinazi	ETH Zurich, Switzerland
Christoph Schlieder	University of Bamberg, Germany
Angela Schwering	University of Munster, Germany
Takeshi Shirabe	Royal Institute of Technology (KTH), Sweden
Barry Smith	State University of New York at Buffalo, USA
John Stell	University of Leeds, UK
Kathleen Stewart Hornsby	University of Iowa, USA
Kristin Stock	University of Nottingham, UK
Thora Tenbrink	Bangor University, UK
Sabine Timpf	University of Augsburg, Germany
David Uttal	Northwestern University, USA
Nico Van de Weghe	Ghent University, Belgium
Maria Vasardani	University of Melbourne, Australia
Jan Oliver Wallgrün	The Pennsylvania State University, USA
Robert Weibel	University of Zurich, Switzerland
Stephan Winter	University of Melbourne, Australia
Diedrich Wolter	University of Bamberg, Germany
Zena Wood	University of Exeter, UK
Michael Worboys	University of Greenwich, UK
May Yuan	University of Texas at Dallas, USA

Doctoral Colloquium Committee

Sarah Battersby	University of South Carolina, USA
Amy Lobben	University of Oregon, USA

Tutorial/Workshop Committee

Michela Bertolotto	University College Dublin, Ireland
Clare Davies	University of Winchester, UK

Additional Reviewers

Frank Dylla	Paola Magillo
Jakub Krukar	Tyler Thrash
Joshua Lewis	Vanessa J. Anacta
Matthew Dube	

Sponsors

University of Saskatchewan Social Sciences Research Laboratories, Canada
University of New Mexico Department of Psychology, USA
University of New Mexico Office of Research, USA
University of New Mexico College of Arts and Sciences, USA

Contents

Formalizing and Modeling Space-Time

Outline of a Formal Theory of Processes and Events, and Why GIScience Needs One

Antony Galton$^{(\boxtimes)}$

College of Engineering, Mathematics and Physical Sciences, University of Exeter,
Exeter, UK
a.p.galton@exeter.ac.uk

Abstract. It has often been noted that traditional GIScience, with its focus on data-modelling functions such as the input, storage, retrieval, organisation, manipulation, and presentation of data, cannot readily accommodate the process-modelling functions such as explanation, prediction, and simulation which it is increasingly acknowledged should form an essential element of the GI scientist's toolkit. Although there are doubtless many different reasons for this seeming incompatibility, this paper singles out for consideration the different views of time presupposed by the two kinds of function: on the one hand, the 'frozen' historical time required by data modelling, and on the other, the 'fluid' experiential time required by process modelling. Whereas the former places an emphasis on events as discrete completed wholes, the latter is concerned with on-going continuous processes as they evolve from moment to moment. In order to reconcile the data-modelling and process-modelling requirements of GIScience, therefore, a formal theory of processes and events is developed, within which their fundamental properties can be made explicit independently of any specific implementation context, and their relationships systematically investigated.

Keywords: Data modelling · Process modelling · Event · Process · Formal theory

1 Introduction

This paper begins (in Sect. 2) with a discussion of the ways in which time enters GIS, through an examination of the various kinds of functions it has been thought desirable for a GIS to perform; here I refer to 'functions' of a GIS in the generic sense of the broad *kinds* of activities that a GIS might enable a user to undertake, as opposed to specific operations such as overlay, interpolation, and generalisation, which are often referred to as GIS functions. I draw a broad distinction between two general classes of functions which, following [26], I call *data-modelling functions* and *process-modelling functions*. These two classes are associated with two distinct ways of viewing time, called 'historical' and 'experiential' after [14], or, perhaps more vividly, 'frozen time' and 'fluid time'.

© Springer International Publishing Switzerland 2015
S.I. Fabrikant et al. (Eds.): COSIT 2015, LNCS 9368, pp. 3–22, 2015.
DOI: 10.1007/978-3-319-23374-1_1

The much-debated question of how data-modelling and process-modelling functions can be integrated into a single system is thus seen to involve as a key component the integration of the two corresponding approaches to time. It appears that while the historical view of time is dominated by events, the experiential view is dominated by processes, and hence a successful integration of the two depends on a correct understanding of processes and events, and how they are related. The main purpose of this paper is, having established the necessity for this in the context of GIS, to undertake the initial development and formalisation of a robust and highly general theory of processes and events. Sect. 3 is devoted to a close analysis of the notions of 'process' and 'event' and the concepts needed to handle their interrelationships. The discusssion is informal, but is backed up by a formal development, outlined in the Appendix. No claims are made for completeness of the theory: it is unashamedly a first step, which will require further detailed elaboration before it can fully serve its purpose as a standard reference benchmark for the proper treatment of time in GIS.

2   GIS Functions, and Two Approaches to Time

What are the functions of a GIS? It is usual to draw a contrast between 'traditional GIS functions' comprising the input, storage, retrieval, organisation, manipulation, analysis and presentation of data, and a range of more advanced capabilities such as explanation, simulation and prediction which engage with the data through some form of theoretical understanding of the real-world systems and processes that the data represent.[1] This contrast has been described in various different ways, all tending to the same (sometimes rather despairing) conclusion that it is high time the data-modelling functions of GIS were integrated with the process-modelling requirements of at least a substantial proportion of GIS users. A representative sample of sources in which such statements can be found is [5–7,12,20,24,26,29,31].

These two sets of functions seem to point to two rather different kinds of system. On the one hand, there are those systems, lying somewhere on a continuum between a digital map and a spatial database, which encompass the descriptive and representational functions of a GIS, and on the other, there are systems which encompass the exploratory functions such as prediction and simulation that are increasingly thought of as natural adjuncts to a GIS. This distinction has appeared in the literature under a variety of different names, including 'map-representation systems' vs 'reality-representation systems' [20], 'data models' vs 'process models' [26], 'history models' vs 'process models' [24], and 'information systems' vs 'modelling systems' [12].

---

[1] The term 'analysis' could perhaps be included with the second set of functions as well: it is a broad term which covers a range of different activities. However, many traditional GIS functions such as interpolation, overlay, and generalisation are often described as 'analytical', and many, though not all, of the functions described by O'Sullivan and Unwin in their book on Geographic Information Analysis [23] belong with the 'traditional GIS functions' rather than the 'more advanced capabilities'.

Traditionally, these two kinds of functions — data-modelling functions and process-modelling functions — have been handled separately, the former by a general-purpose GIS and the latter by a special-purpose system designed to meet the specific requirements of a particular application area such as meteorology, geomorphology, animal ecology, traffic systems, or human population studies. The significant question is how to get the two systems to "talk to each other", so that the results of the calculations performed by the process-modelling system can be made available to be stored, manipulated, and output in human-friendly form by the GIS. Many of the authors cited above have asked this question and explored the ramifications of different ways of answering it; but already 20 years ago, Raper and Livingstone [26] suggested that the question was outdated, and that 'the next step should be the fusion of models and spatial representations within new object-oriented environments and *not* the integration of incompatible systems which force representational compromises'.

Of particular relevance to all this is the role of *time* in geographical representation. Time may, but need not, be involved in spatial data modelling, but it is almost invariably of central importance in spatial process modelling. It is significant, however, that while both kinds of modelling may need to work with time, they work with *different kinds of time*. What do I mean by this?

Considering the data-modelling functions first, it is generally accepted that the basic unit of geographical information is a combination of place (*where?*), time (*when?*), and theme (*what?*): In place p at time t there is X. This is Peuquet's *Triad Framework* [24].[2] The basic schema covers a multitude of variations — for example, p can be a point, a grid square, or a region; t can be an instant, a 'standard' interval such as a calendar month or year, or an 'arbitrary' interval; and X can be a value (of a field), an object (which could be a fixed physical feature, something mobile, a social or political unit, a collective, ...), or a process or event — but in essence a GIS consists of a repository of such triples together with a set of algorithms for manipulating them in accordance with the data-modelling functions listed above.

Time is often said to enter this picture in two different ways [28]. There is the time in which the manipulations referred to above occur, known in the temporal database community as *transaction time*; insofar as it is recorded in the database itself it belongs to the metadata associated with the geographical data, indicating, for instance, when a particular triple was entered into the repository, or when it was superseded. This time is concerned with the history of the database itself, and has nothing to do with the temporal dimension of the geographical reality being described. Much more significant, for our present purposes, is the so-called *valid time*, which is the time referred to by the triples, the time in the real world at which the state of affairs described by the triple actually obtained. This time records the history of the geographical reality that the system is being

---

[2] In some more recent treatments, place and time are amalgamated, and the nature of the theme is made more explicit, as in the *geo-atom* of Goodchild *et al.*, which takes the form $\langle \mathbf{x}, Z, z(\mathbf{x}) \rangle$, where '$\mathbf{x}$ defines a point in space-time, Z identifies a property, and $z(\mathbf{x})$ defines the particular value of the property at that point' [18].

used to record, and the presence of a triple (p, t, X) in the repository amounts to an assertion that there really was X in place p at time t — if this was not the case, then the repository is in error and stands in need of correction.

Turning now to the process-modelling functions, we find a rather different picture. Rather than being a repository of facts and data-processing algorithms, a process-modelling system might be thought of as a repository of *theories*, that is, theoretical models of the laws and regularities that are believed to hold sway in the world. These can be used to *generate* a picture of the world — which usually means a picture of the world as it evolves through time. This is 'spatial process modelling', where I am using this term in a generic sense to cover a multitude of different possible formalisms such as numerical solution of partial differential equations, cellular automata [4,9,30], GeoAlgebra [29], agent-based models [7], geographic automata [32], and doubtless many others as yet unthought of.

Spatial process modelling can be used in several different ways, notably:

- *Prediction*: Starting from known present data X, run the model to predict future data Y.
- *Explanation*: Starting from known past data X, run the model to 'predict' present data Y, comparing the result with known present data Z. If Y and Z agree, the model is accepted as providing an explanation for Z, if not, it is rejected and a new model tried.
- *Retrodiction*: Starting from hypothetical past data X, run the model to predict present data Y — if Y agrees with known present data Z, the hypothesis X is corroborated.
- *Planning*: Starting from hypothetical near-future data X, run the model to predict future data Y. If Y agrees with some desired future goal Z, use X as a plan in order to achieve Z.

It is noteworthy that, throughout these examples, (p, t, X) triples are not being asserted but hypothesised, put up for consideration as possibilities — the grammatical mood here is in effect *subjunctive* rather than, as in the data-modelling system, indicative; and this is related to a key distinction between the ways in which data modelling and process modelling relate to time.

The (valid) time of a data-modelling system is *passive*, exactly comparable to space. Here time is "just another" form of space, another coordinate in the multidimensional presentation of data. Time in such a system is as it were static. Insofar as processes and events are represented in the system, they are portrayed as "frozen" in time, inactive, merely more bits of data. In contrast, the time of a process-modelling system is *active*. When running a simulation, events and processes are *enacted* in symbolic form within the system — this is obvious in the case of a real-time simulation which we see unfolding before our eyes, but even in the case where calculations are performed to derive an end result from data pertaining to some earlier time, simulation is taking place in a more abstract sense, and we can say that here too processes are being enacted. This is "fluid" time in the sense of something that flows, in the way that we customarily (albeit metaphorically) conceive of time as doing, and changes really occur.

The results produced by running a process model in such a system are by nature hypothetical. They are not data in the sense of something *given* from outside (for example, entered by the user in the belief that they truly represent the world out there); rather, they are derived within the system. That is why they can be usefully *compared* with real data, as in the scenarios envisaged above. But time is also fluid for a real-time monitoring system, such as a collection of sensors gathering data about real-world processes to send to a data-modelling system for analysis. From the point of view of the latter, it is immaterial whether the data it is given to work with are factual or hypothetical: the same data-processing techniques can be applied to both — after all, hypothetical data are of little use if they do not resemble factual data to the extent that they can be considered as *possibly* factual.

In process modelling and process monitoring, time is modelled *as* time, that is, the temporal sequence in which the represented events are handled computationally corresponds to the sequence in which they occur in the hypothetical world that is being modelled or the actual world that is being monitored. In the case of process modelling, one might say that this is because that world is *in* the machine, so its time *is* the machine's time. In data modelling, by contrast, the time of the world being modelled is captured, not through temporal sequence in the data processing, but symbolically as values on a coordinate axis, which may be stored and processed in any order, just like the values on the spatial coordinate axes. In process modelling, the processor itself might be said to *experience* the time it is modelling, whereas in data modelling it merely *records* it. In [14], I drew a contrast between the 'experiential' and 'historical' accounts of the world, and related these to the distinction between processes and events, processes being concerned with the low-level goings-on that are the immediate objects of experience, events to more synoptic summaries of salient aggregates of processes as they are recorded in the memory. This distinction closely matches that drawn here between "fluid" time and "frozen" time respectively.

The relationship between processes and events is crucial here: the process-modelling system (or, in a different way, a real-time monitoring system) generates ongoing processes, and information about these processes has to be passed in some form to the data-modelling system. But it is, presumably, the responsibility of the latter to extract from this processual flux those hard nuggets of salience which constitute events, and which from a human perspective represent information rather than mere data.

Simulation systems and real-time monitoring systems are points of contact with fluid time: either the time that actually elapses in the real world (and of course it elapses by virtue of the processes going on in the world — there is no need, here, to invoke a Newtonian notion of absolute time flowing independently of anything that happens), or simulated time, which elapses by virtue of the computations driving the simulation. Our *experience* of the world is like this, too: a direct engagement with fluid temporal processes — hence 'experiential' as the designation for the latter. On the other hand, our representation and reasoning about the world displays a strongly event-oriented bias: events, that is salient discrete "chunks" of happening, which can be labelled as individuals and marshalled into networks of cause and effect, form the basic subject-matter of our temporal discourse, and as such, they, rather than processes, have been the focus of many important proposals for how time should be handled in GIS, here conceived primarily as a data-modelling system [21, 25, 34]. What we most want from a temporal information system is information about *what happens*, and that primarily means information about events. It follows that if a process-modelling system is to be successfully integrated with an event-oriented data-modelling system, then we need a robust account of how processes are related to events.

In the remainder of this paper, I shall start to develop a basic theory of processes and events from an informatic perspective, laying down the fundamentals of a logic of occurrence which underlies the presentation of temporal facts. Whatever other specialised apparatus is employed by a GIS for recording and manipulating such facts, I would maintain that it should be founded on a secure logical bedrock of this kind. Without such a foundation, talk about processes and events will continue to be subject to confusion as different researchers use the terms in their own way without any kind of agreed common standards. One thing that emerges from the work presented below is that even with regard to the simple logic of occurrence, there are already some formidable difficulties which must be overcome by carefully distinguishing different varieties of process and event and how they are related. Either that, or sweep the difficulties under the carpet and risk tripping over them later to fall flat on one's face.

3   Towards a Formal Theory of Processes and Events

A trip is an event, whereas travel is a process. [1]

There is little agreement on how to use the terms 'process' and 'event': Worboys [33] speaks of an 'astonishing variety of usage and definition', and notes that 'One person's process is another's event, and vice versa'. But whereas Worboys, in that paper, declines to pursue the matter further, I believe that we can and should strive to achieve a common understanding of the issues at stake here, which are not just a matter of terminology but strike deeply into the conceptual foundations of how we represent and reason about the world.

One of the problems is that there seem to be two fundamentally different meanings of the word 'process', and I believe that a good deal of the confusion surrounding the use of this term is due to the conflation of these two meanings.

On the one hand, the word 'process' is used to denote an activity that is not intrinsically bounded and which can, at a sufficiently coarse temporal scale, be conceptualised as homogeneous. Processes in this sense include the flowing of a river, cliff erosion along a stretch of coastline, the year-on-year growth of a tree, the gradual encroachment of built-up area into the countryside surrounding a city, the movement of traffic along a street, continental drift, as well as human activities such as walking, talking, eating, swimming, and travelling. This kind of process contrasts strongly with the notion of 'event', which prototypically refers to an intrinsically bounded, discrete occurrence which may, at a sufficiently coarse temporal scale, be conceptualised as point-like.[3] Examples of events in this sense include the collapse of a particular chunk of cliff, the falling of a tree (whether through human or natural agency), the construction (or destruction) of a house, a volcanic eruption or an earthquake, a journey from A to B, and human actions such as a walk from home to the office, utterance of a sentence, eating an egg, or swimming a length.

The second use of the word 'process' is to denote a structured closed routine leading to a specified end point. Processes of this kind typically involve human (or animal) agency, and are often described as 'the process of Xing a Y'. Examples include the processes of making a pot of tea, registering for a conference, booking a train ticket, constructing a window-frame, building a [bird's] nest, and spinning a [spider's] web. Processes of this kind are typically governed by a specific *procedure* which may be specified in advance.[4] Each enactment of the procedure is in fact an event (that is, it is intrinsically bounded and discrete).[5] Moreover, this event is composite: that is, it is built up out of sub-events corresponding to the various phases of the procedure. As such, this kind of process cannot readily be described as homogeneous.

From the above, it is evident that the two kinds of process are radically different from one another. Processes of the first kind are homogeneous (unstructured) and not intrinsically bounded, whereas those of the second kind are structured and intrinsically bounded. On the other hand, it can hardly be regarded as merely coincidental that the same word is used to describe both of them. What is the connection? I shall defer discussion of this until later, but meanwhile, in order to emphasise that there are two very different kinds of phenomenon here, I shall reserve the term 'process' for the first kind — that is, the open-ended ongoing process conceptualised as homogeneous — and call the second kind 'routines'.[6]

Here I want to pick up my earlier use of the phrase 'at a sufficiently coarse temporal scale'. Scale, or *granularity*, is all important here in enabling us to arrive at a worthwhile conception of the relationship between processes and events. One reason for this is that scale is closely related to *aspect*, which is concerned with the different points of view from which one and the same thing can be considered: specifically, in the case of something going on in time (an occurrent), whether we are concerned with the occurrence as a whole, including a beginning and end — and in this case, whether we are primarily interested in the beginning or the end, or perhaps the state resulting once the end has occurred — or, alternatively, with what is going on from moment to moment, how the occurrence presents itself at the point of experience or recording.

---

[3] Cf. [10]: 'An event is an individual episode with a definite beginning and end . . . '.

[4] These are similar to what Aitken and Curtis [3] call Scripts: 'A Script is a typical pattern of events that can be expected to re-occur: "dining in a restaurant" and "brushing one's teeth" being well known examples' (the restaurant example comes from the original exposition of the Script concept by Shank and Abelson [27]).

[5] Cf. [33]: '[C]omputational processes are rather like computer programs, which when executed result in occurrents'. Here it is the program execution itself that is described as an occurrent, not the outputs resulting from it.

[6] In [14], these are called 'open' and 'closed' processes respectively.

To illustrate this with a simple example, consider a succession of bursts of machine-gun fire. Taken together as whole, we might describe this as an event, which begins at the start of the first burst and goes on till the end of the last one. But putting ourselves in the position of someone experiencing this, either as the gunner, his intended victims, or an onlooker, it seems natural to describe it as a process: an ongoing process consisting of one burst after another, with perhaps no indication of when or whether it is going to end. Each individual burst, on the other hand, is clearly an event, with a clear-cut beginning and ending, and the larger-scale process consists of an indefinite number of repetitions of events of this kind.[7] Now turn up the temporal "magnification" to examine the structure of each burst. What it is, is simply a "chunk" of machine-gun fire; here, machine-gun fire is a process. In principle it can go on indefinitely in the same way (hence, unbounded and homogeneous), though in practice any instance of machine-gun fire will have a beginning and an ending, and if we include these in our description of it then what we have is, precisely, a "chunk" of that process; and this is an event. Ignore the beginning and end now, and concentrate on the process of machine-gun fire: on closer examination (i.e., stepping up the temporal magnification again), we see that it consists of a sequence of events, each of which is the firing of an individual cartridge. Each of these firing events can itself be examined more closely to reveal various lower-level processes and events which go to make it up.

This way of looking at the relationship between processes and events can be used to reconcile two rather different views of events that have appeared in the literature. On the one hand, Yuan [35] regards an event as 'a spatial and temporal aggregate of its associated processes', and states that 'a process is measured by its footprints in space and time'. In relation to precipitation, the subject of her case study, she notes that 'an event marks the occurrence of precipitation' whereas 'a process describes how it rains'. Although her understanding of 'process' and 'event' are somewhat different from what is proposed here, the notion that events can be built up (or 'assembled', to use Yuan's word) from processes represents a point of commonality. Contrasted with this is the notion of events as marking points of discontinuity in an otherwise smooth course of history, as expressed, for example, by Langran and Chrisman [21], who portray events as effectively instantaneous transitions between preceding and succeeding states of affairs. Here there is no indication that events may themselves comprise extended episodes within which various processes occur. Our example above, however, shows that these two seemingly very different views of events are quite compatible, and merely reflect different granularities at which events can be portrayed.[8]

---

[7] Note: This must be construed carefully: it is the *type* of event that is repeated, each individual event occurs just once.

[8] It is instructive in this connection to compare Fig. 2 in [21] with Fig. 1 in [35], focussing particularly on the role assigned to the term 'Event' in the two diagrams.

This example has highlighted two general *temporal operators* which can be used to define event-types in terms of processes or vice versa.[9] They are:

- *Chunking*: For a process P, we define an event-type $chunk(P)$, each of whose occurrences consists of P starting, going on for a while, and then stopping. In an information system, an event of type $chunk(P)$ might be constructed by selecting a time-interval $[t_1, t_2]$, and "filling" it with a "texture" corresponding to process P, analogous to selecting a spatial region and filling it with some value such as land-cover type. Note that the *boundary* is not filled: P is not active at the endpoints of an interval on which a chunk of P occurs.
- *Repetition*: For an event-type E, we define a process $rep(E)$, which consists of an indefinite number of occurrences of E in (sufficiently quick) succession. (Here 'sufficiently quick' will depend on the specific event-type involved, and the context in which we are considering it — more on this below).

It should be emphasised here that chunks are to be understood as maximal: as we are using the term, we cannot pick out a day's worth of the Earth's rotation and call this a chunk of rotation.

In the literature, the term 'process' has been used to refer both to processes as we understand them here, and to *chunks* of process. For example, when Yuan [35] describes an event as an 'aggregate of its associated processes', she must mean process chunks, but when she speaks of a process as 'a continuing course of development', the word 'continuing' seems to rule out the chunk interpretation.

What other temporal operations are there? Up to now I may have given the impression that any event-type must be specified as a chunk of some process, but this is not correct. Some events are directly composed of other events, where these constituent events are not sufficiently homogeneous to be regarded as forming a process of type $rep(E)$ for any E. As an example, consider an event in which someone refuels their car: this event consists of a sequence of subevents, namely: drive into the petrol station; if necessary wait in the queue; draw up alongside the petrol pump; switch off the engine, get out of the car, unscrew the cap to the fuel tank; etc., etc. This is, of course, an enactment of a routine, in the sense introduced above. There is no single process of which this event is a chunk. What is needed here is a direct event-composition operator, which combines a sequence of events into a single larger event. It is standard to denote such an operator ';', so the event-type defined as the *sequential composition* of event-types E_1 and E_2 is denoted $E_1; E_2$. If this operator is stipulated to be associative, then any expression of the form $E_1; E_2; \ldots; E_n$ is unambiguous; but as will be shown below, such a stipulation may be problematic, in which case we must distinguish differently-bracketed variants such as $E_1; (E_2; E_3)$ and $(E_1; E_2); E_3$.

---

[9] It is important to note that the general theory has to handle event-types rather than specific unique occurrences. In defining what is meant by a chunk of some process, for example, we are characterising a type of event, not an individual event. There may be many different individual occurrences which come under this description (or only one, or none), whereas an individual event is by nature unique. If we say 'It happened twice' or 'It happened again', by 'it' we can only mean an event-type, of which we are reporting another occurrence.

With events it is natural to consider sequential composition, because events have both beginnings and endings, and therefore one can readily locate the beginning of one event at or just after the end of another, the two together thereby forming a candidate for being considered as constituting a larger event. A process, on the other hand, does not intrinsically have a beginning and an ending; as soon as a process is considered together with its endpoints, it is being treated as an event (that is, a chunk of process). For this reason it does not seem possible to define sequential composition of processes as such. On the other hand, it is natural to consider the *parallel* composition of processes: that is, two processes whose simultaneous operation is regarded as constituting a process in its own right. An example, on a small scale, would be someone driving a car while speaking on their mobile phone — in this case two individually permitted activities become illegal in (parallel) combination — and on a larger scale, the climate becoming both warmer and wetter.

Parallel composition of events is also possible, although it is conceptually more complex, since one has to specify whether the events should begin together, end together, or both — or merely overlap without any coincidence of endpoints. A range of different possible operators might be suggested here; experience with different application contexts might single out some as especially useful.

Various forms of sequential and parallel composition are widely encountered in the literature, forming essential components of algebras or calculi that have been proposed for different purposes. Examples include the event-composition operators used in Active Databases [2,17], Dynamic Logic [19], and Artificial Intelligence [13]. Most such systems only handle events (while sometimes using the term 'process' to denote them, the focus being on routines rather than processes). As a result, the idea of having operators mapping *between* states and processes is more rarely encountered. An exception is in linguistics, where attempts have been made to formalise the semantic relationships amongst different verbs or verb-phrases, and the expression of such relationships through the linguistic phenomenon of aspect (notably perfective *vs* imperfective), as for example 'it is raining' refers to a process but 'it rained three times yesterday' refers to three occurrences of the event-type which we here describe as a chunk of raining. Operators mapping between events and processes or states and vice versa are discussed by, amongst others, [11,22].

Comparable constructs can be found in GIScience, although it is rare to find explicit formalisations of them. Yuan [35] has a section on 'Assembling Events and Processes' which includes a rule that is analogous to chunking (to determine when 'the rain areas in T_1 and T_2 belong to the same rainstorm process'), and a composition rule that builds a precipitation event from overlapping or sequential chunks of precipitation processes which form an unbroken extent. Claramunt *et al.* [8] similarly consider aggregations of process chunks (here just called processes) to form larger 'STP composites'. Worboys and Hornsby [34] discuss the combination of events into 'temporal sequence aggregations', e.g., the sequence of `PlaneLanding` followed by `PlaneTaxiToGate`, followed by `Passenger Deplaning`, but again, no attempt is made to formalise the properties of such sequences. Although the necessity for such operations is frequently acknowledged, and attempts have been made to systematise their definition and behaviour, rigorous formalisations such as we present here seem to be lacking.

The example 'It is raining' cited above leads us to consider a further possible operator, since if we regard the English continuous tense (i.e., the form 'be ...ing') as expressing a process of which the simple form ('It rained') reports an occurrence of a chunk, then it would seem we should be able to apply this to *any* event type to yield a process of which each instance of that event type is a chunk. If we apply this to a routine such as making a pot of tea, we arrive at the idea that each enactment of this routine (that is someone making a pot of tea on one occasion) can be considered to be a chunk of a process called 'making a pot of tea'. In reality, of course, the various stages of making a pot of tea are very different from each other (boiling the water, putting the tea in the pot, pouring the boiling water into the pot, etc.) so we cannot really say there is a single homogeneous activity such that making a pot of tea just involves engaging in that activity for a certain period of time. We can, however, regard the continuous tense as supplying a blanket term to cover all the activities involved at each stage of making a pot of tea, conceptualising them as forming a single notionally homogeneous activity, unified by the fact of their forming part of the larger event: the complete event is then indeed a chunk of that activity. It may be that something like this is at the root of the use of the term 'process' to refer to a closed routine (our second sense above): it refers to the activity involved in executing such a routine, glossing over the fact that this activity may be a complex heterogeneous compound of individually homogeneous subprocesses. This operation, by which a 'higher-level' process is created from an event, can be called *dechunking.*

The operators suggested so far look as though they should form the basis of a formal calculus of processes and events, but the matter is not entirely straightforward. Here I shall discuss some of the issues in an informal way; in the Appendix can be found partial formalisations which highlight where the problems are.

It seems natural to suppose that the *chunk* and *dechunk* operators should be mutually inverse, so that

- for any event type E, $chunk(dechunk(E)) = E$, and
- for any process P, $dechunk(chunk(P)) = P$.

A natural way of defining these two operators is as follows:

- There is an occurrence of event type $chunk(P)$ on interval $[t_1, t_2]$ so long as P is active throughout (t_1, t_2) but not at the endpoints t_1 and t_2.
- The process $dechunk(E)$ is active at time t so long as there is an occurrence of E on some interval $[t_1, t_2]$ such that $t_1 < t < t_2$.

Note the style of these definitions: an event type is defined by providing its *occurrence conditions*, that is, necessary and sufficient conditions that an event of the specified type occurs over a given interval, whereas a process is defined by providing its *activity conditions*, that is, necessary and sufficient conditions that the specified process is active at a given time.

With these definitions we can only obtain the result that *chunk* and *dechunk* are mutual inverses (Appendix, Theorems 3 and 4) if we postulate that:

1. Distinct occurrences of a given event type cannot overlap. In this case we shall call the event type *discrete*.
2. Processes are active on open intervals, i.e., if P is active at time t then there are times $t_1 < t < t_2$ such that P is active throughout (t_1, t_2).
3. A process cannot persist indefinitely into the future or have persisted indefinitely far into the past; that is, at any time there are both earlier and later times at which the process is not active. In this case we shall call the process *locally finite*.[10]

We also have to make a number of assumptions about the temporal ordering; for most purposes it suffices that the ordering is irreflexive, transitive, linear, and dense — but in one place we also need to assume that it is continuous.

Some of these principles are questionable, and we can easily find geographical examples which might lead us to question them. The discreteness principle seems particularly vulnerable. So long as the event type is sufficiently narrowly defined, we can enforce discreteness, but very often if we try to broaden the definition to create a more general event type, overlapping becomes possible. Consider the event type 'flight by A', where A is a particular individual aircraft. Clearly, distinct occurrences of this event type cannot overlap, since an aircraft cannot be engaged in two flights simultaneously. But if we broaden the event type to 'flight by any aircraft', then over the world there are thousands of individual occurrences of this event type in progress at any moment. Similarly, if the event type is 'rainstorm here' (where 'here' denotes any sufficiently small region), then overlapping occurrences are ruled out since it is not possible for two rainstorms to be in progress at the same place at the same time, whereas the broader event type 'rainstorm anywhere' is again one with many temporally overlapping occurrences. If for technical reasons we wish to maintain the discreteness principle, for example in order to preserve the mutually inverse character of the chunking and dechunking operations relating events to the processes they comprise, then we will have to outlaw general event types such as 'flight by an aircraft' or 'rainstorm anywhere'.

Likewise one may wish to question the local finiteness principle for processes. As an example, consider the rotation of the Earth. So long as it has existed, the Earth has rotated, and no doubt it will continue to rotate for as long as it exists. It is of course true that there are times in the distant past when the Earth did not exist, and at these times the Earth's rotation was not active; and in the (we hope distant) future there will be times when the Earth no longer exists. Thus strictly speaking this process is locally (indeed globally) finite; but from the point of view of a GIS, one would naturally wish to ignore this very long-term perspective and treat the rotation of the earth as a constant backdrop to the events and processes one wishes to describe.

---

[10] As distinct from 'globally finite', which would mean there is a time before which the process is never active, and a time after which it is never active.

Turning now to the sequence and repetition operators, we find that they too present problems when we try to formalise them. Consider for example how we might formally specify the conditions of occurrence for an event type of the form $E_1; E_2$ — to be concrete, say, an earthquake followed by a landslide. An occurrence of this event type must consist of an occurrence of type E_1 followed by an occurrence of type E_2. What exactly is meant by 'followed' here? How soon after the earthquake must the landslide occur in order for the two together to count as 'an earthquake followed by a landslide'? The strictest requirement would be that the earthquake must be followed immediately by the landslide; that is, for times t_1, t_2, t_3, the earthquake occurs on the interval $[t_1, t_2]$ and the landslide occurs on the interval $[t_2, t_3]$. But we do not normally insist on this, and indeed it often does not make sense to do so. One of the reasons we may be interested in events of type $E_1; E_2$ is because we are interested in the possibility of a causal connection between the components. Such causal effects may well operate with a delay, which may be rather long in some cases. It is not usually possible to specify a precise upper limit to the length of time that must elapse between an occurrence of E_1 and an occurrence of E_2, though one might specify an imprecise limit and build this into the definition of the compound event type.

There is another difficulty, however. To illustrate this, consider the case:

Here two separate occurrences of E_1 are closely followed by two occurrences of E_2. How many occurrences of $E_1; E_2$ are there? A liberal interpretation might define the occurrence condition for $E_1; E_2$ as follows:

> There is an occurrence of $E_1; E_2$ on interval $[t_1, t_2]$ if and only if there are times $t_1 < t \leq t' < t_2$ such that there is an occurrence of E_1 on $[t_1, t]$ and there is an occurrence of E_2 on $[t', t_2]$.

On this liberal interpretation there are four occurrences of $E_1; E_2$ in the case illustrated above, occupying the intervals $[t_1, t_3]$, $[t_2, t_3]$, $[t_1, t_4]$, and $[t_2, t_4]$, which means that $E_1; E_2$ is not discrete.

If we want to avoid this consequence, we must restrict the occurrence condition for $E_1; E_2$ in some way. If our goal is to secure discreteness then it seems that the least we can do is the following:

> There is an occurrence of $E_1; E_2$ on interval $[t_1, t_2]$ if and only if there are times $t_1 < t \leq t' < t_2$ such that E_1 has an occurrence on $[t_1, t]$ but no other occurrence starting within $[t_1, t_2]$, and E_2 has an occurrence on $[t', t_2]$, but no other occurrence ending within $[t_1, t_2]$.

Under this restricted interpretation, the only occurrence of $E_1; E_2$ in the scenario illustrated above is the one on $[t_2, t_3]$. It can be proved (Appendix, Theorem 6) that thus defined, an event type of form $E_1; E_2$ is discrete so long as E_1 and

E_2 are. This is a desirable consequence if we wish the chunking and dechunking operations to be mutually inverse for all event and process types.

There is a disadvantage to this manoeuvre, however. With our first, simple, definition of $E_1; E_2$ we can prove (Appendix, Theorem 5) that sequential composition is an associative operator, that is, that

$$E_1; (E_2; E_3) = (E_1; E_2); E_3.$$

The advantage of this is that it gives us for free, as it were, a definition of three-fold sequential composition, which we would naturally write as $E_1; E_2; E_3$. The restricted definition of $E_1; E_2$, on the other hand, is not associative. This is easily seen from an example such as the following:

Here, according to the restricted version of sequential composition, there is an occurrence of $E_1; (E_2; E_3)$ on $[t_1, t_2]$, but not of $(E_1; E_2); E_3$. In such a case it is less clear what we might mean by $E_1; E_2; E_3$.

Thus we have a broad interpretation of ';' which is associative but not discrete, and a narrow interpretation which is discrete but not associative. It is natural to ask whether some compromise interpretation can secure both associativity and discreteness, but it would appear that as we move from our broad interpretation in the direction of the narrow interpretation by adding gradually more stringent conditions, we lose associativity before we gain discreteness.[11]

It should be noted, however, that overlapping is not always undesirable — we have already seen that requiring discreteness can restrict the level of generality of allowable event-type descriptions. Another case where overlapping should be allowed concerns repetition. What if we form the sequential composition of an event type with itself? An occurrence of $E; E$ consists of two occurrences of E, one after the other. If now we have *three* consecutive occurrences of E, then the first two constitute one occurrence of $E; E$ and the second and third constitute a second, overlapping with the first. Of course the three occurrences of E together also constitute an occurrence of $E; E; E$. If we allow such composite events, then we can easily define the repetition process, $rep(E)$ as $dechunk(E; E)$ — or if we require a minimum of, say, five occurrences of E to count as a process of repetition, then $dechunk(E; E; E; E; E)$.

In view of these considerations, we should not insist that event types are discrete, say, or that event composition must be associative (and similarly with other properties one may wish to enforce in some cases), but rather use discreteness to define a specific subclass of events, which one may invoke as the occasion demands. Then for example we can state that if E is discrete then so is

---

[11] I have not proved this; it is a conjecture based on experiments with a number of plausible candidate definitions.

$chunk(dechunk(E))$, and that if P is locally finite then $dechunk(chunk(P)) = P$. In other words, the main theorems of the formal theory will be conditional in form. This leaves it open to the user of the theory to decide whether they want all events to be discrete, say, or only events of a certain type. In general, if an event type is precisely defined by means of its occurrence condition, then this is already sufficient to determine whether it is discrete.

A more general moral is this. It seems obviously desirable, for the purposes of building a robust and reliable computational infrastructure for a temporal GIS, to give precise generic specifications of the properties and relationships pertaining to the most basic patterns of temporal phenomena. However, as soon as we try to do this in a rigorous and comprehensive way, we find the enterprise to be fraught with difficulties, and it becomes far from clear exactly how these basic notions should be defined, or what the consequences of defining them in any particular way may be. For this reason, unpalatable as it may be from the perspective of GIS developers eager to proceed quickly to a stage where they can begin working with concrete applications, a good deal of logical or mathematical "spade work" has to be done before we can deliver a product with the required degree of reliability. Here we have been able to describe (and, as outlined in the Appendix, execute) only a small portion of such spade work.

Beyond this, a further requirement is to develop, within the formal framework outlined here, a way of representing causal relations amongst states, processes and events. These causal relations would not in themselves constitute an explanatory theory, but would provide an interface through which the relationships determined by means of causal rules within the process-modelling system can be fed back into the data-modelling system. The basic qualitative causal vocabulary consists of, in addition to 'cause' itself, such terms as 'perpetuate', 'maintain', 'allow', 'enable', 'prevent' and 'disable'. In everyday speech these terms may be used in a variety of ways; for the purpose of a more disciplined treatment we need to determine exactly what relations we want to refer to, and to select appropriate vocabulary to describe them. For more on this see [15, 16].

4   Conclusion

Following a long-standing tradition in GIScience, this paper began by focussing on the distinction between two fundamentally different classes of function that it has been thought desirable for a GIS to accommodate: on the one hand the data-modelling functions that are concerned with tasks such as the input, storage, retrieval, organisation, manipulation, and presentation of data, and on the other hand the process-modelling functions such as explanation, simulation and prediction which, to be successful, must embody a theory of the real-world phenomena which are being modelled.

It was noted that these two kinds of function, and the systems in which they are implemented, treat time in different ways: for the purposes of data modelling, time is regarded as another static dimension like those of space, providing another coordinate dimension for indexing thematic elements. This is what I have called 'frozen' or 'historical' time. By contrast, for the purposes of process modelling, it is necessary to take seriously the dynamic nature of time, enabling data to be collected on the fly or generated hypothetically by means of various kinds of process simulation. This leads to a view of time which I have labelled 'fluid' or 'experiential' time. While historical time places an emphasis on completed events, experiential time is more concerned with ongoing processes.

As has previously been acknowledged, the integration of data-modelling functions and process-modelling functions gives rise to considerable difficulties. There may be many reasons for this, but the one I have focussed on in this paper is precisely the discrepancy between the divergent approaches to time that are required by the two kinds of function, exacerbated by the lack of a principled theory of temporal phenomena from an informatic perspective. In the second part of the paper, the foundations for such a theory were laid down: a clear distinction between events specified by occurrence conditions and processes specified by activity conditions, as well as operations for deriving events from processes (chunking), processes from events (dechunking, repetition), and events from events (sequential composition), thus providing a formal framework within which processes and events can be accommodated within an information system.

The exposition of the formal theory has not addressed the issue of how it might be implemented in a working system. While this may be perceived by some as a weakness, I believe that, on the contrary, it is essential. We are talking about the fundamental structure of some of our most basic temporal concepts; any specific implementation must inevitably include many details, concerning for example the data structures used for representing different elements of the theory, which might be specified in many different ways compatibly with the underlying theory and which, being essentially irrelevant to that theory, would only serve as a distraction and, perhaps, be accorded undue importance. Having been developed in a clean, implementation-independent way, the theory can then stand as a benchmark, or reference standard, against which many different implementations in specific systems may be assessed.

A   Notes Towards a Formal Theory of Processes and Events

All the theorems listed below have been proved, but there is no space here to include the proofs. These may be obtained from the author on request.

We define a many-sorted first-order language with identity, with sorts $\mathcal{P}$ (Processes), $\mathcal{E}$ (Event types), and $\mathcal{T}$ (Time instants). We could have introduced an additional sort for time intervals, but instead we will refer to an interval by means of a pair of instants, representing its beginning and end points.

The primitive predicates are:

– *Active*, of type $\mathcal{P} \times \mathcal{T}$, where $Active(P, t)$ means that P is on-going at t.
– *Occurs*, of type $\mathcal{E} \times \mathcal{T} \times \mathcal{T}$, where $Occurs(E, t_1, t_2)$ means that an event of type E occurs on the interval $[t_1, t_2]$.

– $<$, of type $\mathcal{T} \times \mathcal{T}$, where $t_1 < t_2$ means that t_1 precedes t_2. We assume the ordering $<$ is irreflexive, transitive, linear, and dense; also, in one place, we assume that the order is continuous (a second-order property).

The only axioms we assert here are that the start of an event precedes its end and that processes are active on open intervals:

(**AxOcc**). $Occurs(E, t_1, t_2) \rightarrow t_1 < t_2$
(**AxAct**). $Active(P, t) \rightarrow \exists t_1 t_2 (t_1 < t < t_2 \wedge \forall t'(t_1 < t' < t_2 \rightarrow Active(P, t')))$

We define a number of additional predicates, as follows:

An event-type is **discrete** if distinct occurrences cannot overlap:

$$Discrete(E) =_{\text{def}} \forall t_1 t_2 t_3 t_4 (Occurs(E, t_1, t_2) \wedge Occurs(E, t_3, t_4)$$
$$\rightarrow t_2 \leq t_3 \vee t_4 \leq t_1 \vee (t_1 = t_3 \wedge t_2 = t_4)))$$

A process is **locally finite** if it neither always has been, nor always will be, active:

$$LocFin(P) =_{\text{def}} \forall t \exists t_1 t_2 (t_1 < t < t_2 \wedge \neg Active(P, t_1) \wedge \neg Active(P, t_2)).$$

Subtype: The relation $\sqsubseteq \subset (\mathcal{E} \times \mathcal{E}) \cup (\mathcal{P} \times \mathcal{P})$ is defined by

$$E_1 \sqsubseteq E_2 =_{\text{def}} \forall t_1 t_2 (Occurs(E_1, t_1, t_2) \rightarrow Occurs(E_2, t_1, t_2))$$
$$P_1 \sqsubseteq P_2 =_{\text{def}} \forall t (Active(P_1, t) \rightarrow Active(P_2, t)).$$

Equality for event-types and processes: For $X \in \mathcal{E} \cup \mathcal{P}$,

$$X_1 = X_2 =_{\text{def}} X_1 \sqsubseteq X_2 \wedge X_2 \sqsubseteq X_1$$

Chunking: The function $chunk : \mathcal{P} \rightarrow \mathcal{E}$ is defined contextually, via an occurrence condition for the event-type $chunk(P)$, as follows:[12]

$$Occurs(chunk(P), t_1, t_2) =_{\text{def}} t_1 < t_2 \wedge \forall t(t_1 \leq t \leq t_2$$
$$\rightarrow (Active(P, t) \leftrightarrow t_1 < t < t_2))$$

Dechunking: The function $dechunk : \mathcal{E} \rightarrow \mathcal{P}$ is defined contextually, via an activity condition for the process $dechunk(E)$, as follows:[13]

$$Active(dechunk(E), t) =_{\text{def}} \exists t_1 t_2 (t_1 < t < t_2 \wedge Occurs(E, [t_1, t_2]))$$

Using these axioms and definitions we can prove:

Theorem 1. $Discrete(chunk(P))$.

Theorem 2. $Discrete(E) \rightarrow LocFin(dechunk(E))$.

---

[12] The first conjunct of the definiens is required to ensure that $chunk(P)$ satisfies **AxOcc**.

[13] The legitimacy of this definition depends on the fact, easily proved, that $dechunk(E)$, so defined, satisfies **AxAct**.

The converse of Theorem 2 does not hold: if E has only two occurrences, which overlap, then $dechunk(E)$ is locally finite but E is not discrete.

The next two theorems show that for discrete events and locally finite processes, *chunk* and *dechunk* are mutually inverse.

Theorem 3. $Discrete(E) \rightarrow chunk(dechunk(E)) = E$.

Theorem 4. $LocFin(P) \rightarrow dechunk(chunk(P)) = P$.

Note that even if P is not locally finite we have $dechunk(chunk(P)) \sqsubseteq P$, which holds for any process P.

We define two different flavours of sequential composition operator, which we call *weak* and *strong*. Other definitions are possible.

Weak Sequential Composition:

$$Occurs(E_1; E_2, t_1, t_2) =_{\text{def}}$$
$$\exists t_3 t_4 (t_1 < t_3 \leq t_4 < t_2 \wedge Occurs(E_1, t_1, t_3) \wedge Occurs(E_2, t_4, t_2))$$

The next theorem establishes the associativity of weak sequential composition:

Theorem 5. $E_1; (E_2; E_3) = (E_1; E_2); E_3$.

As a result of this theorem, we can drop the parentheses and write $E_1; E_2; E_3$. As noted in the main text, under Weak Sequential Composition $E_1; E_2$ is not discrete.

Strong Sequential Composition:

$$Occurs(E_1 \hat{;} E_2, t_1, t_2) =_{\text{def}} \exists t_3 t_4 (t_1 < t_3 \leq t_4 < t_2$$
$$\wedge Occurs(E_1, t_1, t_3) \wedge Occurs(E_2, t_4, t_2)$$
$$\wedge \neg \exists t, t' ((t_3 \leq t < t_2 \wedge Occurs(E_1, t, t'))$$
$$\vee (t_1 < t' \leq t_4 \wedge Occurs(E_2, t, t'))))$$

The next theorem establishes that the strong sequential composition of two discrete events is discrete.

Theorem 6. $Discrete(E_1) \wedge Discrete(E_2) \rightarrow Discrete(E_1 \hat{;} E_2)$.

As noted in the main text, the operator $\hat{;}$ is not associative.

For **repetition**, we want to define a process $rep(E)$ which is active during a period in which E repeatedly occurs. The simplest case is where E occurs twice. This can be expressed as an occurrence of the event $E; E$ (but not $E \hat{;} E$, since $E \hat{;} E$ cannot occur). We define:

$$rep(E) =_{\text{def}} dechunk(E; E)$$

This would mean that two occurrences of E suffice for this process to be active. Normally we would expect a larger number (think of our bursts of machine-gun fire). We could arbitrarily decide for some n that we require $rep(E) = dechunk(E; E; \cdots ; E)$, where the right-hand side contains n copies of 'E'. While it would clearly not be feasible to fix an n which will always give satisfactory results, the important thing is that as we increase n we obtain a sequence of processes each of which is special case of the previous one. This is shown by the following theorem:

Theorem 7. $dechunk(E; E; E) \sqsubseteq dechunk(E; E)$.

As well as the indeterminacy as to how many repetitions of E are required before we say that the process $rep(E)$ is active, there is an indeterminacy as to how far apart the individual occurrences of E must be in time. Resolution of both these indeterminacies must depend on the nature of the specific event-type in question and the context in which it is considered.

References

1. Abler, R., Adams, J.S., Gould, P.: Spatial Organization: The Geographer's View of the World. Prentice-Hall International, Englewood Cliffs (1971)
2. Adaikkalavan, R., Chakravarthy, S.: SnoopIB: interval-based event specification and detection for active databases. In: Kalinichenko, L.A., Manthey, R., Thalheim, B., Wloka, U. (eds.) ADBIS 2003. LNCS, vol. 2798, pp. 190–204. Springer, Heidelberg (2003)
3. Aitken, S., Curtis, J.: Design of a process ontology: Vocabulary, semantics, and usage. In: Gómez-Pérez, A. (ed.) Proceedings of the 13th International Conference on Knowledge Engineering and Knowledge Management (EKAW02), pp. 108–113 (2002)
4. Batty, M.: Geocomputation using cellular automata. In: Openshaw, S., Abrahart, R.J. (eds.) GeoComputation, pp. 95–126. Taylor and Francis, London (2000)
5. Bivand, R., Lucas, A.: Integrating models and geographical information systems. In: Openshaw, S., Abrahart, R.J. (eds.) GeoComputation, pp. 331–363. Taylor and Francis, London (2000)
6. Bregt, A.K., Bulens, J.: Integrating GIS and process models for land resource planning. In: Heineke, H., et al. (eds.) European Soil Bureau Research Report No. 4, pp. 293–304. Laboratory of GeoInformation Science and Remote Sensing, Wageningen University, The Netherlands (1998)
7. Brown, D.G., Riolo, R., Robinson, D.T., North, M., Rand, W.: Spatial process and data models: towards integration of agent-based models and GIS. J. Geogr. Syst. **7**, 25–47 (2005)
8. Claramunt, C., Parent, C., Thériault, M.: Design patterns for spatio-temporal processes. In: Spaccapietra, S., Maryanski, F. (eds.) Searching for Semantics: Data Mining, Reverse Engineering, pp. 415–428. Chapman and Hall, New York (1997)
9. Couclelis, H.: Cellular worlds: a framework for modeling micro-macro dynamics. Environ. Plann. A **17**, 585–596 (1985)
10. Reis Ferreira, K., Camara, G., Monteiro, A.M.V.: An algebra for spatiotemporal data: from observations to events. Trans. GIS **18**(2), 253–269 (2014)
11. Galton, A.: The Logic of Aspect: An Axiomatic Approach. Clarendon Press, Oxford (1984)
12. Galton, A.: Dynamic collectives and their collective dynamics. In: Cohn, A.G., Mark, D.M. (eds.) COSIT 2005. LNCS, vol. 3693, pp. 300–315. Springer, Heidelberg (2005)
13. Galton, A.: Eventualities. In: Fisher, M., Gabbay, D., Vila, L. (eds.) Handbook of Temporal Reasoning in Artificial Intelligence, pp. 25–58. Elsevier, New York (2005)
14. Galton, A.: Experience and history: processes and their relation to events. J. Logic Comput. **18**, 323–340 (2008)

15. Galton, A.: States, process and events, and the ontology of causal relations. In: Donnelly, M., Guizzardi, G. (eds.) Formal Ontology in Information Systems: Proceedings of the 7th International Conference (FOIS 2012), pp. 279–292 (2012)
16. Galton, A., Worboys, M.: Processes and events in dynamic geo-networks. In: Rodríguez, M.A., Cruz, I., Levashkin, S., Egenhofer, M. (eds.) GeoS 2005. LNCS, vol. 3799, pp. 45–59. Springer, Heidelberg (2005)
17. Gehani, N.H., Jagadish, H.V., Shmueli, O.: Event specification in an active object-oriented database. ACM SIGMOD Rec. **21**(2), 81–90 (1992)
18. Goodchild, M.F., Yuan, M., Cova, T.J.: Towards a general theory of geographic representation in GIS. Int. J. Geogr. Inf. Sci. **21**(3), 239–260 (2007)
19. Harel, D.: Dynamic logic. In: Gabbay, D., Guenthner, F. (eds.) Handbook of Philosophical Logic. volume II: Extensions of Classical Logic, pp. 497–604. Reidel, Dordrecht (1984)
20. Hazelton, N.W.J., Leahy, F.J., Williamson, I.P.: Integrating dynamic modeling and geographic information systems. URISA J. **4**(2), 47–58 (1992)
21. Langran, G., Chrisman, N.R.: A framework for temporal geographic information. Cartographica **25**(3), 1–14 (1988)
22. Moens, M., Steedman, M.: Temporal ontology and temporal reference. Comput. Linguist. **14**, 15–28 (1988)
23. O'Sullivan, D., Unwin, D.J.: Geographic Information Analysis. Wiley, Hoboken (2003)
24. Peuquet, D.J.: It's about time: a conceptual framework for the representation of temporal dynamics in geographic information systems. Ann. Assoc. Am. Geogr. **84**, 441–461 (1994)
25. Peuquet, D.J., Duan, N.: An event-based spatiotemporal data model (ESTDM) for temporal analysis of geographical data. Int. J. Geogr. Inf. Syst. **9**(1), 7–24 (1995)
26. Raper, J., Livingstone, D.: Development of a geomorphological data model using object-oriented design. Int. J. Geogr. Inf. Syst. **9**, 359–383 (1995)
27. Shank, R.C., Abelson, R.: Scripts, plans, goals and understanding. Erlbaum, Hillsdale (1977)
28. Snodgrass, R.T.: Temporal databases. In: Frank, A.U., Formentini, U., Campari, I. (eds.) GIS 1992. LNCS, vol. 639, pp. 22–64. Springer, Heidelberg (1992)
29. Takeyama, M., Couclelis, H.: Map dynamics: integrating cellular automata and GIS through geo-algebra. Int. J. Geogr. Inf. Sci. **11**, 73–91 (1997)
30. Tobler, W.R.: Cellular geography. In: Gale, S., Olsson, G. (eds.) Philosophy in Geography, pp. 379–386. D. Reidel, Dordrecht (1979)
31. Torrens, P.M.: Process models and next-generation geographic information technology. ArcNews Online, Summer 2009. http://www.esri.com/news/arcnews/summer09articles/process-models.html
32. Torrens, P.M., Benenson, I.: Geographic automata systems. Int. J. Geogr. Inf. Sci. **19**(4), 385–412 (2005)
33. Worboys, M.: Event-oriented approaches to geographic phenomena. Int. J. Geogr. Inf. Sci. **19**, 1–28 (2005)
34. Worboys, M.F., Hornsby, K.: From objects to events: GEM, the geospatial event model. In: Egenhofer, M., Freksa, C., Miller, H.J. (eds.) GIScience 2004. LNCS, vol. 3234, pp. 327–343. Springer, Heidelberg (2004)
35. Yuan, M.: Representing complex geographic phenomena in GIS. Cartography Geogr. Inf. Sci. **28**(2), 83–96 (2001)

Extracting Causal Rules
from Spatio-Temporal Data

Antony Galton[1]([⊠]), Matt Duckham[2], and Alan Both[2]

[1] University of Exeter, Exeter, UK
a.p.galton@exeter.ac.uk
[2] RMIT University, Melbourne, Australia

Abstract. This paper is concerned with the problem of detecting causality in spatiotemporal data. In contrast to most previous work on causality, we adopt a logical rather than a probabilistic approach. By defining the logical form of the desired causal rules, the algorithm developed in this paper searches for instances of rules of that form that explain as fully as possible the observations found in a data set. Experiments with synthetic data, where the underlying causal rules are known, show that in many cases the algorithm is able to retrieve close approximations to the rules that generated the data. However, experiments with real data concerning the movement of fish in a large Australian river system reveal significant practical limitations, primarily as a consequence of the coarse granularity of such movement data. In response, instead of focusing on strict causation (where an environmental event initiates a movement event), further experiments focused on perpetuation (where environmental conditions are the drivers of ongoing processes of movement). After retasking to search for a different logical form of rules compatible with perpetuation, our algorithm was able to identify perpetuation rules that explain a significant proportion of the fish movements. For example, approximately one fifth of the detected long-range movements of fish over a period of six years were accounted for by 26 rules taking account of variations in water-level alone.

1 Introduction

In this paper we address the problem of detecting causality in spatiotemporal data. Broadly speaking one might approach this problem in two different ways, which may be labeled *probabilistic* and *logical*. In a probabilistic approach, such as is exemplified by a substantial body of deep and detailed work associated particularly with researchers such as Pearl [10] and Spirtes *et al.* [11], one looks for patterns of conditional dependence and independence in the data which exhibit the characteristic "signatures" of genuinely causal correlations. A typical outcome from this kind of approach is a list of functional dependencies between the values of observational variables, resulting in statements to the effect that one variable has a specific causal influence on another. These approaches may be described as *data-driven* and in particular are appropriate if one has no prior expectation of the form taken by causal laws.

© Springer International Publishing Switzerland 2015
S.I. Fabrikant et al. (Eds.): COSIT 2015, LNCS 9368, pp. 23–43, 2015.
DOI: 10.1007/978-3-319-23374-1_2

A logic-based approach, in contrast, is driven by a prior conception of the form that causal laws might take, and the process of inferring laws from data is targeted to the discovery of laws of that form. Experiments performed in accordance with this conception may be thought of as investigating the extent to which the given data can be described in terms of laws of a specified form, as opposed to simply trying to discover general causal connections within the data.

We do not here argue for the merits of either approach over the other; the work reported here takes the logical approach rather than the probabilistic, and may be regarded as an investigation into the feasibility of the former.

Specifically, this paper explores the definition of a logical approach to causal analysis of data, capable of generating causal rules of specific logical forms. In Sect. 2 we introduce causation and perpetuation in the context of the past literature. In Sect. 3 we develop the foundations of logical detection of causal rules, leading to the construction of an algorithm for identifying causal rules from data. The performance of this algorithm is explored first with synthetic data (Sect. 4), and subsequently with real data about the environmental context for the movements of fish in the Murray River, Australia (Sect. 5). A statistical analysis of these resulting causal rules in Sect. 6 demonstrates that the causal rules generated do indeed have explanatory structure in several key respects. Finally, Sect. 7 concludes the paper with a look back at the implications for causation in geographic space.

2   Events, Processes, and Causality

The problem with causality, as highlighted by the philosopher David Hume in the 18th century, is that causation is experientially indistinguishable from correlation; that is, causal relations are not themselves overtly present in data but are manifested through correlations which are. But correlations in data can also arise by chance, unconnected with any of the causal mechanisms that in reality gave rise to the data, leading to the appearance of causal dependencies where none are in fact present. And even when a correlation is not the result of chance, it does not mean that there must be a direct causal connection between the correlates, which may instead be independently caused by some common unobserved third element. Thus the problem for anyone seeking to detect causal relations through the analysis of data is how to separate out the genuine cases of causality from such non-causal correlations.

This problem is particularly acute if one takes the kind of "broad brush" view of causality common to many probabilistic approaches. Probabilistic approaches regard causation, appropriately enough, as a relation between events, but then confuse matters by regarding an "event" as anything one can assign a probability to. This is in stark contrast to our normal understanding of events as things that *happen*, i.e., discrete changes in the world. Events in this sense play a central role in causality, but it is important to recognise other forms of causal relation involving processes or states, which differ from events in the manner in which they occupy time [2,9]. The importance of these temporal distinctions for causal analysis was pointed out in [12,13].

In the work reported here we take a more focused view of the logical structure of causal relations, which is sensitive to the aspectual distinctions between states, processes, and events in a way that probabilistic approaches typically are not. The ontological framework of our research is taken from [7]; here we recapitulate the main features of this approach.

We take the view that in its strictest sense the verb 'cause' should be understood as naming a relation between discrete events: that is, one event (such as a ball hitting a window) may be said to cause another (such as the window breaking). Loosely, we may also speak of one *process* causing another (e.g., the action of the tides causing erosion) but properly considered this is a different relation because it does not happen at a discrete moment but continues, in an ongoing, cumulative fashion, over a period. For this reason we prefer the term 'perpetuation' for this: the action of the tides perpetuates the erosion process.

Relations of causation and perpetuation apply to individual events or processes: a particular ball impact, at a definite location and time, causes a particular window breaking; the action of the tides along a particular stretch of coastline over a particular period perpetuates the process of erosion along that stretch during that period. But our understanding of the world expects such individual instances to reflect general *laws* referring not to individual occurrent tokens but to *types*. It is, however, generally impossible to formulate a valid law of the form 'Any event of type E_1 causes an event of type E_2', since the causation of an E_2 instance by an E_1 instance is typically dependent on some appropriate enabling conditions: for example, if I turn the door handle and push on the door, the door will open — but only if it is unlocked. The importance of such conditions in causality was emphasised by [1], following [4], and indeed has been widely recognised in the philosophical literature on causality (cf., [5]).

In the light of this, we prefer to formulate *conditional* causal rules along the lines of 'If events of types $E_1, E_2, E_3, \ldots$ occur and conditions $C_1, C_2, \ldots$ hold, then an event of type F will occur'. The conditions can in general be modelled as *states* which either hold or not at each time; but such states may record the state of a process variable, e.g., given a process of variation in water temperature, an example of a state might be that the water temperature exceeds 15°C. These are the kinds of conditions we use in application examples below.

The distinctive feature of the work reported in this paper is that we are looking not just for patterns in data that might betray the existence of causal relationships, but patterns that arise from causal laws of specific forms. Our search for correlations is thus guided by the forms of laws that we hope to find. In the next section we describe the algorithm we use for this.

3 An Algorithm for Extracting Causal Rules from Data

3.1 The Data

The algorithm takes as inputs one or more *history files*. A history file records the occurrences of events and the values of process variables at every time-step over some period $T = [0, t_{max}]$, where time-steps are represented as non-negative integers. Events and processes are collectively called *occurrents*: so the set of occurrents, $\mathcal{O}$, may be written as $\mathcal{O} = \mathcal{E} \cup \mathcal{P}$, where $\mathcal{E}$ and $\mathcal{P}$ are the sets of events and processes respectively. Note that $\mathcal{E} \cap \mathcal{P} = \emptyset$.

An event (properly an *event-type*) is represented for the purposes of the algorithm as a function from time-steps to non-negative integers, i.e., for $e \in \mathcal{E}$ we have $e : T \to \mathbb{Z}^+ \cup \{0\}$, where $e(t)$ is the number of distinct occurrences (i.e., *event-tokens*) of e at time t. In any actual application, of course, e will have some semantics specifying its meaning in relation to the application domain; but any such semantics is unknown, and irrelevant, to the algorithm. For many applications the events of interest will be such that $e(t)$ is always either 0 or 1, but this is not invariably the case, and in particular it is not the case for the domain of animal movement we describe later.

Similarly, a process is represented as a function from time-steps to real numbers, so that for $p \in \mathcal{P}$ we have $p : T \to \mathbb{R}$. Typically, though not invariably, they are the discrete-time analogues of continuously-varying real-world functions — for example a process might record the variation in water temperature or water level at a particular station along a river.

3.2   Causal Rules

Amongst events, we regard some as possible *causes* and others as *effects*. These refer to the roles they play in *causal rules*. The most general form of causal rule we handle is:

$$R : [\mathsf{Causes}_R \mid \mathsf{Conditions}_R] \Rightarrow \mathsf{effect}_R \text{ after } \mathsf{Delay}_R,$$

where

- $\mathsf{Causes}_R \subset \mathcal{E}$ is a set of events;
- $\mathsf{Conditions}_R$ is a set of *conditions*, where each condition c is an expression of the form '$v_c^- \leq p \leq v_c^+$', where $v_c^-, v_c^+ \in \mathbb{R}$ and $p \in \mathcal{P}$;
- $\mathsf{effect}_R \in \mathcal{E} \setminus \mathsf{Causes}_R$ is an event distinct from any of the causes;
- Delay_R is a *delay interval* $[d_R^-, d_R^+]$, where d_R^-, d_R^+ are integers such that $0 \leq d_R^- \leq d_R^+$.

In a condition, v_c^- and v_c^+ are the limits of a range within which the value of p_c must fall to satisfy it.

Inclusion of a delay interval does not mean that we are contemplating some mysterious "action at a distance" across time, but simply that the transmission of causal power from cause to effect is mediated by some process that is initiated by the cause and culminates in the effect — e.g., at a traffic intersection, I press the button for the pedestrian signal, and some seconds later (or minutes if I am unlucky) the lights change to enable me to cross.

The causal rule R is *activated* at time t if and only if both:

1. For every $e \in \mathsf{Causes}_R$, $e(t) > 0$.
2. For every $c \in \mathsf{Conditions}_R$, $v_c^- \leq p_c(t) \leq v_c^+$.

An activation of the rule at time t is *explanatory* if the effect predicted by the rule does indeed occur, i.e.:

– For some $d \in \mathsf{Delay}_R$, $\mathsf{effect}_R(t + d) > 0$.

Conversely, an occurrence of effect_R at time t is *explained* by rule R if some activation of R is made explanatory by that occurrence of the effect, i.e.,

– For some $d \in \mathsf{Delay}_R$, R is activated at $t - d$.

It is possible for a rule-activation to explain more than one occurrence of its effect, and also for an effect to be explained by more than one rule-activation. These may be regarded as unsatisfactory situations from the perspective of some real-world applications, but in others may be perfectly acceptable, e.g., one and the same environmental event may trigger migration in many individual fish; and migration by a fish may be triggered by two different environmental events each of which would be sufficient on its own to cause it.

From the general form of rule as presented here, a number of special cases can be identified that are of interest. If $\mathsf{Conditions} = \emptyset$, we have an *unconditional* rule, which can be written in simplified form as

$$R : \mathsf{Causes}_R \Rightarrow \mathsf{effect}_R \text{ after } \mathsf{Delay}_R.$$

If $\mathsf{Delay}_R = [d, d]$ we have a *one-delay* rule, which can be written as

$$R : [\mathsf{Causes}_R \mid \mathsf{Conditions}_R] \Rightarrow \mathsf{effect}_R \text{ after } d$$

and in the special case $d = 0$ we have a *simultaneous causation* rule, written

$$R : [\mathsf{Causes}_R \mid \mathsf{Conditions}_R] \Rightarrow \mathsf{effect}_R.$$

For any of these rules it will sometimes be convenient to abbreviate the part before the $\Rightarrow$ as $\mathsf{antecedent}_R$, which is neutral as to its composition out of causes and conditions, e.g.,

$$R : \mathsf{antecedent}_R \Rightarrow \mathsf{effect}_R \text{ after } \mathsf{Delay}_R.$$

3.3   The Problem

The problem which the algorithm is designed to solve may be stated simply as follows: Given a data set in the form described in Sect. 3.1, we seek a set of rules $\mathcal{R}$ which, as nearly as possible, accounts fully for the data, in the following sense:

1. For each $t \in T$ and $R \in \mathcal{R}$, if R is activated at t then it is explanatory, i.e., effect_R occurs after an admissible delay.
2. For each occurrence of each effect f in the data, there is a rule $R \in \mathcal{R}$ which explains it, i.e., $f = \mathsf{effect}_R$ and R is activated within an admissible delay time preceding the occurrence.

These rules can be roughly characterised as "no false positives" and "no false negatives" respectively, though the precise interpretation of these terms in the present context is delicate and will be discussed further below.

With real-world data it is unrealistic to expect to find a rule-set which fully accounts for the data in this sense, which is why we add the caveat 'as nearly as possible' to the problem statement. To interpret this we need to find a measure of *how nearly* a rule-set fully accounts for the data. This is discussed in Sect. 3.4.

3.4   Evaluating a Rule Set

Given some data and a set of rules (however these have been discovered, whether by the algorithm described here or in some other way), we need a principled way of evaluating the rules with respect to the data. For this purpose two commonly used measures are *precision* and *sensitivity*. In general, for a rule of the form 'If P then Q' these are defined as

$$precision = \frac{TP}{TP + FP} \qquad sensitivity = \frac{TP}{TP + FN}$$

where

- TP is the number of *true positives*, i.e., instances satisfying both P and Q,
- FP is the number of *false positives*, i.e., instances satisfying P but not Q,
- FN is the number of *false negatives*, i.e., instances satisfying Q but not P.

Our problem is how to define these quantities for a causal rule of the form

$$R : \mathsf{antecedent}_R \Rightarrow \mathsf{effect}_R \text{ after } [d^-, d^+].$$

In particular, what do we mean by an 'instance'?

In the case of TP, we could take either a *cause-centred* (cTP) or an *effect-centred* (eTP) approach as follows:

- cTP is the number of explanatory activations of R
- eTP is the number of occurrences of effect_R which are explained by R.

In general, these figures will be different. For the other two quantities of interest, it is natural to count FP in the cause-centred way, and FN in the effect-centred way, as follows:

- cFP is the number of non-explanatory activations of R
- eFN is the number of occurrences of effect_R that are not explained by R

We now define *cause-centred precision* and *effect-centred sensitivity* as follows:

$$c\text{-}precision = \frac{cTP}{cTP + cFP} \qquad e\text{-}sensitivity = \frac{eTP}{eTP + eFN}$$

Thus *c-precision* measures what fraction of the rule activations are explanatory, and *e-sensitivity* measures what fraction of occurrences of the effect are explained by the rule.

These definitions can be used to evaluate an individual rule; to evaluate a set of rules $\mathcal{R}$ for the same effect we use

- cTP: the number of explanatory activations of a rule in $\mathcal{R}$
- cFP: the number of non-explanatory activations of a rule in $\mathcal{R}$
- eTP: the number of occurrences of effect explained by at least one rule in $\mathcal{R}$
- eTP: the number of occurrences of effect not explained by any rule in $\mathcal{R}$

The harmonic mean of the *c-precision* and *e-sensitivity* is called the F_1 score and provides a useful single measure against which rules can be ranked:

$$F_1 = 2\left(\frac{c\text{-}precision \cdot e\text{-}sensitivity}{c\text{-}precision + e\text{-}sensitivity} \right).$$

3.5   The Algorithm

The algorithm is presented below as Algorithm 1. Here we give an informal explanation of it to help the reader understand how it works, as well as some pertinent observations. Note that, in the algorithm, we use $\mathcal{F}$ to refer to the set of effects to be explained.

The algorithm is guided in its search for causal rules by the strict form to which any such rule must adhere. For each effect f, and each subset E of the events available to act as causes, we consider whether any of the data for f can be explained by a rule whose cause-set is E.[1] Such a rule could only be activated at those times T_E at which every event in E occurs; we can therefore immediately discard any set E for which there are no such times, along with any supersets of that set (line 5). If on the other hand T_E is non-empty, we need to consider whether each time in T_E is followed by an occurrence of f within $d_{\max}$ time-steps, where $d_{\max}$ is the maximum allowed delay for a rule (set by the user).

Let D_T be the set of all delays d in the range $[0, d_{\max}]$ for which some time in T_E is followed by an occurrence of f after a delay of d time steps. Any causal rule generating some of these occurrences of f from E must have a delay interval encompassing some of the delays in D_T. Hence if D_T is empty we can discard E and all its supersets (line 10).

If E is still not discarded, then we have a set of times T_E at which all the putative causes in E occur, and for each of these times there may or may not be an occurrence of f within a delay in the set D_T. The times for which such an occurrence exists are put in the set T^+, the rest in T^- (line 12).

If T^- is empty, this means that *whenever* all of E occur, f occurs after a suitable delay. Letting d^- and d^+ be the minimum and maximum delays in D_T, we can set up the unconditional rule '$E \Rightarrow f$ after $[d^-, d^+]$' (line 15), and this is guaranteed to generate no false positives for the data, i.e., to satisfy the first condition in Sect. 3.3. (There may of course be false negatives since there may be more than one rule for effect f, with different cause-sets).

---

[1] At line 3 of the algorithm we are required to iterate over the power set of $\mathcal{E}$. Since this leads to combinatorial explosion if $\mathcal{E}$ is too big, we in practice restrict the iteration to subsets of $\mathcal{E}$ up to some predetermined size. In any case we are most likely to be interested in rules with a small number of causes in the antecedent.

Algorithm 1. The rule-detection algorithm

1   Let $\mathcal{R} = \emptyset$;
2   **foreach** $f \in \mathcal{F}$ **do**
3   $\quad$ **foreach** $E \subseteq \mathcal{E}$ **do**
4   $\quad\quad$ Let T_E be the set of $t \in T$ such that $e(t) \geq 1$ for every $e \in E$.;
5   $\quad\quad$ **if** $T_E = \emptyset$ **then** jettison E and all its supersets;
6   $\quad\quad$ **else**
7   $\quad\quad\quad$ **foreach** $t \in T_E$ **do**
8   $\quad\quad\quad\quad$ let D_t be the set of $d \in [0, d_{\max}]$ such that $f(t + d) = 1$;
9   $\quad\quad\quad$ Let $D_T = \bigcup_{t \in T} D_t$;
10 $\quad\quad\quad$ **if** $D_T = \emptyset$ **then** jettison E and all its supersets;
11 $\quad\quad\quad$ **else**
12 $\quad\quad\quad\quad$ Let $T^+ = \{t \in T_E \mid D_t \neq \emptyset\}$ and $T^- = \{t \in T_E \mid D_t = \emptyset\}$;
13 $\quad\quad\quad\quad$ **if** $T^- = \emptyset$ **then** **we have an unconditional rule****
14 $\quad\quad\quad\quad\quad$ Let $d^- = \min(D_T)$ and $d^+ = \max(D_T)$;
15 $\quad\quad\quad\quad\quad$ $\mathcal{R} \leftarrow \mathcal{R} \cup \{[E \mid \emptyset] \Rightarrow f \text{ after } [d^-, d^+]\}$;
16 $\quad\quad\quad\quad$ **else** **we look for conditional rules****
17 $\quad\quad\quad\quad\quad$ **foreach** $p \in \mathcal{P}$ **do**
18 $\quad\quad\quad\quad\quad\quad$ Sort T_E w.r.t. the value of p at each time. Call the sorted list T_E^s;
19 $\quad\quad\quad\quad\quad\quad$ Create a new list T_E^w from T_E^s such that the ith element of T_E^w is the number of elements of T^+ occurring in the subsequence of T_E^s with indices in the range $[i - h, i + h]$ (where h is a pre-determined constant);
20 $\quad\quad\quad\quad\quad\quad$ Now find all maximal subsequences of T_E^w of length greater than $2h + 1$ in which the values are all positive;
21 $\quad\quad\quad\quad\quad\quad$ **foreach** *subsequence covering indices* $i_1, \ldots, i_n$ **do**
22 $\quad\quad\quad\quad\quad\quad\quad$ Let t_0 and t_1 be the i_1th and i_nth elements of T_E^s and put $v^- = p(t_0)$ and $v^+ = p(t_1)$;
23 $\quad\quad\quad\quad\quad\quad\quad$ Let $D_T^p = \bigcup\{D_t \mid p(t) \in [v^-, v^+]\}$ and let $d^- = \min(D_T^p)$ and $d^+ = \max(D_T^p)$;
24 $\quad\quad\quad\quad\quad\quad\quad$ $\mathcal{R} \leftarrow \mathcal{R} \cup \{[E \mid v^- \leq p \leq v^+] \Rightarrow f \text{ after } [d^-, d^+]\}$;
25 Remove from $\mathcal{R}$ any rule that is *covered* (see below) by another rule in $\mathcal{R}$;

If on the other hand T^- is not empty, then not all occurrences of E are followed by f within the acceptable delay time. In this case, we might still find an unconditional rule that admits exceptions (false positives), and if we are interested in these we can relax the condition $T^- = \emptyset$ at line 13. But with this condition in place, we must proceed to the search for conditional rules (lines 16–24). To this end we consider in turn each of the processes available to supply conditions (remember that a condition takes the form $v^- \leq p \leq v^-$, where $[v^-, v^+]$ is the range within which the process variable p must fall for the condition to be satisfied).

Suppose that in fact all the data for f could be accounted for by a single rule '$[E \mid v^- \leq p \leq v^+] \Rightarrow f$ after $[d^-, d^+]$'. Since $T^- \neq \emptyset$, there are occurrences of E that are not followed by f within an appropriate delay. The non-occurrence of f must be explained by the value of p being outside the range $[v^-, v^+]$ at that time. If, therefore, we sort the times in T_E with respect to the value taken by p at those times to give the sequence T_E^s (line 18), marking each time "good" or "bad" according as the effect f does or does not occur then, the "good" times will form a consecutive run within the sequence, with the values of p at the start and end points of this run bracketed by the "true" values v^- and v^+ — and this is the maximal run of consecutive elements within the sequence for which this is the case. If just one rule fully accounts for the data, a close approximation to it (differing only in the precise value-range in the condition) can be discovered by the above procedure.

In general, however, we expect there to be other rules, with the same effect, whose presence prevents the simple procedure above from working, it being unlikely that the "good" times in the value-sorted sequence will form a single consecutive run. In this case, two immediate remedies suggest themselves:

- On the one hand, we could simply take as our v^- and v^+ the smallest and greatest values attained by p on T^+. The resulting rule is guaranteed to exclude any false negatives, since every actual occurrence of f within the delay range is covered, but it may admit many false positives (the gaps in the sequence of "good" points).
- On the other hand, since there is no single run of consecutive "good" values, we could look for *all* such runs in the sequence and construct a new rule for each, using the extreme p values within that run as our v^- and v^+ for that rule. This method will create a set of rules for f which are guaranteed to exclude false positives (since none of the rules will be activated at any of the "bad" points) but at the cost of a proliferation of rules each of which allows many false negatives.

The method actually used in the algorithm is a compromise between these two approaches, and is found in practice to generate rules with fewer false positives than the first and fewer false negatives than the second.

What we do is to run a sliding "window" of length $2h + 1$ along the sequence T_E^s (with suitable adjustments for the first and last h positions in the list), recording in T_E^w the total numbers of "good" points within the window at each position (line 19). This achieves a smoothing effect on the sequence, allowing us to identify the ranges of values for p within which the occurrence of the effect is more frequent than elsewhere. These show up as maximal runs of positive values in T_E^w; they are the ranges we use in the conditions for rules (lines 20–24).

Finally, having collected a set of rules for effect f, we discard any which are superfluous because they are covered by other rules in the set (line 25). Rule R_1 *covers* rule R_2 with respect to the data so long as $\mathsf{effect}_{R_1} = \mathsf{effect}_{R_2}$ ($= f$, say), every occurrence of f in the data that is explained by R_2 is also explained by R_1, and every non-explaining activation of R_2 is also an activation of R_1. In this case R_2 is superfluous and can be dropped from the rule-set.

It should be noted that the algorithm, as currently constituted, can only generate rules with $|\mathsf{Conditions}_R| \leq 1$.

4   Working with Synthetic Data

The algorithm was first tested on synthetic data sets, generated using artificial causal rules. This form of synthetic data set enabled investigation of how well the algorithm could retrieve known rules from the data. For this it was necessary to: (a) define occurrents to feature in the antecedents of the rules; (b) generate histories for those occurrents over an adequate number of time-steps; and then (c) determine the activation history for each rule and thereby generate histories for the effects of the rules. The occurrent histories and effect histories were then used as inputs to the rule-detection algorithm.

Several types of occurrent were defined, as follows:

1. Events:
   - Periodic events, specified by the number of time-steps from one occurrence to the next
   - Random events, specified by the probability of occurrence at any time-step
2. Processes:
   - Sinusoidal processes, specified by the period
   - Gaussian processes, stipulated to have mean 0 and standard deviation 1
   - Markovian processes, in which the differences between the values at consecutive time-steps have a Gaussian distribution with mean 0 and standard deviation 0.1.

4.1   Experiment 1

For this first set of experiments, the occurrents used were those listed in Table 1, and the rules used were those listed in Table 2. It will be noted that some of the occurrents did not feature in any of the rules. This does not mean that they played no role in any of the experiments with this rule-set. They were available to the rule-detection program, which could therefore look for rules featuring these occurrents. Thus these occurrents acted as "red herrings", and indeed it will be seen from the results in Table 3 that in two cases the best rules found did feature occurrents from this set (pGauss2 and pMarkov2).

Three runs were performed with this set of occurrents and rules, each with 1000 time-steps. For each rule found by the algorithm, the *c-precision* and *e-sensitivity* were computed, and from these the F_1 score was derived. For each effect, the rule with highest F_1 score is reported in Table 3. It will be noted that the rule for effect f1 is deterministic, since the antecedent is a periodic event, with the same occurrences in each run, and the delay interval is a single point. This does not mean that the detection program has an identical task for this effect in all three runs, since it does not "know" in advance that it was generated by a rule of that type, and therefore is also looking for rules whose activations may differ between the runs. None the less, both for this effect and the non-deterministic f2 and f3, the program reliably found the correct rule on each occasion; these are, of course, the effects generated by unconditional rules.

With the conditional rules, the program had a less easy time of it, but still managed to find good approximations to the "true" rules in almost every case.

Table 1. Occurrents used in Experiment 1

Occurrent	Type	Parameters
pSineHigh	sinusoidal process	period 10
pSineMedium	sinusoidal process	period 18
pSineLow	sinusoidal process	period 32
pGauss1	Gaussian process	
pGauss2	Gaussian process	
pMarkov1	Markovian process	
pMarkov2	Markovian process	
ePeriHigh	periodic event	period 9
ePeriMedium	periodic event	period 24
ePeriLow	periodic event	period 50
eRandomHigh	random event	probability 0.4
eRandomMedium	random event	probability 0.25
eRandomLow	random event	probability 0.1

Table 2. Rules used to generate data for Experiment 1, with the number of times that each was activated in 1000 time steps for runs 1, 2, and 3.

Rule	Activations
ePeriLow $\Rightarrow$ f1 after 5	20,20,20
ePeriMedium $\Rightarrow$ f2 after $[3,6]$	42,42,42
ePeriHigh, eRandomHigh $\Rightarrow$ f3 after $[0,4]$	42,57,54
[eRandomHigh $\mid$ $0 \leq$ pSineMedium ≤ 1] $\Rightarrow$ f4 after 2	200,238,226
[eRandomMedium $\mid$ $0 \leq$ pSineHigh ≤ 1] $\Rightarrow$ f5 after $[0,3]$	162,149,152
[ePeriHigh, eRandomHigh $\mid$ $0 \leq$ pGauss1] $\Rightarrow$ f6 after 3	21,26,27
[ePeriMedium, eRandomMedium $\mid$ $0 \leq$ pMarkov1] $\Rightarrow$ f7 after $[2,4]$	4,7,9

The exceptions were for f7 in runs 1 and 3, where the program identified the correct causes, but mistook the conditions, attributing the effect to conditions involving the processes pMarkov2 and pGauss2 which in fact figure in none of the correct rules. The poor performance for this effect can be explained by the fact that in each run it occurred at less than 1 % of the time-steps, so there was not enough relevant data for the rule-detection algorithm to work on, with the result that a spurious non-causal correlation happened to provide a better fit to the data than the best approximation to the true rule discoverable by the algorithm.

In summary, these and other experiments performed with synthetic data demonstrated that the algorithm was, in most cases, able to retrieve from synthetic data the causal rules that generated it. In instances where the algorithm failed, there were frequently plausible explanations for that failure, such as a lack of relevant data generated by a specific rule.

Table 3. The best rule discovered by the program for each effect in each run of Experiment 1, with *c-precision* and *e-sensitivity* (expressed as percentages)

Run	Best rules found	CP, ES
1	ePeriLow ⇒ f1 after 5	100, 100
2	ePeriLow ⇒ f1 after 5	100, 100
3	ePeriLow ⇒ f1 after 5	100, 100
1	ePeriMedium ⇒ f2 after $[3, 6]$	100, 100
2	ePeriMedium ⇒ f2 after $[3, 6]$	100, 100
3	ePeriMedium ⇒ f2 after $[3, 6]$	100, 100
1	ePeriHigh, eRandomHigh ⇒ f3 after $[0, 4]$	100, 100
2	ePeriHigh, eRandomHigh ⇒ f3 after $[0, 4]$	100, 100
3	ePeriHigh, eRandomHigh ⇒ f3 after $[0, 4]$	100, 100
1	[eRandomHigh $\mid -0.342 \leq$ pSineMedium $\leq 0.985] \Rightarrow$ f4 after $[0, 4]$	92, 100
2	[eRandomHigh $\mid 0.643 \leq$ pSineMedium $\leq 0.985] \Rightarrow$ f4 after $[0, 4]$	100, 76
3	[eRandomHigh $\mid -0.342 \leq$ pSineMedium $\leq 0.985] \Rightarrow$ f4 after $[0, 4]$	97, 100
1	[eRandomMedium $\mid -0.588 \leq$ pSineHigh $\leq 0.951] \Rightarrow$ f5 after $[1, 5]$	86, 90
2	[eRandomMedium $\mid 0 \leq$ pSineHigh $\leq 0.951] \Rightarrow$ f5 after $[0, 4]$	100, 100
3	[eRandomMedium $\mid -0.588 \leq$ pSineHigh $\leq 0.951] \Rightarrow$ f5 after $[2, 6]$	71, 84
1	[ePeriHigh, eRandomHigh $\mid 0.03 \leq$ pGauss1 $\leq 1.959] \Rightarrow$ f6 after 3	100, 100
2	[ePeriHigh, eRandomHigh $\mid 0.033 \leq$ pGauss1 $\leq 1.96] \Rightarrow$ f6 after 3	100, 100
3	[ePeriHigh, eRandomHigh $\mid 0.065 \leq$ pGauss1 $\leq 2.975] \Rightarrow$ f6 after 3	100, 100
1	[ePeriMedium, eRandomMedium $\mid -4.249 \leq$ pMarkov2 $\leq -0.76] \Rightarrow$ f7 after $[2, 3]$	100, 75
2	[ePeriMedium, eRandomMedium $\mid 0 \leq$ pMarkov1 $\leq 1.761] \Rightarrow$ f7 after $[2, 4]$	100, 100
3	[ePeriMedium, eRandomMedium $\mid 0.482 \leq$ pGauss2 $\leq 1.543] \Rightarrow$ f7 after $[2, 4]$	100, 67

5   Working with Real Data

5.1   Fish Movement Data Set

The real-world data set used for this study was one we had previously worked with, as described in [3]. Lyon and collaborators [8] gathered data on fish movement in the Murray River system in south-eastern Australia. Over 1000 individual fish were tagged with radio transmitters, and their movements were monitored by 18 river-side radio receivers located at strategic positions along the river, which thereby divided the river and its tributaries into 24 zones, labelled a–x (see Fig. 1). The movement of tagged fish between the zones was tracked over a period of six years, during which time a number of environmental

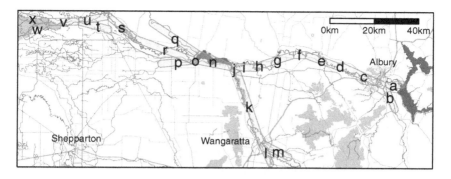

Fig. 1. Map of study area, Murray River, Australia, showing river zones a–x

variables were also monitored, including water temperature, water level, and salinity. The environmental variables were recorded at a coarser spatial granularity than the fish movements, since the recording stations were more widely spaced along the river than the radio receivers: thus the values for these variables in a zone are taken to be those recorded at the nearest station to the zone.

The data thus consisted of records of the following types:

- For each environmental variable, a record of its value at each recording station on each day of the period of study;
- A collection of records of zone-boundary crossings by individual fish, where each record takes the form 'fish i moves from zone z_1 to zone z_2 on day d'.

The aim of our study was to determine to what extent the movement of fish was causally influenced by the variations in the environmental variables.

To this end, fish-movement event types were defined as follows. For each pair z_1, z_2 of adjacent zones, where z_2 is downstream from z_1, the event $z_1 \backslash z_2$ occurs whenever a fish moves from z_1 to z_2, and the event z_2/z_1 occurs whenever a fish moves from z_2 to z_1. Note that it is possible for there to be several occurrences of any one of these events on any given day.

Two sets of experiments were performed using this data, which are reported in the next two sections.

5.2   Experiment 2

For this experiment, we looked for unconditional rules relating fish movement events to a certain set of events defined in terms of the environmental variables. For each environmental variable v and each group of zones G relating to a given recording station for that variable, we defined the event $v3q(G)$ as occurring whenever the value of v recorded at G crossed from the third to the fourth quartile of its range. Thus for example the event $wl3q(cd)$ stands proxy for 'onset of high water level in zones c and d'.

Table 4. The top-ranking rules by F_1 score from Experiment 2 (part)

Rule	F_1
wl3q(efgh) $\Rightarrow$ d\e after $[3, 10]$	0.29
wl3q(cd) $\Rightarrow$ d\e after $[0.10]$	0.26
wl3q(cd) $\Rightarrow$ e\f after $[1, 10]$	0.24
wl3q(efgh) $\Rightarrow$ e\f after $[1, 9]$	0.24
wl3q(ijklm) $\Rightarrow$ i\n after $[0, 9]$	0.24
wl3q(cd) $\Rightarrow$ c\d after $[5, 8]$	0.20
wl3q(cd) $\Rightarrow$ f\g after $[0, 9]$	0.20
wl3q(efgh) $\Rightarrow$ c\d after $[3, 9]$	0.16
wl3q(efgh) $\Rightarrow$ f\g after $[0, 10]$	0.16

The algorithm was asked to look for rules with these quartile-boundary crossing events as causes, and the fish-movement events as effects. We did not pursue this line of enquiry beyond the initial stages as it became clear that the results were somewhat disappointing. Here we present just those results obtained when we looked for rules relating the environmental events wl3q(cd), wl3q(efgh), wl3q(ijklm) and the downstream boundary-crossing events c\d, d\e, e\f, f\g, g\h, h\i, i\n, k\j, m\k. Only 26 rules were found, all with F_1 scores below 30 %. The highest ranking rules, with their F_1 scores, are listed in Table 4.

On the face of it, some of these rules make more sense from a spatial point of view than others. We would expect the strongest causal influence on a fish's movement between two zones to come from the environmental conditions within the zone from which the fish is moving. Thus of the first two rules in the table, the one relating d\e to wl3q(cd) is *prima facie* more "sensible" than the one relating the same effect to wl3q(efgh). In fact the presence of both rules, with comparable F_1 scores, reflects the high correlation between the values of wl3q(cd) and wl3q(efgh) (correlation coefficient 0.9768). This high correlation explains why each rule with the former event as cause is paired with a rule with the latter event, with similar F_1 score. Equally, the *low* correlations between these two values and wl(ijklm) (-0.031 and -0.047 respectively) account for the absence of similar pairings with rules involving that event.

5.3   Discussion

The disappointingly low F_1 scores found in Experiment 2 prompted us to revise our ideas about the kind of causal rule we should be looking for. The initial idea was that initiation of fish movement should be triggered by some environmental event, in accordance with the principle that events are caused by events, so quartile-boundary crossing was used as a way of deriving candidate events from the processes provided in the data. However, there is something rather arbitrary about this choice of events, and coupled with the fact that the crossing of zone-boundaries is also a rather crude proxy for initiation of fish-movement, it is not surprising that the rules discovered, although not implausible, were rather weak.

On reflection, it seemed that rather than looking for rules relating environmental events to the *initiation* of fish movement, it would be more fruitful to look for rules relating environmental processes to the fish movement, considered as a process itself. The kind of causality considered in our third experiment is thus *perpetuation* rather than causation in the narrow sense, the zone-crossing events now being considered as proxies for upstream or downstream movement *processes*.

6 Experiment 3: Exploring Processes and Perpetuation

In order to handle perpetuation, we need to specify rules without causes in the antecedent. To model this, we require rules in which Causes is empty, so that all the burden of causality is borne by the conditions. In order to work with this kind of rule using the algorithm, a "dummy" event was generated which occurred at every time-step. This was achieved simply by defining the event (called always) as a random event with probability 1. For clarity, the rule

[always | Conditions] ⇒ effect after Delay

will be written in shorter form as

Conditions ⇒ effect after Delay.

We shall call these "Always-rules".

This section reports on a systematic exploration of the rules generated by the algorithm when tasked with identifying *perpetuation* in the fish data set. For brevity, we focus on the rules generated from causal analysis of data about water levels and movement. However, the same analysis has been conducted on the water-temperature data with congruent results.

6.1 Support

The algorithm is able to identify a large number of candidate rules from the data set. For example, more than 1000 rules are found for each upstream and downstream movement in response to water level. However, many of these rules are derived from conditions or events that occur only a handful of times. Figure 2 shows a scatterplot of *e-sensitivity* and *c-precision* of rules generated for upstream and downstream movement in response to water levels.

Figure 2 highlights those rules that relate to less than 10 instances of conditions (orange "+") or to less than 10 instances of effects (blue "×"). It is immediately noticeable that rules supported by few condition instances also have lower *e-sensitivity*, and similarly rules supported by few effect instances tend to have lower *c-precision*. Hypothesis testing confirms this visual expectation, significant at the 1 % level. Taking this result as a evidence of overfitting, those rules that were supported by less than 10 condition or effect instances in the data were excluded from the subsequent analyses.

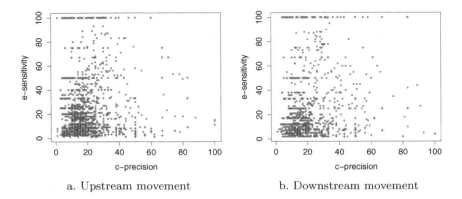

a. Upstream movement   b. Downstream movement

Fig. 2. Scatterplot of rule *e-sensitivity* against *c-precision*, highlighting rules with fewer than 10 instances of conditions (orange "+") and fewer than 10 instances of effects (blue "×") (Color figure online).

6.2   Spatial Coincidence

Next we examined the spatial coincidence between conditions and effects. The causal analysis is agnostic about whether an effect is in any way spatially related to its condition. As in Experiment 2 (Sect. 5.2), it was found that many of the rules generated relate conditions in one zone to effects in a different zone. However, one might hope that "sensible" rules would relate conditions in one zone to effects in the same zone.

We tested whether those rules that related conditions to spatially proximal effects (i.e., where the condition was spatially coincident with the start of movement) tended to have higher F_1 scores than rules that related conditions to spatially distal effects. A non-parametric Wilcoxon rank sum hypothesis test indicated that there was no evidence to support the hypothesis that spatially coincident rules have higher F_1 scores ($p = 0.39$ for upstream and $p = 0.88$ for downstream movement, which leads us to fail to reject the null hypothesis that proximal and distal conditions are drawn from the same population of F_1 scores).

Thus, as in Sect. 5.2, the data do not support the expectation that F_1 scores are higher for spatially proximal effects; indeed it appears that rules that relate conditions to distal effects are just as likely to have good *c-precision* and *e-sensitivity* as those that relate conditions to proximal effects. As we have already seen, such rules can potentially occur both as an effect of spatial autocorrelation in conditions and as a granularity effect (cf. Sect. 5.2). Nevertheless, we restricted our subsequent analyses to examine only "sensible" rules (where condition and effects are spatially proximal) on the grounds that such rules are more meaningful (even if our data did not indicate that they were statistically distinct).

6.3   Shuffled Data

In this context, we examined the degree to which the rules might still relate to meaningful patterns, rather than arbitrary overfitting, by repeating the causal analysis with a "shuffled" data set. In our shuffled data set, observations of environmental variables were arbitrarily reassigned to randomly selected zones (e.g., the water level in zone a might be reassigned to zone f at time t_1 and reassigned to zone p at time t_2, and so on). This process ensures that any structure in the data resulting from causal relationships is lost, while still allowing comparison with the unshuffled data set (since the movements are unchanged, and the distribution of the total set of environmental variables is unchanged).

There are two main reasons in this case for preferring shuffling to more conventional cross-validation (where the algorithm results are applied to a reserved portion of the data). First, cross-validation is sensitive to how the data set is partitioned. Spatial, seasonal, and longer-term variations (including drought conditions in the earlier years of the study) are expected to lead to statistical non-stationarity in the data. Consequently, by partitioning the data, especially with respect to time or space, we would run the risk that the reserved portion exhibits different properties to the training data. Second, cross validation cannot yield information about the "correct" rules, since (unlike in our experiments with synthetic data) we have no ground truth in the form of causal rules with which to compare the results, such rules being manifested only through correlations in the data (as discussed in Sect. 2). Cross-validation will only tell us how sensitive our results are to partitions of the data. This information is already implicitly available in the support for each rule, and indeed rules with low support are discarded anyway (Sect. 6.1). By contrast, shuffling allows us to create a second data set for cross-validation that has identical statistical properties (same numbers, timing, and locations of movement events, same numbers and distributions of process variables) to the original, unshuffled data. Any spatial relationships between causes and effects are thus scrambled in the shuffled data set. As a consequence, any rules inferred from the shuffled data are a priori examples of overfitting, and any difference between the results for the unshuffled and shuffled data sets can be ascribed to underlying spatial patterns in the unshuffled data set.

Figure 3 shows the boxplots of F_1 scores for rules generated from both shuffled and unshuffled data sets. In all cases, the F_1 scores for the unshuffled data set are significantly higher (at the 1 % level, $p < 0.0001$) than for the shuffled data set. Thus we may infer that the rules generated do indeed derive from *some* meaningful patterns of movement, and are not purely overfitting.

6.4   Condition Value Ranges

Looking at the rules themselves, it was noticeable that the size of the value range in the antecedent (condition) for a rule was strongly correlated with the F_1 score for that rule (Fig. 4). In other words, rules with larger ranges for the environmental variables in the antecedent tended to be associated with larger F_1

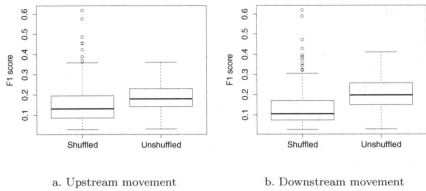

a. Upstream movement   b. Downstream movement

Fig. 3. Boxplot of F_1 scores for rules generated from shuffled and unshuffled data sets.

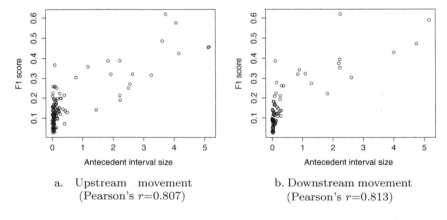

a. Upstream   movement   b. Downstream movement
(Pearson's r=0.807)   (Pearson's r=0.813)

Fig. 4. Scatterplot of condition interval length against F_1 score.

scores. This is encouraging as, in general, such rules can be regarded as stronger: they "say more," since they make assertions about a wider range of instances and therefore it takes less to falsify them. (Conversely, rules with larger delay intervals can be regarded as weaker, since they "say less" about precisely what effects are expected to result from a condition.)

6.5   Top-Ranked Rules

Finally, we looked at the top-ranked rules (in terms of F_1 score) for each effect, in Table 5. In total, these 15 rules accounted for more than 20 % of all upstream movements found in the data set. Combined with the 11 top-ranked rules for downstream movement, which accounted for more than 18 % of all downstream movements, a total of 26 rules accounted for a significant minority (approximately one-fifth) of all movements. Given that water level is but one potential

Table 5. The best rules discovered for each upstream movement effect. *Note that zones i and j meet at a confluence, so it is possible to move upstream into j from i and upstream into i from j

Best rule found	F_1 score
$2.79 \leq$ wl(cd) $\leq 4.72 \Rightarrow$ d/c after $[0, 5]$	0.32
$2.39 \leq$ wl(efgh) $\leq 5.03 \Rightarrow$ e/d after $[0, 5]$	0.32
$2.81 \leq$ wl(efgh) $\leq 5.03 \Rightarrow$ f/e after $[0, 5]$	0.38
$1.77 \leq$ wl(efgh) $\leq 1.92 \Rightarrow$ g/f after $[0, 5]$	0.25
$0.77 \leq$ wl(efgh) $\leq 1.55 \Rightarrow$ h/g after $[0, 5]$	0.30
$126.41 \leq$ wl(ijklm) $\leq 131.53 \Rightarrow$ i/h after $[0, 5]$	0.45
$126.85 \leq$ wl(ijklm) $\leq 128.69 \Rightarrow$ i/j after $[0, 5]*$	0.39
$126.98 \leq$ wl(ijklm) $\leq 128.16 \Rightarrow$ j/i after $[0, 5]*$	0.36
$126.89 \leq$ wl(ijklm) $\leq 126.92 \Rightarrow$ j/k after $[4, 5]$	0.26
$124.67 \leq$ wl(np) $\leq 124.75 \Rightarrow$ n/i after $[0, 5]$	0.26
$1.60 \leq$ wl(or) $\leq 6.75 \Rightarrow$ o/n after $[0, 5]$	0.46
$3.02 \leq$ wl(or) $\leq 6.75 \Rightarrow$ r/o after $[0, 5]$	0.62
$2.24 \leq$ wl(stuv) $\leq 6.40 \Rightarrow$ s/r after $[0, 5]$	0.42
$2.33 \leq$ wl(stuv) $\leq 6.40 \Rightarrow$ u/s after $[0, 5]$	0.58
$2.78 \leq$ wl(stuv) $\leq 6.40 \Rightarrow$ v/u after $[0, 5]$	0.48

driver of movement, and that our rules take but one, simple form, this small set of rules does seem to provide a surprisingly compact representation of almost one fifth of the data set.

7   Conclusions and Further Work

We have developed the foundations of an algorithm that is able to identify the instances of rules of a particular logical form that best describe a given data-set. The approach can handle a range of logical forms, including simple causation, causation with conditional rules, and perpetuation. Our experiments show that for synthetic data, where the underlying causal rule is known, the approach is able to derive close approximations of the underlying causal rules from data.

In the case of real data, however, granularity effects may often confound an attempt to derive strict causation, where one event initiates another. In our example of fish movement, for example, the spatial and temporal granularity of the data (movement between granular zones of tens of kilometers and with a finest temporal granularity of one day), our algorithm struggles to identify strict causal relationships. However, by tasking the algorithm to look instead for perpetuation rules (termed in our system "Always-rules"), the algorithm is able to identify a suite of rules that compactly describe the data. Amongst our key results are included, considering rules relating fish movement to water level alone:

- The rules generated do relate to meaningful structure in the movement data, describing movements that are significantly different from random movements;
- The top-ranked rules in each zone compactly describe approximately 20 % of the fish movements.

While this study has demonstrated the potential of our approach, future work on a much wider range of data sets is needed to further validate our initial results (in particular with finer-granularity information for events and process variables). Beyond this, comparison with probabilistic alternatives would assist both in validating our results and in further elucidating the practical implications of our logical approach. In the longer term, an integration of both probabilistic and logical approaches may be advantageous. It is also likely that, in moving towards operational data-mining tools for identifying causal relationships in movement data, we can complement our algorithm with visualisation capabilities for assisting users with sorting and filtering inferred rules. More broadly, we believe that visualisation of causal spatial rules could be a fruitful area for future research.

Finally, it is worth reflecting on the secondary role played in our account of causation by space, when compared with time. Cause and time are intimately linked through the familiar maxim that an effect cannot precede its cause, a reflection of the asymmetrical directedness of time. Since space exhibits no such directionality, there is no comparable maxim relating cause and space. Space and time do, however, share the attribute of extension, which gives rise to the measures of distance and duration. A general expectation for causality is that causal influence should be proximal with respect to both space and time: that is, we expect an effect to be spatially and temporally close to its cause (compare our remarks in [3] commenting on [6]). Where we find that this is apparently not the case — where a cause at one place and time leads to an effect at a distant place after a time delay — we normally suppose this to be explicable in terms of some unobserved process carrying the causal influence from the cause location to the effect location. But precisely because the process is unobserved, it is not possible for a mechanism that extracts causal rules from data to detect it, with the result that spatial linkages between cause and effect may, at least for some types of data set, show up only weakly, if at all, in the analysis.

Acknowledgments. Antony Galton's work was supported by the EPSRC, project EP/M012921/1. Matt Duckham's work is supported by funding from the Australian Research Council (ARC) under the Discovery Projects Scheme, project DP120100072. Alan Both's work is supported by funding from ARC project DP120103758.

References

1. Allen, E., Edwards, G., Bédard, Y.: Qualitative causal modeling in temporal GIS. In: Kuhn, W., Frank, A.U. (eds.) COSIT 1995. LNCS, vol. 988. Springer, Heidelberg (1995)

2. Allen, J.F.: Towards a general theory of action and time. Artif. Intell. **23**, 123–154 (1984)
3. Bleisch, S., Duckham, M., Galton, A., Laube, P., Lyon, J.: Mining candidate causal relationships in movement patterns. Int. J. Geogr. Inf. Sci. **28**(2), 363–382 (2013)
4. Bunge, M.: Causality. Dover, New York (1966)
5. Davidson, D.: Causal relations. J. Philos. **64**, 691–703 (1967)
6. El-Geresy, B.A., Abdelmoty, A.I., Jones, C.B.: Spatio-temporal geographic information systems: a causal perspective. In: Manolopoulos, Y., Návrat, P. (eds.) ADBIS 2002. LNCS, vol. 2435, pp. 191–203. Springer, Heidelberg (2002)
7. Galton, A.: States, process and events, and the ontology of causal relations. In: Donnelly, M., Guizzardi, G. (eds.) Formal Ontology in Information Systems: Proceedings of the Seventh International Conference (FOIS 2012), pp. 279–292. IOS Press, Amsterdam (2012)
8. Lyon, J.P.: Snags underpin Murray River restoration plan. ECOS **177**, 1 (2012)
9. Moens, M., Steedman, M.: Temporal ontology and temporal reference. Comput. Linguist. **14**, 15–28 (1988)
10. Pearl, J.J.: Causality: Models, Reasoning, and Inference. Cambridge University Press, New York (2000)
11. Spirtes, P., Glymour, C., Scheines, R.: Causation Prediction and Search. Springer, New York (1993)
12. Terenziani, P.: Towards a causal ontology coping with the temporal constraints between causes and effects. Int. J. Hum Comput Stud. **43**, 847–863 (1995)
13. Terenziani, P., Torasso, P.: Time, action-types, and causation: an integrated analysis. Comput. Intell. **11**(3), 529–552 (1995)

Modelling Spatial Structures

Franz-Benjamin Mocnik$^{(\boxtimes)}$ and Andrew U. Frank

Vienna University of Technology, 1040 Vienna, Austria
mail@mocnik-science.net
www.mocnik-science.net

Abstract. Data is spatial if it contains references to space. We can easily detect explicit references, for example coordinates, but we cannot detect whether data implicitly contains references to space, and whether it has properties of spatial data, if additional semantic information is missing. In this paper, we propose a graph model that meets typical properties of spatial data. We can, by the comparison of a graph representation of a data set to the graph model, decide whether the data set (implicitly or explicitly) has these typical properties of spatial data.

Keywords: Space · Spatial structure · Spatial data · Spatial information · Time · Tobler's law · Principle of least effort · Graph model · Spatial network · Scale invariance

1 Introduction

It is widely assumed that information is in large part of a spatial nature [17]. Evidence for exact percentages is rare [21], but the large number of spatial data sets demonstrates the importance of spatial information, e. g. data about public transport, cadastres, maps, and weather data.

Tobler claimed that "everything is related to everything else, but near things are more related than distant things" [43], which is known as Tobler's first law of geography. This law is not universally true but has been proven to be statistically valid for many spatial data sets, e. g. for spatially referenced Wikipedia articles [22]. Tobler's law is, in case of human activities, motivated by the principle of least effort [52]: it claims that people choose the path of least effort, and as movement in space usually requires more effort for longer distances, human activities more often relate near than distant things.

Physical properties of tangible space are very similar at different scales. Classical mechanics holds, for example, for everyday items as well as for solar systems. Representations of human activities and processes that depend on physical properties are thus often independent of scale. This is not true in general, but it applies, for instance, to many transport networks [28, 34]. Both, Tobler's law and scale invariance, are characteristics of spatial data in many cases.

Properties of data can be influenced by the data's relation to space, but also by other reasons: for example, the location of cities depends on properties of

© Springer International Publishing Switzerland 2015
S.I. Fabrikant et al. (Eds.): COSIT 2015, LNCS 9368, pp. 44–64, 2015.
DOI: 10.1007/978-3-319-23374-1_3

space (since transport costs are related to distance, etc.) but also on landscape morphology (shape and structure of water bodies, natural resources, etc.) and others. If the properties of data originating from the properties of space (called spatial structure) predominate, the data is called spatial. This classification is not a binary but rather a fuzzy classification: a data set can expose a spatial structure to a certain extent, and spatial and non-spatial structures may coexist.

Explicit references to space enable us to check whether a data set has typical properties of spatial data, e. g. whether Tobler's law is met [22]. When a data set does not expose explicit references to space, we cannot check in the same way whether Tobler's law is met or whether it has typical properties of spatial data. However, data can meet Tobler's law without explicitly including references to space: some things are more related than others, and relations (within the data set) may have the same structure as relations in spatial information. The issue lies with how to detect a spatial structure without explicit references to space. A very short overview over this topic has been provided by the author at the Vienna Young Scientists Symposium [38].

In this paper, we propose a spatial graph model that has some typical properties of spatial data sets, e. g. Tobler's law and scale invariance. The comparison of a data set with the proposed graph model enables us to determine whether the data set exhibits these typical properties or not.

In the next section, we discuss typical properties of spatial information, including Tobler's law and scale invariance (Sect. 2). Then we outline existing graph models as well as related work and argue why these models are not suitable as general models of spatial data (Sect. 3). We propose a spatial graph model (Sect. 4) and show that it meets the properties discussed in Sect. 2 (Sect. 5). The comparison between spatial data sets and the proposed graph model is discussed and evaluated on several data sets, including data sets about public transport, water distribution networks, formalizations of games, and biological networks (Sect. 6).

2   Typical Properties of Spatial Information

Information which exposes a number of references to space is, by definition, called *spatial information*. These references relate the underlaying data with things that are placed in space. The structure of spatial information as well as the structure of data, which becomes spatial information by interpretation, is based on the properties of space and the entities that constitute space: the existence of distance and the effort of travelling leads to a predominance of relations between near things; the similarity of space and physical processes at different scales of tangible reality leads to scale invariance of spatial data; and non-uniform distributions of objects in space lead to not necessarily uniform but in many cases bounded distributions of relations. We call such a structure of data in this paper a *spatial structure*[1], and we say that data *has a spatial structure* (in which case we also speak of *spatial data*)

---

[1] Time has a similar effect on data, because it can be modelled by one-dimensional Euclidean vector spaces.

if it exposes some of these properties. It is important to note that data can, by the above definition, have a spatial structure without being interpreted and actually without being related to space; we only require that the data's structure *can* be interpreted in such a way.

In this section, we discuss three typical properties of spatial information that the proposed graph model meets. For this discussion, we assume data sets to have representations that explicitly expose references to space as well as relations of objects within the data set. Such representations of data as a graph are called *graph representations* in this paper. Graph representations can be stored in triple stores or graph databases to practically verify properties of the data sets.

2.1 Tobler's First Law

A topological core concept of space is neighbourhood [29]: when things are near, we say that they are located in the same neighbourhood. The concept is relative in the sense that the meanings of near and neighbourhood depend on context, and do not necessarily relate to Euclidean space but to some concept of distance, e. g. travelling time or fuel consumption.

The existence of distance and the concept of neighbourhood (both properties of space), as well as the existence of "costs" to relate distant things, lead to spatial autocorrelation, *Tobler's first law of geography*: "Everything is related to everything else, but near things are more related than distant things" [43]. Many data sets reveal that more distant things are statistically less related for different reasons, e. g. due to transport or communication costs. For example, Tobler's law has been proven true in large part for spatially referenced Wikipedia articles [22].

2.2 Scale Invariance

Space, conceptualized as a Euclidean vector space, has no preferred unit. After rescaling space, it cannot be distinguished from the unscaled one, and physical processes of our tangible world remain (nearly) the same when rescaled. As soon as objects are placed in space, they define a unit and a scale. If interrelations between objects only depend on relative distances and the Euclidean structure, the objects and their interrelations do not change with the rescaling of space. The system of objects and interrelations is, in this case, called *scale-invariant*[2]. The effect of scale invariance can be observed in different data sets, e. g. for metro and railway networks [34], and for road networks [28].

2.3 Bound Outdegree

The average edge degree in a planar graph can be proven to be strictly less than 6, which can be seen by Euler's formula and the fact that a face has at

---

[2] The concept of *scale invariance* of a graph embedded in space should not be confused with the concept of *scale-freeness* of a graph, which is characterized by a power law distribution of the nodes' edge degrees and hence by the invariance of the distribution's shape under rescaling of the total number of edges.

least three edges and each edge has at most two faces. The edges of a planar graph can be oriented such that the outdegree is bounded by 3 [11]. We expect that the outdegree of a graph embedded in space behaves similar, even if it is not completely planar, and we hence expect the outdegree to have an upper bound which is considerably lower than the one of a complete graph.

When nodes are non-uniformly distributed in space, we could expect the outdegree to be non-uniform for different nodes as well. Following the above argument, we however expect the outdegree to be bounded, as is true in the example of public transport: nodes representing stops of public transport are usually more dense in city centres than in the countryside, but there exist edges in the countryside, and the outdegree is not arbitrarily high in city centres.

We have discussed three typical properties of spatial information, as well as the structure that spatial data in consequence has. In the next section, we discuss existing graph models and argue why they are not suitable as general models of spatial data.

3 Existing Graph Models

Existing graph models have been developed in order to model different aspects of information. An overview of space-related graphs and their properties has been provided by Barthélemy [8]. In this section, we review well-known graph models and argue why they are not suitable to model spatial data in general.

- The *Erdős-Rényi model* is a randomly chosen graph with a given number of nodes and edges [15]. The *Gilbert model* is a graph where edges between pairs of nodes exist with a given probability [19]. Both models are not suitable for modelling spatial data in general, because spatial data is not completely random: the structure of spatial data is influenced by space, and some configurations of edges are expected to occur more often than others.
- *Barabási and Albert* proposed a graph model where nodes are added incrementally [4]. Each time a node is added, edges are more likely to be introduced between the new node and nodes that already have a high number of edges. Thus, the majority of nodes is joined by a very low number of edges, whereas only a very low number of nodes is joined by a high number of edges, resulting in a power-law degree distribution. Graphs with power-law distributions are called *scale-free*, because the distribution of the edge degree is scale-invariant. This model and similar ones have been used to model internet links [5,42], citation networks [3], and social networks [40]. The construction of Barabási-Albert models however does not reflect Tobler's law.
- The family of *exponential random graph models* consist of graphs whose edges follow the distribution of the exponential family [23]. Exponential random graph models have been used to model social networks [24]. These graphs are tailored to fit statistical properties, but they do not refer to spatial properties.
- *Hierarchical network models* relate duplicates of small graphs at different hierarchies [6]. These models are suitable for hierarchical aspects which spatial

data, in principle, can have. Spatial data however is at the core not solely characterized by hierarchies but primarily by Tobler's law and other properties.

– *Watts and Strogatz* proposed a model with a very short typical path length and high clustering [48]. Spatial graphs usually have longer path length than this model, because relations tend to exist only in neighbourhoods.
– *Planar graphs*, i. e. graphs that can be embedded in two-dimensional space such that their edges do not intersect, have been studied widely [2, 18, 30, 35, 41, 45, 46, 50]. Graphs have been proven to be planar if and only if they neither contain the complete graph K_5 with 5 nodes nor the complete bipartite graph $K_{3,3}$ with 6 nodes as a subgraph after the contraction of edges [46]. Spatial data sets usually cannot be represented by planar graphs, because they remain spatial after local modifications (in particular after the introduction of K_5 or $K_{3,3}$ as subgraphs).
– *Spatial generalizations of existing models* have been discussed, e. g. by considering only planar graphs during construction [14, 37], or other modifications [27, 36]. For example, the Barabási-Albert model has been modified by taking distance between nodes into account [7, 42, 51]. These generalizations share aspects of spatial data, but as most of their characteristics originate from the non-generalized models, they are not suitable as models for spatial data in general.
– A class of *geometric graph models* assumes nodes to have explicit locations in space, and edges to be modelled by the distance between nodes: an edge between two nodes with distance $d(p, q)$ is introduced with probability $f(d(p, q))$, where $f : \mathbb{R} \to [0, 1]$ is some probability function. For example, radio transmitters (with constant transmitting power) have been modelled [25] using the function

$$f(l) = \begin{cases} 1 & \text{if } l < r \\ 0 & \text{otherwise.} \end{cases} \tag{1}$$

A similar model was discussed by Waxman [49], who proposed a smoothened, continuous probability function

$$f(l) = \beta \exp\left(-\frac{l}{r}\right). \tag{2}$$

Both models depend on the absolute distance between points and thus are not scale-invariant. A scale-invariant variant of this model was discussed by Aldous [1]: edges to the k nodes with minimal distance are introduced for each node (for a given $k > 0$). This model does not reflect the fact that for a spatial data set, the distribution of the relations, in particular the number of relations per node, usually depends on the locations of the nodes in space.

As argued in this section, existing graph models are tailored to model specific types of data but not spatial data in general. In the next section, we introduce a model of spatial data in general. The construction of the model is motivated by the three typical properties of spatial information of Sect. 2.

4   A Scale-Invariant Spatial Graph Model

In this section, we introduce a graph model that has numerous properties of spatial data, including the three ones discussed in the last section. The graph model does not aim at modelling particular types of spatial data, e.g. data about public transport or communication data. It aims, instead, at having typical properties of spatial data, and thus at sharing similarities with many spatial data sets. In the following, we motivate that the proposed model meets the properties of Sect. 2. The proof is given later in Sect. 5.

For the construction of a graph model, we ask which edges have to be introduced for a given set of nodes in space in order to model spatial data. To be more exact, we ask which configuration of edges would produce the properties of Sect. 2.

Taking the second part of Tobler's law "near things are more related than distant things" literally means that, as soon as a point p is related to a point q by an edge, every edge q' with a shorter distance, i.e. $d(p, q') < d(p, q)$, is with a high probability also related to p by an edge. In the proposed model, an edge between p and q' does not only exist with a high probability, but with a probability of 1.[3] This means that edges to a number of nearest points are introduced, and the number can vary for different nodes.

As Tobler's law describes things in space, the number of edges depends on the distribution of the nodes, in particular on the distance between nodes. If the number depends on the absolute distance between points, the model is not scale-invariant. The following model only uses the relative distance between nodes, and is therefore scale-invariant:

Definition 1 (Scale-Invariant Spatial Graph Model – SISG Model). *Let V be an n-dimensional Euclidean vector space with metric d. To a finite set of points $S \subset V$ and a real number $\rho > 1$, we associate the abstract[4] (directed and simple) graph $\mathcal{M}_\rho(S, V)$ consisting of*

(i) a node for every point $p \in S$, and
(ii) a directed edge (p, q) if and only if

$$d(p, q) \leq \rho \cdot \min_{q_0 \in S \setminus \{p\}} d(p, q_0)$$

where q_0 denotes the nearest neighbour of p.

The graph $\mathcal{M}_\rho(S, V)$ is called the scale-invariant spatial graph model (SISG model) of the generating set $S \subset V$ of dimension $\dim V$ and density parameter ρ. We call $\mathcal{M}_\rho(S, V)$ to be generated by the set S.

---

[3] This choice is made because it enables us to analytically compute some properties of the model. A variant of the model which introduces edges only with a certain probability is left to future work. As long as this effect is less dominant than other ones, we expect the properties of the proposed model and its variant to be similar.

[4] A graph is called *abstract* if its nodes and edges contain no additional semantics. An abstract graph is, in particular, not embedded in space, and the nodes have no location.

In Fig. 1(a), a graph representation of data from the national railway operator in Sweden is depicted[5]. Nodes in the graph representation relate to stops, and edges to pairs of successive stops, i. e. stops p and q such that at least one train travels from p to q without stopping in between. Figure 1(b) shows a SISG model with the stops of the data set used in (a) as generating set. As expected, edges exist only between near nodes. The SISG model is not expected to model the graph representation exactly, because it is not a model of public transport but of spatial data in general. We hence expect the model to share the properties discussed in Sect. 2 with the graph representation. The Gilbert model (a random graph) with the stops of the data set used in (a) does not share significant spatial properties with the graph representation in (a). In particular, edges exist between nodes independent of their distance.

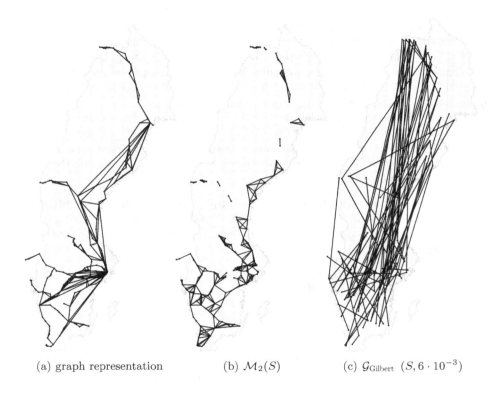

(a) graph representation   (b) $\mathcal{M}_2(S)$ (c) $\mathcal{G}_{\text{Gilbert}}\ (S, 6 \cdot 10^{-3})$

Fig. 1. Graphs whose nodes S are the stops of the national railway operator SJ in Sweden; (a) graph representation of the data set, (b) a SISG model and (c) a Gilbert model; the parameters are chosen such that similarities and dissimilarities visually stand out

---

[5] The data is publicly accessible in the General Transit Feed Specification (GTFS) format [44].

The definition of a SISG model is based on a generating set of points placed in space. If a model is to be computed without prior choice of a generating set, a set of randomly placed points can be used. Without further assumptions, however, it is not clear which distribution of the points should be used, and as physical space is uniform, there is no preferred distribution. A uniform distribution[6] of points is hence a convenient choice:

Definition 2 (Uniform Scale-Invariant Spatial Graph Model). *For a given dimension m, we denote a (uniform) SISG model with a generating set consisting of s randomly distributed points with uniform distribution in an m-dimensional ball by $\mathcal{M}_\rho^m(s)$.*

The SISG model as well as the uniform SISG model are constructed to have some typical properties of spatial data. In the next section, we discuss some basic properties of the model and formally prove that the SISG model has the typical properties of spatial data that were discussed in Sect. 2.

5   The Model has Typical Properties of Spatial Data

We have constructed SISG models such that they have some typical properties of spatial data. In this section, we examine the properties that were discussed in Sect. 2, and we formally prove that SISG models have these properties.

5.1   Tobler's First Law

Tobler claimed that "near things are more related than distant things". In accordance with this, we are able to prove:

Theorem 1. *If there is an edge between two nodes p and q in a SISG model $\mathcal{M}_\rho(S, V)$, then there exists an edge between p and any node q′ that has distance smaller than d(p, q).*

Proof. We have

$$d(p, q') \overset{(a)}{\leq} d(p, q) \overset{(b)}{\leq} \rho \cdot \min_{q_0 \in S \setminus \{p\}} d(p, q_0)$$

where (a) is due to the presumption of the theorem and (b) due to the definition of a SISG model. The equation proves that p and q' are connected by an edge according to the definition of a SISG model.   □

5.2   Scale Invariance

The change of scale in a vector space V can be described by transformations $\tau: V \to V$ that scale distances between arbitrary points p and q by a constant factor $\sigma > 0$:

$$d(\tau(p), \tau(q)) = \sigma \cdot d(p, q).$$

---

[6] A uniform distribution is a distribution where a point is placed at each location in space with the same probability.

As such transformations change scale, they are called *scale transformations* of relative scale σ. SISG models can be proven to be invariant under such transformations:

Theorem 2. *SISG models are invariant under scale transformations, i. e.*

$$\mathcal{M}_\rho(S, V) = \mathcal{M}_\rho(\tau(S), V)$$

(as abstract graphs) for every scale transformation $\tau : V \to V$.

Proof. The definition of the SISG model does not explicitly include the location of nodes, just their distances. In particular, the inequality in Definition 1 stays invariant because both sides are multiplied by the relative scale $\sigma > 0$. □

5.3 Bound Outdegree

The outdegree of a node has a lower bound:

Theorem 3. *If a SISG model has at least two nodes, each node has outdegree of at least* 1.

Proof. As at least two nodes exist, each node p has a nearest node p_0, and thus an outgoing edge to p_0. □

As SISG models are simple by definition (i. e. there exists not more than one edge for each pair of nodes), the outdegree of a node is bound by $(s-1)$ where s is the number of nodes. This upper bound however is meaningless, because it is met for every simple graph and not only for SISG models. It can be shown that the expected outdegree of a node is much lower than this upper bound, when the number of nodes approaches infinity[7]:

Theorem 4. *In a SISG model* $\mathcal{M}_\rho(S, V)$ *with* S *uniformly distributed, the expectation value of the outdegree of a node converges to* $\rho^{\dim V}$ *for* $|S| \to \infty$.

Proof. Consider points to be uniformly distributed in a vector space of dimension $m = \dim V$. For an arbitrarily chosen point p and a real number $L > 0$, let S be the set of all points in the m-dimensional ball $B_m(p, L)$ of radius L centred in p. We denote the minimal distance between p and the remaining points by $r = \min_{q \in S \setminus \{p\}} d(p, q)$.

If for an $R < L$ the m-dimensional open ball $B_m(p, R)$ does not contain any point of S apart from p, the points of $S' = S \setminus \{p\}$ are in $B(L, R) = B_m(p, L) \setminus B_m(p, R)$. Denoting the volume of the m-dimensional ball of radius

---

[7] Analytical results are much easier to derive when the number of nodes approaches infinity and hence only inner regions of the graphs have to be considered. The results can, however, be expected to approximately hold for finite graphs as well, when the number of nodes is sufficiently high.

L by $\text{Vol}_m(L)$, the density of points in the ball $B_m(p, L)$ equals $s/\text{Vol}_m(L)$ with $s = |S|$ for $L \gg R$. Thus, we expect

$$\mu = \frac{s}{\text{Vol}_m(L)} \cdot [\text{Vol}_m(\rho R) - \text{Vol}_m(R)] + 1 = s(\rho^m - 1)\frac{\text{Vol}_m(R)}{\text{Vol}_m(L)} + 1 \quad (*)$$

points in $B(\rho R, R)$, namely the one at minimal distance r (the cases where more than one point is at minimal distance is a null set) and the ones in the inner of $B(\rho R, R)$. (The second equality is due to the fact that $\text{Vol}_m(\rho R)$ equals $\rho^m \text{Vol}_m(R)$.) If $R \leq r$, we expect at least μ points in $B(\rho r, r)$, i.e. at least μ edges starting in p.

For a given R, the probability of $R \leq r$, i.e. the probability that all $s - 1$ points S' have distance greater than R to the point p, is

$$\left(1 - \frac{\text{Vol}_m(R)}{\text{Vol}_m(L)}\right)^{s-1}.$$

Inserting Eq. $(*)$ proves that the probability that at least μ edges are starting at p is

$$\nu(\mu) = \left(1 - \frac{\mu - 1}{s(\rho^m - 1)}\right)^{s-1}.$$

The probability that at most μ edges are starting at p equals $1 - \nu(\mu)$, and the corresponding probability density function is given by $-\frac{d}{d\mu}\nu(\mu)$. To compute the expectation value for the number of edges starting at p, we first compute

$$\pi(\mu) = -\int \mu \frac{d}{d\mu}\nu(\mu)\, d\mu$$

$$= -\mu \cdot \nu(\mu) + \int \nu(\mu)\, d\mu$$

$$= \left[(\mu - 1)\left(\frac{1}{s} - 1\right) - \rho^m\right] \cdot \nu(\mu).$$

The expectation value of the number of edges starting at p can be computed as

$$\pi(\mu)\big|_1^{s-1} = \rho^m + \left[(s - 2)\left(\frac{1}{s} - 1\right) - \rho^m\right] \cdot \left(1 - \frac{s - 2}{s} \cdot \frac{1}{\rho^m - 1}\right)^{s-1}.$$

When $s \to \infty$, the second summand vanishes.   □

We have proven that SISG models have the typical properties of spatial information which were discussed in Sect. 2. In the next section, we use this fact to test data sets for spatial structures.

6   Application: Testing Data for Spatial Structures

SISG models have typical properties of spatial data and can thus serve as prototypes of spatial data. In this section, we discuss an application of SISG models: the comparison of data sets with the model enables us to discover spatial structures.

6.1   The Problem of Testing Data for Spatial Structures

Space can lead to a spatial structure, i. e. typical properties of spatial data, as was discussed in Sect. 2. If a data set contains explicit references to space, it can be checked which typical properties the data set has. The situation is different when a data set only implicitly contains references to space, e. g. when semantics is missing: the data set contains references to space but we are not aware of them, and the references can, in consequence, not be used to check which typical properties the data set has. This raises the question of *how to check for a spatial structure without explicit references to space.*

When a data set is similar to a uniform SISG model, we can conclude that the data set has, by and large, the properties discussed in Sect. 2 because the SISG model has them. In this case, the data set has a spatial structure, i. e. a structure that data typically has when it can be interpreted as spatial information. We can, by no means, conclude that the data set becomes spatial information when it is interpreted[8], but there is a good chance that the data set *can* become spatial information by interpretation if the data set is similar to a SISG model. The information of the data set is, in this case, a good candidate for spatial information.

Data sets very rarely *equal* a SISG model, because the properties discussed in Sect. 2 are not exact laws but rather loose properties. It is however sufficient that a data set and a uniform SISG model are *similar* in order to conclude that they have similar properties. Spatial data sets can even have properties very different to the ones discussed in Sect. 2, depending on how representations are constructed, and on to what extent non-spatial information is included in the representation. Such a data set cannot be detected when it is compared to SISG models.

Many examples of spatial information have, in addition to a spatial structure, further structures. Examples are manifold: the structure of a town, in particular the configuration of rivers and bridges, has an impact on timetable information; the importance of controlling a number of persons with clear responsibilities leads in many organizations to hierarchical structures, even if the organizations are spatially organized, e. g. by affiliates; and the preference of nodes with a large number of edges during the growth of a network can lead to a power law distribution of the nodes' edge degrees, e. g. in case of social networks. The comparison of a data set to the SISG model can gradually determine how similar the data set set is to the model, and thus to what degree the data set has a spatial structure.

We have discussed that we can, with some limitations, approach the question of whether a data set has a spatial structure by comparing it to the SISG model. In the next section, we approach the question of how we can compare a data set with a SISG model.

---

[8] There cannot exist any way to conclude whether a certain interpretation of a data set is spatial without knowledge of the interpretation, because there can exist spatial and non-spatial interpretations of the same data set.

6.2   The Problem of Comparing Data to the SISG Model

In Sect. 6.1, we discussed the importance of comparing a data set to the SISG model in order to approach the question whether the data set is spatial. The question of how to conduct this comparison is approached in this section.

The uniform SISG model depends on a density parameter ρ, a dimension m and a number of nodes s. For given parameters, we can compute an explicit model $\mathcal{M}_\rho^m(s)$. The SISG model $\mathcal{M}_\rho^m(s)$ can be understood as an abstract graph, i. e. as a set of abstract nodes and edges, ignoring the fact that we know the parameters ρ, m, and s that were used to generate the model. We will discuss two methods (Theorems 5 and 6) that enable us to approximately recover ρ^m. As both methods estimate the same value, we expect their results to be approximately equal for any SISG model.

As the two methods of estimating ρ^m do computationally not depend on the fact that the graph is a SISG model, we can apply them to any data set that is represented as a graph. This enables us to check whether a graph representation has a spatial structure: if the computations of both estimates result different values for a data set, it does not have the properties discussed in Sect. 2. If the estimates are approximately equal, we can conclude that the graph representation shares some properties with a SISG model.

Both methods of estimating ρ^m assume that the number of nodes approaches infinity. Data sets are necessarily finite, and the used analytical results are hence, in general, not valid for data sets. If the size of a data set is sufficiently large, we will assume that the estimations are good enough for a reasonable comparison to the SISG model. We will see in Sect. 6.4 that the results are reasonable for the examined data sets.

Theorem 4 provides a first method of estimating ρ^m for a given SISG model (as an abstract graph): we expect each node to have an outdegree of ρ^m, and as the outdegree equals the number of edges divided by the number of nodes, we can immediately conclude:

Theorem 5. *For a graph $\mathcal{M}_\rho^m(s)$ with e edges, we expect $\rho^m = e/s$ for $s \to \infty$.*

A second approach to estimate ρ^m compares the density of subgraphs. The density of a graph is defined as:

Definition 3 (Density of a Graph [12]). *The density[9] of a graph G consisting of $n > 1$ nodes and e edges is defined as*

$$c_{\text{density}}(G) = \frac{e}{n \cdot (n-1)}.$$

By Theorem 5, we expect the density of $\mathcal{M}_\rho^m(s)$ to be $\rho^m/(s-1)$ for $s \to \infty$. As every subgraph of a SISG model is again a subgraph of the SISG model generated with the same density parameter and the nodes of the subgraph as generating set, we expect an induced subgraph[10] of $\mathcal{M}_\rho^m(s)$ with t nodes to have

---

[9] The notation of density of a graph should not be confused with the notation of density of elements distributed in space.

[10] A subgraph H of a graph G is called *induced* if every edge (p, q) of G with p and q nodes in H is also an edge of H.

density $\rho^m/(t-1)$ for $t \to \infty$. For small subgraphs, the estimation may be worse than for large ones as the limit of $t \to \infty$ is not considered. If the computation of a subgraph's density does not only include the edges between nodes of the subgraph but also includes all edges that start (and not necessarily end) in the subgraph, the result is similar to the one of an infinite graph because it cannot be distinguished from a fragment of an infinite graph. We define:

Definition 4 (Total Density of a Subgraph). *The* total density *of a subgraph $H \subset G$ consisting of $n > 1$ nodes is defined as*

$$c_{\text{total density}}(H, G) = \frac{e}{n \cdot (n-1)}$$

where G has e edges starting at a node in H.

Using this definition, we can conclude:

Theorem 6. *A subgraph H of $\mathcal{M}_\rho^m(s)$ with t nodes is expected to approximately have total density $\rho^m/(t-1)$ for $H \ll \mathcal{M}_\rho^m(s)$, i. e. in case that the number of nodes in the subgraph is much smaller than the number of nodes in the SISG model.*

An estimate of ρ^m can be found by computing the total density of subgraphs and fitting by the function $c_{\text{total density}}(t) = \rho^m/(t-1)$ where t is the number of nodes in the subgraph. This estimate of ρ^m by Theorem 6 should approximately equal the estimate by Theorem 5 for uniform SISG models. In the next section, we introduce spatial and non-spatial data sets that are used in Sect. 6.4 to evaluate the considerations of this section.

6.3   Data Sets for the Comparison

In this section, we introduce data sets from different domains that have explicit references to space resp. time, and some data sets that have no explicit references to space nor time. These data sets are used in the next section to evaluate the methods proposed in Sect. 6.2.

As examples for spatial data, we examine data about public transport in Sweden which was already used in Sect. 4. Each of these data sets contains data from one transport agency [44]. These networks are explicitly related to space by the coordinates of the stops.

In addition to these spatial data sets, we examine the high-voltage power grid in the Western States of the USA [48], and the network of airports in the USA [13]. In the latter one, airports are represented as nodes, and an edge exists between two nodes if a flight was scheduled between the corresponding airports in 2002.

Water distribution networks are further examples for spatial data sets. They consist of one or more sources and a number of sinks. Pipes are represented by edges, and as the aim of the network is to distribute the water, there is a flow direction and the pipes can hence be represented by directed edges. Walski

Table 1. Estimates of ρ^m for different data sets

Graph	$|N|$	$|E|$	$\widehat{\rho^m}^N$	$\widehat{\rho^m}^D$	χ^2	Ref.
⊗ $\mathcal{M}_2^2(1000)$	1 000	3 939	3.94	4.69	$4.53 \cdot 10^{-3}$	def. 2
⊕ $\mathcal{M}_{2.3}^2(1000)$	1 000	5 111	5.11	5.73	$2.80 \cdot 10^{-3}$	def. 2
⊙ $\mathcal{M}_{2.5}^2(1000)$	1 000	5 947	5.95	6.84	$6.85 \cdot 10^{-3}$	def. 2
□ SJ (national railway provider)	176	544	3.09	4.30	$9.46 \cdot 10^{-3}$	[44]
Länstrafiken Sörmland	2 100	4 382	2.09	2.37	$9.32 \cdot 10^{-4}$	[44]
Östgötatrafiken	2 643	5 960	2.26	2.08	$1.07 \cdot 10^{-3}$	[44]
Blekingetrafiken	1 215	2 643	2.18	2.15	$8.81 \cdot 10^{-4}$	[44]
⊠ Hallandstrafiken	1 503	3 331	2.22	2.36	$4.92 \cdot 10^{-4}$	[44]
Värmlandstrafiken	1 682	3 743	2.23	2.59	$2.53 \cdot 10^{-3}$	[44]
Västmanlands Lokaltrafik	1 491	3 223	2.16	2.41	$5.32 \cdot 10^{-4}$	[44]
Dalatrafik	3 359	7 366	2.19	2.48	$3.92 \cdot 10^{-3}$	[44]
⊞ Karlstadsbuss	251	530	2.11	2.30	$4.52 \cdot 10^{-4}$	[44]
⊟ Luleå Lokaltrafik	205	459	2.24	2.73	$2.51 \cdot 10^{-3}$	[44]
⊡ Stadsbussarna Östersund	247	526	2.13	2.54	$5.68 \cdot 10^{-4}$	[44]
⊡ Swebus	158	435	2.75	3.78	$1.21 \cdot 10^{-2}$	[44]
◇ Power grid in the USA	4 941	13 188	2.67	3.74	$1.62 \cdot 10^{-3}$	[48]
◈ Network of airports in the USA	500	5 960	11.92	45.52	$1.26 \cdot 10^{-3}$	[13]
❖ Water distr. netw. Anytown	24	43	1.79	2.07	$2.31 \cdot 10^{-2}$	[33]
⬦ Water distr. netw. W.-C. Ranch	1 785	1 983	1.11	1.88	$4.30 \cdot 10^{-3}$	[33]
☆ Pizza Napoletana	2 291	3 687	1.61	2.16	$1.04 \cdot 10^{-2}$	[16]
△ Tic-tac-toe (2x2 board)	30	44	1.47	1.53	$4.57 \cdot 10^{-3}$	
⬠ Tic-tac-toe (3 moves, 3x3 board)	3 890	74 169	19.07	24.85	$4.11 \cdot 10^{-3}$	
⬭ Rubik's Cube (3 rot., 2x2x2 size)	1 417	1 644	1.16	5.15	$4.44 \cdot 10^{-2}$	
⬢ Rubik's Cube (3 rot., 3x3x3 size)	4 602	5 364	1.17	9.59	$3.63 \cdot 10^{-2}$	
▽ p2p Gnutella network 09	8 114	26 013	3.21	6.03	$9.40 \cdot 10^{-4}$	[39]
▽ Met. network of A. fulgidus	1 567	3 631	2.32	10.92	$2.95 \cdot 10^{-3}$	[26]
▽ Met. network of C. elegans	1 469	3 447	2.35	7.68	$1.97 \cdot 10^{-1}$	[26]
▼ Met. network of E. coli	2 897	7 104	2.45	12.03	$1.32 \cdot 10^{-1}$	[26]
▽ Graph of Wikipedia votes	7 115	103 689	14.57	78.82	$3.35 \cdot 10^{-3}$	[31, 32]

The following types of data sets are examined: ○ SISG models, □ transport networks, ◇ other spatial graphs, ☆ recipes, △ games, ▽ other data sets

$|N|$ number of nodes

$|E|$ number of edges

$\widehat{\rho^m}^N$ estimate of ρ^m by the number of nodes and edges (according to Theorem 5)

$\widehat{\rho^m}^D$ estimate of ρ^m by density (fit of the arithmetic mean of the total density of 10 (randomly chosen) connected induced subgraphs consisting of n nodes ($n = 1, \ldots, 50$), by $\rho^m/(t-1)$, according to Theorem 6)

χ^2 residuals for $\widehat{\rho^m}^D$

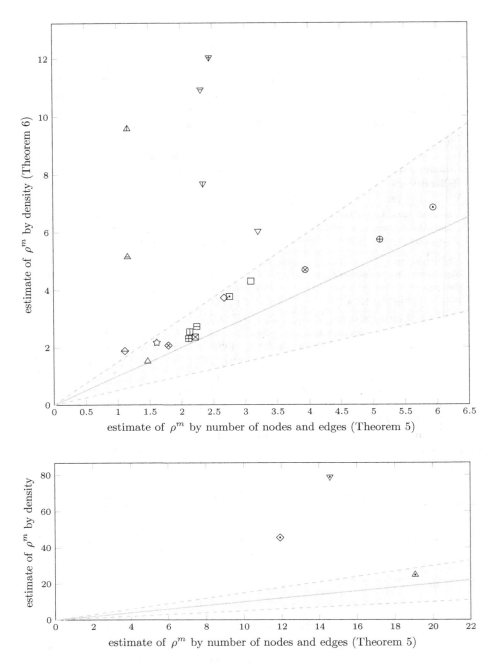

Fig. 2. Estimates of ρ^m for different data sets (see Table 1); if a data set has the typical properties of spatial data discussed in Sect. 2, both estimates coincide; for the grey area, the estimates differ by less than a factor of 1/2

introduced the hypothetical water distribution network of Anytown which has been used as a prototypical example in many studies [47]. Another example is the water distribution network of the Wolf-Cordera Ranch which distributes water to about 370,000 persons [33].

Recipes and Games are examples of activities that are related to time. We examine a formalization of a pizza recipe [16] where we represent the state of the ingredients as nodes and the actions as edges. Similarly, we examine the Rubik's Cube and Tic-tac-toe games. For both games, we consider different sizes of the cube resp. board, and we restrict the number of rotations resp. moves in the game to restrict the size of the network.

Data about a computer network (p2p Gnutella network 09) [39], metabolic networks of different cellular organisms [26], and a graph of Wikipedia votes [31,32] have no explicit references to space (in the used representations as abstract graphs).

In addition to these data sets, we consider three uniform SISG models to validate the hypothesis that both estimates coincide.

The data sets introduced in this section are very different in their structure. In Sect. 6.2, the relation between spatial structure and properties of SISG models was discussed. In the next section, we evaluate the considerations about this relation for the introduced data sets and discuss in how far we can distinguish between data with spatial structure, temporal structure, and data exposing none of these structures.

6.4 Results of the Comparison

In Sect. 4, we proposed a graph model for spatial data. As an application of the model, we compare it with data sets in order to evaluate whether the data sets can be categorized by their spatial structure. In particular, we argued in Sect. 6.2 that a data set can only have the properties discussed in Sect. 2 if the two estimates of ρ^m (Theorems 5 and 6) are approximately equal. In this section, we compare both estimates for the data sets that we introduced in the previous section.

The estimates[11] of ρ^m for the different data sets can be found in Table 1 and Fig. 2. In case both estimates coincide, the data sets are placed on the diagonal. If a data set has the properties which are discussed in Sect. 2, we expect it to have approximately equal estimate of ρ^m. If a data set has a low number of edges and is only temporal, we expect it to also have approximately equal but lower estimate of ρ^m, because time has one dimension whereas space has more. The estimates are not expected to coincide for data sets without spatial structure, but the values could coincide by chance.

As expected, both estimates and the value ρ^m used during the generation of the model are of similar size for all three SISG models. The differences between the two estimates can be shown to be an effect of the finiteness of the models.

---

[11] Note that the estimate by density can slightly differ for each computation because it depends on the random choice of subgraphs.

As transport networks have explicit references to space, they have a spatial structure, and both estimates are as expected of similar size, as can be seen in Fig. 2. The estimates are between about 2 and 4.5. As stops of public transport are placed in two-dimensional space and the density parameter is larger than 1 (but not much larger, because the networks are far from being complete graphs), the estimates are within a reasonable range.

Both estimates approximately coincide for the power grid in the USA, as is expected for a spatial graph. The estimates are between 2.7 and 3.7, which is within a reasonable range compared to the values for transport networks.

The network of airports in the USA can be embedded in space by the natural location of the airports, but both estimates are, nevertheless, very different. This effect is caused by the high number of non-spatial aspects which are influencing the network: the importance of a low average number of connections separating two airports, cultural aspects leading to more connections, legal restrictions (night flight restrictions, ban on unsafe airlines, taxes), etc. These aspects cause a number of structural properties that SISG models do not have, because these properties are not typical for spatial data in general: a non-uniform distribution of the airports in space, a high number of long-distance connections, a high number of hubs, a strong hierarchical organization (domestic and intercontinental), communities of strongly related airports, and many more [8].

The formalization of the pizza recipe as well as the game Tic-tac-toe are strongly influenced by time: most steps in pizza baking and all steps in Tic-tac-toe are irreversible, and time therefore induces a partial order on the nodes of the network. As can be seen in Fig. 2, both estimates are of the same size for each of these examples, which is expected because spatial and temporal data share some typical properties. The estimates for the pizza recipe and for Tic-tac-toe (for a board of size 2×2) are lower than those of transport networks, as is expected because time is one-dimensional (and space usually two- or higher-dimensional). For Tic-tac-toe with a board of size 3×3, there are many more choices in each step (with equal probability) resulting in a much less dominant temporal structure.

Water distribution networks can be embedded in space by the natural locations of the junctions. Both estimates approximately coincide for the water distribution network of Anytown, as is expected for a spatial graph. The estimates for the water distribution network of the Wolf-Cordera Ranch do not coincide but the values are not very different either. For both data sets, the estimates are between 1.1 and 2.1, which is reasonable due to the existence of a flow direction and the resulting low number of edges compared to the number of nodes.

The estimates for the remaining data sets differ by more than a factor of $1/2$.[12] These data sets do not have the typical properties of spatial data that were discussed in Sect. 2, suggesting that neither a spatial nor a temporal structure is *decisive* for these data sets.

---

[12] The factor of $1/2$ is chosen to visually illustrate how near data sets are depicted to the diagonal in Fig. 2. This choice is arbitrary and has no relevance for the fact that some data sets are depicted much nearer to the diagonal than others.

We have seen in this section that the argument of Sect. 6.2 is also valid (with minor deviations) for finite SISG models. The comparison of the estimates for different data sets has been shown to provide a meaningful characterization of the examined data sets according to their spatial structure.

7   Conclusion

Information with references to space exposes in many cases some typical properties, including Tobler's law and scale invariance. We introduced the scale-invariant spatial graph (SISG) model and showed that it meets some of the typical properties of spatial data. The model can therefore serve as a model of spatial data.

As an application of the SISG model, we showed how to determine whether a data set shares some typical properties of spatial data with the SISG model, even if the data set does not contain any *explicit* references to space. The evaluation of this consideration showed that spatial and temporal structures could be detected for the examined data sets.

The construction of SISG models introduces edges between two points if their distance is smaller than a certain value, which depends on the distribution of the points in space. This construction is a scale-invariant variant of a model proposed by Huson (Eq. 1 in Sect. 3). Waxman has introduced a smoothened variant of Huson's model by introducing edges with a probability that depends on the distance between two points (Eq. 2 in Sect. 3). Future research may, in a similar way, introduce and analyse a smoothened variant of the SISG model.

Properties of the SISG model were analytically computed only for models of infinite size. For finite models, these properties are different, and the influence of the "boundary region" has to be examined. Future research may analytically compute properties of finite models as well as result in algorithms that are taking the influence of the "boundary region" into account, e. g. when testing data sets for spatial structures.

In Sect. 6.2, we discussed two methods that enable us to estimate the value ρ^m for a given data set such that the data set is similar to the uniform SISG model $\mathcal{M}_\rho^m(s)$. Separate estimations of ρ and m would enable a more specific classification of data sets. In particular, it could be examined which influence generalization of data has on spatial structure, and which methods of generalization leave which parameters of the model invariant.

Spatial dependency is only one of the factors that influence the structure of data, and most data sets are characterized by additional aspects. Transport networks, for example, are usually connected (or have very few connected components), but SISG models with low dimension m and low density parameter ρ are in many cases disconnected; the outdegree equals the indegree for most nodes in a transport network; and the number of edges joining a node is usually between 2 and 4 for road networks. Future research may discuss how the SISG model can be modified in order to model specific types of spatial data, e. g. data about public transport or communication, and how other structures of data can be modelled.

Many processes are characterized by a hierarchical system, e. g. in Christaller's central place theory [10]. A combination of SISG models in different hierarchies could be used to model this effect. Such a hierarchical SISG model could be used to model and identify different hierarchies without assuming additional semantics. This would enable us to distinguish between local transport networks and nation-wide ones.

Algorithms are efficient if they take advantage of the data's structure. Future research may show how knowledge gained about the spatial structure of a data set can be used to improve and optimize algorithms, and SISG models, as a prototype for spatial data, could be used to gain insights into how these improvements and optimizations could be carried out. In particular, spatial indexes like R-trees [20] and R*-trees [9] could possibly be generalized to graphs that follow Tobler's law.

References

1. Aldous, D.J., Shun, J.: Connected spatial networks over random points and a route-length statistic. Stat. Sci. **25**(3), 275–288 (2010)
2. Archdeacon, D., Bonnington, C.P., Little, C.H.C.: An algebraic characterization of planar graphs. J. Graph Theor. **19**(2), 237–250 (1995)
3. Barabási, A.L., Jeong, H., Néda, Z., Ravasz, E., Schubert, A., Vicsek, T.: Evolution of the social network of scientific collaborations. Phys. A **311**, 590–614 (2002)
4. Barabási, A.L., Albert, R.: Emergence of scaling in random networks. Science **286**, 509–512 (1999)
5. Barabási, A.L., Albert, R., Jeon, H.: Scale-free characteristics of random networks: the topology of the world-wide web. Phys. A **281**(1–4), 69–77 (2000)
6. Barabási, A.L., Ravasz, E., Vicsek, T.: Deterministic scale-free networks. Phys. A **299**(3–4), 559–564 (2001)
7. Barthélemy, M.: Crossover from scale-free to spatial networks. Europhys. Lett. **63**(6), 915–921 (2003)
8. Barthélemy, M.: Spatial networks. Phys. Rep. **499**(1–3), 1–101 (2011)
9. Beckmann, N., Kriegel, H.P., Schneider, R., Seeger, B.: The R*-tree: an efficient and robust access method for points and rectangles. In: Proceedings of the International Conference on Management of Data (SIGMOD), pp. 322–331 (1990)
10. Christaller, W.: Die zentralen Orte in Süddeutschland: Eine ökonomisch-geographische Untersuchung über die Gesetzmässigkeit der Verbreitung und Entwicklung der Siedlungen mit städtischen Funktionen. Fischer, Jena (1933)
11. Chrobak, M., Eppstein, D.: Planar orientations with low out-degree and compaction of adjacency matrices. Theor. Comput. Sci. **86**, 243–266 (1991)
12. Coleman, T.F., Moré, J.J.: Estimation of sparse Jacobian matrices and graph coloring problems. SIAM J. Numer. Anal. **20**(1), 187–209 (1983)
13. Colizza, V., Pastor-Satorras, R., Vespignani, A.: Reaction-diffusion processes and metapopulation models in heterogeneous networks. Nature **3**, 276–282 (2007)
14. Denise, A., Vasconcellos, M., Welsh, D.J.A.: The random planar graph. Congressus Numerantium **113**, 61–79 (1996)
15. Erdős, P., Rényi, A.: On random graphs I. Publicationes Math. Debrecen **6**, 290–297 (1959)

16. European Commission: Commission Regulation (EU) No 97/2010 of 4 February 2010 entering a name in the register of traditional specialities guaranteed [Pizza Napoletana (TSG)]. Official J. Eur. Union **53**(L34), 7–16 (2010)
17. Franklin, C.: An introduction to geographic information systems: linking maps to databases. Database **15**(2), 12–21 (1992)
18. de Fraysseix, H., Rosenstiehl, P.: A depth-first-search characterization of planarity. Ann. Discret. Math. **13**, 75–80 (1982)
19. Gilbert, E.N.: Random graphs. Ann. Math. Stat. **30**(4), 1141–1144 (1959)
20. Guttman, A.: R-trees: a dynamic index structure for spatial searching. In: Proceedings of the International Conference on Management of Data (SIGMOD), pp. 47–57 (1984)
21. Hahmann, S., Burghardt, D., Weber, B.: 80% of all information is geospatially referenced? Towards a research framework: using the semantic web for (in)validating this famous geo assertion. In: Proceedings of the 14th AGILE Conference on Geographic Information Science (2011)
22. Hecht, B., Moxley, E.: Terabytes of Tobler: evaluating the first law in a massive, domain-neutral representation of world knowledge. In: Hornsby, K.S., Claramunt, C., Denis, M., Ligozat, G. (eds.) COSIT 2009. LNCS, vol. 5756, pp. 88–105. Springer, Heidelberg (2009)
23. Holland, P.W., Leinhardt, S.: An exponential family of probability distributions for directed graphs. J. Am. Stat. Assoc. **76**(373), 33–50 (1981)
24. Hunter, D.R., Goodreau, S.M., Handcock, M.S.: Goodness of fit of social network models. J. Am. Stat. Assoc. **103**(481), 248–258 (2008)
25. Huson, M.L., Sen, A.: Broadcast scheduling algorithms for radio networks. In: Proceedings of the Military Communications Conference (MILCOM), vol. 2, pp. 647–651 (1995)
26. Jeong, H., Tombor, B., Albert, R., Oltvai, Z.N., Barabási, A.L.: The large-scale organization of metabolic networks. Nature **407**, 651–654 (2000)
27. Jespersen, S.N., Blumen, A.: Small-world networks: links with long-tailed distributions. Phys. Rev. E **62**(5), 6270–6274 (2000)
28. Kalapala, V., Sanwalani, V., Clauset, A., Moore, C.: Scale invariance in road networks. Phys. Rev. E **73**, 026130 (2006)
29. Kuhn, W.: Core concepts of spatial information for transdisciplinary research. Int. J. Geogr. Inf. Sci. **26**(12), 2267–2276 (2012)
30. Kuratowski, C.: Sur le problème des courbes gauches en Topologie. Fundamenta Math. **15**(1), 271–283 (1930)
31. Leskovec, J., Huttenlocher, D., Kleinberg, J.: Predicting positive and negative links in online social networks. In: Proceedings of the 19th International Conference on World Wide Web (WWW), pp. 641–650 (2010)
32. Leskovec, J., Huttenlocher, D., Kleinberg, J.: Signed networks in social media. In: Proceedings of the 28th Conference on Human Factors in Computing Systems (CHI), pp. 1361–1370 (2010)
33. Lippai, I.: Water system design by optimization: Colorado Springs utilities case studies. In: Proceedings of the ASCE Pipeline Division Specialty Conference (Pipelines), pp. 1058–1070 (2005)
34. Louf, R., Roth, C., Barthelemy, M.: Scaling in transportation networks. PLoS ONE **9**(7), e102007 (2014)
35. MacLane, S.: A combinatorial condition for planar graphs. Fundamenta Math. **28**(1), 22–31 (1937)
36. McDiarmid, C.: Random graphs on surfaces. J. Comb. Theor. Ser. B **98**(4), 778–797 (2008)

37. McDiarmid, C., Steger, A., Welsh, D.J.A.: Random planar graphs. J. Comb. Theor. Ser. B **93**(2), 187–205 (2005)
38. Mocnik, F.-B.: Modelling spatial information. In: Proceedings of the 1st Vienna Young Scientists Symposium (VSS) (2015)
39. Ripeanu, M., Foster, I., Iamnitchi, A.: Mapping the Gnutella network: properties of large-scale peer-to-peer systems and implications for system design. IEEE Internet Comput. **6**(1), 50–57 (2002)
40. Sala, A., Cao, L., Wilson, C., Zablit, R., Zheng, H., Zhao, B.Y.: Measurement-calibrated graph models for social network experiments. In: Proceedings of the 19th International World Wide Web Conference (WWW), pp. 861–870 (2010)
41. Schnyder, W.: Planar graphs and poset dimension. Order **5**(4), 323–343 (1989)
42. Soon-Hyung, Y., Jeong, H., Barabási, A.L.: Modeling the internet's large-scale topology. Proc. Natl. Acad. Sci. U.S.A. **99**(21), 13382–13386 (2002)
43. Tobler, W.R.: A computer movie simulating urban growth in the detroit region. Econ. Geogr. **46**, 234–240 (1970)
44. Trafiklab: GTFS Sverige. https://www.trafiklab.se. Accessed 28 April 2013 (2013)
45. de Verdière, Y.C.: Sur un nouvel invariant des graphes et un critère de planarité. J. Comb. Theor. Ser. B **50**(1), 11–21 (1990)
46. Wagner, K.: Über eine Eigenschaft der ebenen Komplexe. Math. Ann. **114**(1), 570–590 (1937)
47. Walski, T.M., Brill, E.D., Gessler, J., Goulter, I.C.: Battle of the network models: epilogue. J. Water Resour. Plann. Manag. **113**(2), 191–203 (1987)
48. Watts, D.J., Strogatz, S.H.: Collective dynamics of small-world networks. Nature **393**, 440–442 (1998)
49. Waxman, B.M.: Routing of multipoint connections. IEEE J. Sel. Areas Commun. **6**(9), 1617–1622 (1988)
50. Whitney, H.: Non-separable and planar graphs. Proc. Nat. Acad. Sci. U.S.A. **17**(2), 125–127 (1931)
51. Xulvi-Brunet, R., Sokolov, I.M.: Evolving networks with disadvantaged long-range connections. Phys. Rev. E **66**, 026118 (2002)
52. Zipf, G.K.: The hypothesis of the minimum equation as a unifying social principle: with attempted synthesis. Am. Sociol. Rev. **12**(6), 627–650 (1947)

Strong Spatial Cognition

Christian Freksa[✉]

University of Bremen, Bremen, Germany
freksa@uni-bremen.de

Abstract. The ability to perform spatial tasks is crucial for everyday life and of great importance to cognitive agents such as humans, animals, and autonomous robots. Natural embodied and situated agents often solve spatial tasks without detailed knowledge about geometric, topological, or mechanical laws; they directly relate actions to effects enabled by spatio-temporal affordances in their bodies and their environments. Accordingly, we propose a cognitive processing paradigm that makes the spatio-temporal substrate an integral part of the problem-solving engine. We show how spatial and temporal structures in body and environment can support and replace reasoning effort in computational processes: physical manipulation and perception in spatial environments substitute formal computation, in this approach. The *strong spatial cognition* paradigm employs affordance-based object-level problem solving to complement knowledge-level computation. The paper presents proofs of concept by providing physical spatial solutions to familiar spatial problems for which no equivalent computational solutions are known.

Keywords: Cognitive architecture · Spatial cognition · Pervasive computing · Ubiquitous computing · Spatial affordance · Knowledge representation

1 Introduction

The philosopher John Searle distinguishes "strong" AI from "weak" or "cautious" AI (Artificial Intelligence): "According to weak AI, the principal value of the computer in the study of the mind is that it gives us a very powerful tool. … But according to strong AI, the computer is not merely a tool in the study of the mind; rather, the appropriately programmed computer really *is* a mind, in the sense that computers given the right programs can be literally said to *understand* and have other cognitive states. In strong AI, because the programmed computer has cognitive states, the programs are not merely tools that enable us to test psychological explanations; rather, the programs are themselves the explanations." (Searle 1980, p. 417).

Research in spatial cognition and the formalization of commonsense reasoning by means of logic has made substantial progress in reasoning about space and time in the past 25 years (Egenhofer and Franzosa 1991; Freksa 1991b; Cohn and Hazarika 2001; Renz and Nebel 2007; Ligozat 2011; Wolter and Wallgrün 2012; Dylla et al. 2013). For example, approaches of qualitative spatial reasoning permit the computation of spatial relations that correspond to real or potential configurations in the physical environment. These approaches employ AI tools (weak AI) to describe states of affairs in physical environments and to manipulate these descriptions in such a way that the

© Springer International Publishing Switzerland 2015
S.I. Fabrikant et al. (Eds.): COSIT 2015, LNCS 9368, pp. 65–86, 2015.
DOI: 10.1007/978-3-319-23374-1_4

resulting descriptions correspond to new states of affairs as obtained by physical operations in the spatial environment.

While weak AI models may achieve the same final configurations as the systems they model *(weak generative capacity, result equivalence, informational equivalence)*, only in certain cases they achieve these configurations in the same way *(strong generative capacity, strong equivalence* (Chomsky 1963), *computational equivalence* (Simon 1978)) as the corresponding operations in physical space. The reason is that only some spatial structures can be faithfully replicated in formal representations such as lists, trees, or arrays; other spatial structures and operations must be simulated in terms of the structures available in formal systems. As a consequence, computational operation sequences may be quite different from the sequences of mental and spatial operations modeled.

Whereas purely formal approaches employ sophisticated knowledge about a domain in order to infer new knowledge about spatial configurations, they may be only partially useful when it comes to modeling cognitive processes, their dynamics, their complexity, and their scalability (cf. Dreyfus 1979). Here we require models that operate on the restricted domain knowledge of a cognitive agent and preserve the relevant structures of the system to be modeled; in spatial problem solving these are in particular spatial and temporal structures.

As spatial cognition does not exclusively take place in the mind – it also involves severe interactions between mind, body, and the spatial environment – a full model of spatial cognition must take the roles of the body and the spatial environment for spatial problem solving into account (see Chandrasekaran 2006). In analogy to Searle's notion of *strong AI* we call embodied and situated models of spatial cognition that maintain the structural and functional properties of physical space *strong spatial cognition*. This paper expands on the work presented in (Freksa 2014; Freksa and Schultheis 2014; Freksa 2015).

2   Real-World Spatial Problem Solving

We focus on a special class of real-world problems that are of particular significance for cognitive agents such as humans, animals, and autonomous robots: spatial and temporal problems in physical environments. These problems share basic structural properties that have been intensively studied in spatial cognition research (Freksa 1997) and are quite well understood today on the information processing level, i.e. the level that is directly accessible to computers and computer science. One outcome of this research is the insight that the best solutions to different types of spatio-temporal problems require a considerable variety of approaches and tools (Descartes 1637; Sloman 1985). Some of the best approaches for human spatial problem solving make heavy use of the physical object level, i.e. they manipulate and perceptually inspect physical spatial configurations rather than solving problems entirely on the abstract information level.

In the following sections we will present a number of examples that illustrate different types of approaches to spatial problem solving.

Example 1. When a cognitive agent instantiates the route instruction *turn left at the next intersection*, he, she, or it does not require a detailed mental representation of the intersection or the environment; it also does not need to know and specify a precise turning angle before being able to follow the instruction.

The instruction provides a coarse guideline that permits the agent to move in an unfamiliar environment by perception-based route following and to select a *left* route from several alternatives that it may perceive in the vicinity of the intersection. The information about the turning angle is implicitly present in the spatial configuration that consists of the pose of the agent in relation to the route. The route instruction can be followed by means of a short perception-action loop. This does not require that the turning angle ever be made explicit in a cognitive representation.

In other situations cognitive agents may prefer to have detailed spatial knowledge before starting a spatial action as it may be easier to solve the problem by reasoning than by spatial interaction, as the following example illustrates:

Example 2. When I lost my keys that I last used during a trip some while ago, it may be worthwhile to reconstruct the preceding sequence of events on the trip mentally on the basis of remembered information about my preceding actions. Exploring the environment perceptually also might work to find my keys, but it could be more difficult or laborious, in the particular situation.

Embodied and situated cognitive agents are capable to operate on both the information processing and the physical object level. Perception and action operations serve as interfaces between the two levels; a memory serves to make information about the environment available to information processing in the absence of perceptual information and to store results of information processing for carrying out actions.

The classical cognitive science view treats cognition as a pure information processing activity (Simon 1978) that takes place entirely in the brain of a cognitive agent (respectively in a computer). But it was pointed out early on that the bodies of cognitive agents and the environments in which they perceive and act have significant effects on the types of solutions they pursue (Norman 1993; Clancey 1997; Wilson 2002; Wintermute and Laird 2008).

Furthermore, the information processing approach presupposes that a real-world problem has been comprehensively abstracted into an information processing task before cognitive processing can start. This assumption may be reasonable for routine tasks for which all necessary information is provided and which can be performed according to pre-existing standard patterns; however for novel problems, where a considerable part of the problem solving effort goes into identifying the information needed and finding appropriate representations, approaches, and tools, physical spatial configurations as in diagrams, maps, or other perceivable and/or manipulable spatial configurations typically play an important role for solving problems (cf. Polya 1945). In many cases, problem-specific approaches may be more appropriate and more efficient than general approaches. Specific approaches can take into account particular features of the problem domain to a larger extent than general approaches that abstract from specific characteristics.

In general, cognitive agents have a considerable variety of approaches and tools they can use to attempt solving a given spatial problem. In real-world cognitive problem solving we are confronted with problems where the identification of a suitable approach is more difficult than the computation of a solution on the basis of a given approach; once an appropriate approach has been selected, the problem solving procedure itself may be straightforward. This is illustrated in the following example:

Example 3. Suppose an agent's task is to determine visually (without a depth sensor) whether a tree is on this or on the other side of a fence (Fig. 1a).

A classical image analysis approach could use depth clues in the 2D projection of the 3D configuration to infer whether fence or tree is closer to the agent. Problem: the essential depth dimension is only poorly represented in the 2D projection. Spatial approach: Select a spatial reference frame that highlights the essential dimension; this can be achieved by relocating the agent such that the essential dimension is projected prominently onto the image of the configuration (Fig. 1b); now the task can be solved by considering only one image dimension, as the previous depth dimension has been mapped to the perceptually better accessible width dimension by spatial transformation in the problem domain.

(a)   (b)

Fig. 1. a. Hard visuo-spatial decision problem **b.** Same problem in a more suitable spatial reference frame

This specific example employs physical action only to modify perceptual acquisition of information from the environment without changing the 3-dimensional scene of interest. Note, however, that the 2-dimensional projection of the scene (that is typically the only information available for visual scene analysis) has considerably changed. Other kinds of manipulations would actually change the physical scene of interest in order to simplify or solve the spatial problem at hand.

The discussion shows that we need to address the question of *how to find* a suitable approach to solve a given spatial problem. Finding a suitable approach to solve a novel problem is one of the most interesting and challenging problems for cognitive agents. For the specific domain of spatio-temporal problems, we have good reasons to believe that the time is ripe to tackle this challenge; our belief is rooted in the fact that today we have a much better understanding of the properties of spatial and temporal relations and structures than twenty years ago.

There even is hope that once the challenges of identifying suitable spatio-temporal reasoning approaches to a given problem are tackled, we will be able to make use of the resulting approaches for addressing non-spatial problems, as well. This hope is rooted in the insight that human cognitive agents understand many problems through analogies (Gentner 1983) and metaphors (Lakoff and Johnson 1980). Non-spatial problems may be solved by mapping them onto spatially constrained structures; here they may be solved more easily before mapping the solution back to the problem domain. This procedure would be in contrast to generalizing spatial approaches to unconstrained domains where we would employ highly general approaches.

The spatial (and to a lesser extent the temporal) domain is particularly accessible to autonomous mobile agents with visual, haptic, and auditory perception and memory, as well as with moving, turning, and grasping capabilities. These capabilities enable agents to flexibly interact with their environments. Specifically, agents can actively influence which parts and aspects of their environment they perceive and they can modify spatial configurations in the environment through their actions. So far, these capabilities have not been systematically investigated and exploited for cognitive systems architectures. We therefore propose to develop proof of concept implementations and demonstrations for solving spatio-temporal problems strategically by making use of spatio-temporal affordances.

A main motivation for studying physical operations and processes in spatial and temporal form in comparison to formal or computational structures is that spatial and temporal structures in the body and the environment can substantially support (and even replace) reasoning effort in computational processes (Dewdney 1988). When we compare the use of different forms of representation (see Marr 1982), we observe that the processing structures of problem solving processes differ and facilitate different processing mechanisms (Sloman 1985). Spatial structures that resemble the problem domain may result in a lower complexity than structurally deviating abstract representations, as they can make direct use of the inherent structural properties without a need for describing them (Nebel and Bürckert 1995).

A main objective of our work is to explore the scope of application of this principle. This will involve a representation-theoretic assessment of representational equivalence and similarity, on the level of both result equivalence and strong equivalence. We develop a framework to relate physical actions and perception activities (Bajcsy 1988; Lungarella et al. 2002) to information processing activities, in order to assess the trade-off between physical and mental operations. Such a framework has long been missing in the debate surrounding diagrammatic vs. analytic reasoning (Glasgow et al. 1995).

Our approach builds on well-established paradigms from cognitive science (e.g. *knowledge representation theory* (Palmer 1978), *affordances* (Gibson 1979), *knowledge in the world* (Norman 1980), *conceptual neighborhood* (Freksa 1991a)) and on research carried out in the collaborative research center SFB/TR 8 Spatial Cognition at the University of Bremen over the past twelve years.

3 Background and Motivation

AI research initially was concerned exclusively with mental aspects of cognitive systems, specifically with operations and processes that take place in the brain (respectively computer) (Feigenbaum and Feldman 1963). Advances in robotics and knowledge representation have extended the scope of AI research to model perception and action processes, the (physical) bodies of agents, and the agents' spatial environments (e.g. Davis 1990). The rather general structures of abstract formalisms used for knowledge representation in computers allow describing arbitrary aspects of bodies and environments in detail and to reason about them, including spatial and temporal aspects.

While abstract reasoning about the world can be considered the most advanced level of cognitive ability (see Freksa and Schultheis 2014), this ability requires a comprehensive understanding of mechanisms responsible for the behavior of bodies and environments. But many natural cognitive agents (including adults, children, and animals) lack a detailed understanding of their environments (Piaget 1929) and still are able to interact with them rather intelligently.

Example 4. Children and dogs may be able to open and close doors in a goal-directed fashion without understanding the mechanisms of doors or locks on a functional level.

This suggests that knowledge-based reasoning is not the only way to implement problem solving in cognitive systems. Other systems of perceiving and moving goal-oriented autonomous agents have been proposed in biocybernetics and AI research to model aspects of cognitive agents (e.g. Braitenberg 1984; Brooks 1991; Pfeifer and Scheier 2001). These models implement perceptual and cognitive mechanisms that follow physical laws rather than formal representations that follow the laws of logics. Such systems are capable of reacting to their environments intelligently without encoding knowledge about the mechanisms behind the actions and without the associated computational cost.

In our spatial cognition research we have investigated the potential of qualitative spatial relations, of structure-preserving schematic maps, and of the role of intrinsically spatial structures for spatial reasoning and spatial problem solving (Freksa 1991b, 2013; Schultheis et al. 2014). A main result of this work is that structure-preserving representations can make direct use of spatial relations (e.g. spatial neighborhood, conceptual neighborhood, spatial order, and spatial orientation). Without structure-preservation, these relations would have to be derived through knowledge-based processes in more abstract formal representations of space. Thus, spatial calculi that exploit structure-preserving representations can avoid the necessity of performing certain computational derivations, as they represent crucial relations intrinsically rather than extrinsically (Palmer 1978; Dirlich et al. 1983).

Spatial cognition research also has been concerned with issues of resolution and granularity, both on a physical and on a conceptual level (Hobbs 1985; Freksa and

Barkowsky 1996; Mossakowski and Moratz 2012; Schultheis and Barkowsky 2011). In knowledge representation, we must deal with the issue of level of detail on which we represent objects and configurations in order to solve certain problems. The finer the level of representation, the more problems we will be able to solve, in principle. But this comes at a cost: the more details we have to deal with, the more computation we have to invest. Cognitive processes frequently process information from coarse to fine rather than from fine to coarse. These processes are directly supported by physical and spatial properties of their environments. An illustration is given in the following example:

Example 5. In vision, *coarse* corresponds to *distant* and *fine* corresponds to *close*. The same sensor adapts its 'representation' of the world simply by physically moving towards an object or away from it.

The field of diagrammatic reasoning (Glasgow et al. 1995; Chandrasekaran 2006; Goel et al. 2010) is concerned with problem solving by means of diagrams, a special form of spatial representations. A key issue here is the comparison between formal and diagrammatic representations and reasoning processes. Of particular interest is the equivalence between the reasoning procedure operating on the corresponding formal structure and the problem solving procedure operating on the spatial structure. Strong equivalence has been mainly studied in comparing different formal systems. Comparing processes operating on physical spatial structures with processes operating on formal structures poses an interesting challenge, as we will require a reference framework that includes information processing and re-configuration of spatial configurations.

Our research builds on work in the areas of spatial and temporal reasoning and simulation, data structures, diagrammatic reasoning, mental representations, theories of knowledge representation and computation, and related areas. In the following I will sketch some of the issues that have been particularly relevant for our work.

In 1983, James Allen published a widely referenced paper on Maintaining knowledge about temporal intervals (Allen 1983). In this paper the author developed a calculus for temporal relations based on the set of 13 jointly exhaustive and pairwise disjoint (JEPD) relations that can hold between two temporal intervals. Allen's approach became the role model for numerous calculi for qualitative spatial reasoning (QSR) (e.g. Guesgen 1989; Egenhofer and Franzosa 1991; Randell et al. 1992; Freksa 1992; Ligozat 1993; Zimmermann 1995; Moratz et al. 2000; Van de Weghe et al. 2005). Whereas a single calculus is sufficient for reasoning about temporal relations, a multitude of calculi are required to cover all relevant aspects of spatial relations. In particular, calculi for topological relations, for orientation relations, and for distance relations have been developed (Freksa and Röhrig 1993; Cohn and Hazarika 2001).

Attempts to integrate the different aspects of spatial calculi in a single calculus failed. The reason for this is rooted in the fact that the different aspects of space are strongly intertwined as indicated by the following example:

Example 6. A topological relation between two objects constrains the distance between them; a set of distance relations between several objects constrains their relative orientations, etc.

An integrated calculus would have to compute and update all relations that are affected by a single change in order to maintain consistency among the relations. This would be computationally expensive and not useful if only a single aspect of the spatial configuration is of interest.

This computational dilemma points to a property of space that makes spatial structures particularly interesting: properties of spatial objects and configurations are intrinsically highly interdependent. If we modify one spatial aspect (e.g. distance, orientation, or topological relation) in a spatial structure, other spatial aspects will change 'automatically', as well. We call such a structure a *spatial substrate* (Freksa 2013). If we move an object in space, the spatial locations of all its parts as well as their relations to other objects will change. If we change a single spatial aspect in a spatial substrate, all these changes take place ('for free'); no computing (or otherwise) effort is required.

In other words: The computational dilemma described above mutates into a special feature of spatial substrates: If suitable operations for spatial simulations in spatial substrates are available, we may be able to avoid an enormous amount of computation, whereby consistency is intrinsically guaranteed.

The use and exploitation of structural properties of representations makes up the core idea of data structures: a data structure supports a particular way of organizing information such that it can be used efficiently (Knuth 1997). Depending on the problem structure and on the tasks to be performed on a representation, certain data structures may be better suited than others. Some data structures share structural properties with spatial substrates, in particular lists, trees, and arrays.

Example 7. Binary trees are particularly useful data structures for sorting and searching linearly ordered information, as the linear order can be mapped to the leaves of the tree while the branching structure of the tree corresponds to decisions to be taken in the sorting/searching process.

For other data, in which no order is implied in their appearance, an ordered data structure may be detrimental to the task as the structure may impose an unintended bias; therefore we employ structures that avoid such a bias, in these cases.

Example 8. Hash tables avoid a correspondence with spatial substrates and processes and thus permit data access independent of an ordering.

Unfortunately, we do not have suitable data and access structures for all operations that we would like to perform on computational representations of space, as the following example suggests:

Example 9. Zooming and perspective transformation operations involve computation on all elements of the domain and tend to be computationally expensive.

In spatial substrates, in contrast, we have more flexibility in accessing data. We can perform zooming and perspective transformation by manipulating the data access (perception) apparatus while leaving the data itself untouched. In the performance assessment of computer algorithms, data access operations usually are considered cheap in comparison to computational transformations.

Efficient use of information also depends on the operations we permit on data and knowledge structures. Unrestricted relations and operations may result in high complexity and in unfavorable scaling behavior while having no particular advantages, in many cases.

Example 10. In qualitative reasoning, we can relate each spatial (or temporal) relation to each other in the set of JEPD relations. When reasoning about a given domain, this leads to exponential complexity in the number of objects related to one another.

On the other hand, we know that in the spatial (respectively temporal) domain most transitions between relations cannot occur due to substrate-inherent reasons. This insight allows us to restrict a calculus to take into account only transitions between conceptually neighboring relations (Freksa 1991a). This restriction on the representational level has no negative implications on the object level, as other transitions cannot occur in the represented domain. But it has great advantages: the calculus becomes tractable, as it results in only polynomial complexity (Nebel and Bürckert 1995).

Although spatial calculi have been applied with considerable success to a number of spatial problems (e.g. Wolter et al. 2008; Kreutzmann et al. 2013; Falomir et al. 2013; Dubba et al. 2015), there are at least two aspects of such calculi that seem in need of improvement. First, calculi mainly represent information about an abstract spatial problem, i.e., they largely fail to systematically exploit the constraints and affordances provided by the physical structure of the domain. As a result, spatial problems as represented by calculi may easily become computationally too complex to be solved efficiently, if spatial constraints (or subsets thereof) are not introduced on top of the spatial calculi – for example in the form of conceptual neighborhood (Freksa 1991a; Nebel and Bürckert 1995; Balbiani et al. 2000). Second, there is currently only a poor understanding concerning which calculi are best suited for solving a given spatial problem: When faced with a specific problem it is not clear how to select among the many available formalisms for solving the problem.

In AI, early attempts to exploit spatial structures for reasoning purposes are found in the subfield of diagrammatic reasoning (Larkin and Simon 1987; Glasgow et al. 1995). The idea was to make use of the spatial arrangement of objects on a (simulated) two-dimensional medium to optimize search and decision processes in reasoning. This idea was inspired by the way humans utilize diagrams employing their visual perceptual capabilities. Consequently, applications are found in geometric reasoning, in (physical) problem solving, and in the simulation of physical motion.

From a basic research perspective in cognitive science, spatial substrates play an important role in mental spatial reasoning using visual mental images (Kosslyn 1980, 1994) and spatial mental models (Johnson-Laird 1983, 1995). In these mental representation formats, entities under consideration are dealt with in a similar way as they would be perceived and interacted with in the real world. The properties of these types of mental representations in mental spatial reasoning have been investigated and modeled from a computer science perspective by Schultheis and Barkowsky (2011) and Schultheis et al. (2014), among others.

In general, when a cognitive information processing system is analyzed from an informatics perspective, this can be addressed on three distinct levels (Marr 1982): as a computational theory; from the perspective of knowledge representations and the processes operating on them; and with respect to a specific hardware implementation. It is an essential feature of our approach that we operate on all three levels: we are interested in spatial substrates that are physically realized and utilized by specific hardware; we operate on those substrates employing suitable representation structures and cognitive processes; and we aim at establishing a new paradigm of theoretically investigating cognitive systems from an information processing point of view.

In summary, we can identify numerous results which indicate that adaptations to specific task requirements may be cognitively and computationally more adequate (Sloman 1985) than the previously pursued goal of a 'General Problem Solver' (Newell et al. 1959). In particular, exploiting the restrictions of spatial and temporal substrates may have considerable advantages over employing general abstract approaches. This insight also leads us to question the appropriateness of employing highly expressive representation languages for reasoning about the severely constrained spatial and temporal domains.

4   Objectives of this Work

Cognitive agents such as humans, animals, and autonomous robots comprise brains (respectively computers) connected by powerful interfaces to the environment: sensors and actuators. The sensors and actuators are arranged in their (species-specific) bodies to interact with their (species-typical) environments. All of these components need to be well tuned to one another to function in a fully effective manner. For this reason, we view the entire aggregate (cognitive agent including body and environment (Wilson 2002)) as a *full cognitive system* (see Fig. 2). This is in contrast to classical AI systems, which have focused on the structures and processes within the confinements of the computer.

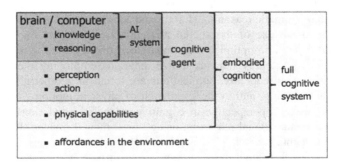

Fig. 2.   Structure of a full cognitive system

Our research is concerned with the investigation of cognitive principles that govern the interaction between these high-level cognitive system components. Although the research is motivated by the capabilities we observe in natural cognitive systems, our goal is not to replicate or characterize a particular natural system. The project aims at investigating and analyzing the distribution, coordination, and execution of tasks among the components of embodied and situated spatial cognitive agents on the system level from a systems engineering/system analysis perspective (Pylyshyn 1988).

In a classical information processing/artificial intelligence approach, we would describe the relevant components outside the brain or computer formally in some knowledge representation language or scheme in order to allow the computer to perform formal reasoning or other computational processing (Russell and Norvig 2010). Physical, topological, and geometric relations would be transformed into abstract information about these relations and the tasks would be performed entirely on the information processing level, where physical, topological, and geometric relations are replaced by descriptions of their properties.

As we are dealing with spatial problems that originate from a spatial substrate and whose solutions eventually are relevant in a spatial substrate (e.g. we solve wayfinding tasks in order to apply the identified route in physical space), we will investigate under which conditions and to which extent we will be able to take advantage of the specific spatial properties and the structure of the problem domain. The goal is to use abstraction only in as far as it is useful for the given problem and to maintain the spatial structure when we can profit from its intrinsic spatial properties (Palmer 1978). In this way we may be able to avoid effort, difficulties, and losses due to the problem abstraction process, the reasoning process (in a possibly not optimally adapted formalism), and the concretion process that maps the abstract problem solution back into the spatial problem domain.

When we talk of spatial structures and spatial reasoning in the context of spatial cognition, we implicitly include temporal structures and temporal reasoning, as we are concerned with cognitive processes and the dynamics of space which must take into account the structure and constraints of time.

Spatial and temporal structures are not as expressive as general abstract languages. Abstract languages can transcend the limitations of the physical realm. In comparison, spatial and temporal structures have the advantage of representing precisely those configurations that we deal with in space and time. Limitations of spatial representation media have been explored in art (e.g. by Pablo Picasso and Maurits Cornelis Escher); their advantages come to bear when representing concrete spatial entities, events, or concepts that we imagine in terms of spatial relations or structures (e.g. abstract hierarchies). Here, the limitations of spatial structures are useful (e.g. when representing spatial configurations or abstract hierarchies on a spatial substrate like a piece of paper) as we do not have to make relations explicit that are implicitly provided by the spatial structure.

Thus, the question is not whether more general abstract representations or more specific concrete representations are better; rather, we need to consider in which form the problem is given and exactly which operations we want to perform. In the end, the question is how we can combine the advantages of a general abstract language with the advantages of a specific representation structure (Freksa and Barkowsky 1999).

As an example, consider geographic maps: they employ a spatial medium – paper or a 2D display of some sort. The spatial structure by itself is rather useless; a map requires symbols that add semantics and establish an abstract correspondence to entities in the real world. In principle, of course, everything that can be represented on a map could be described in terms of an expressive abstract language: all spatial relations that are implicitly expressed through the spatial medium could be made explicit in terms of linguistic or other descriptions.

For perceiving and acting cognitive agents who heavily rely on spatial interdependencies, such descriptions of maps would be difficult to use. Thus, we need to find the right balance between implicitly and explicitly available knowledge – or equivalently: between specific and general representation structures. In maps, this balance varies depending on the purpose of use: some features are expressed spatially (configurations, relative distances, and shapes) while others are expressed symbolically (type of road, type of land use) (Freksa 1999). As both types of representation share the same spatial medium, they compete for kinds of interpretation, as the following example shows:

Example 11. In interpreting a road map, can I multiply the width of a road symbol on the map with the scaling factor of the map to determine the width of the corresponding road in the environment or is the width of the road symbol a constant related to that symbol?

The answer to this question depends on the design decision for the particular map type, namely which aspects of the environment are to be represented pictorially and which symbolically. In our research, we investigate distributions between implicit and explicit knowledge in intrinsic vs. extrinsic representations from a cognitive processing perspective: when we employ cognitive offloading of information processing from the mind (respectively computer) to the environment (Wilson 2002), we will have to add new information processing structures (on a more abstract level) to control the cognitive agent's use of the externalized knowledge. This creates interesting trade-offs that we will investigate in the framework of the spatial substrate processing paradigm.

Just as the *logic programming* paradigm is designed to generate inferences about truth values by employing the laws of logics, we develop a spatial processing paradigm to generate inferences about spatial relations by employing the laws of space and time (Freksa and Schultheis 2014). Logics is an excellent language on the meta-level, for describing states of affairs and for reasoning about them; our objective for developing the spatial processing paradigm is to produce an object level representation for processes of spatial cognition.

Against this background, our objective is to initiate a paradigm shift in representation and problem solving by placing emphasis on representational approaches and solutions that exploit knowledge in the world (Norman 1993) (and the affordances (Gibson 1979) and constraints (Freuder and Mackworth 1994) that come with it) as well as systematically investigating how to best distribute problem solving effort between abstract (knowledge *about* – meta-level) and concretely embodied (knowledge in the world – object level) modes of representation and processing. More specifically, our objectives comprise:

- Characterizing the division of labor between abstract knowledge representation and knowledge in the world.
- Devising representation and processing structures that facilitate the exploitation of knowledge in the world.
- Devising selection and control structures to identify a promising problem solving approach from a set of alternatives.
- Determining how (a) the effort required for building up an abstract representation; (b) the accessibility of knowledge in the world; (c) the frequency with which certain

information is reused; and (d) the availability of computational and memory resources influence the division of labor.

- Developing measures that allow comparing effectiveness and efficiency of different combinations.
- Realizing the proposed approach in a simulation environment and on a robotics platform.

5 Approach

We focus on spatial and spatio-temporal tasks that are directly accessible by perception and allow for manipulation by physical action. This is the domain we understand best in terms of computational structures; we have well established and universally accepted reference systems to describe and compute spatial and temporal relations. The limitation to spatial tasks may turn out less severe than it may appear initially: numerous non-spatial problems can be transformed into equivalent spatial problems where the spatial structure helps to support the problem solving process. Human problem solvers make use of problem spatialization, for example, when visualizing a linguistically specified problem in the form of a diagram in order to better grasp the problem and/or to be better able to formalize it for formal problem solving. Depending on the spatial representation chosen for the diagram, it may be easier or harder to grasp or formalize the problem.

The main hypothesis of the *strong spatial cognition* paradigm is that the 'intelligence' of cognitive systems is grounded not only in specific abstract problem solving approaches, but also – and perhaps more importantly – in the capability of recognizing characteristic problem structures and of selecting particularly promising problem solving approaches for given tasks. Formal representations generally do not facilitate the recognition of such structures due to a bias inherent in the abstraction. This is where *mild abstraction* can help as it abstracts only from few aspects while preserving important structural properties.

The insight that spatial relations and physical operations are strongly connected to cognitive processing will lead to a different division of labor between the perceptual, the representational, the computational, and the locomotive parts of cognitive interaction than the one we have been pursuing with AI systems: rather than putting all the 'intelligence' of the system into the computer, the proposed approach aims at putting more intelligence into the interactions between components and structures of a cognitive system as well as into the structure of the problem representation. More specifically, we aim at exploiting intrinsic structures of space and time to reduce the complexity of computation.

We argue that a flexible assignment of physical and computational resources for cognitive problem solving is closer to natural cognitive systems than the almost exclusively computational approach. For example, when we as cognitive agents search for certain objects in our environment, we have at least two different strategies at our disposal: we can represent the object in our mind and try to imagine and mentally reconstruct where it could or should be – this would correspond to the classical AI approach; or we can visually search for the object in our spatial environment. Which approach is better (or more promising) depends on a variety of factors including memory and physical effort required. Frequently a clever combination of both approaches will be best.

We develop and implement a proof of concept for the proposed approach to spatial problem solving through simulations of the perception and manipulation processes as well as through physical agent models, e.g. as generated by a 3D printer (Freksa 2013). The research is primarily conceived as basic research in cognitive systems engineering: we identify and relate an inventory of cognitive principles and ways of combining them to obtain cognitive performance in spatio-temporal domains.

For this project, we can build on extensive research on spatial and temporal relations, their representation in memory, and with qualitative spatial reasoning in the framework of international interdisciplinary spatial cognition research. Naturally, the proposed approach will not be as broadly applicable as some of the approaches we have pursued in classical AI research, as it is intentionally restricted to spatio-temporal structures; but the approach promises to discover broadly applicable cognitive engineering principles for the design of tomorrow's intelligent agents.

Our philosophy is to understand and exploit pertinent features of space and time as modality-specific properties of cognitive systems that enable powerful specialized approaches in the specific domain of space and time. Since space and time are most basic for perception and action and ubiquitous in cognitive processing, we believe that understanding and utilizing their specific structures will be particularly beneficial.

There are at least two general approaches towards studying cognitive systems: (1) by analysis or (2) by synthesis. A standard empirical approach would be to analyze an existing system in order to understand its functionality. When we are dealing with complex systems whose components interact in multiple ways, it becomes difficult to derive a single theory that explains the functionality of the underlying system.

In our research, we investigate cognitive systems by synthesizing components whose functions we understand. The objective is to provide a proof of concept in order to discuss and compare various system architectures. Our method follows the approach that Braitenberg called 'experiments in synthetic psychology'. As Braitenberg (1984) argues in his book *Vehicles*, this approach may lead in a straightforward way to a well-understood model that can be scrutinized by empirical methods.

Further, we can distinguish at least two types of models of systems that may or may not pursue different objectives: (1) models that aim at reconstructing the functional components, relationships, and performance of a system on a given level of abstraction and granularity (Zadeh 1979; Hobbs 1985) as closely as possible through replicating properties and functionality of their components (object-level models); and (2) models that make the scientists' knowledge about properties, relations, and functions of the modeled system explicit through descriptions or prescriptions (knowledge-level models). Both types of models are suited to enhance our understanding of cognitive systems and both have advantages and disadvantages.

Example 12. Well-known examples of the two types of models from a non-cognitive domain are wind channel models of automobiles or airplane wings (object level) and finite-element algorithms to power virtual wind tunnel simulations (knowledge level).

Before the theory of aerodynamics was mature enough to convey all relevant parameters that influence aerodynamic properties and computers powerful enough to simulate the aerodynamic interactions, engineers placed physical models (full sized or spatially scaled) that maintained crucial features such as shape and surface finish into the air flow generated by physical wind channels. In these wind channels, the engineers could measure physical forces to evaluate their designs under varying environmental conditions. As the theory of aerodynamics advanced (partly due to empirical testing of various aerodynamic shapes), the physical interactions between design and surrounding airflow were better understood; they could be characterized by finite element models, a numerical approximation approach to describing physical field behavior. Supercomputers were employed to calculate aerodynamic properties of cars and airplane wings, and other objects, as the computation of the interactions between all the parameters involves massive computation.

While the finite-element simulation has clear advantages over the wind tunnel simulation, it also has decisive disadvantages. Advantages are: the finite-element model reflects scientific understanding of aerodynamic interactions; furthermore wind tunnel experiments required huge labs with special equipment that required personnel and consumed considerable amounts of energy. Disadvantages are: the mathematical simulation of the aerodynamic processes computes the effects of physical interaction by iterative numerical approximation processes; these processes have no direct correspondence to the physical interaction processes they represent and are not performed in real time. Considerable computation is required to integrate the results from a multitude of interactions.

For building cars and airplane wings the disadvantages of computational simulation are not significant; the simulations can run 'off-line' to compute the required characteristics of the design. In cognitive modeling, however, we are not only interested in the result of applying the model; we are interested in the dynamics of the cognitive processes themselves. Thus we may benefit from an object-level model that intrinsically guarantees certain domain properties – until we understand their significance sufficiently well to simulate them in a cognitive process model in real time.

In modeling computational problem solving in spatial substrates, we are confronted with the spatial substrate (object level) and at least two levels of abstraction: the knowledge level and a meta-knowledge level (Fig. 3).

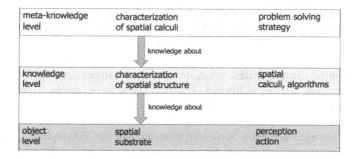

meta-knowledge level	characterization of spatial calculi	problem solving strategy
	↓ knowledge about	
knowledge level	characterization of spatial structure	spatial calculi, algorithms
	↓ knowledge about	
object level	spatial substrate	perception action

Fig. 3. Levels of cognition in spatial problem solving

The knowledge level makes aspects of the spatial structure on the object level explicit in the form of spatial relations and calculi. The meta-knowledge level makes knowledge about the relations and calculi on the knowledge level explicit in the form of knowledge about their use.

A fundamental contribution of computer science and artificial intelligence is to enable algorithmic interpretation of formally described knowledge such that formal representations of knowledge can be executed and thus perform as a model. In this way, the knowledge-level model can turn into a performance model whose behavior resembles that of the object-level system. To what extent it will be possible to reproduce the behavior of the object level system will depend on the structures of the representation and of the interpretation processes. It is not necessarily the most general or most powerful interpretation mechanism that will yield the best correspondence; rather, the best adapted knowledge structures and interpreters will win.

Object-level models do not make knowledge explicit; they maintain system properties and relations implicitly in their system structure. If the system structure and the process structure of the model closely match the corresponding structures of the cognitive system on the object level, its behavior can be expected to closely resemble that of the modeled system. Typically, the structures and processes of the modeled system are only partially known; the model designer fills in other parts on the basis of 'informed speculation'. Matching behavior provides no proof but may provide strongly suggestive evidence for the appropriateness of the model. Running and testing such object-level models and observing their global and detailed behavior can provide very useful information for further theoretical and/or empirical exploration of the cognitive system of interest.

Knowledge-level modeling can be considered the more sophisticated of the two methods, as it starts with knowledge that can be used for formal reasoning about scientific findings along established routes. A weakness of the knowledge-based approach, however, is that it forces the structure of the formalism and the reasoning procedures onto the system under investigation. This may not be a problem as long as we are only interested in the (static) initial and final states of problem solving processes of the cognitive system under investigation; however, if we are interested in modeling the dynamics of a cognitive system, a good match of module and process structures between model and target system become essential. This is where object-level models may score; they start with an engineering approach that initially focuses on system architecture and behavior. From here, valuable scientific knowledge on principles of cognitive processing can be derived. Braitenberg (1984) refers to this as the "law of uphill analysis and downhill synthesis".

In Sect. 2 I presented an example where spatial problem solving is supported by a suitable reference frame for a problem given in a spatial substrate (Example 3). In this example, the objects in the spatial substrate were not manipulated; only the reference frame was changed. In the following, I will present an example of how intrinsic properties of a spatial substrate can be exploited for spatial problem solving by manipulating spatial configurations. Manipulation for spatial problem solving constitutes a more severe interaction with the environment. In Example 13, the spatial problem configuration itself is modified in order to obtain a configuration that is equivalent with respect to the problem to be solved, but easier to analyze by spatial procedures.

Example 13. Suppose an agent's task is to identify the shortest route that connects a location A with a location B given several possible paths that can be chosen.

A classical knowledge-level approach would represent the lengths of the route sections, compute various alternatives of configuring these sections to connect A and B and determine the option with the smallest overall length. Observation: the lengths of the route sections need to be known and several alternatives have to be computed and compared before the one route of interest is identified.

Spatial approach (Dewdney 1988): Here we use a mildly abstracted version of the street network: a paper map in which all paper regions which do not correspond to routes are missing. We obtain a deformable map consisting only of route representations which preserve the relative lengths of the route sections (Fig. 4a).

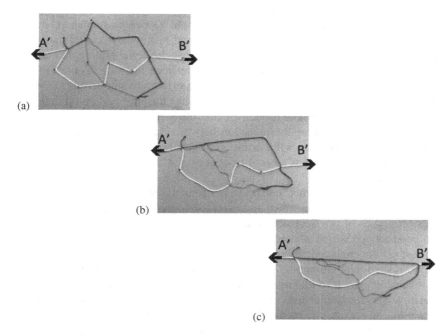

Fig. 4. Determining the shortest route from point A to point B by physical manipulation of a mildly abstracted representation of a route network. (a) The (non-elastic) strings corresponding to route segments preserve the relative distance relations of the original route segments; the distance relations are invariant wrt. physical manipulations (pulling apart strings at A' and B') which distort angles and shapes of the route network (b) and (c). The shortest route is identified as the route corresponding to the straight connection between A' and B' in (c).

The map permits certain spatial reconfigurations of the network through deformation while preserving topology and important geometric constraints; in particular, an agent can (carefully) pull apart the positions A' and B' on the map (Fig. 4b) that correspond to locations A and B until a string of route sections forms a straight line between these positions (Fig. 4c); due to the geometric properties of the representation, the route sections corresponding to the sections on the straight line manifest the shortest route between A and B.

This approach avoids computation by reducing the problem to the relevant single dimension of length on which a basic geometric principle *straight line is shortest connection* can be directly applied. In both, Examples 3 and 13, computational problem solving operations are augmented by spatial operations.

6   Conclusions and Outlook

Our project sets out to investigate a new architecture of artificial cognitive systems that more closely resembles natural cognitive systems than purely knowledge-based AI approaches to cognitive processing. This is to be achieved by involving interaction with space through perception and action.

With today's availability of 3D printers, Example 13 can be implemented in the framework of a robotic system that interacts with and manipulates configurations in a spatial substrate: a route network can be extracted from an aerial photograph or from a map database and be printed (without the regions between the routes) on a 3D printer using a non-elastic and non-rigid printing substrate (comparable to paper). On the printer output, the robot's perception system identifies the start and end points on the route and grasps both points. The robot then cautiously pulls apart the two points until it can identify an almost straight connection between the start and end points of the route network; this connection will correspond to the shortest connecting route in the network. The example serves as a proof-of-concept for our project from which further explorations will follow.

The project brings together perspectives from a variety of disciplines: (1) the cognitive systems perspective, which addresses the cognitive architecture and trade-offs between properties of physical structures and properties of their descriptions; (2) the formal perspective, which characterizes and analyzes the resulting structures and operations; (3) the engineering perspective, which constructs and explores varieties of cognitive system configurations; and (4) the psychological-empirical perspective, which relates the effects of different system behaviors to those of natural agents. In the long term, we see potential technical applications of physically supported cognitive configurations, for example, in the development of future *intelligent materials* (e.g. 'smart skin' where spatially distributed computation is required that needs to be minimized with respect to computation cycles and energy consumption, and more robust and adaptable artificial agents, which can deal with unknown environments).

Acknowledgements. Heated discussions with members of the Bremen Cognitive Systems group, in particular Thomas Barkowsky, Ana-Maria Olteteanu, Holger Schultheis, Frank Dylla, Jasper van de Ven, Zoe Falomir, and Loai Ali, as well as with Werner Kuhn promoted this work. Excellent comments and suggestions for improvement by numerous anonymous reviewers are highly appreciated. The German Research Foundation (DFG) supported this work through generous funding for the SFB/TR 8 Spatial Cognition. This paper is dedicated to Gerhard Dirlich and Ulrich Furbach, who set the foundations for this project with me more than thirty years ago.

References

Allen, J.F.: Maintaining knowledge about temporal intervals. Commun. ACM **26**, 832–843 (1983)

Bajcsy, R.: Active perception. Proc. IEEE **76**, 996–1005 (1988)

Balbiani, P., Condotta, J.F., Ligozat, G.: Reasoning about generalized intervals: Horn representability and tractability. In: TIME 2000, pp. 23–29 (2000)

Braitenberg, V.: Vehicles: Experiments in Synthetic Psychology. MIT Press, Cambridge (1984)

Brooks, R.A.: Intelligence without representation. Artif. Intell. **47**, 139–159 (1991)

Chandrasekaran, B.: Multimodal cognitive architecture: Making perception more central to intelligent behavior. In: Proceedings of the AAAI, pp. 1508–1512 (2006)

Chomsky, N.: Formal properties of grammar. In: Luce, R.D., Bush, R.R., Galanter, E. (eds.) Handbook of Mathematical Psychology II, pp. 323–418. John Wiley and Sons, London (1963)

Clancey, W.: Situated Cognition: on Human Knowledge and Computer Representations. Cambridge University Press, Cambridge (1997)

Cohn, A.G., Hazarika, S.M.: Qualitative spatial representation and reasoning: An overview. Fundamenta informaticae **46**(1), 1–29 (2001)

Davis, E.: Representations of Commonsense Knowledge. Morgan Kaufmann, San Mateo (1990)

Descartes, R.: Discourse on the method, Part VI. In: Newby, I., Newby, G. (prods.) The Project Gutenberg EBook 2008, #59 (1637)

Dewdney, A.K.: The Armchair Universe. W.H. Freeman and Company, San Francisco (1988)

Dirlich, G., Freksa, C., Furbach, U.: A central problem in representing human knowledge in artificial systems: The transformation of intrinsic into extrinsic representations. In: Proceedings of the 5th Cognitive Science Conference, Rochester (1983)

Dreyfus, H.L.: What Computers Can't Do. The Limits of Artificial Intelligence, Revised edn. Harper and Row, New York (1979)

Dubba, K.S.R., Cohn, A.G., Hogg, D.C., Bhatt, M., Dylla, F.: Learning relational event models from video. J. Artif. Intell. Res. **53**, 41–90 (2015)

Dylla, F., Mossakowski, T., Schneider, T., Wolter, D.: Algebraic properties of qualitative spatio-temporal calculi. In: Tenbrink, T., Stell, J., Galton, A., Wood, Z. (eds.) COSIT 2013. LNCS, vol. 8116, pp. 516–536. Springer, Heidelberg (2013)

Egenhofer, M., Franzosa, R.: Point-set topological spatial relations. Intern. J. Geogr. Inf. Syst. **5**(2), 161–174 (1991)

Falomir, Z., Museros Cabedo, L., Castelló, V., González Abril, L.: Qualitative distances and qualitative image descriptions for representing indoor scenes in robotics. Pattern Recogn. Lett. **34**(7), 731–743 (2013)

Feigenbaum, E., Feldman, J.: Computers and Thought. McGraw-Hill, New York (1963)

Freksa, C.: Conceptual neighborhood and its role in temporal and spatial reasoning. In: Singh, M., Travé-Massuyès, L. (eds.) Decision Support Systems and Qualitative Reasoning, pp. 181–187. North-Holland, Amsterdam (1991a)

Freksa, C.: Qualitative spatial reasoning. In: Mark, D.M., Frank, A.U. (eds.) Cognitive and Linguistic Aspects of Geographic Space, pp. 361–372. Kluwer, Dordrecht (1991b)

Freksa, C.: Using orientation information for qualitative spatial reasoning. In: Frank, A.U., Campari, I., Formentini, U. (eds.) GIS 1992. LNCS, vol. 639, pp. 162–178. Springer, Heidelberg (1992)

Freksa, C.: Spatial and Temporal Structures in Cognitive Processes. In: Freksa, C., Jantzen, M., Valk, R. (eds.) Foundations of Computer Science. LNCS, vol. 1337, pp. 379–387. Springer, Heidelberg (1997)

Freksa, C.: Spatial aspects of task-specific wayfinding maps: A representation-theoretic perspective. In: Gero, J.S., Tversky, B. (eds.) Visual and Spatial Reasoning in Design, pp. 15–32. Key Centre of Design Computing and Cognition, University of Sydney, Sydney (1999)

Freksa, C.: Spatial computing – How spatial structures replace computational effort. In: Raubal, M., Mark, D., Frank, A. (eds.) Cognitive and Linguistic Aspects of Geographic Space. Springer, Heidelberg (2013)

Freksa, C.: Strong spatial cognition (ext. abstract). In: Stewart, K., Pebesma, E., Navratil, G., Fogliaroni, P., Duckham, M. (eds.) Ext. Abstr. Proc. GIScience 2014. GEOinfo 40, 282-285, Vienna. Rev. version in Cognitive Processing 15 (Suppl 1): 103-105 (2014)

Freksa, C.: Computational problem solving in spatial substrates – A cognitive systems engineering approach. Int. J. Softw. Inf. 9(2), 279–288 (2015)

Freksa, C., Barkowsky, T.: On the relation between spatial concepts and geographic objects. In: Burrough, P., Frank, A. (eds.) Geographic objects with indeterminate boundaries, pp. 109–121. Taylor and Francis, London (1996)

Freksa, C., Barkowsky, T.: On the duality and on the integration of propositional and spatial representations. In: Habel, C., Rickheit, G. (eds.) Mental Models in Discourse Processing and Reasoning, pp. 195–212. Elsevier, Amsterdam (1999)

Freksa, C., Röhrig, R.: Dimensions of qualitative spatial reasoning. In: Piera Carreté, N., Singh, M.G. (eds.) Qualitative Reasoning and Decision Technologies, pp. 483–492. CIMNE, Barcelona (1993)

Freksa, C., Schultheis, H.: Three ways of using space. In: Montello, D.R., Grossner, K.E., Janelle, D.G. (eds.) Space in Mind: Concepts for Spatial Education. MIT Press, Cambridge (2014)

Freuder, E., Mackworth, A. (eds.): Constraint-Based Reasoning. MIT Press, Cambridge (1994)

Gentner, D.: Structure-mapping: A theoretical framework for analogy. Cogn. Sci. 7(2), 155–170 (1983)

Gibson, J.J.: The Ecological Approach to Visual Perception. Lawrence Erlbaum Associates, New Jersey (1979)

Glasgow, J., Narayanan, N.H., Chandrasekaran, B. (eds.): Diagrammatic Reasoning: Cognitive and Computational Perspectives. AAAI Press, Menlo Park (1995)

Goel, A.K., Jamnik, M., Narayanan, N.H. (eds.): Diagrammatic Representation and Inference. Springer, Berlin (2010)

Guesgen, H.W.: Spatial reasoning based on Allen's temporal logic, TR-89-049. International Computer Science Institute, Berkeley (1989)

Hobbs, J.: Granularity. In: International Joint Conference Artificial Intelligence, pp. 432–435 (1985)

Johnson-Laird, P.N.: Mental Models. Harvard University Press, Cambridge (1983)

Johnson-Laird, P.N.: Mental models, deductive reasoning, and the brain. In: Gazzaniga, M.S. (ed.) The Cognitive Neurosciences, 65, pp. 999–1008. MIT Press, Cambridge (1995)

Knuth, D.: The Art of Computer Programming. Fundamental Algorithms, vol. 1, 3rd edn. Addison-Wesley, Reading (1997)

Kosslyn, S.M.: Image and Mind. Harvard University Press, Cambridge (1980)

Kosslyn, S.M.: Image and Brain - The Resolution of the Imagery Debate. MIT Press, Cambridge (1994)

Kreutzmann, A., Wolter, D., Dylla, F., Lee, J.H.: Towards safe navigation by formalizing navigation rules. Intern. J. Marine Navig. Saf. Sea Transp. 7(2), 161–168 (2013)

Lakoff, G., Johnson, M.: Metaphors We Live By. University of Chicago Press, Chicago (1980)

Larkin, J.H., Simon, H.A.: Why a diagram is (sometimes) worth ten thousand words. Cogn. Sci. 11, 65–99 (1987)

Ligozat, G.: Qualitative triangulation for spatial reasoning. In: Frank, A.U., Campari, I. (eds.) COSIT 1993. LNCS, vol. 716, pp. 54–68. Springer, Heidelberg (1993)

Ligozat, G.: Qualitative Spatial and Temporal Reasoning. Wiley, London (2011)

Lungarella, M., Hafner, V., Pfeifer, R., Yokoi, H.: An artificial whisker sensor for robotics. In: IEEE Conference on Intelligent Robots and Systems (IROS), pp. 2931–2936 (2002)

Marr, D.: Vision. MIT Press, Cambridge (1982)

Moratz, R., Renz, J., Wolter, D.: Qualitative spatial reasoning about line segments. In: Horn, W. (ed.) ECAI 2000, pp. 234–238. IOS Press, Amsterdam (2000)

Mossakowski, T., Moratz, R.: Qualitative reasoning about relative direction of oriented points. Artif. Intell. 180, 34–45 (2012)

Nebel, B., Bürckert, H.J.: Reasoning about temporal relations: A maximal tractable subclass of Allen's interval algebra. JACM 42(1), 43–66 (1995)

Newell, A., Shaw, J.C., Simon, H.A.: Report on a general problem-solving program. In: Proceedings of the International Conference on Information Processing, pp. 256–264. UNESCO, Paris (1959)

Norman, D.A.: The Psychology of Everyday Things. Basic Books Inc., New York (1980)

Norman, D.A.: Cognition in the head and in the world: An introduction to the special issue on situated action. Cogn. Sci. 17, 1–6 (1993)

Palmer, S.E.: Fundamental aspects of cognitive representation. In: Rosch, E., Lloyd, B.B. (eds.) Cognition and Categorization, pp. 259–303. Lawrence Erlbaum, Hillsdale (1978)

Pfeifer, R., Scheier, C.: Understanding Intelligence. MIT Press, Cambridge (2001)

Piaget, J.: The Child's Conception of the World. Routledge and Kegan Paul Ltd., London (1929)

Polya, G.: How to Solve It. Princeton University Press, Princeton (1945)

Pylyshyn, Z.: The role of architecture in theories of cognition. In: VanLehn, K. (ed.) Architectures for Intelligence. Erlbaum, Hillsdale (1988)

Randell, D.A., Cui, Z., Cohn, A.G.: A spatial logic based on regions and connection. In: KR 1992, pp. 165–176 (1992)

Renz, J., Nebel, B.: Qualitative spatial reasoning using constraint calculi. In: Aiello, M., Pratt-Hartmann, I.E., van Benthem, J.F. (eds.) Handbook of Spatial Logics. Springer, The Netherlands (2007)

Russell, S.J., Norvig, P.: Artificial Intelligence: A Modern Approach, 3rd edn. Prentice Hall, Upper Saddle River (2010)

Schultheis, H., Barkowsky, T.: Casimir: An architecture for mental spatial knowledge processing. Top. Cogn. Sci. 3, 778–795 (2011)

Schultheis, H., Bertel, S., Barkowsky, T.: Modeling mental spatial reasoning about cardinal directions. Cogn. Sci. 38(8), 1521–1561 (2014)

Searle, J.: Minds, brains, and programs. Behav. Brain Sci. 3, 417–457 (1980)

Simon, H.A.: On the forms of mental representation. In: Savage, W. (ed.) Perception and Cognition, pp. 3–18. University of Minnesota Press, Minneapolis (1978)

Sloman, A.: Why we need many knowledge representation formalisms. In: Bramer, M. (ed.) Research and Development in Expert Systems, pp. 163–183. Cambridge University Press, New York (1985)

Van de Weghe, N., Kuijpers, B., Bogaert, P., De Maeyer, P.: A qualitative trajectory calculus and the composition of its relations. In: Rodríguez, M., Cruz, I., Egenhofer, M., Levashkin, S. (eds.) GeoS 2005. LNCS, vol. 3799, pp. 60–76. Springer, Heidelberg (2005)

Wilson, M.: Six views of embedded cognition. Psychon. Bull. Rev. 9(4), 625–636 (2002)

Wintermute, S., Laird, J.E.: Bimodal spatial reasoning with continuous motion. In: Proceedings of the AAAI, pp. 1331–1337 (2008)

Wolter, D., Dylla, F., Wölfl, S., Wallgrün, J.O., Frommberger, L., Nebel, B., Freksa, C.: SailAway: Spatial cognition in sea navigation. Künstliche Intelligenz 22(1), 28–30 (2008)

Wolter, D., Wallgrün, J.O.: Qualitative spatial reasoning for applications: New challenges and the SparQ toolbox. In: Hazarika, S.M. (ed.) Qualitative Spatio-Temporal Representation and Reasoning: Trends and Future Directions, pp. 336–362. IGI Global (2012)

Zadeh, L.A.: Fuzzy sets and information granularity. In: Gupta, M., Ragade, R., Yager, R. (eds.) Advances in Fuzzy Set Theory and Applications, pp. 3–18 (1979)

Zimmermann, K.: Measuring without measures: The Δ-calculus. In: Frank, A.U., Kuhn, W. (eds.) COSIT 1995. LNCS, vol. 988, pp. 59–67. Springer, Heidelberg (1995)

Qualitative Spatio-Temporal Reasoning and Representation I

A Conceptual Quality Framework for Volunteered Geographic Information

Andrea Ballatore[1][✉] and Alexander Zipf[2]

[1] Center for Spatial Studies, University of California, Santa Barbara,
Santa Barbara, CA, USA
`aballatore@spatial.ucsb.edu`
[2] Geoinformatics Research Group, Department of Geography,
University of Heidelberg, Heidelberg, Germany
`zipf@uni-heidelberg.de`

Abstract. The assessment of the quality of volunteered geographic information (VGI) is cornerstone to understand the fitness for purpose of datasets in many application domains. While most analyses focus on geometric and positional quality, only sporadic attention has been devoted to the interpretation of the data, i.e., the communication process through which consumers try to reconstruct the meaning of information intended by its producers. Interpretability is a notoriously ephemeral, culturally rooted, and context-dependent property of the data that concerns the *conceptual quality* of the vocabularies, schemas, ontologies, and documentation used to describe and annotate the geographic features of interest. To operationalize conceptual quality in VGI, we propose a multi-faceted framework that includes accuracy, granularity, completeness, consistency, compliance, and richness, proposing proxy measures for each dimension. The application of the framework is illustrated in a case study on a European sample of OpenStreetMap, focused specifically on conceptual compliance.

Keywords: Data quality · Interpretability · Conceptual quality · Volunteered geographic information

1 Introduction

The importance of data quality has been noted since the disciplinary inception of geographic information science (GIScience) [13]. The quality of geographic information has been framed along the spatial, temporal, and thematic dimension, in terms of accuracy, precision (or resolution), consistency, and completeness [32]. Because any discussion on data quality assumes the presence of a producer who encodes some information and a consumer who has to interpret it and use it, the *conceptual quality* of the data is crucial to enable the semantic decoding of data. For example, a dataset can contain highly accurate geometries, but if the description of the entities and their attributes is not clear, articulate, rich, and complete enough, the value of the data for consumers will be severely curtailed.

© Springer International Publishing Switzerland 2015
S.I. Fabrikant et al. (Eds.): COSIT 2015, LNCS 9368, pp. 89–107, 2015.
DOI: 10.1007/978-3-319-23374-1_5

In past decades, the assurance of conceptual quality was facilitated by the fact that both producers and consumers tended to belong to professional circles, and shared to some degree a semantic ground, i.e., the conceptualization of the domain and its entities, and a common vocabulary to describe them. The advent of volunteered geographic information (VGI), with its less centralized production models, leads to a novel state of affairs, with important consequences for conceptual quality. In VGI, different actors generate data for a variety of purposes, interpreting and consuming data produced from other actors. These processes of informal and loosely constrained *prosumption* [10] usually result in data with higher heterogeneity and fragmentation than traditional datasets, creating new, unforeseen barriers to data interpretation. Despite the growth of interest in VGI in both academia and industry, recurring issues make the application of traditional approaches to data quality problematic. For instance, a crowdsourced set of points of interest might possess sufficient quality to enrich spatial social media, but could fail to capture the changes in businesses in rural areas studied by economists.

The problem of quality assessment is intimately linked to the quantification of several orthogonal or correlated dimensions. In fact, it is impossible to state anything about data quality without well-defined criteria to measure it. To date, many researchers have tackled the issue of quality in VGI [5,14,19,20,23]. Unlike the expert-controlled data generated by government agencies, crowd-sourced resources do not come with systematic documentation about their production protocols, biases, and shortcomings [11]. In this sense, the context of VGI is open to many questions concerning conceptual quality, some of which are also relevant to traditional datasets, while others are novel and specific. In VGI, the data is often characterized by loose application of standards and a lack of thorough documentation. The traditional distinction between schema and records in databases does not always hold. Contributors produce data and then re-define and update its schema in an open process, which leads to uneven conceptualizations. In this sense, another peculiar difficulty lies in the associations between classes and instances—or, alternatively, universals and individuals.

How, then, is it possible to operationalize the conceptual quality of VGI, taking into account semantic aspects in the data that are so central to its production? To answer this question, it is important to take an ecological viewpoint on the environments in which VGI producers and consumers operate. Rather than designing the data production as a deterministic process with clear inputs and outputs, in VGI diverse actors use a combination of natural language, data sources, schemas, vocabularies, and software tools to generate the data they are interested in, through many feedback loops. In a semiotic sense, contributors need to develop shared conceptualizations that constrain the intended meaning of the symbols to enable the interpretation of the data, enabling the data as a medium for communication [22]. While the centrality of documentation and semantics is often acknowledged by researchers [e.g., [2]], conceptual quality has not yet been reduced to tractable measures and deployed within the relevant communities. Effective conceptual quality assessment techniques would benefit project owners, contributors, as well as end users, informing the production, evaluation, and consumption of data sources.

To fill this gap, in this article we present a conceptual quality framework that can be adapted and applied to any VGI source, tapping indirect and intrinsic proxy measures. The framework moves a first step towards the operationalization of the following difficult questions: Is the schema appropriate to describe the domain? How current and clear is the documentation? Is there consensus and consistency in the terms used in the data? Is the data prone to foster divergent interpretations? Are the descriptions of classes sufficiently detailed? What are missing elements that need to be described? To what degree are the users interpreting the terms correctly? How intuitive are the terms for the users? Is the data internally coherent? What geographic areas present differences in conceptual quality and why? Does the data conform to an external reference or standard? Is there conformance within specific groups of contributors or within geographic areas?

The remainder of this article is organized as follows. Section 2 surveys the literature on geographic information quality, focusing particularly on VGI. Section 3 illustrates the core ideas of our framework and proposes formal measures to operationalize it. Subsequently, Sect. 4 reports on a case study in which we illustrate the applicability of the framework on OpenStreetMap, a prominent example of VGI. Finally, Sect. 5 draws conclusions and indicates directions for future research.

2   Related Work

This quality framework for VGI lies at the intersection of several research areas, including GIScience, conceptual modeling, and ontology engineering. This section provides an overview of the notions of quality discussed in these inter-related disciplines.

Geographic Information Quality. Because all geographic information is produced through measurement with some level of uncertainty, the debate on quality has been central to geography and GIScience for a long time [17,21,27]. As pointed out by Goodchild and Li [14], broad consensus was established in the 1980s along five dimensions: positional accuracy, attribute accuracy, logical consistency, completeness, and lineage (p. 111), embedded in the US Spatial Data Transfer Standard (SDTS).[1] A pioneering theoretical discussion about the semantic dimension of geographic information quality was provided by Salgé [26]. Assuming that any description of reality is inevitably a reduction to a model, Salgé defined semantic accuracy as the "quality with which geographical objects are described in accordance with the selected model," as well as the "pertinence of the meaning of the geographical object rather than to the geometrical representation" (p. 139).

In practical terms, producers can enforce minimum quality standards in their data collection process, and consumers can assess the quality of a dataset

---

[1] http://mcmcweb.er.usgs.gov/sdts.

through metadata standards [32]. Improvements in web technologies have tightened the feedback loop between consumers and producers, providing mechanisms to improve quality in a targeted way based on users' needs. In the early 2000s, aspects of quality were defined in the International Organisation for Standards (ISO) 19113:2002 for quality principles, and in ISO 19114:2003 for quality evaluation procedures, then superseded by ISO 19157:2013.[2] While progress in theorization and standardization of data quality has been made, particularly in the context of public agencies, many challenges remain to be met. As Hunter et al. [21] pointed out, the communication, visualization, and description of data quality and its application to decision making are far from having satisfactory solutions for the many actors involved.

Conceptual and Ontological Quality. Conceptual modeling and ontology engineering are concerned with the quality of models, schemas, and ontologies [4]. The operationalization of such conceptual dimension of quality offers an important tool to facilitate the adoption, correct interpretation, and re-use of conceptualizations by practitioners. Agent-based approaches have been proposed to model the correctness of spatial information [12]. In recent years, applied ontologists have designed formal semantic approaches to assess the quality of an ontology, epitomized by the OntoClean method [15]. Other approaches are grounded in semiotics: Tartir et al. [30] outlined a triangular model where quality can be assessed in the mappings between the real world and the schema, between the real world and the data, and between the schema and data. In their formulation, metrics for schema quality include dimensions such as relationship, attribute, and inheritance richness, while instance metrics should reflect connectivity, cohesion, and readability of the data.

Along similar lines, Burton-Jones et al. [8] defined ontological quality from four facets: syntactic quality (richness of lexicon and correctness); semantic quality (interpretability, consistency and clarity); pragmatic quality (comprehensiveness, accuracy and relevance); and social quality (authoritativeness and history). An overall indicator of quality is obtained with a linear combination of these four dimensions. In the context of conceptual modeling, Cherfi et al. [9] defined a framework for conceptual quality, outlining metrics applicable to entity-relationship schemas, and to UML diagrams. While these methods inform the foundations of our framework, they are hard to apply directly to the semantically weak folksonomies and tagging models used in VGI [28,31].

Quality in VGI and OpenStreetMap. The emergence of VGI deeply re-configured the geographic information landscape, raising immediate concerns about quality assurance [20]. Analogously to Wikipedia, crowdsourced geographic information can be often of higher quality than authoritative sources, but shows considerable spatial and temporal variability, and is affected by gender and socio-economic contribution inequalities [29]. Vandalism in open mapping platforms has also been identified as a multi-faceted, complex challenge for VGI [1]. More generally, VGI communities have several strategies to ensure data quality [14]. These

---

[2] http://www.iso.org.

include crowdsourcing, based on the assumption that more people working on the same area will tend to result in higher quality [19], social approaches that rely on surveillance and control, and geographic approaches that exploit knowledge from geography to detect unlikely or impossible configurations in the data. The quality of OpenStreetMap (OSM), one of the leading VGI projects, has been studied along different lines of investigation. Several studies compared a sample from the OSM vector dataset against the corresponding data from more traditional and authoritative sources [18,33], showing high variability in the data quality, and identifying several geographical divides, particularly between rural and urban areas, and natural and man-made features.

The conceptual and semantic dimension of the data presents many specific challenges to ensuring quality. OSM has a lightweight semantic model that relies primarily on user-defined tags [2]. While the positional accuracy of features can be measured with standardized methods, the annotation process has no stable ground truth, as it is rooted in alternative conceptualizations of geographic world. Problems identified in OSM semantic set up are the flexibility of the tagging process and the lack of a strict mechanism for checking semantic compliance, even for core elements of the data, often resulting in tag wars [23,24]. In the project's forum and mailing lists, contributors often debate data quality, pointing out a strong need for "consistency in tagging, editor improvements, better documentation, better training materials."[3]

To date, the most substantial attempt at quantifying quality in OSM has been carried out by Barron et al. [5]. Their *iOSMAnalyzer* tool generates a range of intrinsic quality indicators, focusing on the spatio-temporal evolution of the data, including geometric and thematic quality. The framework adopts a fitness-for-purpose perspective, grouping indicators by application area, such as geocoding, routing, and points of interest (POI) search. Despite its comprehensiveness, this framework is narrowly focused on OSM, and could not be easily applied to other datasets. Moreover, the semantic aspects of the vector data are discussed only tangentially. Our proposed quality framework, outlined in the next section, aims at overcoming these limitations.

3   A Framework for VGI Conceptual Quality

To establish a framework to operationalize the conceptual quality of VGI, we analyze and revise each dimension of data quality, comparing it with traditional views on geographic information quality and proposing indirect indicators. In VGI, heterogeneous communities produce information for a variety of purposes, relying on a combination of tools, vocabularies, and data sources [3]. The main purpose of this framework is to enable the measurement of conceptual quality of a VGI dataset. Understanding what information producers meant to express in their data is a crucial, and yet ephemeral aspect of spatial information quality [22]. In traditional database theory, the quality of a database includes the quality

---

[3] http://goo.gl/W0rmVU.

of its conceptual schema, metadata description, and provenance of data [6]. Similarly, in VGI, *conceptual quality* should answer questions about the conceptual schema and its relations with the data. Conceptual quality is intimately intertwined with *interpretability*, the fundamental communication problem between the data creators and consumers.

In the context of VGI, we frame the production and consumption of information semiotically as the interaction of semantic agents in an information community. Hence, we refer to symbols (e.g., words, icons, or images) pointing to concepts, psychological models used by semantic agents to produce and interpret information about domain entities, called referents. The mappings between symbols and concepts is dynamic, and are established through social agreements [3]. To clarify our notion of interpretability, we distinguish between interpretability of the *data format* (i.e., file formats, formal languages, conceptual schemas), and the interpretability of the *domain content* (i.e., the concepts prior to their encoding into data).

The Semantic Gulf. The role of conceptual quality is essential to overcome the semantic gulf that exists between producers and consumers. As shown in Fig. 1, agent A describes a concept in his/her worldview with the symbol 'mountain' and encodes it into a dataset D. Because of cultural, linguistic, and individual variations, when agent B interprets and decodes the symbol S, his/her interpretation (Θ_B) overlaps with that of agent A (Θ_A) only to some degree. Following the notion of intended models by Guarino et al. [16], we refer to the overlap between the interpretations of symbol S for the two users ($\Theta_A \cap \Theta_B$). This overlap is an indicator of the quality of D in the sense that the highest quality would result in equivalent interpretations ($\Theta_A \equiv \Theta_B$). By contrast, the lowest quality leads to totally different interpretations of D ($\Theta_A \cap \Theta_B = \emptyset$). As low conceptual quality causes friction in the interpretation process, the operationalization of conceptual quality is essential to improve the interpretability of the data.

Conceptual quality questions include: how can the meaning of a term be assessed? To what degree is it possible to reconstruct the context and intentions of the producers from the data? How many alternative interpretations exist for a term? Does the compliance to external resources help the interpretation of the data? Is the usage of a term widespread or is it unusual? How easily can a consumer decode the terms? How clear are the constraints on terms? How ambiguous are the terms? To what degree are the agents interpreting the symbols correctly? How intuitive are the symbols for the agents? Are different agents mapping the symbols to different concepts? Are the symbols internally coherent?

Measuring Conceptual Quality. The measurement of conceptual quality can be carried out through some form of psychological testing in a controlled environment, asking human subjects to perform tasks on a piece of information, and measuring the cognitive load and other observable outcomes of the interpretation process. Less formally, ratings about the quality of resources can be collected directly from users of an online platform, identifying issues in the conceptualization. However, these approaches are impractical for large, decentralized projects

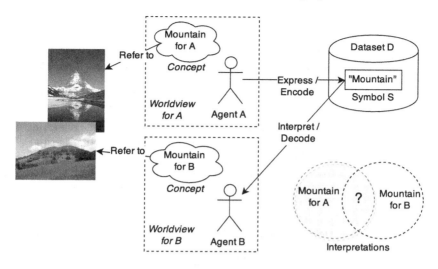

Fig. 1. The semantic gulf

such as OSM. On the other hand, measures of conceptual quality can rely on a number of indirect indicators, used as proxies to unobservable variables. For example, in ontology engineering, a measure of interpretability has been designed on the number of terms defined in external linguistic resources such as WordNet [8]. This extrinsic approach relies on the interlinking of a resource, assuming that connected resources are easier to interpret than isolated ones.

When measuring the conceptual quality of domain content, the documentation of a given term is crucially important. Indicators for this dimension of quality include the number of users who contributed to the definition, the stability of the definition over time, and the amount of discussions generated about it. Measures of VGI interpretability should be applied not only to schemas, as done traditionally in information systems [6], but also to the data itself, which might differ considerably from the documentation in local contexts. From the perspective of interpretability of data formats, different versions of the same piece of information can be evaluated, comparing traditional GIS formats like Esri's shapefile, and semantically richer formats such as RDF. A complementary issue lies in the interpretability of information about quality, which suffers from lack of standardization and from technical complexity [21].

Formalization and Symbols. To formalize the core ideas in the framework, we adopt the following terminology, which we believe can describe the bulk of VGI. A geographic feature τ is an instance of a class C, and has a set A of attributes a. Attributes have values. Feature τ also has a geometric attribute g. Features can be aggregated in a set F. Our framework operationalizes conceptual quality with indicators either at the feature level $I(\tau)$, on a single τ, or at the aggregate level on a set of features $I(F)$.

Table 1. Dimensions of conceptual quality for VGI

Dimension	Sym.	Purpose & indicators
Accuracy	I_{ac}	Distance between conceptualization and domain knowledge. It can be seen as the degree of correctness in the classification of features τ into classes C. *Indicators:* Number of features with multiple classifications; number of contributors.
Granularity	I_{gr}	Level of thematic description present in the data, moving from very abstract to very specific concepts. *Indicators:* Depth of classes in the class hierarchy (if applicable).
Completeness	I_{cl}	Coverage in the conceptualization of the features of interest. A distinction exists between class completeness and attribute completeness. *Indicators:* Number of classes; number of attributes
Consistency	I_{cn}	Degree of homogeneity in the descriptions of geographic features. *Indicators:* Number of features in a class described with the same attributes; ratio between consistent features or attributes to all others, weighted against the absolute number of attributes.
Compliance	I_{cm}	Degree of adherence of an attribute, a feature, or a set of features to a given source S, ranging from non compliance to full compliance. *Indicators:* Ratio between the number of classes and attributes defined in an external source S and the total number of classes and attributes.
Richness	I_{ri}	Amount and variety of dimensions that are included in the description of the real-world entity. *Indicators:* Number of attributes describing a feature.

For example, Lake Tahoe τ belongs to class C_{lake}, and has an attribute a_{name} whose value is set to "Lake Tahoe," and a footprint g that is represented as a polygon. A crucial difference between VGI and the traditional geo-database approach lies in the flexibility and instability of the schema definition. Rather than a clear distinction between schema and records, VGI communities produce datasets and their schemas in an open-ended way. Classes, instances, and their attributes tend to be fluid and mutable, rather than centrally defined and controlled. The remainder of this section discusses the notion of interpretability, followed by several complementary dimensions that need to be considered, summarized in Table 1.

3.1 Conceptual Accuracy

The notion of accuracy is central to the definition of geographic information quality. Accuracy answers questions about the correctness of the information with respect to a measurable phenomenon in the real world, for which there is a true value can that in principle be assessed. Accuracy is perhaps the best understood dimension of quality [22]. Because of the strong spatiality of geographic data, positional accuracy is a core preoccupation, complemented by temporal accuracy (also known as currency) [27].

More relevant for conceptual accuracy, notions of thematic or attributional accuracy indicate the degree of correctness of attribute values grounded in a spatio-temporal region, typically in the context of classification, for example indicating an area as industrial where it is in fact residential. Thematic information relies on a conceptualization that defines salient domain entities, categories, and their attributes. In this sense, conceptual accuracy concerns the distance between the concepts and the real world entities that they are supposed to describe according to an observer. Low conceptual accuracy indicates that the instances encountered by contributors are not intuitively or easily described with the selected conceptualization, resulting in semantic noise.

The measurement of spatio-temporal accuracy relies on the assumption that a true value can be obtained at a higher accuracy using appropriate measurement techniques. This assumption cannot always be performed in thematic information, and can clash with the multi-authored, choral nature of VGI. Conceptual accuracy should answer several questions: To what degree are the classes and attributes capturing the underlying domain knowledge? In the case of categorical variables, are the categories reflecting the domain knowledge? Are there many observations that do not fit the categories? How good is the agreement on the classification of instances when performed by different actors?

For conceptual accuracy, the heterogeneity of VGI presents new and peculiar challenges. Contributors describe the objects of interest using loosely defined vocabularies that present high lexical and semantic variation [2]. The inconsistencies in the attributional data makes the measurement of the global conceptual accuracy very hard, prompting, again, local measures. New measures of accuracy for VGI should include a social dimension that plays a huge role in the production process. Intrinsic, local measures include that by Haklay et al. [19], who suggested that the number of active users in an area shows a non-linear relationship with positional accuracy. Along similar lines, Bishr and Kuhn [7] tapped the social dimension in VGI by using trust as a proxy measure of quality.

Given a set of classes C, we define conceptual accuracy I_{ac} as the degree of correctness in the classification of features τ into classes C. Although such an assessment of classification accuracy needs some form of extrinsic ground truth (i.e., a classification having higher accuracy), it is possible to devise indirect indicators of conceptual accuracy I_{ac}. The core impact of conceptual accuracy occurs in the definition of the schema and the application of the schema on the instances. When the classification of a feature is difficult, contributors tend to classify it in multiple, incompatible ways [23]. Hence, one indicator consists of the number of features that have been classified in different ways at different points in time:

$$I_{ac}(F) = 1 - \frac{|\exists \tau \in F : \tau \in C_1 \wedge \tau \in C_2|}{|F|}; \quad I_{ac} \in [0, 1] \qquad (1)$$

High values of I_{ac} indicate low level of negotiation in the classification of features. While high $I_{ac}(F)$ might indicate that too few people worked on a classification to evaluate its quality, high $I_{ac}(F)$ signals that the contributors did not encounter problems in the classification of F.

3.2   Conceptual Granularity

While accuracy generally concerns the distance between a measurement and the true value, granularity answers questions about the precision of information, i.e., the repeatability of measurements, regardless of their true value [22]. Accuracy and granularity are orthogonal: a piece of information can present high accuracy and low granularity, and vice-versa. The term resolution is a synonym of granularity. The notion of scale is indeed related to granularity, as different scales require higher or lower granularity. Geographic information quality standards prominently include granularity as a fundamental element to evaluate fitness for purpose. For example, satellite imagery can be described as having "10 m resolution."

In VGI, the heterogeneity in the production process results in varying granularity. As observed for accuracy, the assessment of granularity loses meaning when performed at the global level. Spatial granularity in VGI is bound by the technical apparatus available to mappers (e.g., GPS sensors), and by the pre-existing geospatial infrastructure, such as the quality of satellite imagery that OSM contributors rely on to draw roads [18]. While the notion of granularity is well understood at the spatial and temporal level, thematic information presents deeper challenges. The notion of thematic resolution has been defined as the precision of the scalar or nominal variables [32].

In an open process of negotiation, VGI contributors express quantitative and qualitative measurements about a wide range of phenomena, usually based on a loosely defined conceptualization. For this reason, the conceptual structure of information is rather fluid, and its granularity is hard to assess. As in the case of OSM, contributors define hierarchies of classes C and their attributes to describe the concepts of interest, such as *university*, *park*, and *river*. Using the categorization by Rosch [25], such taxonomical hierarchies span from the superordinate level (e.g., built environment), to the basic level (e.g., house), and to the subordinate level, which includes more specific concepts, rarely used in day-to-day language (e.g., detached single-unit house).

Given a VGI dataset, *conceptual granularity* should answer questions about the level of thematic description is present in the data, moving from very abstract to very specific concepts. Are the objects described simply as buildings or are they categorized in sub-types? Hence, for a feature τ in class C, we want to devise a measure that quantifies the thematic granularity in the data. Among all classes defined in the data, how specific is C? To achieve this goal, measure of level of generality of a term in a hierarchy. This approach is meaningful only if the classes are organized in a subsumption hierarchy, which is not always the case. An indicator is the depth of a class C in the class hierarchy, for features τ:

$$I_{gr}(F) = \sum_{i=1}^{|F|} depth(C) : \tau_i \in C \wedge \tau_i \in F; \quad I_{gr} \in [0, \infty) \qquad (2)$$

As concepts can be organized in alternative ways in a taxonomy, caution is needed when comparing different datasets that adopt radically different approaches (such as OSM and the GeoNames gazetteer). The amount of details included in the description is captured by conceptual richness (see Sect. 3.6).

3.3   Conceptual Completeness

Geographic data can be evaluated in terms of the coverage of the entities of interest in the real world. Given some mapping rules, completeness answers questions about how many objects are included or missing from the dataset. Completeness can be measured spatially (is the target space surveyed in its entirety?), temporally (how well is the target space covered at a given time?), and thematically (are all relevant types of features included?) [32]. To be assessed, completeness needs an external reference that can be used as ground truth, and for this reason its measurement tends to be extrinsic.

Completeness in VGI is challenging as the mapping rules are either loosely defined or left implicit. As in traditional datasets, VGI completeness can be assessed extrinsically, using higher-quality data as ground truth [18]. In many instances, such as in the case of disaster management, the ground truth does not exist in the first place, and extrinsic measures might not be applicable. Therefore, intrinsic measures appear as particularly valuable. For example, Barron et al. [5] suggest that, if the growth of additions to the dataset is slowing down for a given feature type, in spite of general growth, that might indicate high completeness.

Depending on the degree of openness and structure of a data collection procedure, contributors decide what they want to include in the data from the potentially infinite knowledge about a geographic area. For this reason, *conceptual completeness* concerns the coverage in the conceptualization of the features of interest. Conceptual completeness can be further specialized in *class completeness* (e.g., how many building types are present in the dataset) and *attribute completeness* (e.g., how many streets have a name attribute).

To support the measurement of conceptual completeness in VGI, intrinsic measures can use various social and semantic signals as indirect indicators of completeness. Absolute and global completeness measures are doomed to be not very meaningful for VGI, because of the non-parametric distribution of features in the geographic space. By contrast, local and relative measures of completeness should answers questions about completeness with respect to a given type of features and attribute by comparing spatial, temporal, or thematic subsets. As simple indicators I_{cl}, we adopt the number of classes and the number of attributes in a set of features:

$$I_{cl\ C}(F) = |C : \tau \in F \wedge \tau \in C|; \tag{3}$$
$$I_{cl\ A}(F) = |A : a \in \tau \wedge \tau \in F|; \quad I_{cl} \in [0, \infty)$$

Geo-statistical approaches can support the formulation of intrinsic measures of conceptual completeness based on these indicators. The automatic detection of missing attributes is a proxy to attributional completeness, rooted in the distribution of attributes over space, highlighting statistically anomalous regions. A similar approach can be applied to class completeness, exploiting geographic knowledge about an area to identify areas with unusually low number of classes being instantiated. From a social dimension, collective or individual activity patterns cannot be reliable indicators, but they might provide a crude proxy indicators to conceptual completeness.

3.4   Conceptual Consistency

For each piece of geographic information, many alternative representations are possible. *Conceptual consistency* answers questions on the degree of homogeneity in the descriptions of geographic features. Are a set of features described with the same classes and attributes? Are synonyms used in the data? Are there multiple names for the same features? Are there individual and regional variations in the usage of terms or concepts? While in formal systems consistency usually refers to logical contradictions, VGI rarely relies on highly formal languages, favoring simple vocabularies, folksonomies, or tagging mechanisms.

Measures of consistency can focus on the use of classes and attributes in the data, with the advantage that no knowledge about the conceptual schema is needed. Given a set of features, pair-wise comparison can be used to identify clusters of features described similarly, both within the same spatial unit and between different spatial units. The ratio between consistent features or attributes to all others, weighted against the absolute number of attributes, can be used as a simple indicator of consistency, applicable at different granularities. As consistency is an intrinsic characteristic of the data, it is possible to devise an indicator I_{cn} based on a feature set F in class C containing attributes A:

$$I_{cn}(F, C) = \frac{|\forall(A_{\tau_i}, A_{\tau_j}) : A_{\tau_i} \equiv A_{\tau_j} \wedge \tau_i, \tau_j \in C|}{|\tau : \tau \in C \wedge \tau \in F|}; \quad I_{cn} \in [0, 1] \qquad (4)$$

High (low) values of I_{cn} indicate that the description of class C tends to be (in)consistent. This measure captures the homogeneity of attributes across different features. Effective measures of consistency are useful to identify communities that adopt different representational conventions and terms, going beyond the global binary classification as correct or incorrect. The measures of consistency are also useful to analyze consistency over time, and not only in space, detecting regional trends.

3.5   Conceptual Compliance

Compliance can be seen as an orthogonal dimension to consistency. Unlike consistency, compliance is extrinsic, as it refers to an external resource, such as documentation, meta-data, standards, or guidelines. *Conceptual compliance* answers questions about the degree of adherence of an attribute, a feature, or a set of features to a given source S, ranging from non-compliance to full compliance. In VGI, contributors rely on a combination of sources to produce the data, intrinsically (using resources defined within the same project), and extrinsically (adopting external sources). Measuring conceptual compliance I_{cm} would increase the homogeneity of data, facilitating its interpretability. The quality of these resources S, such as the readability and completeness of the documentation, is out of the scope of conceptual compliance.

VGI projects define formats, schemas, vocabularies, and conventions to be used in the data, usually indicating a hierarchy of reference sources. For instance, OSM indicates its wiki website as the most authoritative source of documentation, and other sources such as the map editors as less reliable and possibly non-compliant. These pieces of documentation indicate at different levels of detail how to describe buildings, what spatial and temporal reference systems should be adopted, how street addresses should be encoded, etc. Indicators of conceptual compliance I_{cm} can be applied to attributes $a \in A$, to features τ, or to sets of features F, with respect to a given source S, such as a conceptual schema:

$$I_{cm}(A, S) = \frac{|A \in S|}{|A|}; \quad I_{cm}(F, S) = \frac{|\tau : \tau \in F \wedge \tau \in S|}{|F|}; \quad I_{cm} \in [0, 1] \quad (5)$$

These indicators enables the measurement of conceptual compliance, distinguishing it from conceptual consistency. A set of features F can be consistent and non compliant, and vice-versa. In projects like OSM, the detection of consistent and non compliant subset of the data can also help contributors identify suspect deficiencies in the documentation S.

3.6 Conceptual Richness

A facet that is rarely mentioned in current quality frameworks concerns the richness of the data. By *conceptual richness*, we mean the amount and variety of dimensions that are included in the description of the real-world entity. For example, a building can be described as a simple point or footprint in space, and this description can be enriched by a unbounded set of observations about its architecture, usage, materials, infrastructure, ownership, functions, history, etc. A measure of richness I_{ri} therefore needs to quantify the dimensions of a feature τ, enabling comparison with other features (or sets of features).

Richness can concern either the conceptual schema or the data, bearing in mind that in VGI the alignment between classes and instances cannot be taken for granted. This facet of conceptual quality is orthogonal to conceptual completeness, in the sense that a dataset can possess high richness but low completeness, and vice-versa. To measure richness of the conceptual schema, we can rely on number of classes and attributes defined in a dataset. At the feature level, richness $I_{ri}(\tau)$ can be quantified as the number of attributes. The richness of a set of features F can be computed as the mean of number of attributes defined in the features:

$$I_{ri}(\tau) = |a \in \tau|; \quad I_{ri}(C) = \sum_{n=1}^{|C|} |a \in C_n|; \quad I_{ri}(F) = \frac{\sum_{n=1}^{|F|} I_{ri}(\tau)}{|F|} \quad (6)$$

These measures enable the comparison of different datasets and regions with respect to their richness, highlighting disparities in the data as well as in the conceptualization. However, the measurement of richness faces many challenges. In heterogeneous datasets, different attributes can be describing the same dimensions inconsistently, making it challenging to distinguish between emergent richness and noise. Moreover, the assumption that a higher number of attributes leads to better conceptual quality does not always hold true, for example in the case of machine-generated default attributes in OSM. In such cases, measures of information content might be helpful.

4   A Case Study on Conceptual Compliance

To illustrate our quality framework, we choose the measurement of conceptual compliance on real crowdsourced data as a case study. As a data source, we selected OpenStreetMap (OSM), the collaborative mapping project. For reasons of space, we focus on conceptual compliance I_{cm}, one of the most critical dimensions for OSM, leaving a more thorough and comprehensive evaluation of the approach as future work. In OSM, geographic features are encoded in the form of vector data, with geometries (points, polylines, and polygons) described with attributes called tags (e.g., *place=city, name=Berlin*).

The intended meaning of the attributes are documented in the OSM Wiki website,[4] which hosts the definitions of the intended meaning and usage of tags. As OSM contains a wide range of feature types, we restrict the analysis to road-related features, described with the *highway* tag. The rationale for this choice lies in the centrality of the road network in the project: producers and consumers alike are particularly concerned about its quality for routing applications, where conceptual compliance is particularly important.

OSM contributors choose the attributes to describe a feature based on a number of compliance sources S. The official documentation is hosted on the OSM Wiki website, but the map editing tools, such as JOSM and iD,[5] are particularly central to the tagging process. Hence, our case study aims at answering the following questions: How compliant is the road network with the attributes defined in the OSM Wiki website? What is the compliance of data with respect to the most popular map editing tools? What is the spatial variation in conceptual compliance?

Selection of Regions. To explore conceptual compliance, we identified a sample of areas in Germany and the UK, which are expected to present geographic and cultural variability in the European context. A densely populated and highly developed region was selected for each country (respectively Upper Bavaria and the South East region of England), contrasted with regions characterized by relatively low population density and economic development (Mecklenburg-West Pomerania and the North East of England). Because the size of administrative

---

[4] http://wiki.openstreetmap.org.

[5] http://wiki.openstreetmap.org/wiki/Comparison_of_editors.

Table 2. The European regions included in the study, with estimated population, area (source: *freebase.com*), OSM highway objects, conceptual compliance I_{cm} for three sources, and averages.

Region	Pop. (M)	Area (km² K)	Highway objects	Wiki I_{cm}	JSOM I_{cm}	iD I_{cm}
Germany						
Upper Bavaria	4.47	17.5	59,531	0.92	0.96	0.88
Mecklenburg-West Pomerania	1.61	23.2	40,775	0.96	0.92	0.88
England						
South East	8.64	19.1	13,796	0.92	0.9	0.92
North East	2.59	8.6	831	0.81	1	1
Average	–	–	–	0.9	0.96	0.92

units varies considerably between these countries in the Nomenclature of Territorial Units for Statistics (NUTS), we selected regions from the NUTS2 for Germany (*Regierungsbezirk*) and regions from NUTS1 for England, resulting in comparable units, summarized in Table 2. These regions provide a small sample of European OSM data with respect to population, size, and culture. The OSM data was downloaded in January 2015.

In this study, we restricted the analysis to the objects tagged with at least one *highway* tag. These tags are used to describe not only highways, as the name would suggest, but all road-related information, which is of particular importance to the OSM community and users.[6] The selected regions are summarized in Table 2, including an estimate of the current population, area, and their total number of highway objects in OSM.

Computation of Conceptual Compliance. After having extracted the OSM data, we calculated the conceptual compliance I_{cm} as defined in Sect. 3.5. The indicator was computed at the aggregated level on the regions, as well as on a 10 km grid, in order to be able to observe the spatial variation at a higher granularity. As compliance sources S, we included the OSM Wiki website, and the two most popular editors that have a set of predefined tags (*JOSM* and *iD*). For each source, we consider compliant a tag that is explicitly defined and documented, and non-compliant all the others. A distinction was made between keys that should only accept a set of values (*highway=residential, highway=primary,* etc.), and open-value tags that accept any value (*name=*, ref=**), relaxing the compliance definition for the latter cases.

Results and Discussion. The conceptual compliance for OSM Wiki website and the two editors for each of the four regions is displayed in Table 2. The conceptual compliance ranges from 0.81 to 1, for an average of 0.93, indicating that 7 % of tags are not compliant, and their interpretation is problematic. The compliance

---

[6] http://wiki.openstreetmap.org/wiki/Key:highway.

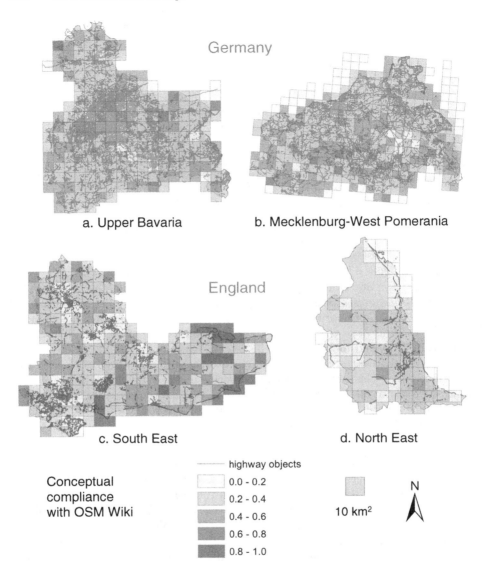

Fig. 2. Conceptual compliance I_{cm} with OSM Wiki on four European regions, calculated on a $10\,\text{km}^2$ grid.

average for the Wiki (0.9) is lower than for editors JOSM (0.96) and iD (0.92), confirming the misalignment between the different sources of compliance that OSM users complain about (see Sect. 2).

Non-compliant tags include for example *highway=no* and *highway:historic=-primary*. Some tags appear to be deprecated (e.g., *highway=byway*), and their status with respect to compliance is hard to assess. The conceptual compliance was then calculated on a $10\,km^2$ grid. Figure 2 shows choropleth maps of the four selected regions, as well as the locations of the highway objects, highlighting the high spatial variability of conceptual compliance. This simple measure already enables contributors and users to quantify the amount of non-compliant tags, and localize them spatially. This information can be used for fitness-for-purpose by consumers, and for quality assurance by producers. Moreover, the incongruities between the OSM Wiki website and the map editors can be identified and resolved systematically with our approach.

5 Conclusions

In this paper, we outlined a multi-dimensional framework for the assessment of conceptual quality, tailored for the context of VGI. Conceptual quality answers questions about the quality of conceptualization and its relationship with the data. This notion is strongly related to interpretability, the communication problem between the data creators who encode information according to their explicit and implicit knowledge, and consumers who need to interpret the data, reconstructing its intended meaning. Conceptual quality is essential to facilitate the communication over the semantic gulf that separates producers and consumers.

As conceptual quality is a complex, multi-faceted notion, six dimensions were identified: accuracy, granularity, completeness, consistency, compliance, and richness. Each dimension of conceptual quality was defined as complementary to traditional notions of quality developed in GIScience, proposing indicators to compute it and operationalize it. As an initial illustration of the framework, we explored a case study on four regions in Europe in OSM, focusing on the conceptual compliance of the tags. The case study highlights the wide applicability of conceptual quality to real data, and its potential to identify semantic and modeling issues in VGI.

Operationalizing conceptual quality is essential to increase the usability of VGI, adding a semantic facet to traditional notions of spatial, temporal, and thematic quality. The current state of the framework has several limitations that need addressing before deployment in realistic settings. The indicators described in this article need to be applied to OSM and other datasets in order to assess their strengths and weaknesses. Some dimensions of conceptual quality, such as conceptual granularity, will certainly prove harder than others to operationalize in different contexts. Without doubt, much empirical work is needed to deploy the framework effectively, with the goal of increasing the value and interpretability of VGI.

The core future direction for this work involves the application of the six dimensions to different datasets, comparing and contrasting the results, and tailoring more sophisticated and alternative indicators. For example, the relationship between how many contributors work on a region and its conceptual completeness needs further investigation [19]. A more mature version of the framework will be implemented into actual tools for VGI contributors and users, particularly for OSM. Conceptual quality, in its many empirically unexplored facets, will play an important role in overcoming the barriers to the usage of data as communication medium, mitigating the friction encountered when crossing the semantic gulf.

Acknowledgments. The authors thank Sophie Crommelinck and Sarah Labusga (University of Heidelberg) for the implementation of the case study, and the OpenStreetMap community for supplying the data.

References

1. Ballatore, A.: Defacing the map: cartographic vandalism in the digital commons. Cartographic J. **51**(3), 214–224 (2014)
2. Ballatore, A., Bertolotto, M.: Semantically enriching VGI in support of implicit feedback analysis. In: Tanaka, K., Fröhlich, P., Kim, K.S. (eds.) W2GIS 2011. LNCS, vol. 6574, pp. 78–93. Springer, Heidelberg (2010)
3. Ballatore, A., Bertolotto, M., Wilson, D.: Computing the semantic similarity of geographic terms using volunteered lexical definitions. Int. J. Geograph. Inf. Sci. **27**(10), 2099–2118 (2013a)
4. Ballatore, A., Wilson, D.C., Bertolotto, M.: A survey of volunteered open geoknowledge bases in the semantic web. In: Pasi, G., Bordogna, G., Jain, L.C. (eds.) Quality Issues in the Management of Web Information. ISRL, vol. 50, pp. 93–120. Springer, Heidelberg (2013b)
5. Barron, C., Neis, P., Zipf, A.: A comprehensive framework for intrinsic OpenStreetMap quality analysis. Trans. GIS **18**(6), 877–895 (2014)
6. Batini, C., Scannapieco, M.: Data Quality: Concepts, Methodologies and Techniques. Springer, Berlin (2006)
7. Bishr, M., Kuhn, W.: Trust and reputation models for quality assessment of human sensor observations. In: Tenbrink, T., Stell, J., Galton, A., Wood, Z. (eds.) COSIT 2013. LNCS, vol. 8116, pp. 53–73. Springer, Heidelberg (2013)
8. Burton-Jones, A., Storey, V., Sugumaran, V., Ahluwalia, P.: A semiotic metrics suite for assessing the quality of ontologies. Data Knowl. Eng. **55**(1), 84–102 (2005)
9. Si-said Cherfi, S., Akoka, J., Comyn-Wattiau, I.: Conceptual modeling quality - from EER to UML schemas evaluation. In: Spaccapietra, S., March, S.T., Kambayashi, Y. (eds.) ER 2002. LNCS, vol. 2503, pp. 414–428. Springer, Heidelberg (2002)
10. Dodge, M., Kitchin, R.: Crowdsourced cartography: mapping experience and knowledge. Environ. Plan. A **45**(1), 19–36 (2013)
11. Flanagin, A.J., Metzger, M.J.: The credibility of volunteered geographic information. GeoJournal **72**(3–4), 137–148 (2008)
12. Frank, A.U.: Spatial communication with maps: defining the correctness of maps using a multi-agent simulation. In: Habel, C., Brauer, W., Freksa, C., Wender, K.F. (eds.) Spatial Cognition 2000. LNCS (LNAI), vol. 1849, pp. 80–99. Springer, Heidelberg (2000)
13. Goodchild, M.F., Gopal, S.: The Accuracy of Spatial Databases. CRC Press, Boca Raton (1989)
14. Goodchild, M.F., Li, L.: Assuring the quality of volunteered geographic information. Spat. Stat. **1**, 110–120 (2012)
15. Guarino, N., Welty, C.A.: An overview of OntoClean. In: Staab, S., Studer, R. (eds.) Handbook on Ontologies, 2nd edn, pp. 201–220. Springer, Berlin (2009)

16. Guarino, N., Oberle, D., Staab, S.: What is an ontology? In: Staab, S., Studer, R. (eds.) Handbook on Ontologies, 2nd edn, pp. 1–17. Springer, Berlin (2009)
17. Guptill, S., Morrison, J. (eds.): Elements of Spatial Data Quality. Elsevier, Oxford (1995)
18. Haklay, M.: How good is volunteered geographical information? A comparative study of OpenStreetMap and ordnance survey datasets. Environ. Plan. B Plan. Des. **37**, 682–703 (2010)
19. Haklay, M., Basiouka, S., Antoniou, V., Ather, A.: How many volunteers does it take to map an area well? The validity of Linus' law to volunteered geographic information. Cartographic J. **47**(4), 315–322 (2010)
20. Heipke, C.: Crowdsourcing geospatial data. ISPRS J. Photogrammetry Remote Sens. **65**(6), 550–557 (2010)
21. Hunter, G., Bregt, A., Heuvelink, G., Bruin, S., Virrantaus, K.: Spatial data quality: problems and prospects. Research Trends in Geographic Information Science, LNGC, pp. 101–121. Springer, Berlin (2009)
22. Kuhn, W.: Core concepts of spatial information for transdisciplinary research. Int. J. Geogr. Inf. Sc. **26**(12), 2267–2276 (2012)
23. Mooney, P., Corcoran, P.: Characteristics of heavily edited objects in OpenStreetMap. Future Internet **4**(1), 285–305 (2012a)
24. Mooney, P., Corcoran, P.: The annotation process in OpenStreetMap. Trans. GIS **16**(4), 561–579 (2012b)
25. Rosch, E.: Principles of categorization. In: Margolis, E., Laurence, S. (eds.) Concepts: Core Readings, pp. 189–206. MIT Press, Cambridge (1999)
26. Salgé, F.: Semantic accuracy. In: Guptill, S., Morrison, J. (eds.) Elements of Spatial Data Quality, pp. 139–151. Elsevier, Oxford (1995)
27. Shi, W., Fisher, P., Goodchild, M.F. (eds.): Spatial Data Quality. CRC Press, Boca Raton (2003)
28. Solskinnsbakk, G., Gulla, J.A., Haderlein, V., Myrseth, P., Cerrato, O.: Quality of hierarchies in ontologies and folksonomies. Data Knowl. Eng. **74**, 13–25 (2012)
29. Stephens, M.: Gender and the GeoWeb: divisions in the production of user-generated cartographic information. GeoJournal **78**(6), 981–996 (2013)
30. Tartir, S., Arpinar, I., Moore, M., Sheth, A., Aleman-Meza, B.: OntoQA: metric-based ontology quality analysis. In: IEEE Workshop on Knowledge Acquisition from Distributed, Autonomous, Semantically Heterogeneous Data and Knowledge Sources, at the 5th IEEE International Conference on Data Mining 2005, ICDM 2005, pp. 1–9. IEEE (2005)
31. Van Damme, C., Hepp, M., Coenen, T.: Quality metrics for tags of broad folksonomies. In: Proceedings of International Conference on Semantic Systems (I-SEMANTICS), Graz, Austria, pp. 118–125 (2008)
32. Veregin, H.: Data quality measurement and assessment. NCGIA Core Curriculum in Geographic Information Science (1998). http://www.ncgia.ucsb.edu/giscc/units/u100/u100_f.html
33. Zielstra, D., Zipf, A.: A comparative study of proprietary geodata and volunteered geographic information for Germany. In: Painho, M., Santos, M.Y., Pundt, H. (eds.) Proceedings of the 13th AGILE International Conference on Geographic Information Science, pp. 1–15 (2010)

A Coq-Based Axiomatization of Tarski's Mereogeometry

Richard Dapoigny$^{(\boxtimes)}$ and Patrick Barlatier

LISTIC/Polytech Annecy-Chambéry, University of Savoie,
Po. Box 80439, 74944 Annecy-le-vieux cedex, France
`{richard.dapoigny,patrick.barlatier}@univ-savoie.fr`
`http://www.polytech.univ-savoie.fr/index.php?id=listic-accueil&L=1`

Abstract. During the last decade, the domain of Qualitative Spatial Reasoning, has known a renewal of interest for mereogeometry, a theory that has been initiated by Tarski. Mereogeometry relies on mereology, the Leśniewski's theory of parts and wholes that is further extended with geometrical primitives and appropriate definitions. However, most approaches (i) depart from the original Leśniewski's mereology which does not assume usual sets as a basis, (ii) restrict the logical power of mereology to a mere theory of part-whole relations and (iii) require the introduction of a connection relation. Moreover, the seminal paper of Tarki shows up unclear foundations and we argue that mereogeometry as it is introduced by Tarski, can be more suited to extend the whole theory of Leśniewski. For that purpose, we investigate a type-theoretical representation of space more closely related with the original ideas of Leśniewski and expressed with the Coq language. We show that (i) it can be given a more clear foundation, (ii) it can be based on three axioms instead of four and (iii) it can serve as a basis for spatial reasoning with full compliance with Leśniewski's systems.

Keywords: Mereology · Mereogeometry · Point · Solids · Balls · Coq

1 Introduction

In Knowledge Representation and Reasoning (KRR), and especially in the subdomain of Qualitative Spatial Reasoning (QSR) the very nature of topology and its relation to how humans perceive space stems from what is called mereotopology [2,22,32,45]. The term mereotopology refers (i) to mereology, i.e., the theory of parts and wholes (see e.g., [35] for a complete analysis) and (ii) to the addition of topological relation(s) to mereology (e.g., the "Connected" relation) in order to get sufficient expressiveness for reasoning about space. Notice that mereotopology is general enough since it can be applied in many other fields where the spatial character is not the primary purpose. An alternative approach for the modeling of spatial regions which relies on geometrical primitives is known as mereogeometry [43]. It has known some gain of interest during the last decade [5,8,11,21] and appears as a promising research area. Mereo-geometrical

© Springer International Publishing Switzerland 2015
S.I. Fabrikant et al. (Eds.): COSIT 2015, LNCS 9368, pp. 108–129, 2015.
DOI: 10.1007/978-3-319-23374-1_6

relations are required to be invariant to the strength of the type of geometry, e.g., for affine geometry they should be invariant under affine transformations. Mereogeometry has been applied mainly in the area of control tasks for mobile robotics [25, 26] and in the semantics of spatial prepositions [3].

The development of such region-based theories can be seen as an appealing alternative to set theory, point-set topology and Euclidian geometry. The quest for such theories of space are primarily motivated by human cognition, i.e., how humans perceive their spatial environment. It is also well-known that these region-based theories are able to draw topological or mereological conclusions even in the absence of precise data. Furthermore, they bridge the gap between low-level and high-level representations of space by providing a way to understand the nature of points, e.g., offering a structure that is not evident in Euclidian geometry [18].

Region-based theories using topological properties have an expressive power which is much more restricted than point-based geometry. From that perspective, mereogeometry and more especially, Tarski's mereogeometry, appears as a powerful alternative. However, it has not be fully formalized, and several authors have provided a fully formal system based on Tarski'work, either using a first-order language [10] or set theory [21], or using parthood together with a sphere predicate [5]. As far as we know, the totality of these approaches consider mereology as the theory of parts and wholes and restrict the mereological contribution to the theory, as a single *part-of* relation. This assumption leads to many difficulties such as the addition of the so-called Weak/Strong Supplementation principle whose assumption is highly debatable [35, 45]. Furthermore, mereogeometry is not compelled to provide a connection primitive [5, 21, 43].

We observe that all the mereogeometrical theories developed so far (i) depart from the classic mereology of Leśniewski which does not assume usual sets as a basis, (ii) perceive mereology as a mere theory of part-whole relations[1] and (iii) restrict the underlying logic to first-order logic whereas the original theory of Leśniewski is clearly higher-order. We argue that mereogeometry as it is introduced by Tarski, appears more suited to extend the foundations of Leśniewski's mereology for building a sound theory of space.

The first objective of the paper is to motivate for a mereogeometry based on Leśniewski's work with the purpose of providing a well-founded region-based theory of space [22]. The second objective is to develop a type-theoretical theory of space more closely related with the original ideas of Leśniewski [28]. Finally, besides expressing Leśniewski's mereology with the Coq language [6], we are able to provide a version of Tarski's mereogeometry in Coq. After a short presentation of motivations in Sect. 2, we describe the basis of Leśniewski's mereology in Sect. 3. The type-theoretical account of Leśniewski's mereology is summarized in Sect. 4. Section 5 discusses the foundations of mereogeometry which (i) extends the type-theoretical mereology and (ii) builds upon Tarski's work.

---

[1] Most descriptions suffer from the misunderstanding that mereology is formally nothing more than a particular elementary theory of partial ordering.

2   Motivations for a Leśniewski's Approach to Mereogeometry

We first present some arguments which advocate for using mereogeometry as an appealing alternative to mereotopology. Besides, we show how a detailed analysis of Tarski's papers will motivate the Leśniewskian approach as a sound basis for mereogeometry.

In mereotopology as well as in mereogeometry, the common theoretical core is the so-called mereology basis. However, in the quasi-totality of approaches the term mereology has little connection (if any) with the Leśniewski's Mereology. The notable exception is the mereogeometry of Tarski [42, 43]. The strong point for using Leśniewski's mereology stands in the fact that it can be built from algebraic structures, and more precisely from a quasi boolean algebra (i.e., without the zero element). It was stated by Tarski [44] and further proved by Clay [14] that every mereological structure can be transformed into complete Boolean lattice by adding zero element (its non-existence is a consequence of axioms for mereology) and conversely, that every complete Boolean lattice can be turned into a mereological structure by deleting the zero element. Furthermore, a major difficulty we see in mereotopological theories stems from the addition of a purely topological primitive (i.e., connection) to a mereological framework. This commitment results in hybrid systems whose theoretical foundations are unclear. As underlined in [5], mereotopological theories are limited to describe topological properties and therefore, their expressive power is more restricted than point-based geometry. Another benefits of using mereogeometry is its ability to reconstruct points from region-based primitives [19]. For instance, in Tarski's mereogeometry, the domain of discourse is classified into spheres and thus one can recover points from spheres i.e., metrical ones. For these reasons, we argue for a mereogeometrical framework using the Leśniewski's mereological deductive basis. The best candidate for such a framework is the geometry of solids first proposed in [42] and modified in [43].

However, a detailed analysis of Tarski's papers shows up some unclear methodological aspects [8]. The authors point out that Tarski's theory suffers from (i) some methodological divergences from Leśniewski's mereology and (ii) a free mixing of Leśniewski's mereology and the type-theoretical approach of Whitehead and Russel's *Principia mathematica*. Point (i) refers to the facts that in Leśniewski's theories, axioms precede definitions and that axioms should not contain defined notions [9]. Whereas in the early paper [42] Tarski assumes the availability of Leśniewski's mereology, in the second paper [43] he provides a minimal mereology based on three definitions and two axioms. As advocated in [33], the axiomatization of mereology given by Tarski is deficient in that one of the Leśniewski's axioms of mereology is not provable from his axioms. By adopting the original version of Leśniewski as axiom system, we are able to prove that this assertion is irrelevant here. However, while the proposed version fails to come up to the aesthetic rules of "well-constructed axiom system", it has the virtue of intuitive simplicity and it expresses faithfully Leśniewski's mereological principles. Point (ii) requires the clarification of two aspects: (a) the exact semantics behind the notions of "class" and "sum" and (b) the relation between the domain of discourse and the range of quantifiers. To make (a) precise, a footnote in the second paper of Tarski [43] claims that the sense in which class should be interpreted is Russellian. While the approach of Bennett [5] develops a region-based version of Tarski's theory by restricting the mereology to first-order axioms without involving set-theory, it results in an undecidable model. Alternatively, the analysis detailed in [21] relies on an algebraic framework which supports higher-order logic but uses set theory which is incoherent if one expects to work with Leśniewski's mereology. Departing from these approaches, the contributions of the paper turn out (i) to formalize Tarski's geometry of solids w.r.t. Leśniewski's mereology in a strict methodological sense and (ii) to provide a type-theoretical account for the whole theory based on the Calculus of Inductive Constructions.

3   Leśniewski's Mereology

3.1   Foundations

In the early 20th century, S. Leśniewski first proposed a higher-order logical theory (called protothetic) based on a single axiom, the equivalence axiom. This logical system was the support of a theory called mereology, whose purpose was the description of the world with collective classes and mereological relations such as the so-called *part-of* relation. While set theory relies on the opposition between element and set, mereology is rather based on the opposition between part and whole. Mereology primarily assumes the distinction between the distributive and collective interpretations of a class. A distributive class (a usual set) is the extension of a concept such as for example, the solar system, $P = \{Mercury, Venus, Earth, Jupiter, Mars, Saturn, Neptune, Pluto\}$ in which P is a distributive class which contains nine elements and nothing else. By contrast, the notion of collective class involves both the previous planets and a lot of entities such as the moon, the oceans, water, the fishes in it, the red Jupiter spot, the rings of Saturn and the like. It should also be noted that Leśniewski's mereology assumes the interplay between distributive and collective classes. The ontology introduces names, where name is the distributive notion, while the mereology relies on classes, where class is the collective notion. The theory also admits two kinds of variables which are either of the type propositional and belong to the semantic category S or to the category of names N. Any name may refer to a singular name, a plural name or the empty name. Notice that the introduction of plural names gives mereology an expressive power that goes beyond the capabilities of first-order logic (e.g., predicates may be introduced which have plural subjects). The logical system relies on the equivalence relation which allows to introduce new symbols as names of arbitrary elements of any proposition. Leśniewski's idea was to construct entirely new foundation for mathematics. For that purpose, its theory consists of three subsystems:

- prototethics which corresponds roughly to a higher-order propositional calculus introduces the equivalence, the category S and extensionality,

- ontology referred to as Leśniewski's Ontology (LO) which includes protothetic, could be described as a theory of particulars. It introduces the copula "ε" (distinct from the set-theoretic $\in$) and the category N,
- mereology (which includes LO) whose axioms are not deducible from logical principles is rather a theory which introduces collective classes, name-forming functor "class of", "part of" and "element of" relations together with their properties.

In the following upper case letters will denote singular names while lowercase will refer to plural names. Leśniewski's mereology can be presented in many settings depending on the adopted set of primitive terms. We refer here to the original system (1916). Notice that different versions of mereology have been proposed, but it has been proved that each of these versions are inferentially equivalent to the original system.

Protothetic. Protothetic is a logically rigorous system, a propositional calculus with quantifiers and semantic categories developed for supporting LO, which itself supports mereology. The construction of computable systems of protothetic (referred as S_1 to S_5) allowed Leśniewski to prove that S_5 is a consistent and complete system [29,38]. In protothetic, a single category S is used as the universe of propositions. It is a higher-order two-valued logic based on the equivalence relation denoted $\equiv$ between terms which can be propositions or propositional expressions (e.g., a propositional function having two propositional arguments such as S/SS^2). Without getting into the details, the formalization of protothetic obeys a succession of refinements that could broadly divided in two steps: (i) a quantified equivalential calculus and (ii) a logic of bivalent propositions which should support quantification on any variable whatever its category, and should admit the law of extensionality for propositions. The extensionality principle for propositional types is written as:

$$\forall p\, q\, f, (p \equiv q) \equiv (f(p) \equiv f(q))$$

where p, q and f denote propositional variables.

Leśniewski's Ontology. LO can be perceived as an extension of the traditional Aristotelian formal logic and interpreted as a theory a certain kind of general principles of existence. Given two names, a and b, the sentence $a\varepsilon b$ is true iff a denotes exactly one object, and this object is named by b. The distributive meaning of classes is captured by ε since "A is an element of the extension of the objects a", is identical to "A is a", i.e., $A\varepsilon a$. Intuitively it means that one object (individual) falls under the scope of a collective notion (a sum of individuals). A single axiom states the properties of the copula ε:

$$\forall Aa,\ A\varepsilon a \equiv ((\exists B, B\varepsilon A) \wedge (\forall CD, (C\varepsilon A \wedge D\varepsilon A) \to C\varepsilon D) \wedge (\forall C, C\varepsilon A \to C\varepsilon a))$$

The first conjunction of the right side of the equivalence prevents A from being an empty name, the second conjunction states the uniqueness of A while the last

---

[2] The two argument categories are on the right side and the resulting category on the left side.

conjunction refers to a kind of convergence (anything which is A is also an a). The only functor of LO provides the meaning of ε via a single axiom and rules of the system. Many definitions are given on the form of equivalences, that is using primitives of the language rather than meta-level primitives which do not belong to the language. We only describe definitions which are of interest for the purpose of the paper. The first one defines the *singular_equality* which holds when two singular names denote the same object.

$$\forall A\ B, singular\_equality\ A\ B \equiv (A\ \varepsilon\ B\ \wedge\ B\ \varepsilon\ A)$$

This relation is symmetric and transitive, but not reflexive. The second one is the *weak_equality* between plural names (symmetric, reflexive and transitive), i.e., an equivalence relation.

$$\forall a\ b, weak\_equality\ a\ b \equiv \forall A, (A\ \varepsilon\ a) \equiv (A\ \varepsilon\ b)$$

The extensionality principle is involved to prove the following theorem of LO where A, B, C are members of the category N, while ϕ belongs to N/N:

$$\forall A\ B\ C\ \phi,\ A\ \varepsilon\ (\phi\ B)\ \wedge\ singular\_equality\ B\ C\ \rightarrow\ A\ \varepsilon\ (\phi\ C).$$

Mereology. Mereology is an axiomatic theory of parts built on LO which relies on the concept of (collective) classes. For that purpose, it can be formalized in terms of different primitives. The most usual formalization introduces a mereological element called part (pt), as a primitive. Mereology is developed on a minimal collection of axioms and new primitive functors are defined whose the most important are that of (collective) class, denoted Kl and that of element of a class referred to as el. From now, it should be clear that the notions of "collective class", "mereological sum" and Kl are equivalent. Besides, it has been demonstrated that Leśniewski's mereology is consistent [13,27].

3.2   Contribution of Leśniewski's Mereology to Mereogeometry

Representing all the Leśniewski's apparatus in Coq will add some value to the current work. The first benefit comes from the underlying higher-order logic (prototethic) which is coherent with the choice of a higher-order type theory on which Coq relies. The higher-order framework of Coq offers much more expressiveness than first-order provers. For example, relations can be composed at the meta level from existing relations. The usual critique concerns the lack of automation inherent to higher-order provers, however, Coq comes with a proof-search mechanism and with several decision procedures that enable the system to automatically synthesize lumps of proof. Mereo-geometrical definitions (see Sect. 5) rely on user's requests that can be solved using lemmas and tactics in a quasi-automatic mode. The library proved in [17] uses more than a hundred theorems that are all available for reasoning in mereogeometry.

The second benefit relates to the conceptualization. First, using Leśniewski's mereology simplifies Tarski's approach to mereogeometry and clarifies many semantic issues (see Sect. 5). Using LO is a way to provide sound and simple ontological foundation for spatial reasoning. Second, the theory is able to aggregate in a single framework distributive and collective classes giving rise to an expressive and coherent theory. For example, instead of representing the relation of "proper-part of" as a relation between two names x and y, Leśniewski's mereology provides a more expressive relation [35] using the two-place functor ε and the one-place functor pt with: $x \, \varepsilon \, pt \, y$ resulting in a stronger system. Not only we get more information but, we also avoid to eliminate the notion of distributive class [14] and consequently the interplay between the two notions of class (distributive and collective).

An alternative to mereogeometry is the so-called mereotopology [45]. However, there is some foundational difficulty here. On the one hand, the original mereology of Leśniewski is rather unknown[3] while on the other hand, most authors view his work as a mere theory of parts and mix it inside a set-based framework. In fact, Leśniewski's theory is incompatible with a set-based approach and it is one of the underlying objectives of our work to provide a coherent theory independent from set theory. While mereology-based axiomatizations of Tarskis mereogeometry were given previously, the paper has a number of new contributions: (i) it incorporates Lesniewski's mereology (with its prototethic and ontology components) together with Tarski's mereogeometry in a coherent system, (ii) it is fully formalized and (iii) it is computer verified.

4   The Type-Theoretical Mereology

4.1   Using Coq as a Logical Foundation

Tarski has pointed out that propositional connectives can be defined in terms of logical equivalence and quantifiers in the context of higher-order logic [41]. Then, Quine further showed how quantifiers and connectives can be expressed in terms of equality and the abstraction operator in the context of Church's type theory [31]. These results led Henkin to define the propositional connectives and quantifiers as they appear in prototethic, in type theory having at least function application, function abstraction, and equality [23]. Versions of dependent type theory in computer science such as the Calculus of Inductive Constructions (CIC) are largely inspired from Church-style (constructive) logics. As a consequence, there is a semantic connection between the logical content of CIC and prototethic since the former can be seen as an extension of Church's type theory which itself is able to represent propositional connectives and quantifiers of prototethic. As underlined in [1,7], there is a significant advantage for using a dependently typed functional language such as CIC and its implementation with the Coq language, mainly because its programming language combines powerful logical capabilities and reasonable expressive power.

---

[3] This is partly due to the destruction of most of his work during the second world war and to the difficulty to assess prototethic.

We have to prove that the system of axioms and rules of inference governing sentential connectives and quantifiers of protothetic is expressible in Coq. The qualitative measure suggested here is relative to the system $S5$ of protothetic and the CIC which is, with some variants, the one implemented in the theorem prover Coq.

Proposition 1. *CIC is at least as expressive as protothetic.*

Proof. The first step is to prove that CIC is at least as expressive as Church Simple Type Theory (STT) [12]. Whereas STT has variables ranging over functions, together with binders for them, System F^4 supports a mechanism of universal quantification over types which results in variables ranging over types, and binders for them. It yields that generic data types such as list, trees, etc. can be encoded. It follows that system F is more expressive than STT. CIC adds universes to the system F, which leads to an improvement of its consistency strength. Adding dependent types in CIC enhances the computational power but does not affect its consistency strength. As a result, the expressive power of CIC is higher than STT.

In the second step we have to prove that Church Simple Type Theory is at least as expressive as protothetic. In protothetic quantification is allowed on propositional variables and variables of propositional functors to any degree. It follows that protothetic is equivalent in expressive potential to a theory of propositional types [23,36]. Now, if we consider LO which includes protothetic, all symbols of LO can be substituted with variables that can be quantified, then LO is equivalent in expressive power to STT. It yields that protothetic which is a sub-theory of LO is at most as expressive as STT. Combining the two claims, we derive that CIC is at least as expressive as protothetic.   □

It turns out that Leśniewski's mereology can be expressed in CIC and thus, encoded in Coq. CIC [15,16] is an intensional type theory using a universe for logic (*Prop*) and a hierarchy of universes for types whose low-level universe is *Type* reflecting merely the Leśniewski's categorization with the respective categories of LO, S (propositions) and N (names). The Calculus of constructions (and therefore Coq) also includes an inference rule for equality between types called conversion rule. However, it does not include primarily an extensional equivalence and requires a way of defining an extensional conversion within an intensional system [30,34]. For that purpose setoids provide an efficient way to introduce extensionality which maintains the difference between identity and equivalence i.e., with an interpretation of intensional equality (the equality on the original set) and extensional equality (the equivalence relation on the new set). Setoids have been early introduced into the Coq theorem prover [4,24], and refined using type classes [40]. The encoding of setoids for extensionality in the paper will refer to the latter. The basic idea is to express mereology in Coq with

---

[4] Known as Girard–Reynolds polymorphic lambda calculus. It extends STT by the introduction of a mechanism of universal quantification over types (second-order) and is itself extended in CIC.

some constraints: (i) mapping the category S onto $Prop$ and the category of names to the type N, (ii) using dedicated equivalence relations for expressing functors and (iii) expressing extensionality through setoids.

4.2 A Short Introduction to the Coq Theorem Prover

The Coq language is a tool for developing mathematical specifications and proofs. As a specification language, Coq is both a higher order logic (quantifiers may be applied on natural numbers, on functions of arbitrary types, on propositions, predicates, types, etc.) and a typed lambda-calculus enriched with an extension of primitive recursion (further details are given in [6]). The underlying logic of the Coq system is an intuitionist logic. This means that the proposition $A \vee \neg A$ is not taken for granted and, if it is needed, the user has to assume it explicitly. This allows to clarify the distinction between classical and constructive proofs.

Its building blocks are terms and the basic relation is the typing relation. Coq is designed such that it ensures decidability of type checking. All logical judgments are typing judgments. The type-checker checks the correctness of proofs, that is, it checks that a data structure complies to its specification. The language of the Coq theorem prover consists in a sequence of declarations and definitions. A declaration associates a name with a qualification. Qualifications can be either logical propositions which reside in the universe[5] $Prop$ or abstract types which belong to the universe $Type$ (the universe $Type$ is stratified but this aspect is not relevant here).

Conversion rules such as β-reduction allow for term reductions which in that particular case, will formalize the substitution rule of protothetic. Standard equality in Coq is the Leibniz equality where propositionally equal terms are meant to be equivalent with respect to all their properties. Leibniz equality will be replaced by appropriate equivalence relations that are defined in setoids. The proof engine also provides an interactive proof assistant to build proofs using specific programs called tactics. Tactics are the cornerstone of proof-search in the process of theorem proving.

4.3 Expressing Mereology in Coq

Models of mereology that have been constructed so far focus on the interpretation of part-whole relations letting aside the interpretation of the copula ε [13]. We depart from this assumption and encode the Leśniewski's systems in Coq based on a single axiom for the copula. Two semantic categories are defined, $Prop$ for propositional variables and N for name variables. Any constant of any semantic category is well typed by means of formerly introduced symbols. The principle of extensionality for propositional types is defined as the morphism (iff refer here to the usual first-order condition):

---

[5] Also called *sort*.

Notation "a ≡≡ b" := (iff a b) (at level 70).

The copula ε is introduced by a single axiom.

Class **N**.
Parameter *epsilon* : $N \rightarrow N \rightarrow$ Prop.

Notation "A 'ε' b" := (*epsilon* A b) (at level 70).

Axiom *isEpsilon* : $\forall~A~a$, $A~\varepsilon~a$ ≡≡ $((\exists~B, B~\varepsilon~A) \wedge (\forall~C~D, (C~\varepsilon~A \wedge D~\varepsilon~A \rightarrow$
$$C~\varepsilon~D)) \wedge (\forall~C, C~\varepsilon~A \rightarrow C~\varepsilon~a)).$$

The axiom includes three sub-axioms. The first one requires the existence of A, the second assumes the uniqueness of A while the last one expresses the convergence of all which can be A. From the full ontology, we retain two kinds of equality between names:

Parameter *singular_equality* : $N \rightarrow N \rightarrow$ Prop.
Parameter *D4* : $\forall~A~B$, *singular_equality* $A~B$ ≡≡ $(A~\varepsilon~B \wedge B~\varepsilon~A)$.

Parameter *weak_equality* : $N \rightarrow N \rightarrow$ Prop.
Parameter *D6* : $\forall~a~b$, *weak_equality* $a~b$ ≡≡ $(\forall~A, (A~\varepsilon~a)$ ≡≡ $(A~\varepsilon~b))$.

Whereas D4 defines the identity of singulars A and B in terms of ε, D6 formalizes the weak equality between a and b. In LO, the law of extensionality for identities written $\forall AB\Phi$, $A = B \wedge \Phi(A) \rightarrow \Phi(B)$ which requires the equality between singular names, states the generalization of the singular equality with the higher-order functor Φ [39]. Such an higher-order extensionality can be captured in Coq on the basis of name equivalence relations. Unlike the singular equality (which is irreflexive), the functor for weak equality has the properties of an equivalence relation. It is easily derivable from definition $D6$, the iff morphism and basic tactics.

Lemma OntoT23 : reflexive _ weak_equality.
Lemma OntoT24 : symmetric _ weak_equality.
Lemma OntoT25 : transitive _ weak_equality.

It follows that the weak equality is an equivalence relation. Using setoids, we can state the extensionality lemma MereoT16 using higher-order categories including names:

Definition Phi $(E1: N \rightarrow N \rightarrow$Prop$)(E2:N)(E3:N \rightarrow N)(E4:N) := E1~E2~(E3~E4)$.
Lemma MereoT16 : $\forall(A:N)(B:N)(C:N)(Phi:N \rightarrow N)$, $A\varepsilon Phi~B \wedge$ *singular_equality* $B\,C$
$$\rightarrow A\varepsilon Phi~C.$$

The ontology described so far consists in an axiom and a collection of lemmas currently introduced. The syntactic domain of the ontology is the set of well-formed expressions which can be expressed w.r.t. all semantic categories that have been defined. The semantic domain include a collection of objects together with a collection of names, each collection being possibly empty.

The mereology adds a number of functors such as the KI functor which describes the collective classes. For that purpose, two name forming functors are required, i.e., pt (for "part of") and el (for "element of") which belong to the semantic category $N \rightarrow N$.

Parameter *pt*   : $N \rightarrow N$.
Parameter *el*   : $N \rightarrow N$.

With such declarations, a partitive relation $A \varepsilon (pt\ b)$ (read "A is a part of b") and a membership relation $A \varepsilon (el\ b)$ (read "A is an element of b") are introduced. The former is taken as a primitive of mereology and two axioms govern its properties. They respectively state the asymmetry and transitivity of the ε pt relation. Notice that ε pt, ε el and ε Kl respectively correspond to what are termed "proper-part-of", "part-of" and "sum" in Tarski's mereogeometry. If we were eliminating ε we would remove the notion of distributive class [14] and consequently the interplay between the two notions of class, aspects of mereology which Leśniewski thought were so important that he constructed LO in order to describe them clearly. For the sake of simplicity, definition ProperPart_of represents an alias for handling *proper part of* relations and should not be confused with a mereological definition using equivalence ($\equiv\equiv$).

Definition ProperPart_of   := fun $A\ B \Rightarrow A \varepsilon$ pt B.

Axiom *asymmetric_ProperPart*   : $\forall A\ B,\ A \varepsilon$ pt $B \rightarrow B \varepsilon$ (*distinct* (pt A)). (*A1*)

Axiom *transitive_PoperPart*   : **Transitive** ProperPart_of.   (*A2*)

The term distinct whose type is $N \rightarrow N$ is such that from b, the type distinct b will capture the meaning of "to be different from b". The next mereological definition follows the structure of ontological definitions and expresses "being an element of" as a name-forming functor.

Parameter $MD1$: $\forall A\ B,\ A \varepsilon$ el $B \equiv\equiv (A \varepsilon A \wedge$ (*singular_equality* $A\ B\ \vee$
$\qquad\qquad\qquad\qquad\qquad\qquad\qquad\qquad$ ProperPart_of $A\ B$)).

To formulate the next axioms, we first introduce the mereological class, which is also defined as a name forming functor in the usual way:

Parameter $MD2$: $\forall A\ a,\ A \varepsilon$ Kl $a \equiv\equiv (A \varepsilon A \wedge (\exists B,\ B \varepsilon a) \wedge (\forall B,\ B \varepsilon a \rightarrow$
$\quad B \varepsilon$ el $A) \wedge (\forall B,\ B \varepsilon$ el $A \rightarrow \exists C\ D,\ (C \varepsilon a \wedge D \varepsilon$ el $C \wedge D \varepsilon$ el $B)))$.

The third axiom states that a mereological class is unique.

Axiom *Kl_uniqueness* : $\forall A\ B\ a,\ (A \varepsilon$ Kl $a \wedge B \varepsilon$ Kl $a) \rightarrow$
$\qquad\qquad\qquad\qquad\qquad\qquad\qquad$ singular_equality $A\ B$. (*A3*)

The fourth one postulates the class existence.

Axiom *Kl_existence* : $\forall A\ a,\ A \varepsilon a \rightarrow \exists B,\ B \varepsilon$ Kl a.   (*A4*)

No additional rules of inference are added in mereology. Many lemmas can be proved on the basis of mereological axioms. We only mention the following lemmas since they illustrate the crucial difference between set theory and Leśniewski's mereology.

Lemma MereoT26 : $\forall A\ a,\ A \varepsilon$ Kl $a \rightarrow A \varepsilon$ Kl (Kl a).
Lemma MereoT27 : $\forall A\ a,\ A \varepsilon$ Kl (Kl a) $\rightarrow A \varepsilon$ Kl a.
Lemma MereoT29 : $\forall A,\ \neg(A \varepsilon$ Kl *empty*).

Lemmas MereoT26 and MereoT27 show that in mereology, the class of a and the class of the class of a stand for the same object, which is in contrast with set theory. Lemma MereoT29 asserts that the class resulting from the empty name does not exist. The system of mereology is consistent which can be proved by many ways (e.g., with an appropriate model for mereology such as [13] or from prototethic [27]). Additional name-forming functors can be added to the theory to enhance its expressiveness. Leśniewski added the *coll* (collection of) functors, while Lejewski added the *ov* (overlap) functor.

Parameter *coll* $: N \to N.$ (* collection of *)
Parameter *ov* $: N \to N.$ (* overlap *)

Parameter $MD3$: $\forall\ P\ a,\ P\ \varepsilon\ coll\ a \equiv\equiv (P\ \varepsilon\ P \wedge \forall\ Q,\ Q\ \varepsilon\ el\ P \to \exists\ C\ D,\ (C\ \varepsilon\ a$
$\wedge\ C\ \varepsilon\ el\ P \wedge D\ \varepsilon\ el\ C \wedge D\ \varepsilon\ el\ Q)).$
Parameter $MD7$: $\forall\ P\ Q,\ P\ \varepsilon\ ov\ Q \equiv\equiv (P\ \varepsilon\ P \wedge \exists\ C,\ (C\ \varepsilon\ el\ P \wedge C\ \varepsilon\ el\ Q)).$

A useful lemma in the following states that if an object P is an a, then P is a collection of objects a.

Lemma XIII : $\forall\ a\ P,\ P\ \varepsilon\ a \to (P\ \varepsilon\ coll\ a).$

In addition, it is provable that if an object is a sub-collection of another, then it is equivalent to say that it is a part of this object.

Lemma $MereoT44$: $\forall\ A\ B,\ B\ \varepsilon\ el\ A \equiv\equiv (B\ \varepsilon\ subcoll\ A).$

To clarify the notion of class we describe an example extracted from [20] in Fig. 1 which represents a rectangle labeled R including geometric parts. Let us suppose that the generic name a describes the squares of R. In other words, we get: $A\ \varepsilon\ Kl(square\ of\ R)$ with the obvious meaning "A is the class of squares of R". The four parts of the class definition MD2 are easily provable:

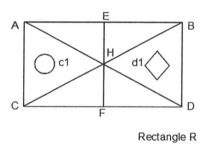

Rectangle R

Fig. 1. Example of a collective class.

1. the name A denotes an object, i.e., rectangle R,
2. there exists a B such that $B\ \varepsilon\ (square\ of\ R)$ (e.g., $AECF$),
3. any square of R is an element of A e.g., $EBFD\ \varepsilon\ el\ A$,
4. for any element of R, e.g., the triangle BHD, an object C which is a square of R should exists, while another object D should also exist as both an element of C and an element of B. If C denotes $EBFD$ and D, the diamond $d1$, then we easily show that $d1$ is both element of $EBFD$ and BHD. We can also consider a more complex object. Let us define B as the collective object including $c1$ (a circle) and BHD. Then, there exists C which is a square of R e.g., $EBFD$ and D which is both element of C and B, e.g., the diamond $d1$.

As a result, the content of collective classes appears much more expressive than distributive classes of set theory. The syntactical aspect of Leśniewski's

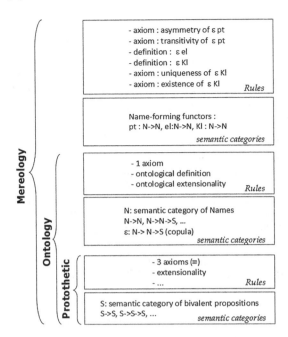

Fig. 2. An overview of Leśniewski's mereology.

Mereology is summarized in Fig. 2. Each sub-theory defines (i) available categories and well-formed expressions, (ii) the assumed set of axioms and (ii) the set of inference rules.

5   The Type-Theoretical Mereogeometry

We first recall the set of definitions suggested by Tarski to support the mereogeometry of solids. In the second subsection, we discuss some issues about Tarski's work while the last subsection presents the type-theoretical account of Tarski mereogeometry.

5.1   Tarski's Definitions

Tarski starts out from some mereological notions such as being a part of and sum and then, provides a minimal set of axioms constituting the mereological part on which the rest of the theory relies. This part will be replaced in the present work by the mereology developed in Sect. 4. Apart from this mereological part, Tarski takes *sphere* (which will be called ball here) as the only primitive notion specific to the geometry of solids. Assuming these notions suffices to formulate a set of nine geometrical definitions which constitutes the basis of mereogeometry.

Definition 1. *The ball A is externally tangent to the ball B if (i) the ball A is disjoint from the ball B and (ii) given two balls X and Y containing as part the ball A and disjoint from the ball B, at least one of them is part of the other.*

Definition 2. *The ball A is internally tangent to the ball B if (i) the ball A is a proper part of the ball B and (ii) given two balls X and Y containing the ball A as a part and forming part of the ball B, at least one of them is a part of the other.*

Definition 3. *The balls A and B are externally diametrically tangent to the ball C if (i) each of the balls A and B is externally tangent to the ball C and (ii) given two balls X and Y disjoint from the ball C and such that A is part of X and B a part of Y, X is disjoint from Y.*

Definition 4. *The balls A and B are internally diametrically tangent to the ball C if (i) each of the balls A and B is internally tangent to the ball C and (ii) given two balls X and Y disjoint from the ball C and such that the ball A is externally tangent to X and B to Y, X is disjoint from Y.*

Definition 5. *The ball A is concentric with the ball B if one of the following conditions is satisfied: (i) the balls A and B are identical, (ii) the ball A is a proper part of B and besides, given two balls X and Y externally diametrically tangent to A and internally tangent to B, these balls are internally diametrically tangent to B and (iii) the ball B is a proper part of A and besides, given two balls X and Y externally diametrically tangent to B and internally tangent to A, these balls are internally diametrically tangent to A.*

Definition 6. *A point is the class of all balls which are concentric with a given ball.*

Definition 7. *The points A and B are equidistant from the point C if there exists a ball X which belongs as element to the point C and which satisfies the following condition: no ball Y belonging as element to the point A or to the point B is a part of X or is disjoint from X.*

Definition 8. *A solid is an arbitrary sum of balls.*

Definition 9. *The point P is an interior point of the solid B if there exists a ball A which is at the same time an element of the point P and a part of the solid B.*

The 2D version of these definitions is illustrated in Fig. 3.

5.2   Some Issues in Tarski's Work

While in the sketchy paper of Tarski [42] the logical background seems rather unclear [8,21] we depart from all approaches stemming from a set-based framework and rather argue that an axiomatization of Tarski's work based solely on

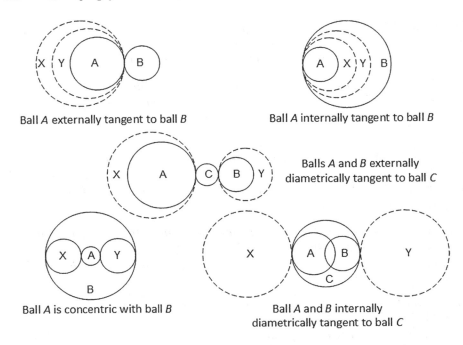

Fig. 3. Tarski's primitive definitions.

Leśniewski's Mereology is possible. Instead of introducing axioms for part-whole relations, we rather rely on the Leśniewski's mereology together with its set of axioms and theorems. In fact Tarski applies Leśniewski's mereology and does not extend it. We will justify these claims by discussing the successive issues highlighted in the recent survey about Tarski's work [8] where the author points out that the role of Leśniewski's systems is marginal.

As argued by the author, Tarski's work can be given a meaningful account to some otherwise logically unclear parts of it, only if one assume a type theoretical approach. We have shown in Sect. 4 that using a type-theoretical background substituting prototheticic in Leśniewski's mereology is a possible approach that is coherent with the work of Henkin [23]. Another remark concerns the notions of being a part of and sum. While being a part of is said to be a primitive notion in Lesniewski's Mereology, but sum is not. We advocate that the references of Tarski to Leśniewski's work are syntactically exact (e.g., the word "sum" in definition 8 refers to the definition of "collection" in mereology [37]).

Then, the author explains that the notion of domain of discourse remains unclear, i.e., that the range of the quantifiers can be that of solids or that of balls. We rather advocate for a more uniform solution in which balls and solids are names, and more precisely plurals in the sense of Leśniewski. Such an assumption yields a quantification over variables that are always of type N (name) and avoids the issues concerning the domain of discourse. She also says that Tarski is freely mixing notions from Leśniewski's mereology (e.g., collective classes) with those of the system of *Principia mathematica*, e.g., distributive classes. However, in Leśniewski's systems collective and distributive classes coexist [14]. It follows that instead of expressing Tarski's mereogeometry in a set-based framework [21], a more coherent picture can be proposed using Leśniewski's axiomatization. Collective classes are addressed with the ε *Kl* functor while distributive classes require new categories and definitions (in the form of equivalences). The Russellian notion of class that appears in Definition 6, is solved using a definition in the sense of Leśniewski by constraining the quantification through concentricity (see next subsection).

A first benefit of the proposed type-theoretical formulation is that some axioms of Mereogeometry can be proved as theorems. This fact enhances the soundness and simplifies the whole theory. A second advantage is that many critiques of Tarski's paper can be avoided such as the difficulty to assess the domain of discourse (see, e.g., [8]).

5.3   Expressing s of Mereogeometry in Coq

The first commitment concerns balls[6] and solids. These concepts are subject to discussion for they are seen as distinct distributive classes. Using the nominalist view of Leśniewski, these names are considered as constant plurals, since (i) they denote a constant plurality and (ii) they refer to collections of objects. In other words, they are instances of names (N) which are constrained to appear as the right argument of the copula (ε). In LO, constant plurals (e.g., "empty name" or "universal name") are defined in a similar way: their category must be defined first (i.e., N) and then, a definition explains their property in the form of an equivalence. Since "balls" is a primitive, no specific property is required. Alternatively, solids does not refer to a primitive name. As a consequence, it belongs to the category N and it is defined as a particular collection of balls (see Definition 8 below).

```
Parameter balls          : N.
Parameter solids         : N.
```

In such a way the domain of discourse is that of truth values and names (basic categories) with a restriction to balls and solids. Notice that the restriction is not given by types but rather by theorems which directly constrain their use. Using the ε *pt* functor, the four axioms of mereology and derived theorems, then it is easy to prove that the relation εel is a partial order:

```
Parameter PartOf : relation N.
Parameter MD11 : ∀ A a, PartOf A a ≡≡ (A ε el a).

Lemma Reflexive_Element_of        : Reflexive PartOf.
Lemma AntiSymmetric_Element_of    : Antisymmetric PartOf.
Lemma Transitive_Element_of       : Transitive PartOf.

Theorem PartOf_is_partial_order   : POrder PartOf.
```

We first introduce the short-hand symbol $\leqslant$ which stands for ε *el*. Then, several relations among balls are defined such as concentricity, relying on the

---

intended interpretation of the primitives. Points are defined as (mereological) collections of concentric balls. Equidistance among points makes use of properties of concentric balls while Euclidean axioms are able to constrain equidistance. We introduce successively:

Parameter *et*	: $N \rightarrow N$.
Parameter *it*	: $N \rightarrow N$.
Parameter *edt*	: $N \rightarrow N \rightarrow N$.
Parameter *idt*	: $N \rightarrow N \rightarrow N$.
Parameter *con*	: $N \rightarrow N$.
Parameter *point*	: $N \rightarrow N$.
Parameter *equid*	: $N \rightarrow N \rightarrow N$.
Parameter *ipoint*	: $N \rightarrow N$.

The relations of external tangency (ε *et*), internal tangency (ε *it*), external diametricity (ε *edt*), internal diametricity (ε *idt*), concentricity (ε *con*), point (*point*), equidistance (ε *equi*) and interior point (ε *ipoint*) are defined using the already defined name functors.

Definition 1: external tangency.

Parameter *ET* : $\forall A\ B,\ A\ \varepsilon\ et\ B \equiv\equiv ((A\ \varepsilon\ balls) \wedge (B\ \varepsilon\ balls) \wedge (A\ \varepsilon\ ext\ B) \wedge$
$\forall X\ Y, ((X\ \varepsilon\ balls) \wedge (Y\ \varepsilon\ balls) \wedge (A \leqslant X \wedge X\ \varepsilon\ ext\ B) \wedge$
$(A \leqslant Y \wedge Y\ \varepsilon\ ext\ B)) \rightarrow (X \leqslant Y \vee Y \leqslant X)).$

Definition 2: internal tangency.

Parameter *IT* : $\forall A\ B,\ A\ \varepsilon\ it\ B \equiv\equiv ((A\ \varepsilon\ balls) \wedge (B\ \varepsilon\ balls) \wedge (A < B) \wedge$
$\forall X\ Y, ((X\ \varepsilon\ balls) \wedge (Y\ \varepsilon\ balls) \wedge (A \leqslant X \wedge X \leqslant B) \wedge$
$(A \leqslant Y \wedge Y \leqslant B)) \rightarrow (X \leqslant Y \vee Y \leqslant X)).$

Definition 3: external diametrical tangency.

Parameter *EDT* : $\forall A\ B\ C,\ A\ \varepsilon\ edt\ B\ C \equiv\equiv ((A\ \varepsilon\ balls) \wedge (B\ \varepsilon\ balls) \wedge$
$(C\ \varepsilon\ balls) \wedge (B\ \varepsilon\ et\ A) \wedge (C\ \varepsilon\ et\ A) \wedge \forall X\ Y, ((X\ \varepsilon\ balls) \wedge$
$(Y\ \varepsilon\ balls\ ) \wedge (B \leqslant X \wedge X\ \varepsilon\ ext\ A\ ) \wedge (C \leqslant Y \wedge Y\ \varepsilon\ ext\ A))$
$\rightarrow (X\ \varepsilon\ ext\ Y)).$

Definition 4: internal diametrical tangency.

Parameter *IDT* : $\forall A\ B\ C,\ A\ \varepsilon idt\ B\ C \equiv\equiv ((A\ \varepsilon\ balls) \wedge (B\ \varepsilon\ balls) \wedge (C\ \varepsilon\ balls)$
$\wedge (B\ \varepsilon\ it\ A) \wedge (C\ \varepsilon\ it\ A) \wedge \forall X\ Y, ((((X\ \varepsilon\ balls) \wedge (Y\ \varepsilon\ balls)$
$\wedge (X\ \varepsilon\ ext\ A) \wedge (Y\ \varepsilon\ ext\ A) \wedge (B\ \varepsilon\ ext\ X) \wedge (C\ \varepsilon\ ext\ Y))$
$\rightarrow (X\ \varepsilon\ ext\ Y)).$

Definition 5: concentric balls.

Parameter *CON* : $\forall A\ B,\ A\ \varepsilon\ con\ B \equiv\equiv ((A\ \varepsilon\ balls) \wedge (B\ \varepsilon\ balls)) \wedge singular\_$
$equality\ A\ B \vee (A < B \wedge \forall X\ Y, ((X\ \varepsilon\ balls) \wedge (Y\ \varepsilon\ balls) \wedge$
$(A\ \varepsilon\ edt\ X\ Y) \wedge (X\ \varepsilon\ it\ B) \wedge (Y\ \varepsilon\ it\ B)) \rightarrow (B\ \varepsilon\ idt\ X\ Y)) \vee$
$(B < A \wedge \forall X\ Y, (X\ \varepsilon\ balls) \wedge (Y\ \varepsilon\ balls) \wedge (B\ \varepsilon\ edt\ X\ Y) \wedge$
$(X\ \varepsilon\ it\ A) \wedge (Y\ \varepsilon\ it\ A) \rightarrow (A\ \varepsilon\ idt\ X\ Y))).$

Definition 6: point.

Parameter *POINT* : $\forall P\ B,\ P\ \varepsilon\ (point\ B) \equiv\equiv ((P\ \varepsilon\ P) \wedge (B\ \varepsilon\ balls) \wedge$
$\forall B', (B'\ \varepsilon\ balls) \wedge B'\ con\ B).$

Definition 7: equidistance.

Parameter *EQUID* : $\forall A\ B\ C,\ A\ \varepsilon\ equid\ B\ C \equiv\equiv ((A\ \varepsilon\ balls) \wedge (B\ \varepsilon\ balls) \wedge$

$(C \ \varepsilon \ balls) \land \exists \ X, ((X \ \varepsilon \ balls) \land (X \ \varepsilon \ con \ A) \land \neg \exists \ Y,$
$((Y \ \varepsilon \ balls) \land Y \ \varepsilon \ (union \ B \ C) \land (Y \leqslant X) \lor (Y \ \varepsilon \ ext \ X))))$.

Definition 8: solids.

Parameter *TarskiD8* : $\forall \ A, A \ \varepsilon \ solids \equiv\equiv \exists \ B, (B \ \varepsilon \ B \land (B \ \varepsilon \ coll \ balls) \land$
$(A \ \varepsilon \ subcoll \ B))$.

Definition 9: interior point.

Parameter *IPOINT* : $\forall \ P \ X \ C, P \ \varepsilon \ (ipoint \ X) \equiv\equiv (X \ \varepsilon \ solids \land P \ \varepsilon \ (point \ C) \land$
$\exists \ A', ((A' \ \varepsilon \ balls) \land (A' \ \varepsilon \ P) \land (A' \leqslant X)))$.

5.4 Revisiting the Axiom System

The axiom system of Tarski for the geometry of solid can be broadly divided in three parts, (i) axioms stating the existence of a correspondence between notions of the geometry of solids and notions of ordinary point geometry, (ii) two axioms establishing a correspondence between notions of the geometry of solids and topology and (iii) internal axioms that are derivable from Leśniewski's mereology. Axioms of the former part are:

Axiom 1. *The notions of point and equidistance of two points to a third satisfy all axioms of ordinary Euclidean geometry of three dimensions.*

More specifically, this axiom states that (i) points as they are introduced in definition 6 correspond to points of an ordinary point-based geometry and (ii) the relation *EQUID* corresponds to an ordinary equidistance relation. With Π standing for mereogeometrical points, the structure $\langle \Pi, \text{EQUID} \rangle$ is a Pieri's structure [42]. Then, it can be proved that $\langle \Pi, \text{EQUID} \rangle$ is isomorphic to ordinary Euclidian geometry $\langle \mathbb{R}^3, \text{EQUID}^{\mathbb{R}^3} \rangle$ (see [21] p. 500).

Axiom 2. *If A is a solid, the class α of all interior points of A is a non-empty regular open set.*

Axiom 3. *If the class α of points is a non-empty regular open set, there exists a solid A such that α is the class of all its interior points.*

The second axiomatic part relies on the structure $\langle \Pi, \text{EQUID} \rangle$. It follows that we are able to define the family of open balls $\mathcal{O}b_\Pi$ in it and then introduce in Π the family $\mathcal{O}_\Pi$ of open sets together with appropriate topological operations of closure and openness such that $\langle \Pi, \mathcal{O}_\Pi \rangle$ is a topological space. If in $\langle \Pi, \mathcal{O}_\Pi \rangle$, we introduce the family $\mathcal{O}r_\Pi^0$ of all regular open sets excluding the empty set, then it can be proved that $\langle \mathcal{O}r_\Pi^0, \mathcal{O}b_\Pi, \subseteq \rangle$ is isomorphic to $\langle \mathcal{O}r_{\mathbb{R}^3}^0, \mathcal{O}b_{\mathbb{R}^3}, \subseteq \rangle$ (see [21] for more details).

The third axiomatic part of the geometry of solid is derivable from Leśniewski's mereology as follows. The first axiom of Tarski is stated as:

Axiom 4. *If A is a ball and B a part of A, there exists a ball C which is a part of B.*

Using previous lemmas and definitions from Leśniewski's mereology, it can be proved in Coq as the theorem:

Theorem *TA4* : ∀ *A B*, (*A ε balls* ∧ *B ε el A*) → ∃ *C*, (*C ε balls* → *C ε el B*).
    Proof.
        intros *A B H1*.
        destruct *H1* as [*H1 H2*].
        assert (*H0:=H1*);apply *OntoT5* in *H0*.
        apply *XIII* in *H1*.
        apply *MereoT44* in *H2*.
        assert (*H3*:(*B ε solids*)).
        apply *TarskiD8*;exists *A*.
        split;[assumption | split;assumption].
        clear *H0 H1 H2*.
        assert (*H1:=H3*);apply *OntoT5* in *H3*.
        apply *TarskiD8* in *H1*.
        destruct *H1* as [*C H1*].
        decompose [and] *H1*;clear *H1*.
        apply *MD3* in *H2*;apply *MereoT44* in *H4*.
        destruct *H2* as [*H0 H2*];clear *H0*.
        apply *H2* in *H4*;clear *H2*.
        destruct *H4* as [*E H4*];destruct *H4* as [*F H4*].
        decompose [and] *H4*;clear *H4*.
        exists *F*;intro;assumption.
    Qed.

The second axiom says that:

Axiom 5. *If A is a solid and B a part of A, then B is also a solid.*

The proof in Coq is detailed below.

Theorem *TA4'* : ∀ *A B*, (*A ε solids* ∧ *B ε el A*)→ *B ε solids*.
    Proof.
        intros *A B H*.
        destruct *H* as [*H1 H2*].
        apply *TarskiD8* in *H1*.
        destruct *H1* as [*C H3*].
        destruct *H3* as [*H3 H4*];destruct *H4* as [*H4 H5*].
        apply *MereoT44* in *H2*.
        apply *TarskiD8*;exists *C*.
        split;[assumption | split;[assumption |
            apply (*Transitive_subcoll B A C*);split;assumption
            ]]. Qed.

Another axiom is provided by Tarski which relies on the definition of *interior points* and is formulated as follows.

Axiom 6. *If A and B are solids, and all the interior points of A are at the same time interior points of B, then A is a part of B.*

As advocated by the author, Axiom 6 (i) relies on the interplay between *interior points* and the set-based definition of regular open sets, and thus requires to state the relation between a solid and its interior points and (ii) is merely an alternative of any axiom among Axioms 4 and 5. It is well-known that the more axioms one assumes in a formal system, the harder it becomes to preserves its soundness. If we use one of the Axioms 4 or 5 we get a more stronger system than the one obtained with Axiom 6 (see Theorem 1.3 in [21]). It follows that the resulting axiom system obtained by blurring Axiom 6 provides a minimal system that can serve as a basis for constructing spatial theories.

6 Conclusion

Generalizing solids to spatial regions, geometrical theories based on mereology present an appealing impact on spatial theories. As underlined in [11], they provide formal theories adequate for different tasks. Among their benefits, (i) they make possible a direct mapping from empirical entities and laws to theoretical entities and formulas, (ii) they have the ability to formalize human learning, conceptualization, and categorization of spatial entities and relations and (iii) they have received a particular emphasis in the field of formal ontology with mereogeometrical notions. The theory of Tarski, has been proved to be semantically complete with regards to the models expressed in terms of R^n and has been axiomatized by Bennett [5].

Major problems are that (i) the set-based interpretation (e.g., interpreting A ε pt B as $A \subseteq B$) considerably weakens the logical power of Leśniewski's framework and (ii) the approach described in [21] in which the fourth axiom is replaced with a new postulate asserting that the domain of discourse of Tarski's theory coincides with arbitrary mereological sums of balls, does not simplify the work of Tarski either. What we have proposed so far to avoid these problems, is a logical foundation having the following properties: (i) the proposed set of structures featuring geometrical entities and relations relies on Tarski's mereogeometry, (ii) it has a model in ordinary three-dimensional Euclidian geometry [43], (iii) it is based on three axioms instead of four (iv) it is coherent with Leśniewski's mereology and does not suffer the defects cited in [8] and (v) it will serve as a basis for spatial reasoning with full compliance with Leśniewski's systems. These systems have precisely the property to be scalable, which is a significant argument for extending the theory with new definitions in appropriate applications. Future work will develop this last aspect.

References

1. Appel, A.W., Felty, A.P.: Dependent types ensure partial correctness of theorem provers. J. Funct. Program. **14**(1), 3–19 (2004). Cambridge University Press
2. Asher, N., Vieu, L.: Toward a geometry of common sense: a semantics and a complete axiomatization of mereotopology. In: Proceedings of the International Joint Conference on Artificial Intelligence (IJCAI-95), Montreal (1995)
3. Aurnague, M., Vieu, L.: A theory of space-time for natural language semantics. In: Korta, K., Larrazabal, J.M. (eds.) Semantics and Pragmatics of Natural Language: Logical and Computational Aspects. ILCLI Series I, pp. 69–126. Universidad Pais Vasco, San Sebastian (1995)

4. Barthe, G., Capretta, V., Pons, O.: Setoids in type theory. J. Funct. Program. **13**(2), 261–293 (2003)
5. Bennett, B.: A categorical axiomatisation of region-based geometry. Fundam. Informaticae **46**(1–2), 145–158 (2001)
6. Bertot, Y., Castéran, P.: Interactive Theorem Proving and Program Development. Coq'Art: The Calculus of Inductive Constructions. Texts in Theoretical Computer Science. An EATCS Series. Springer, Heidelberg (2004)
7. Bertot, Y., Théry, L.: Dependent types, theorem proving, and applications for a verifying compiler. In: Meyer, B., Woodcock, J. (eds.) VSTTE 2005. LNCS, vol. 4171, pp. 173–181. Springer, Heidelberg (2008)
8. Betti, A., Loeb, I.: On Tarski's foundations of the geometry of solids. Bull. Symbolic Logic **18**(2), 230–260 (2012)
9. Betti, A.: Leśniewski, Tarski and the axioms of mereology. Chap. 11. In: Mulligan, K., et al. (eds.) The History and Philosophy of Polish Logic: Essays in Honour of Jan Woleński, pp. 242–258. Palgrave Macmillan, Basingstoke (2013)
10. Borgo, S., Guarino, N., Masolo, C.: A pointless theory of space based on strong congruence and connection. In: Proceedings of 5th International Conference on Principle of Knowledge Representation and Reasoning (KR 1996), pp. 220–229. Morgan Kaufmann (1996)
11. Borgo, S., Masolo, C.: Full mereogeometries. Rev. Symbolic Logic **3**(4), 521–567 (2010)
12. Church, A.: A formulation of the simple theory of types. J. Symbolic Logic **5**, 56–68 (1940)
13. Clay, R.E.: The consistency of Leśniewski's mereology relative to the real number system. J. Symbolic Logic **33**, 251–257 (1968)
14. Clay, R.E.: Relation of Leśniewski's mereology to boolean algebra. J. Symbolic Logic **39**(4), 638–648 (1974)
15. Coquand, T., Huet, G.: The calculus of constructions. Inf. Comput. **76**(2–3), 95–120 (1988)
16. Coquand, T., Paulin, C.: Inductively defined types. In: Martin-Löf, P., Mints, G. (eds.) COLOG-88. LNCS, vol. 417, pp. 50–66. Springer, Heidelberg (1990)
17. Dapoigny, R., Barlatier, P.: Tarski_Mereogeometry, University of Savoie (2015). http://www.polytech.univ-savoie.fr/index.php?id=listic-logiciels-lib-spatial&L=1
18. Eschenbach, C., Heydrich, W.: Classical mereology and restricted domains. Int. J. Hum. Comput. Stud. **43**(56), 723–740 (1995)
19. Gerla, G.: Pointless geometries. In: Buekenhout, F. (ed.) Handbook of Incidence Geometry, pp. 1015–1031. Elsevier, North-Holland (1995)
20. Gessler, N.: Introduction à l'oeuvre de S. Leśniewski. Part III: La méréologie. CdRS, Université de Neuchâtel (2005)
21. Gruszczyński, R., Pietruszczak, R.: Full development of tarski's geometry of solids. Bull. Symbolic Logic **14**(4), 481–540 (2008)
22. Hahmann, T., Grüninger, M.: Region-based theories of space: mereotopology and beyond. In: Hazarika, S.M. (ed.) Qualitative Spatio-Temporal Representation and Reasoning: Trends and Future Directions, pp. 1–62. IGI Publishing, USA (2012)
23. Henkin, L.: A theory of propositional types. Fundam. Math. **52**, 323–334 (1963). Errata 53, 119
24. Hofmann, M.: Extensional concepts in intensional type theory. Ph.d. thesis, University of Edinburgh (1995)
25. Kortenkamp, D., Bonasso, R.P., Murphy, R. (eds.): Artificial Intelligence and Mobile Robots: Case Studies of Successful Robot Systems. MIT Press, Cambridge (1998)

26. Kuipers, B.J., Byun, Y.T.: A qualitative approach to robot exploration and map learning. In: Proceedings of the IEEE Workshop on Spatial Reasoning and Multi-Sensor Fusion, pp. 390–404. Morgan Kaufmann, San Mateo (1987)
27. Lejewski, C.: Consistency of Leśniewski's mereology. J. Symbolic Logic **34**(3), 321–328 (1969)
28. Leśniewski, S.: Podstawy ogólnej teoryi mnogosci. I, Moskow: Prace Polskiego Kola Naukowego w Moskwie, Sekcya matematyczno-przyrodnicza (1916). (English translation by Barnett, D.I.: Foundations of the general theory of sets. I. In: Surma, S.J., Srzednicki, J., Barnett, D.I., Rickey, F.V. (eds.) S. Leśniewski, Collected Works, vol. 1, pp. 129–173. Kluwer, Dordrecht (1992))
29. Leśniewski, S.: Einleitende Bemerkungen zur Fortsezung meiner Miteilung u.d.T. "Grundzüge eines neuen Systems der Grundlagen der Mathematik". Collectanea Logica **I**, 1–60 (1938)
30. Oury, N.: Extensionality in the calculus of constructions. In: Hurd, J., Melham, T. (eds.) TPHOLs 2005. LNCS, vol. 3603, pp. 278–293. Springer, Heidelberg (2005)
31. Quine, W.: Unification of universes in set theory. J. Symb. Logic **21**, 216 (1956)
32. Randell, D.A., Cui, Z., Cohn, A.G.: A spatial logic based on regions and connection. In: Proceedings of 3rd International Conference on Knowledge Representation and Reasoning, pp. 165–176. Morgan Kaufmann, San Mateo (1992)
33. Rickey, V.F.: A survey of Leśniewski's logic. Stud. Logica **36**, 407–426 (1977)
34. Seldin, J.P.: Extensional set equality in the calculus of constructions. J. Log. Comput. **11**(3), 483–493 (2001)
35. Simons, P.: Parts: A Study in Ontology. Clarendon Press, Oxford (1987)
36. Simons, P.M.: Nominalism in Poland. In: Srzednicki, J.T.J., Stachniak, Z. (eds.) Leśniewski's Systems Prototetic. Nijhoff International Philosophy Series, vol. 54, pp. 1–22. Springer, Netherlands (1998)
37. Sinisi, V.F.: Leśniewski's foundations of mathematics. Topoi **2**(1), 3–52 (1983)
38. Slupecki, J.: S. Leśniewski's protothetics. Stud. Logica **1**, 44–112 (1953)
39. Sobociński, B.: On the single axioms of protothetic I. Notre-Dame J. Formal Logic **1**, 52–73 (1960)
40. Sozeau, M.: A new look at generalized rewriting in type theory. J. Formalized Reasoning **2**(1), 41–62 (2009)
41. Tarski, A.: Sur le terme primitif de la Logistique. Fundam. Math. **4**, 196–200 (1923)
42. Tarski, A.: Les fondements de la géométrie des corps (Foundations of the geometry of solids). In: Ksiega Pamiatkowa Pierwszego Polskiego Zjazdu Matematycznego, vol. 7, pp. 29–33 (1929)
43. Tarski, A.: Foundations of the geometry of solids. In: Tarski, A. (ed.) Logics, Semantics Metamathematics. Papers from 1923–1938 by Alfred Tarski. Clarendon Press, Oxford (1956)
44. Tarski, A.: On the foundation of Boolean algebra. In: Woodger, J.H. (ed.) Logic, Semantics, Metamathematics: Papers from 1923 to 1938, pp. 320–341. Clarendon Press, Oxford (1956)
45. Varzi, A.C.: Parts, wholes, and part-whole relations: the prospects of mereotopology. Data Knowl. Eng. **20**(3), 259–286 (1996)

Shape Similarity Based on the Qualitative Spatial Reasoning Calculus eOPRAm

Christopher H. Dorr[1]($^{(\boxtimes)}$), Longin Jan Latecki[2]($^{(\boxtimes)}$), and Reinhard Moratz[1]($^{(\boxtimes)}$)

[1] NCGIA and School of Computing and Information Science,
University of Maine, Orono, USA
{christopher.h.dorr,reinhard.moratz}@maine.edu
[2] Department of Computer and Information Sciences, Temple University,
Philadelphia, USA
latecki@temple.edu

Abstract. In our paper we investigate the use of qualitative spatial representations (QSR) about relative direction and distance for shape representation. Our new approach has the advantage that we can generate prototypical shapes from our abstract representation in first-order predicate calculus. Using the conceptual neighborhood which is an established concept in QSR we can directly establish a conceptual neighborhood between shapes that translates into a similarity metric for shapes. We apply this similarity measure to a challenging computer vision problem and achieve promising first results.

Keywords: Qualitative spatial reasoning · Qualitative shape representation · Computer vision

1 Overview

Qualitative spatial reasoning (QSR) abstracts metrical details of the physical world and enables computers to make predictions about spatial relations even when precise quantitative information is unavailable [4]. From a practical viewpoint QSR is an abstraction that summarizes similar quantitative states into one qualitative characterization. A complementary view from the cognitive perspective is that the qualitative method *compares* features within the object domain rather than by *measuring* them in terms of some artificial external scale [7]. This is the reason why qualitative descriptions are quite natural for humans.

The two main directions in QSR are topological reasoning about regions [25, 27, 31] and positional reasoning about point configurations, like reasoning about orientation and distance [3, 7, 10, 33].

There is also considerable work about using positional reasoning to describe the qualitative shape of 2D regions [17, 28, 29]. Many of these approaches represent qualitative shape by listing the relative positions of the adjacent vertices of polygons enumerating the outline of the polygon [9]. However, this work only

---

R. Moratz—The Principal Investigator and responsible lab author.

S.I. Fabrikant et al. (Eds.): COSIT 2015, LNCS 9368, pp. 130–150, 2015.
DOI: 10.1007/978-3-319-23374-1_7

makes very limited use of concepts of qualitative distance. Based on the recent work by Moratz and Wallgrün [22] there is a candidate for a finer resolution positional QSR calculus called $e\mathcal{OPRA}_m$ which is suited to describe outlines of polygons at different levels of granularity.

The motivation for using qualitative shape descriptions is as follows: qualitative shape descriptions can implicitly act as a schema for measuring the similarity of shapes, which has the potential to be cognitively adequate. Then, shapes which are similar to each other would also be similar for a pattern recognition algorithm. There is substantial work in pattern recognition and computer vision dealing with shape similarity [24]. Here with our approach to qualitative shape descriptions and shape similarity, the focus is on achieving a representation in first-order predicate calculus that can be integrated with ontological reasoning in a straight-forward manner [5].

To enable verification of our shape representation by visually comparing shapes with similar descriptions, the qualitative shape representation must be reversible. That means it must be possible to take the qualitative shape representation and generate prototypical shapes that match the description. In previous work about QSR-based shape description, it was only possible to take shapes and generate their QSR-based descriptions. It was not possible to take a QSR-based shape description and let an automatic algorithm generate a sample shape matching this description. Our work described in this paper presents the first QSR-based shape description capable of generating prototypical shapes based on the QSR-based representation.

In this paper, we discuss the steps taken to reconstruct simple polygons using their $e\mathcal{OPRA}_m$ descriptions. For our purposes, polygons are defined as a simple closed polylines, or a non self-intersecting chain of line segments in the Cartesian plane $\mathbb{R}^2$. Inputs are converted into qualitative $e\mathcal{OPRA}_m$ descriptions, which are then reconstructed as polygons through a combination of state-space searching and constraint propagation.

We show that given an appropriate level of granularity, the $e\mathcal{OPRA}_m$ calculus can be used to represent and reconstruct similar approximations of simple polygons. Additionally, we have applied our approach to a challenging real-world application in computer vision.

Results presented in this paper are produced by a small set of Python programs developed to perform the deconstruction and reconstruction tasks. Roughly, the deconstruction and reconstruction is a three-step process: (1) compute the vertex-pairwise $e\mathcal{OPRA}_m$ direction and distance descriptions of the input polyline; (2) perform an initial reconstruction by "tracing" the qualitative hull description; (3) refine the results of (2) via a greedy search.

2   The eOPRAm Spatial Calculus

Fully fledged qualitative positional reasoning requires representational formalisms, i.e., qualitative spatial calculi, that combine direction information – either in absolute or relative form – with distance information. We start our description with an overview of the developments that led to the $\mathcal{OPRA}_m$ calculus: $\mathcal{OPRA}_m$ supports reasoning about relative directions, but does not address distances[1].

2.1   Developments Leading to the $\mathcal{OPRA}_m$ calculus

Figure 1 shows three examples of spatial calculi capable of expressing relative direction, in chronological order of invention. Arguably, the most natural way to express relative direction is by using ternary relations describing where object C is wrt. object B when seen from A (where A, B, C are points in the 2D plane). This approach has been used in Freksa's Doublecross calculus [7] shown in Fig. 1a. Other calculi employing this approach include the Flipflop calculus [16] and its successors [30], as well as the Singlecross calculus [7].

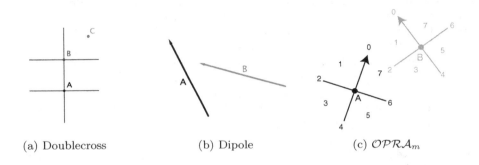

(a) Doublecross   (b) Dipole   (c) $\mathcal{OPRA}_m$

Fig. 1. Examples of direction calculi.

The Dipole calculus [19,20] shown in Fig. 1b is able to express the relative direction of extended objects with an intrinsic direction represented by oriented straight line segments called dipoles. Details about the dipole relations can be found in [19]. It is important to note that the Dipole calculus is a *binary* calculus, and can only be applied when the objects involved can be adequately represented by dipoles. In many potential application scenarios, objects do not have a meaningful length and as such are better represented as points.

The $\mathcal{OPRA}_m$ calculus [18,23] (see Fig. 1c) addresses this issue by attaching intrinsic orientations to point objects, yielding the concept of *oriented points*. $\mathcal{OPRA}_m$ is also an example of a multi-granular calculus in which a granularity parameter m can be used to instantiate relational schemas with different resolutions. Given the granularity parameter m, a partitioning of the plane into $4m$ direction sectors is established for each of the two oriented points A and B involved in a relation. The respective relations are written as $A \; _m\angle_a^b \; B$ where a is the direction sector of A that contains B and b is the direction sector of B that

---

[1] Material in this section is presented as an abridged summary of previous work by Moratz and Wallgrün [22].

contains A (cmp. again Fig. 1c). Oriented points can be seen as dipoles with an infinitely small length. In this conceptualization, the length of the objects involved no longer has any importance, and objects without an intrinsic orientation can be provided with one: for instance, pointing towards some other particular object.

Essentially, this method attaches a feature which is used as a local reference to an object in the 2D-plane which, geometrically, is still a featureless point. We call this principle *hidden feature attachment* and the extended point objects *augmented points*. This approach can be also be adapted to modalities other than direction. In the next section, we employ this approach to introduce the concept of *elevated points* able to represent distance information.

2.2 Elevated Point and Basic Distance Relations

In analogy to the attachment of a reference direction to a featureless point, we can extend point-based binary direction calculi with a local reference distance and then express relative distance relations by comparing these reference distances. We refer to these reference distances as *elevations* due to an analogy with local basic perceptions of a cognitive agent forming the basis for establishing distance categories: We imagine that the point standing for a particular location is *elevated* above the 2D-plane (see point A in Fig. 2) representing, for instance, the viewpoint of a human observer who visually perceives the environment. From this observer perspective, Gibson's insights about natural perspective [8] motivate the availability of depth clues which let the observer distinguish local distances based on a comparison of their distance with this elevation or height value.[2]

We can interpret the elevation as describing a circle around the point providing a reference distance, when projected to the ground plane. As an abstraction, every salient location/point (e.g., the second point B in Fig. 2) can be assigned such a reference distance. This elevation can also be unspecified or unknown, leading to disjunctions in the distance relations defined in the following. We refer to such points with an attached elevation feature as *elevated points*, or simply e-points.

Distances between two e-points A and B can now be locally compared using the projections of their elevations onto the 2D plane, and then considering the resulting partitioning of the ground plane into three partitions called *close, equal,* and *distant* for each of the two points (Fig. 2).

To formally specify the e-point relations, we use two-dimensional continuous space, in particular $\mathbb{R}^2$. Every e-point S on the plane is an ordered pair of a point $\mathbf{p}_S$ represented by its Cartesian coordinates x_S and y_S, with $x_S, y_S \in \mathbb{R}$, and an internal reference distance $\delta_S \in \mathbb{R}^+$ which corresponds to the elevation height in the cognitive motivation.

$$S = (\mathbf{p}_S, \ \delta_S), \qquad \mathbf{p}_S = (x_S, \ y_S)$$

---

[2] Note that while this analogy makes use of three-dimensional space, our model refers to the 2D plane.

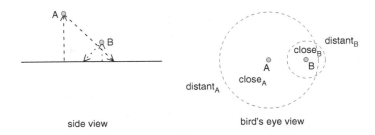

Fig. 2. Qualitative spatial distance relations between two points based on the notion of elevation.

The metric distance between e-points A and B is simply the Euclidean distance:

$$|A - B| = \sqrt{(x_A - x_B)^2 + (y_A - y_B)^2}$$

We now introduce a granularity parameter m to allow for the distance modality to have an adjustable granularity (for the concept of scalable granularity, compare [23]). We define the basic schema from Fig. 2 with three partitions to correspond to $m = 2$.

The notation scheme for the relations of this purely distance-based calculus uses the symbol $A \,_m\bigcirc_i^j\, B$ allowing numbers $1, 2, 3$ to be used in the relation names instead of the mnemonic names *close*, *equal*, and *distant* (e.g., $A \,_2\bigcirc_2^1\, B$ instead of $A \,_2\bigcirc_{\text{equal}}^{\text{close}}\, B$. Figure 3 shows the corresponding schema for granularity $m = 4$. The intuition behind this schema is the empirical observation that subjective distances can be modeled as functions of objective distances with the property that larger distances are represented in a coarser resolution [2]. Another design criterion was the intended property that base relations should have single base relations as converses (strong converse operation). As a result, distances smaller than *equal* have equidistant dividing borders and the distances greater than *equal* are multiplicative inverses to guarantee single base relations as converses in the case of **same** relations. The granularity parameter m corresponds to the number of dividing borders including the **same** case.

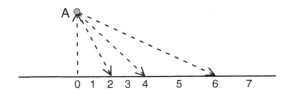

Fig. 3. Qualitative spatial distances for granularity parameter $m = 4$: Four projections corresponding to $\delta * 0$, $\delta * 1/2$, δ, and $\delta * 2$ are used to divide the plane into eight $(2m)$ partitions numbered 0 to 7.

To define the schema for qualitative relative distances $A \ {}_m\bigcirc_i^j \ B$ $(0 < i,$ $j < 2m)$ between e-point pairs with $\mathbf{p}_A \neq \mathbf{p}_B$ formally, we first define the boundaries of the distance partitioning around A to be given by a function $b_A(i)$ with $0 \leq i \leq 2m$ and i being an even number (written as $i \equiv_2 0$ for i mod $2 = 0$):

$$b_A(i) = \begin{cases} \infty & \text{if } i = 2m \\ \frac{i\,\delta_A}{m} & \text{if } i \leq m \\ \frac{m\,\delta_A}{2m-i} & \text{otherwise} \end{cases} \tag{1}$$

The distance component i in $A \ {}_m\bigcirc_i^j \ B$ with $0 < i < 2m$, is then defined as

$$A \ {}_m\bigcirc_i^* \ B \iff (i \equiv_2 0 \wedge |A - B| = b_A(i)) \tag{2}$$
$$\vee$$
$$(i \equiv_2 1 \wedge b_A(i-1) < |A - B| < b_A(i+1))$$

where the $*$ stands for an arbitrary value for j. Analogously, for the distance partition around B and parameter j in $A \ {}_m\bigcirc_i^j \ B$:

$$b_B(j) = \begin{cases} \infty & \text{if } j = 2m \\ \frac{j\,\delta_B}{m} & \text{if } j \leq m \\ \frac{m\,\delta_B}{2m-j} & \text{otherwise} \end{cases} \tag{3}$$

$$A \ {}_m\bigcirc_*^j \ B \iff (j \equiv_2 0 \wedge |A - B| = b_B(j)) \tag{4}$$
$$\vee$$
$$(j \equiv_2 1 \wedge b_B(j-1) < |A - B| < b_B(j+1))$$

For the **same** relations in which the two points have the same position (i.e., $\mathbf{p}_A = \mathbf{p}_B$), the formal schema $A \ {}_m\bigcirc_{same}^i \ B$ (where $0 < i < 2m$) for a general granularity parameter m is given by:

$$A \ {}_m\bigcirc_{same}^i \ B \iff (i \equiv_2 0 \wedge \delta_B = b_A(i)) \tag{5}$$
$$\vee$$
$$(i \equiv_2 1 \wedge b_A(i-1) < \delta_B < b_A(i+1))$$

where $b_A(i)$ is the boundary function already defined above.

2.3   The $e\mathcal{OPRA}_m$ Spatial Calculus

We now combine the idea of hidden feature attachment to represent distance information, which led to the concept of elevated points, with the $\mathcal{OPRA}_m$ calculus.

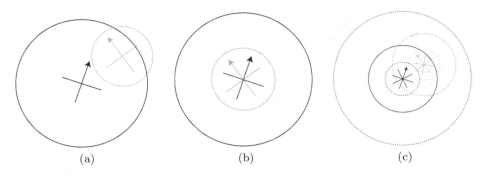

(a)   (b)   (c)

Fig. 4. (a) $e\mathcal{OPRA}_2$ relation ${}_2\angle_7^1 \, {}_{close}^{distant}$, (b) $e\mathcal{OPRA}_2$ relation ${}_2\angle_{same}^{1 \, close}$, and (c) $e\mathcal{OPRA}_4$ relation ${}_4\angle_{15}^{3} \, {}_3^{5}$.

$\mathcal{OPRA}_m$ deals with the relative direction of oriented points (points with an intrinsic reference direction) in the plane. Hence, one would need to combine the concepts of oriented point and elevated point to define points with an intrinsic orientation and elevation as the domain for the combined calculus.

Formally, the domain is $\mathbb{R}^2 \times [0...360) \times \mathbb{R}^+$ (the set of oriented and elevated points) and the resulting eo-points can be written as triples $((x_A, y_A), \theta_A, \delta_A)$ with the first component being Cartesian coordinates x_A and y_A of the point, θ_A giving the orientation, and δ_A the elevation. Figure 4 illustrates the resulting distance and direction partitions using again a single granularity parameter.

Relations of the extended calculus, which we refer to as $e\mathcal{OPRA}_m$ [3], are written in the following way: when the two involved points do not coincide, the relations are written as ${}_m\angle_{a_1 \, d_1}^{a_2 \, d_2}$ with m, d_1, and d_2. Where, d_1 is the distance partition of A which contains B, and d_2 is the distance partition of B which contains A.

Relations for coinciding points (**same** cases) are analogously written as ${}_2\angle_{same}^{a \, d}$. Figure 4 (a) shows the relation ${}_2\angle_7^1 \, {}_{close}^{distant}$, (b) shows ${}_2\angle_{same}^{1 \, close}$, and (c) shows ${}_4\angle_{15}^{3} \, {}_3^{5}$. For further technical details, see Moratz and Wallgrün [22].

2.4   Conceptual Neighborhoods

Whereas qualitative spatial constraint reasoning deals with static aspects of space, conceptual neighborhood-based spatial reasoning (CNH spatial reasoning) can be used to represent spatio-temporal processes. CNH reasoning uses the base relations of a qualitative spatial constraint calculus, but is independent from its operations (e.g. composition table etc.).

Two base relations of the underlying calculus are conceptually neighbored if there exist a continuous sequence of metric instances that starts in one relation and ends in the neighboring relation without resulting in a third relation in between.

---

[3] The first three letters of the symbol $e\mathcal{OPRA}_m$ stand for elevated oriented point.

The concept of general relations typically is not used in CNH reasoning. The information about which base relations are neighbors is stored in a CNH graph that has the base relations as vertices and has edges between base relations that are conceptually neighbored. A transition between two neighboring relations can then represent an object movement in a short period of time. Longer trajectories are represented as sequences of such transitions.

A qualitative spatial CNH calculus can be defined as a triple $C = (\mathcal{D}^n, \mathcal{B}, \mathcal{N})$. Again $\mathcal{D}$ is the domain and $\mathcal{B}$ is a set of base relations. $\mathcal{N}$ is the CNH graph of $\mathcal{B}$. CNH graphs corresponding to the base relations of $e\mathcal{OPRA}$ calculi can be derived using the methods which were developed in [6] for the $\mathcal{OPRA}$ calculus.

With our application of CNH graphs to shape representation we omit the same relations. Then two $e\mathcal{OPRA}$ relations are conceptually neighbored if their parametric expressions only differ by 1 in only one of the parameters[4]. And in our application, we use the conceptual neighborhood for representing shape deformation and measurement errors or relatedness/similarity in a wider context rather then limiting to the representation of trajectories of moving objects.

3 Shape Preprocessing with Discrete Curve Evolution (DCE)

Discrete curve evolution (DCE) was first introduced in [12]. This section is based on [15], where it was part of computing a shape similarity measure of planar shapes. DCE has also been used in [1] for schematizing maps by simplification of geographic shapes, and in [11] for automatic extraction of key frames in videos.

Since contours of objects in digital images are distorted due to digitization noise and segmentation errors, it is desirable to neglect the distortions while at the same time preserving the perceptual appearance at the level sufficient for object recognition. An obvious way to neglect the distortions is to eliminate them by approximating the original contour with one that has a similar perceptual appearance. To achieve this, an appropriate approximation method is necessary. We perform the approximation with discrete curve evolution (DCE).

Since any digital curve can be regarded as a polygonal curve without loss of information (with possibly a large number of vertices), it is sufficient to study evolutions of polygonal shapes. A polygonal curve is a sequence of line segments such that any two consecutive segments share a vertex (which is an intersection of their endpoints). The basic idea of applying DCE to polygonal curves (or polygons) is very simple:

- In every evolution step, a pair of consecutive line segments s_1, s_2 is substituted with a single line segment joining the endpoints of $s_1 \cup s_2$.

---

[4] Note, that our parameters are elements of a *cyclic* group so that no modulo operation is required.

The key property of this evolution is the order of the substitution. The substitution is done according to a relevance measure K given by

$$K(s_1, s_2) = \frac{\beta(s_1, s_2)l(s_1)l(s_2)}{l(s_1) + l(s_2)}, \tag{6}$$

where $\beta(s_1, s_2)$ is the turn angle at the common vertex of segments s_1, s_2 and l is the length function normalized with respect to the total length of a polygonal curve C. The main property of this relevance measure is the following

- The higher the value of $K(s_1, s_2)$ the larger is the contribution to the shape of the curve of arc $s_1 \cup s_2$.

A cognitive motivation of this property is given in [13]. A detailed description of our discrete curve evolution can be found in [13,14]. A few example stages of the discrete curve evolution are shown in Fig. 5.

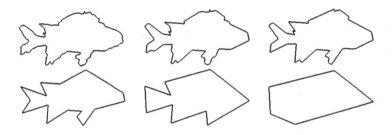

Fig. 5. A few stages of the discrete curve evolution (DCE).

DCE does not require any control parameters to achieve the task of shape simplification, i.e., there is no parameters involved in the process of the discrete curve evolution. However, we clearly need a stop parameter, which is the number of iterations the evolution is performed[5].

4   Generating Qualitative Shape Descriptions with $e\mathcal{OPRA}_m$

In this section we discuss the steps involved in creating qualitative $e\mathcal{OPRA}_m$ descriptions of simple polylines. First, we examine the task of translating quantitative angular measures into qualitative $e\mathcal{OPRA}_m$ direction intervals. Next, we address the problem of converting qualitative polyline edge lengths to qualitative distances.

---

[5] The stop parameter can also be defined with respect to the number of desired vertices, i.e., some pre-specified resolution. This is the approach taken in this paper to enable comparison between polylines with the same number of vertices.

4.1 $e\mathcal{OPRA}_m$ Descriptions of Polygons

Qualitative $e\mathcal{OPRA}_m$ descriptions of polygons are comprised of three primary components: a granularity measure m, a pairwise set of qualitative directions, and a set of qualitative distances.

The first component, m, defines the granularity of our representation. Although $e\mathcal{OPRA}_m$ supports the use of different granularity measures for direction and distance, here we use one value for both measures. Once the value of m is specified, it is used to create the pairwise qualitative direction matrix and to transform edge lengths into qualitative distances.

Qualitative Directions. To compute the pairwise qualitative direction matrix, one must first find the quantitative counter-clockwise turn angles between each pair of connected edges, or ordered triple of vertices. Given two edges $\mathbf{v}_1$ and $\mathbf{v}_2$ sharing a common vertex, the positive (counter-clockwise) turn angle β (from Eq. 6) is defined as:

$$\beta = \arctan2(\mathbf{v}_{y_2}, \mathbf{v}_{x_2}) - \arctan2(\mathbf{v}_{y_1}, \mathbf{v}_{x_1})$$

$$\beta = \begin{cases} \beta & \textbf{if } \beta \geq 0 \\ \beta + 2\pi & \textbf{if } \beta < 0 \end{cases} \tag{7}$$

After all turn angles are computed, the next step is to convert them into relative directions. This is done simply by dividing each turn angle (reference direction) by the angular resolution specified by the $e\mathcal{OPRA}_m$ granularity measure m. The angular resolution of a given $e\mathcal{OPRA}_m$ representation is obtained by dividing the set of possible directions $[0, 2\pi]$ radians into $4m$ partitions.

$$e\mathcal{OPRA}_m \text{ angular resolution} = 2\pi/4m = \pi/2m \tag{8}$$

$$\text{turn angle } \theta \text{ as } e\mathcal{OPRA}_m \text{ direction} = \theta/(\pi/2m) = (\theta \times 2m)/\pi \tag{9}$$

$$e\mathcal{OPRA}_m \text{ direction } i \text{ as turn angle} = i \times \text{angular resolution} \tag{10}$$

For example, given $m = 2$, a reference direction of $\pi/2$ translates to an $e\mathcal{OPRA}_2$ direction partition of $(\pi/2 \times 4)/\pi = 2$. When the result of Eq. 9 is not an exact integer, the direction partition is assigned as follows:

$$e\mathcal{OPRA}_m \text{ direction interval } i = \begin{cases} \lfloor i \rfloor + 1 & \textbf{if } \lfloor i \rfloor \text{ is even} \\ \lfloor i \rfloor & \textbf{if } \lfloor i \rfloor \text{ is odd} \end{cases} \tag{11}$$

Given the input shape from Fig. 6, the counter-clockwise turn-angle hull (starting at vertex A) is computed to be:

Counter-clockwise turn-angle hull: $[90°, 90°, 90°, 270°, 90°, 0°, 90°, 270°, 90°]$

Given an $e\mathcal{OPRA}_m$ granularity parameter of $m = 8$, these angles are translated to $e\mathcal{OPRA}_8$ directions of: $[8, 8, 8, 24, 8, 0, 8, 24, 8]$.

Fig. 6. A simple eight-sided polygon.

Qualitative Distances. The next step in creating our qualitative $e\mathcal{OPRA}_m$ description is to translate the absolute distances between adjacent vertices into qualitative reference distances. As with directions, $e\mathcal{OPRA}_m$ reference distances are computed by partitioning a plane into $2m$ sections, and assigning quantitative values a qualitative representation (see Fig. 3). Qualitative distances are computed by comparing each edge's length with the previous edge's length as a "control" length.[6] Under $e\mathcal{OPRA}_m$, a control length δ is used to turn a series of distance ratios into a series of qualitative distances.

The sequence of distance ratios is defined by the granularity parameter m such that there are $2m$ partitions: the first half of these ratios represent distances $\leq \delta$ marked at even intervals of $0, \frac{1}{m}, \frac{2}{m}, \ldots, \frac{m-1}{m}, 1$. The second half of the ratios represent distances $> \delta$, and are reciprocals of the first half (reversed): $\frac{m}{m-1}, \frac{m}{m-2}, \ldots, m$. For example, given $m = 4$, these ratios would be $\left(0, \frac{1}{4}, \frac{1}{2}, \frac{3}{4}, 1, \frac{4}{3}, 2, 4, \infty\right)$.

Given these ratios, reference distances are calculated by multiplying each of the ratios by a control length to create quantitative distance markers.

In our implementation, the qualitative distance ratios are created once given m, and applied over each edge of the input polygon. Using the previous edge as a control length δ, each distance ratio is turned into a quantitative distance. Once the pair of distances which bound the current edge length is identified, a qualitative distance is assigned as the average of the two bounding distances.

For example, to encode the value for a $e\mathcal{OPRA}_4$ target edge of length 5 using a control length of 4, distances are computed as:

$$4 \times \left(0, \frac{1}{4}, \frac{1}{2}, \frac{3}{4}, 1, \frac{4}{3}, 2, 4, \infty\right) = \left(0, 1, 2, 3, 4, \frac{16}{3}, 8, 16, \infty\right)$$

Our target length of 5 falls in the 5th partition, between $\left[4, \frac{16}{3}\right]$, yielding a length of $\frac{14}{3}$, or ≈ 4.67. This value is used as the qualitative edge length during reconstruction, and as the control length for computing the next edge length. In cases where the target length falls between $[n, \infty)$, the resultant value is assigned to be n.

5   Initial Reconstruction and Refinement

The first step in the reconstruction process is to "trace" the hull of the $e\mathcal{OPRA}_m$ descriptions. To start, we turn the hull list of $e\mathcal{OPRA}_m$ direction intervals back

---

[6] For the first edge, the last edge is used as the control.

into turn angles (by multiplying each value by the angular resolution) and use the qualitative distances as edge lengths to establish the vertices. This process gives us a rough representation which we will further refine in the next stages.

5.1   Initial Reconstruction

Tracing the $e\mathcal{OPRA}_m$ hull is performed like:

Algorithm 1. Tracing the Qualitative Hull of an n-sided Polygon

---

$T \leftarrow$ list of n $e\mathcal{OPRA}_m$ directions converted back to angles
$D \leftarrow$ list of n $e\mathcal{OPRA}_m$ qualitative edge lengths
$P \leftarrow$ empty list to hold reconstructed points

$P_0 \leftarrow (0,0)$ ▷ assign arbitrary origin

for $i \leftarrow 0, \ldots, n$ **do**
    $p_{i+1} \leftarrow \text{NEXTXY}(p_i, \ D_i, \ T_i)$
end for

return P ▷ return list P of points generated by NEXTXY

The NEXTXY method simply returns the next x, y coordinates using the standard vector equations of:

$$x_{n+1} = x_n + (distance * \cos \theta)$$
$$y_{n+1} = y_n + (distance * \sin \theta) \tag{12}$$

Once completed, the process returns a sequence of points representing our first attempt at reconstructing the input polygon. Given an input polygon with n points, our initial reconstruction will have $n + 1$ points: this is because our reconstructed polygon has two points representing p_0. The first of these points is p_0 from the input polygon, and the second is the p_n (the last coordinate) we generated when tracing the hull. In a perfect world, these two values would be identical at this stage, however we often need to adjust the initial reconstruction such that we can achieve closure and remove the duplicate point.

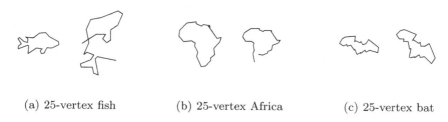

(a) 25-vertex fish   (b) 25-vertex Africa   (c) 25-vertex bat

Fig. 7. Three examples of input polylines and their initial reconstructions.

To this extent, we begin the refinement process by creating qualitative descriptors of our initial reconstructed polygon just as we did with the original input polygon. This allows us to guide the refinement process by comparing our reconstructions with the input shape via comparing qualitative descriptors. Mainly, we look to the pairwise $e\mathcal{OPRA}_m$ direction matrices as measures of similarity.

5.2   Refinement

In most cases, the initial reconstruction will produce a polyline which is not closed, and which only matches the qualitative hull descriptors. The goal of the refinement stage is two-fold: (1) achieve closure and (2) minimize the difference between the initial polyline's pairwise $e\mathcal{OPRA}_m$ direction matrix and the reconstructed polyline's pairwise $e\mathcal{OPRA}_m$ direction matrix.

The refinement process operates by iteratively generating and selecting valid "successor polylines," with the two goals above in mind. Successor polylines are created by applying a set of four adjustments to each vertex of the reconstructed polyline (minus the endpoints). These adjustments are: (1) extend outgoing edge length (2) contract outgoing edge length (3) increase turn angle, and (4) decrease turn angle.

Successor polylines are only passed on to the next steps if: the resulting polyline does not increase either the closure gap or the $e\mathcal{OPRA}_m$ direction matrix difference[7]. These checks ensures that only "good" (closer to closure) and "legal" successors (those which do not break any existing matches between $e\mathcal{OPRA}_m$ direction matrices) are returned.

Using the initial reconstructed polyline, its closure gap, and its pairwise $e\mathcal{OPRA}_m$ direction matrix difference from the initial shape as a baseline, the refinement begins by returning all possible valid polylines which are "one step," or one adjustment away from the current state.

When complete, the method returns a list of valid successor polylines, sorted by their closure gap. The "best" of these valid successors (as measured by minimizing the closure gap and $e\mathcal{OPRA}_m$ direction matrix difference) is then used to repeat the process.

Initially, the angular and distance adjustments are fairly coarse. This allows for large or otherwise obvious adjustments to be made as soon as possible. However, if the algorithm determines that either: (1) there are no valid "successors" that improve the polyline, or (2) the differences are no longer moving towards zero, it will decrease the magnitude of the adjustments.

The largest angular adjustment (one-quarter of the angular resolution) is determined by the $e\mathcal{OPRA}_m$ granularity m, and the minimum angular adjustment is arbitrarily set to $1/20^{th}$ of the initial adjustment. Given the nature of the edge length mutator, the distance adjustments are defined slightly differently (as percent of current length), but effectively range from $currentLength \times 1.1$ to $currentLength \times 0.9$.

---

[7] In 6.2, we present a more detailed look at the $e\mathcal{OPRA}_m$ direction matrix comparison metric.

Algorithm 2. Generating Successor Polylines

Input: initial reconstruction s_0, goal polyline $e\mathcal{OPRA}_m$ direction matrix $goal$

$\Delta_0 \leftarrow \text{DIFF}(s_0, goal)$ ▷ initial $e\mathcal{OPRA}_m$ matrix difference

$gap_0 \leftarrow \text{ClosureGap}(s_0)$ ▷ initial closure gap

$S \leftarrow$ empty list to hold successors

$S_{valid} \leftarrow$ empty list to hold valid successors

for $i \leftarrow 1, 2, \ldots, n - 1$ **do**
    for $adjustment, mutator$ in $adjustments, mutators$ **do**
        $S_i \leftarrow \text{MUTATOR}(p_i, adjustment)$
    end for
end for

for $i \leftarrow 0, 1, \ldots, n - 1$ **do**
    if $\text{DIFF}(S_i, goal) < \Delta_0$ **then**
        $\Delta_0 \leftarrow \text{DIFF}(S_i, goal)$
        $S_{valid_i} \leftarrow S_i$
    end if
end for

return S_{valid} ▷ return list of valid successor polylines

Output: list of valid successors "one-step" away from s_0

Algorithm 3. Refining Successor Polylines

Input: successor polyline q_0, input polyline p_0

$\Delta_0 \leftarrow \text{DIFF}(q_0, p_0)$ ▷ initial $e\mathcal{OPRA}_m$ matrix difference

$gap_0 \leftarrow \text{ClosureGap}(q_0)$ ▷ initial closure gap

$S \leftarrow \text{Successors}(q_0, p_0)$ ▷ sorted successor polylines

loop
    $gap_{current} = \text{ClosureGap}(S_0)$
    if $gap_{current} \leq$ minimum appreciable gap **then**
        break ▷ gap is negligible, break
    end if
    if $gap_{current} \geq gap_0$ **then**
        if scores not improving **then**
            break ▷ no improvements left
        end if
        if adjust factor < 20 **then**
            adjust factor $+= 1$ ▷ try increasing adjustment resolution
        end if
    end if

    $gap_0 \leftarrow gap_{current}$ ▷ otherwise we have a better gap
    $S \leftarrow \text{Successors}(S_0, p_0)$ ▷ so use best state to get more successors
end loop

The refinement process will stop when either: (1) the closure gap becomes negligible[8], or (2) the score stops moving even after changing the adjustment resolution 20 times.

As this stage, the algorithm will try to "snap" the polyline shut by setting the endpoint vertices as equal (as long as doing so does not change any qualitative relations). Snapped or not, the refinement process then returns the final state.

6   Experimental Results

6.1   Qualitative Reconstruction

Given an $e\mathcal{OPRA}_m$ granularity of $m = 8$, many simple polylines can successfully be translated to $e\mathcal{OPRA}_8$ qualitative descriptions, and reassembled back into approximately similar polylines which achieve both closure and 0-difference $e\mathcal{OPRA}_8$ direction hull descriptions, with near-0 $e\mathcal{OPRA}_8$ pairwise direction matrix differences.

Following is a sample result, using an input polyline with 10 vertices. Three shapes are shown in Fig. 8: from left to right, the input polyline, the initial reconstruction, and the final adjusted polyline. In the case below, the input and final polylines have identical $e\mathcal{OPRA}_8$ hull direction descriptions.

Fig. 8. 10-vertex representation of a bird. The final polyline (right) exhibits both some skew and scale deviations from the initial polyline (left). However, both the input and final polylines have the same $e\mathcal{OPRA}_8$ direction hull descriptions.

(a) 25-vertex fish   (b) 25-vertex Africa   (c) 25-vertex bat

Fig. 9. Three examples of input polylines and their final refined reconstructions. Compare to Fig. 7, showing the input polylines and their initial reconstructions.

---

[8] Currently, this is defined as $gap \leq 10\%$ of the shortest hull edge length.

6.2   Shape Comparison and Error

In this section we look at the notion of error as it relates to pairs of $e\mathcal{OPRA}_m$ shape descriptions sharing a common granularity parameter m. Figure 10 depicts a pair of shapes which are identical with the exception of a single vertex. Although we can compare the two shapes' $e\mathcal{OPRA}_m$ pairwise direction matrices by strict element matching, a more flexible approach would be to compute several statistics regarding the element-wise difference between direction matrices with respect to the $e\mathcal{OPRA}_m$ granularity m.

To that extent, we implement a pair of $e\mathcal{OPRA}_m$ direction matrix difference measures which account for m as well as the cyclic nature of direction intervals[9]. Specifically, we look at the element-wise median and mean difference. These values are computed by finding the cyclic difference matrix between the two $e\mathcal{OPRA}_m$ direction description matrices and computing the resultant matrix mean and median.

Given m, we can express this difference error in quantitative angular units, "raw" $e\mathcal{OPRA}_m$ units, or as a percentage of the $e\mathcal{OPRA}_m$ angular domain $[0, 2\pi]$ with respect to the number of possible directions. To demonstrate these concepts, we examine a pair of polylines which are identical, with the exception of one vertex (Fig. 10).

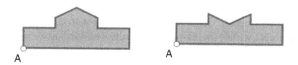

Fig. 10. Two simple shapes with a single different vertex. Vertices are labeled counterclockwise from the start vertex A. Here, the vertex which differs the most is vertex F on the "top" of the shape.

In the Fig. 10, the median and mean error between the two $e\mathcal{OPRA}_8$ direction matrices is $(0.0, 1.247)$ $e\mathcal{OPRA}_8$ direction interval units, or $(0.0\%, 3.9\%)$ with respect to $m = 8$. We can also express the difference visually, as either the element-wise error square matrix, or in a condensed per-vertex format, which simply sums each row and column pair into a single value representing the sum error associated with a given vertex as shown in Fig. 11.

It is important to note that before a pair of shapes can be compared using these measures, they must first both be aligned such that the start vertex is the same for both shapes. In the cases presented here, this has been done ahead of time, but in real-world cases, this must be done by generating error matrices considering each possible vertex as the start vertex, and selecting the one which minimizes the error.

---

[9] Given the cyclic property of direction intervals in $e\mathcal{OPRA}_m$, we are interested in the shortest-path distance from one interval to another instead of the raw absolute difference. I.e., any error greater than $2m$ can be expressed as $4m - error$.

Fig. 11. A square $e\mathcal{OPRA}_m$ direction absolute-minimum-error matrix between two shapes condensed down to a single row, showing the error per vertex. Given our example from Fig. 10, we can indeed see that the culprit is vertex F, the vertex which differs in the two original shapes. The neighboring vertex G also carries some of the sum error, as its relation to F is largely different.

6.3   Shape Neighborhoods

In Fig. 12 is a plot of several simple polygons: the first two polygons are identical minus a 180 degree rotation, while the 3rd and 4th shapes are variants on the initial shape. All of the shapes below are similar within a 5 % mean-error measure.

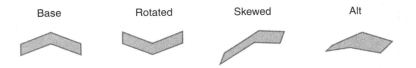

Fig. 12. Four sample shapes. Both the base and rotated versions have identical $e\mathcal{OPRA}_m$ descriptions, while the skewed and alt. versions are similar within a 5 % error.

6.4   Real-World Applications

In a new project with our cooperation partners Biodiversity Research Institute (BRI), HiDef Aerial Surveying, and SunEdison we are sponsored by the U.S. Department of Energy to further develop technology to understand how birds and bats avoid wind turbines. The collaboration will refine a stereo-optic, high-definition camera system already under development by HiDef Aerial Surveying. The collaboration will deploy systems in order to track flying animals in three dimensions. The technology will use two ultra high-definition cameras that are offset to create a three dimensional view of a wind turbine, the horizon, and an area surrounding the turbine.

Using the camera system depicted above, we have captured several sample images of a small quadcopter. These images have then been used as real-world data to test the $e\mathcal{OPRA}_m$ similarity measures described in Sect. 6.2.

Before comparing the images above with respect to their $e\mathcal{OPRA}_m$ descriptions, each image must be processed and transformed into a single polyline representing the shape of the quadcopter. This is relatively simple, and involves the following steps:

- Convert the source image to grayscale
- Apply a bilateral filter to the grayscale image to remove noise

Fig. 13. Stereo-optic camera system with small quadcopter on top left corner.

- Perform an adaptive Gaussian thresholding on the filtered image
- Detect keypoints in the thresholded image
- Compute the alpha-shape surrounding the keypoints.

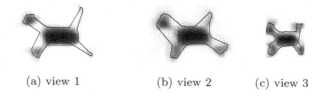

(a) view 1   (b) view 2   (c) view 3

Fig. 14. Three views of a quadcopter, post initial processing with alpha-shape overlaid.

The DCE process is then applied to each of the resulting alpha-shapes (or, concave hulls) to compress the polylines down to the same number of vertices; in this case, we have chosen 18 vertices.

For all three of the images in Fig. 14, the filtering, thresholding, and alpha-shape computation use the same parameters. It is worth noting that tuning these parameters for each individual image would likely yield better results regarding the final alpha-shape.

The next stage involves simplifying the three alpha-shapes down to the same number of vertices; as mentioned earlier we have chosen 18 for this test. After this, we can compare the three views of the quadcopter by examining the similarity or difference between their respective $e\mathcal{OPRA}_8$ representations. As with our previous examples, we use a granularity parameter $m = 8$.

Below is a presentation of shape similarity and comparison error of the alpha-shapes from Fig. 15 in numeric format, expressed as a percent of the $e\mathcal{OPRA}_8$ direction domain as explained in Sect. 6.2.

– Between the 1st and 2nd images, the median, mean error is 9.36 %, 13.03 %.

Fig. 15. Automatic comparison of the three simplified alpha-shapes computed from the quadcopter images in Fig. 14a, b, and c. Each final alpha-shape has 18 vertices, and is specified in the counter-clockwise direction, starting from the vertex labeled A.

– Between the 1st and 3rd images, the median and mean error is 6.25 %, 10.31 %.
– Finally, comparing the 1st and 3rd images, the median and mean error is 12.5 %, 17.34 %.

All four of the shapes can be considered similar within an $e\mathcal{OPRA}_8$ tolerance of 20 %.

7   Conclusion and Outlook

We have developed a qualitative shape description schema based on the qualitative relative direction and distance calculus $e\mathcal{OPRA}_m$. With this new method, we implicitly have a schema for similarity of shapes which has the potential to be cognitively adequate. To enable tests for this cognitive adequacy, the qualitative shape representation must be reversible. With our approach, we can take the qualitative shape representation and generate prototypical specific shapes that match the abstract description. Our work as described in this report is the very first QSR-based shape description capable of generating prototypical shapes based on the abstract QSR-based representation.

Future work includes performing empirical studies with human test subjects to test the cognitive adequacy of our new qualitative shape representation. We also will investigate the application of our shape representation to the formalization of affordances. Affordances are the perceived potential function of everyday objects like chairs, desks, stairs, tables etc. These affordances can assist in the task of object categorization based on 3D shape information [21, 26, 32].

Acknowledgment. The authors would like to thank Jan Oliver Wallgrün for helpful discussions related to the topic of this paper. Our work was supported in part by the National Science Foundation under Grant Nos. CDI-1028895, OIA-1027897 and IIS-1302164. The information, data, or work presented herein was funded in part by the Office of Energy Efficiency and Renewable Energy (EERE), U.S. Department of Energy, under Award Number DE-EE0006803. (Disclaimer: This report was prepared as an account of work sponsored by an agency of the United States Government. Neither the United States Government nor any agency thereof, nor any of their employees, makes any warranty, express or implied, or assumes any legal liability or responsibility for the accuracy, completeness, or usefulness of any information, apparatus, product, or process disclosed, or represents that its use would not infringe privately owned rights.

Reference herein to any specific commercial product, process, or service by trade name, trademark, manufacturer, or otherwise does not necessarily constitute or imply its endorsement, recommendation, or favoring by the United States Government or any agency thereof. The views and opinions of authors expressed herein do not necessarily state or reflect those of the United States Government or any agency thereof.)

References

1. Barkowsky, T., Latecki, L.J., Richter, K.-F.: Schematizing maps: simplification of geographic shape by discrete curve evolution. In: Habel, C., Brauer, W., Freksa, C., Wender, K.F. (eds.) Spatial Cognition 2000. LNCS (LNAI), vol. 1849, pp. 41–53. Springer, Heidelberg (2000)
2. Berendt, B.: Modelling subjective distances. In: Brewka, G., Habel, C., Nebel, B. (eds.) KI 1997. LNCS, vol. 1303, pp. 195–206. Springer, Heidelberg (1997)
3. Clementini, E., Felice, P.D., Hernández, D.: Qualitative representation of positional information. Artif. intell. **95**(2), 317–356 (1997)
4. Cohn, A.G.: Qualitative spatial representation and reasoning techniques. In: Brewka, G., Habel, C., Nebel, B. (eds.) KI 1997. LNCS, vol. 1303, pp. 1–30. Springer, Heidelberg (1997)
5. Dubba, K., Cohn, A., Hogg, D., Bhatt, M., Dylla, F.: Learning relational event models from video. J. Artif. Intell. Res. (JAIR) (to appear)
6. Dylla, F., Wallgrün, J.O.: Qualitative spatial reasoning with conceptual neighborhoods for agent control. J. Intell. Robot. Syst. **48**(1), 55–78 (2007)
7. Freksa, C.: Using orientation information for qualitative spatial reasoning. In: Frank, A.U., Formentini, U., Campari, I. (eds.) GIS 1992. LNCS, vol. 639, pp. 162–178. Springer, Heidelberg (1992). doi:10.1007/3-540-55966-3_10
8. Gibson, J.: The Ecological Approach to Visual Perception. Houghton Mifflin, Boston (1979)
9. Gottfried, B.: Tripartite line tracks. In: International Conference on Computer Vision and Graphics, pp. 288–293 (2002)
10. Isli, A., Moratz, R.: Qualitative spatial representation and reasoning: algebraic models for relative position. Univ, Bibliothek des Fachbereichs Informatik (1999)
11. Latecki, L.J., DeMenthon, D., Rosenfeld, A.: Automatic extraction of relevant frames from videos by polygon simplification. In: Sommer, G., Krüger, N., Perwass, C. (eds.) Mustererkennung, pp. 412–419. Springer, Heidelberg (2000)
12. Latecki, L.J., Lakämper, R.: Discrete approach to curve evolution. In: Levi, P., Schanz, M., Ahlers, R.-J., May, F. (eds.) Mustererkennung. Heidelberg, pp. 85–92. Springer, 1998 (1998)
13. Latecki, L.J., Lakämper, R.: Convexity rule for shape decomposition based on discrete contour evolution. Comput. Vis. Image Underst. **73**(3), 441–454 (1999)
14. Latecki, L.J., Lakämper, R.: Polygon evolution by vertex deletion. In: Nielsen, M., Johansen, P., Fogh Olsen, O., Weickert, J. (eds.) Scale-Space 1999. LNCS, vol. 1682, pp. 398–409. Springer, Heidelberg (1999)
15. Latecki, L.J., Lakamper, R.: Shape similarity measure based on correspondence of visual parts. IEEE Trans. Pattern Anal. Mach. Intell. **22**(10), 1185–1190 (2000)
16. Ligozat, G.: Qualitative triangulation for spatial reasoning. In: Campari, I., Frank, A.U. (eds.) COSIT 1993. LNCS, vol. 716, pp. 54–68. Springer, Heidelberg (1993)
17. Lovett, A., Forbus, K.D.: Shape is like space: modeling shape representation as a set of qualitative spatial relations. In: AAAI Spring Symposium: Cognitive Shape Processing (2010)

18. Moratz, R.: Representing Relative Direction as a Binary Relation of Oriented Points. In: Brewka, G., Coradeschi, S., Perini, A., Traverso, P. (eds.) Proceedings of ECAI-06. Frontiers in Artificial Intelligence and Applications, vol. 141, pp. 407–411. IOS Press, The Netherlands (2006)

19. Moratz, R., Renz, J., Wolter, D.: Qualitative spatial reasoning about line segments. In: Proceedings of ECAI 2000, pp. 234–238 (2000)

20. Moratz, R., Lücke, D., Mossakowski, T.: A condensed semantics for qualitative spatial reasoning about oriented straight line segments. Artif. Intell. **175**(16–17), 2099–2127 (2011). doi:10.1016/j.artint.2011.07.004

21. Moratz, R., Tenbrink, T.: Affordance-based human-robot interaction. In: Rome, E., Hertzberg, J., Dorffner, G. (eds.) Towards Affordance-Based Robot Control. LNCS (LNAI), vol. 4760, pp. 63–76. Springer, Heidelberg (2008)

22. Moratz, R., Wallgrün, J.O.: Spatial reasoning with augmented points: extending cardinal directions with local distances. J. Spat. Inf. Sci. **5**, 1–30 (2014)

23. Mossakowski, T., Moratz, R.: Qualitative reasoning about relative direction of oriented points. Artif. Intell. **180–181**, 34–45 (2012). doi:10.1016/j.artint.2011.10.003

24. Prince, S.J.: Computer vision: models, learning, and inference. Cambridge University Press, New York (2012)

25. Randell, D.A., Cui, Z., Cohn, A.G.: A spatial logic based on regions and connection. In: KR 1992, pp. 165–176 (1992)

26. Raubal, M., Moratz, R.: A functional model for affordance-based agents. In: Rome, E., Hertzberg, J., Dorffner, G. (eds.) Towards Affordance-Based Robot Control. LNCS (LNAI), vol. 4760, pp. 91–105. Springer, Heidelberg (2008)

27. Renz, J., Nebel, B.: On the complexity of qualitative spatial reasoning: a maximal tractable fragment of the region connection calculus. Artif. Intell. **108**(1), 69–123 (1999)

28. Schlieder, C.: Reasoning about ordering. In: Kuhn, W., Frank, A.U. (eds.) COSIT 1995. LNCS, vol. 988. Springer, Heidelberg (1995)

29. Schlieder, C.: Qualitative shape representation. Geogr. Objects Indeterminate Boundaries **2**, 123–140 (1996)

30. Scivos, A., Nebel, B.: The finest of its class: The practical natural point-based ternary calculus lr for qualitative spatial reasoning. In: Freksa, C., Knauff, M., Krieg-Brückner, B., Nebel, B., Barkowsky, T. (eds.) Spatial Cognition IV. LNCS (LNAI), vol. 3343. Springer, Heidelberg (2005)

31. Worboys, M.F., Clementini, E.: Integration of imperfect spatial information. J. Vis. Lang. Comput. **12**(1), 61–80 (2001)

32. Wunstel, M., Moratz, R.: Automatic object recognition within an office environment. In: Proceedings of the First Canadian Conference on Computer and Robot Vision, pp. 104–109. IEEE (2004)

33. Zimmermann, K., Freksa, C.: Qualitative spatial reasoning using orientation, distance, and path knowledge. Appl. Intell. **6**(1), 49–58 (1996)

From Metric to Topology: Determining Relations in Discrete Space

Matthew P. Dube[1,2(✉)], Jordan V. Barrett[2,3], and Max J. Egenhofer[1]

[1] School of Computing and Information Science, University of Maine,
5711 Boardman Hall, Orono, ME 04469-5711, USA
matthew.dube@umit.maine.edu
[2] Upward Bound Math-Science, University of Maine, 5713 Chadbourne Hall,
Orono, ME 04469-5713, USA
[3] Department of Physics, Syracuse University, Physics Building,
Syracuse, NY 13244, USA

Abstract. This paper considers the nineteen planar discrete topological relations that apply to regions bounded by a digital Jordan curve. Rather than modeling the topological relations with purely topological means, metrics are developed that determine the topological relations. Two sets of five such metrics are found to be minimal and sufficient to uniquely identify each of the nineteen topological relations. Key to distinguishing all nineteen relations are regions' margins (i.e., the neighborhood of their boundaries). Deriving topological relations from metric properties in $\mathbb{R}^2$ vs. $\mathbb{Z}^2$ reveals that the eight binary topological relations between two simple regions in $\mathbb{R}^2$ can be distinguished by a minimal set of six metrics, whereas in $\mathbb{Z}^2$, a more fine-grained set of relations (19) can be distinguished by a smaller set of metrics (5). Determining discrete topological relations from metrics enables not only the refinement of the set of known topological relations in the digital plane, but further enables the processing of raster images where the topological relation is not explicitly stored by reverting to mere pixel counts.

Keywords: Discrete spatial regions · Spatial reasoning · Model interoperability · Topological relations · Spatial queries · Geographic information systems

1 Introduction

We live in a world where metric and discrete data abound. The proliferation of geo-sensor networks [37] has dramatically increased the amount of spatial data available to decision makers in the forms of location-based sensor readings as well as spatially correlated imagery. While data are a critical foundation for understanding the world, it is important to move up the data, information, knowledge, and wisdom hierarchy [1] to gain more insights than what mere measurements provide. Metric and discrete data *per se* yield at times information in their own right, but many real-world applications require making sense of these data in some larger context. As metric data can be communicated effectively, such data collections increase their value.

Much of spatial communication is based on topological concepts [8, 17, 28, 29, 34]. Capitalizing on that insight, topological relations can serve as a vehicle for

© Springer International Publishing Switzerland 2015
S.I. Fabrikant et al. (Eds.): COSIT 2015, LNCS 9368, pp. 151–171, 2015.
DOI: 10.1007/978-3-319-23374-1_8

translating spatial language [8, 28]. Topological information is also often considered the default, preferred source for decision-making [17, 29]. Given that so much of people's spatial thought and communication inherently reverts to topological properties, the quest of converting metric and discrete data into topological information is paramount in a digital world. Mathematics shows that metric spaces are underneath the umbrella of topology [42], yet knowledge about some of the finer elements of interaction within digital spaces [39, 43] is still missing.

Modern spatial information systems will benefit from a better transition from measured spatial data to extracted spatial information. The advantages may be in finding a format that can be better communicated verbally. They may also come in system-internal qualitative models that better reflect a user's conceptualization when querying about spatial configurations, for instance in the form of sketches [4, 7, 11]. Critical in both tasks is the frequent topological nature [6] of spatial concepts.

The digital plane, a discrete space that relates to most spatial data acquired by imagery sensors, differs from the continuous plane in some fundamental spatial properties. While some formalizations of spatial concepts apply equally to continuous and discrete spaces, it is their distinct properties that have formed an impediment for a seamless treatment of spatial information across continuous and discrete spaces. Achieving the conversion from metric data to topological information allows for interoperability between information systems designed in either context, a desired result that has been attempted in both continuous and discrete settings.

This paper considers the topological relations exhibited between discrete regions in the digital plane that are bound by digital Jordan curves [19, 43]. Using metrics similar to those developed for continuous spaces [14, 18, 40], the topological relations are systematically studied relative to the qualitative values that the regions produce. While these metrics originating for continuous space have been developed and used for the specification of natural-language terms [40], these exact refinements are inappropriate for discrete spaces, insofar as typical digital spaces are constructed of countably infinite rectangular pixels, changing the boundary from 1-dimensional to 2-dimensional. This paper develops a set of metric refinements specifically for discrete topological relations between regions bound by digital Jordan curves and analyzes them for the ability to derive topological relations, independent of topological knowledge.

We demonstrate that several combinations of these metrics are sufficient for identifying the explicit topological relation, and furthermore that exactly two combinations of metrics are minimal, reducing the amount of necessary calculations. A previous approach to deriving topological relations from metrics in discrete space [41] relies on the construction of six circles for video analysis through the minimum bounding rectangle. Video is a pixelated medium, however, whereas circles are continuous objects, which are not achievable in such a domain. In an effort to achieve a cleaner model by remaining in the same embedding space at all times, this paper provides a metric encapsulation for topological relations, which is defined by the properties of the embedding discrete space.

The remainder of this paper is structured as follows. Section 2 details the literature relative to topological relations and reasoning within metric and discretized environments, followed by a summary of available measures that relate topology to metrics [12, 15, 16, 19] (Sect. 3). Section 4 develops, analog to continuous-space metrics,

metrics for discrete topological relations, for which the transition backwards to topological relations is analyzed in Sect. 5. Since the metrics projected from continuous space are found to be insufficient to distinguish all topological relations in the discrete environment, a modified set of metrics with new operators, unique to discretized environments, are introduced (Sect. 6). Section 7 analyzes these metrics for how well they capture discrete topological relations. Section 8 analyzes the distribution of these metrics relate to the conceptual neighborhood graph of the relations. Section 9 draws conclusions and sketches opportunities for future work.

2 Models for Discrete Topological Relations

The digital plane $\mathbb{Z}^2$ has been studied in two very different constructions relative to discretized topological relations. One is built upon the digital topology [39], whereas the other is built upon the use of digital Jordan curves [43]. Both approaches have their foundations in the underlying concepts of *interior*, *boundary*, and *exterior*, defined rigorously within point-set topology [2] and utilized throughout the 9-intersection [16] and 9^+-intersection [31, 32].

Let A be a subset of a topological space X.

Definition 1. The *interior* of A, denoted by A°, is the union of all open sets contained in A [2].

Definition 2. The *closure* of A, denoted by $\bar{A}$, is the intersection of all closed sets containing A [2].

Definition 3. The *boundary* of A, denoted by ∂A, is the difference between A's closure and A's interior (i.e., $\partial A = \bar{A} \backslash A^\circ$) [2].

Definition 4. The *exterior* of A, denoted by A^-, is the difference between the embedding space X and A's closure (i.e., $A^- = X \backslash \bar{A}$) [2].

The interior of an object represents the points that are not accessible from points outside of the object. Similarly, points in the exterior are not accessible from points inside of the object. The boundary represents the middle ground, that is, points that can access both inside and outside directly without going through an additional medium.

The digital topology itself exploits the space *between* pixels as the boundary for an object [39], motivating one particular approach to discretized topological relations [44]. For two discrete regions, both with simply connected interiors and exteriors, this approach identifies the same eight candidate relations in the digital plane $\mathbb{Z}^2$ as are found using either the 4-intersection [15], the 9-intersection [16], or RCC-8 [38] within a continuous space, such as $\mathbb{R}^2$ (Fig. 1). These digital topological relations (DTRs) with a digital-topology boundary (dtb) are referred to as DTR$_{\text{dtb}}$. The two principal advantages of this approach are that it directly mirrors continuous space and that only few restrictions must be enforced to make the terms of the model applicable. It is of note that there are many forms of the digital topology [30], but for the purposes of this study, the standard digital topology is the one reviewed [44].

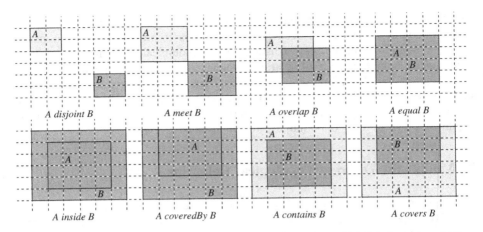

Fig. 1. The set of eight DTR$_{dtb}$, which use the boundary of standard digital topology [44].

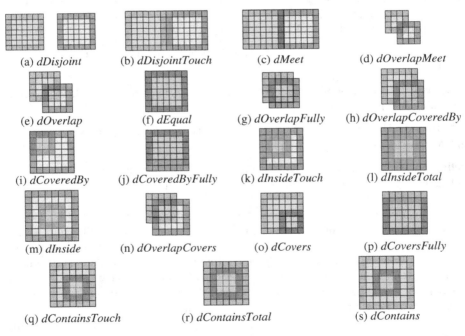

Fig. 2. The set of 19 DTR$_{Jcb}$, which use the digital Jordan curve, when embedded in a planar digital space [19].

On the other hand, the approach of the digital Jordan curve [10, 27, 43] maintains the philosophy of a continuous embedding space. It portrays the boundary as a part of the displayed manifestation of the space and separates interior from exterior. The approach with a Jordan-curve boundary (Jcb) yields a set of 19 digital topological relations between two simply connected discrete regions in $\mathbb{Z}^2$ [19], referred to as DTR_{Jcb} (Fig. 2).

The DTR_{Jcb} have a finer granularity of topological relations than DTR_{dtb}. This gain in expressive power, however, comes at a price. Regions in DTR_{Jcb} require all pixels in the interior of a region to have at least three 8-neighbor pixels that are also part of the interior. Likewise, every exterior pixel is required to have three 8-neighbor pixels that are also part of the exterior. The digital Jordan curve also constrains the boundary in such a manner that every boundary pixel must have exactly two 4-neighbor pixels as also part of the boundary [19]. If such constraints on the interior or exterior were lifted, more relations would exist.

The 19 DTR_{Jcb} can also be interpreted in terms of the 9^+-intersection [26, 27], which is a refinement of the vanilla 9-intersection where particular subsets of the space are identified as important (by being separate). While the 9-intersection alone is insufficient to distinguish *dDisjoint* from *dDisjointTouch* (Fig. 2), the 9^+-intersection achieves this differentiation by considering a discrete region to have three mutually exclusive parts of the boundary—the *inner boundary*, the *boundary core*, and the *outer boundary*. In this context, inner and outer boundaries can be seen as the discrete topology boundaries. In another context, these boundaries could be further considered as a marginal approach, having pixel depth [23].

3 Metric Refinements of Topological Relations in Continuous Space

Based on Naive Geography's premise "Topology Matters, Metric Refines" [17], a set of eleven metrics, applicable to a region, provide further details about topological relations in a continuous space. Such metric refinements apply to topological relations between a region and a line [18, 36] as well as between two regions [14]. Another set of metric refinements, not considered in the context of this paper, applies to line-line relations [36] as lines in $\mathbb{R}^2$ have co-dimension 1, yielding more degrees of freedom. Each of the eleven metrics for region-region relations is normalized by the relative sizes of the regions' interiors or boundaries. Through the normalization, the metrics typically yield a value between 0 and 1. Figure 3 summarizes, for region-region relations, each metric's name, gives its formal specification, and provides a graphic that highlights the parts that are the numerators in each metric refinement.

In addition to offering more details about topological relations from the perspective of the 9-intersection, these metrics also allow for the determination of topological relations from metric information by accounting for all 9-intersection cells. In continuous space, six of the eleven metrics are sufficient to determine any of the eight topological region-region relations [14]. In discrete space, this insight about deriving topological information from metric properties would be particularly useful as counting pixels is a much easier task than trying to identify the discrete topological relation [3, 5, 35]. The behavior of the region-region relations in $\mathbb{R}^2$ applies immediately to the eight DTR_{dtb}(Fig. 1), because their boundary retains a linear form. The 19 DTR_{Jcb}, however, have an extended boundary of uniform size. In order to utilize these metrics in such a

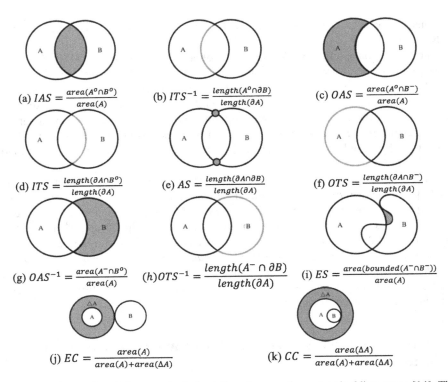

Fig. 3. Metrics used to define topological relations in a continuous embedding space [14]. The metrics' acronyms stand for *inner area splitting* (IAS), *inner traversal splitting inverse* (ITS^{-1}), *outer area splitting* (OAS), *inner traversal splitting* (ITS), *alongness splitting* (AS), *outer traversal splitting* (OTS), *outer area splitting inverse* (OAS^{-1}), *outer traversal splitting inverse* (OTS^{-1}), *exterior splitting* (ES), *expansion closeness* (EC), and *contraction closeness* (CC). Metrics (a)–(i) reflect topological intersections from the 9-intersection, while metrics (j) and (k) are controls for point intersections of a boundary (having no length), with ΔA representing the growth or diminishment of the object to change the relation. Two constraints apply to these metrics: (1) IAS + OAS = 1 and (2) ITS + AS + OTS = 1.

situation, they must be adapted to reflect pixels that have an area, as opposed to points that have no area. Interiors and boundaries now are represented by a finite number of pixels, whereas the exterior (while itself countably infinite) may have bounded portions that are finite.

4   The d*-Metric Refinements for DTR$_{Jcb}$

Among the eleven refinements in $\mathbb{R}^2$ are *expansion closeness* (Fig. 3j) and *contraction closeness* (Fig. 3k), which relate to making the figure region larger or smaller until it changes its topological relation. While these two metrics are necessary in $\mathbb{R}^2$ to derive some topological relations, we hypothesize that these two metrics are not needed in $\mathbb{Z}^2$.

In a continuous space, a boundary-boundary intersection can have length 0 (i.e., it occurs at a point), making *alongness splitting* 0, the same value as if there is no boundary-boundary intersection. As a consequence, certain varieties of *meet* can be confused for *disjoint* (and similarly for *covers* and *contains* as well as *coveredBy* and *inside*). The corresponding feature in a discrete space, however, is comprised of finitely many pixels so that any non-empty boundary-boundary intersection would make alongness splitting always greater than 0. Since both area and boundary of the DTR$_{Jcb}$ have an extent, neither expansion closeness nor contraction closeness should meaningfully contribute to determining any of the 19 DTR$_{Jcb}$

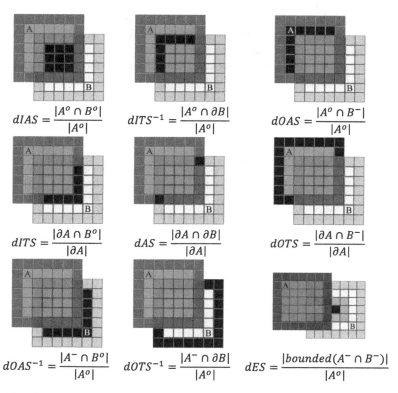

$$dIAS = \frac{|A^o \cap B^o|}{|A^o|} \qquad dITS^{-1} = \frac{|A^o \cap \partial B|}{|A^o|} \qquad dOAS = \frac{|A^o \cap B^-|}{|A^o|}$$

$$dITS = \frac{|\partial A \cap B^o|}{|\partial A|} \qquad dAS = \frac{|\partial A \cap \partial B|}{|\partial A|} \qquad dOTS = \frac{|\partial A \cap B^-|}{|\partial A|}$$

$$dOAS^{-1} = \frac{|A^- \cap B^o|}{|A^o|} \qquad dOTS^{-1} = \frac{|A^- \cap \partial B|}{|A^o|} \qquad dES = \frac{|bounded(A^- \cap B^-)|}{|A^o|}$$

Fig. 4. The d*-metrics, which are the continuous space metrics (Fig. 3) adapted to discrete embedding spaces.

The other nine metrics from continuous space systematically transform to corresponding metrics in $\mathbb{Z}^2$. Determining area and length of a boundary resort to counting pixels. Since length and area have the same domain (i.e., pixel counts), artifacts in the denominator patterns across Fig. 3 can be rectified when moving to a discrete space. The measures ITS^{-1} and OTS^{-1} are thus normalized by A^o rather than by ∂A. The same normalization would have been impossible in $\mathbb{R}^2$ as the domains of length and area in a continuous space are mismatched. The set of digital metrics for DTR_{Jcb} is referred to as the *d*-metrics*. Figure 4 displays d*-metrics, together with a graphical depiction of the portion of the relation that each metric captures.

5   Analysis of d*-Metrics for Discrete Topological Relations

Given the cardinality basis of the discrete metrics (Sect. 4), the possible values for each DTR_{Jcb} can be inferred directly from their corresponding 9-intersection matrix. If an interior row or boundary row only has one non-empty entry, the value of the corresponding metric x must be $x = 1$. If either an interior or boundary row has multiple non-empty entries, these metrics must be $0 < x < 1$. If any intersection is empty, the corresponding metric must be $x = 0$. For the exterior row, any non-empty intersection must have a metric value $x > 0$, with the exception of the exterior-exterior intersection, which can always be $x = 0$, but in other cases can split the exterior, producing $x > 0$. Table 1 summarizes the distribution of the d*-metrics for each of the 19 DTR_{Jcb} (Fig. 2).

Table 1. Relevant metric values for each of the d*-metrics relative to the 19 DTR_{Jcb} (Fig. 2).

Relation	dIAS	dITS^{-1}	dOAS	dITS	dAS	dOTS	dOAS^{-1}	dOTS^{-1}	dES
dDisjoint	0	0	1	0	0	1	0 < x	0 < x	0
dDisjointTouch	0	0	1	0	0	1	0 < x	0 < x	0 or 0 < x
dMeet	0	0	1	0	(0,1)	(0,1)	0 < x	0 < x	0 or 0 < x
dOverlapMeet	0	(0,1)	(0,1)	(0,1)	(0,1)	(0,1)	0 < x	0 < x	0 or 0 < x
dOverlap	(0,1)	(0,1)	(0,1)	(0,1)	(0,1)	(0,1)	0 < x	0 < x	0 or 0 < x
dEqual	1	0	0	0	1	0	0	0	0
dOverlapFully	(0,1)	(0,1)	0	(0,1)	(0,1)	(0,1)	0	0 < x	0
dOverlapCoveredBy	(0,1)	(0,1)	0	(0,1)	(0,1)	(0,1)	0 < x	0 < x	0 or 0 < x
dCoveredBy	1	0	0	(0,1)	(0,1)	0	0 < x	0 < x	0
dCoveredByFully	1	0	0	(0,1)	(0,1)	0	0	0 < x	0
dInsideTouch	1	0	0	1	0	0	0 < x	0 < x	0
dInsideTotal	1	0	0	1	0	0	0	0 < x	0
dInside	1	0	0	1	0	0	0 < x	0 < x	0
dOverlapCovers	(0,1)	(0,1)	(0,1)	(0,1)	(0,1)	(0,1)	0	0 < x	0 or 0 < x
dCovers	(0,1)	(0,1)	(0,1)	0	(0,1)	(0,1)	0	0	0
dCoversFully	(0,1)	(0,1)	0	0	(0,1)	(0,1)	0	0	0
dContainsTouch	(0,1)	(0,1)	(0,1)	0	0	1	0	0	0
dContainsTotal	(0,1)	(0,1)	0	0	0	1	0	0	0
dContains	(0,1)	(0,1)	(0,1)	0	0	1	0	0	0

Table 1 reveals some insights about the dependency between topological relations and the metric values. While most relations are different from each other on a metric level, *dDisjoint* and *dDisjointTouch* are identical, as are *dContains* and *dContainsTouch*, and similarly *dInside* and *dInsideTouch*. This assertion is not surprising in that the matrices for each pair of relations are the same. Exterior splitting (dES) contributes to determining *dDisjointTouch* only in cases when the exterior has at least one separation, the exterior. Some *dDisjointTouch* configurations, however, do not have such a separation; therefore, dES is not a consistent measure for *dDisjointTouch*.

While the set of d*-metrics is not enough to discern all distinct relations, it suffices to discern all unique 9-intersection matrices of the DTR_{Jcb}. Using a sieve designed in Java 1.6.0 and Microsoft Excel, we determined for the 16 unique 9-intersection matrices the minimal combination of metrics that can specify each matrix (Table 2). Some relations are specified in exactly one way (such as *dOverlap*), whereas the relation *dMeet* (the relation with the most identifiable sets) can be identified in 11 distinct ways. Any other identifying set of metrics has at least one of the corresponding metric sets for the particular relation as a subset.

Using Table 2, it is possible to discern the minimal set to distinguish all possible discrete matrices from one another, finding the smallest possible union of metrics that carries at least one of the options from each row. There are exactly three smallest sets, each of cardinality 5: (1) dIAS, dOAS, dAS, $dOAS^{-1}$, $dOTS^{-1}$, (2) dIAS, dOAS, dITS, dOTS, $dOAS^{-1}$, and (3) dIAS, dOAS, dITS, dAS, $dOAS^{-1}$. Since *dOverlap* has only one possible set that identifies it, that condition is a subset of all three of these metric combinations.

Though the worst-case scenario (*dOverlapFully*) requires four metrics, only one additional metric is necessary to achieve full matrix differentiation. If one were to use a derivative of Expansion Closeness (EC) or Contraction Closeness (CC) (Fig. 3), it is possible to distinguish *dDisjoint* from *dDisjointTouch* (and likewise for their *dInside* and *dContains* counterparts), though not all *dDisjoint* instances have Expansion Closeness greater than one pixel. As such, the 19 DTR_{Jcb} cannot always be distinguished by d*-metrics, even if discrete versions of EC and CC were added to the d*-metrics. Instead, a different structure is required.

6 Metrics Unique to a Discrete Topological Space

While the d*-metrics cannot identify uniquely all 19 DTR_{Jcb}, the metrics can differentiate thirteen 9-intersection matrices, lacking a mechanism only for distinguishing the three pairs of relations with the same matrix (*dDisjoint* and *dDisjointTouch*; *dInside* and *dInsideTouch*; and *dContains* and *dContainsTouch*). To disambiguate these cases, we consider two additional concepts, reminiscent of the margin approach applied to mereotopological relations [23].

Definition 5. Let X be a discrete region bounded by a digital Jordan curve Y. Consider the set Z of all pixels z such that $z \in X^o$ and z has an edge neighbor $y \in Y$. The set Z is called the ***Jordan inner margin*** of X, denoted X^{JI}. The same relation regarding X^- is called the ***Jordan outer margin*** of X, denoted X^{JO}.

Both Jordan margins help to position the boundary of these same matrix relations, effectively capturing the information passed along about objects in continuous space by Expansion Closeness (EC) and Contraction Closeness (CC) (Fig. 3). These two parts of an object can be used to refine these relations in a method concurrent with the 9^+-intersection [26, 27], where these relations would be distinct.

Table 2. Uniquely identifying minimal subsets of the d*-metrics for the DTR$_{Jcb}$ matrices.

Relation	Minimal subsets
dDisjoint	(1) dOTS dOTS^{-1}; (2) dOTS dOAS^{-1}; (3) dOAS dOTS; (4) dOAS dAS; (5) dITS^{-1} dOTS; (6) dIAS dOTS; (7) dIAS dAS; (8) dITS, dAS, dOTS^{-1}; (9) dITS, dAS, dOAS^{-1}; (10) dITS^{-1}, dITS, dAS
dMeet	(1) dOAS dOTS; (2) dOAS dAS; (3) dITS^{-1} dOTS; (4) dITS, dOTS, dOTS^{-1}; (5) dITS, dOTS, dOAS^{-1}; (6) dITS, dAS, dOTS^{-1}; (7) dITS, dAS, dOAS^{-1}; (8) dITS^{-1}, dITS, dAS; (9) dIAS, dITS, dOTS; (10) dIAS, dITS, dAS; (11) dIAS, dITS^{-1}, dAS
dOverlapMeet	(1) dIAS dITS; (2) dIAS dOAS; (3) dIAS dITS^{-1}
dOverlap	(1) dIAS, dOAS, dOAS^{-1}
dEqual	(1) dAS; (2) dOTS dOTS^{-1}; (3) dITS dOTS; (4) dITS^{-1} dOTS^{-1}; (5) dIAS dOTS^{-1}; (6) dIAS dITS; (7) dITS^{-1}, dITS, dOAS^{-1}; (8) dITS^{-1}, dOAS, dITS
dOverlapFully	(1) dOAS, dOTS, dOAS^{-1}, dOTS^{-1}; (2) dOAS, dITS, dOTS, dOAS^{-1}; (3) dITS^{-1}, dOAS, dOAS^{-1}, dOTS^{-1}; (4) dITS^{-1}, dOAS, dITS, dOAS^{-1}; (5) dIAS, dOAS, dOAS^{-1}, dOTS^{-1}; (6) dIAS, dOAS, dITS, dOAS^{-1}
dOverlapCoveredBy	(1) dOAS, dOTS, dOAS^{-1}; (2) dITS^{-1}, dOAS, dOAS^{-1}; (3) dIAS, dOAS, dOAS^{-1}
dCoveredBy	(1) dAS, dOTS, dOAS^{-1}; (2) dITS, dOTS, dOAS^{-1}; (3) dITS^{-1}, dITS, dOAS^{-1}; (4) dIAS, dAS, dOAS^{-1}; (5) dIAS, dITS, dOAS^{-1}
dCoveredByFully	(1) dAS, dOTS, dOAS^{-1}; (2) dITS, dOTS, dOAS^{-1}; (3) dITS^{-1}, dAS, dOAS^{-1}; (4) dITS^{-1}, dITS, dOAS^{-1}; (5) dIAS, dAS, dOAS^{-1}; (6) dIAS, dITS, dOAS^{-1}
dInsideTotal	(1) dITS dOAS^{-1}; (2) dAS, dOAS^{-1}, dOTS^{-1}; (3) dAS, dOTS, dOAS^{-1}; (4) dITS^{-1}, dAS, dOAS^{-1}; (5) dIAS, dAS, dOAS^{-1}
dInside	(1) dITS dOAS^{-1}; (2) dAS, dOTS, dOAS^{-1}; (3) dOAS, dAS, dOAS^{-1}; (4) dIAS, dAS, dOAS^{-1}
dOverlapCovers	(1) dOAS, dOAS^{-1}, dOTS^{-1}; (2) dOAS, dITS, dOAS^{-1}
dCovers	(1) dOAS, dOTS, dOAS^{-1}; (2) dOAS, dAS, dOTS^{-1}; (3) dOAS, dITS, dOTS; (4) dOAS, dITS, dAS
dCoversFully	(1) dOAS, dOTS, dOAS^{-1}; (2) dOAS, dAS, dOTS^{-1}; (3) dOAS, dITS, dOTS; (4) dOAS, dITS, dAS
dContainsTotal	(1) dOAS dOTS; (2) dOAS, dAS, dOTS^{-1}; (3) dOAS, dITS, dAS; (4) dITS^{-1}, dOAS, dAS; (5) dIAS, dOAS, dAS
dContains	(1) dOAS dOTS; (2) dOAS dAS

Since the area of concern is only the intersection of the Jordan margins with the boundaries of the other object, only two metrics apply, called *Jordan inner margin boundary splitting* (JIMBS) and *Jordan outer margin boundary splitting* (JOMBS). While it is possible to create 25 metrics, one for each intersection of the 9$^+$-intersection when applied to this purpose, we subsequently show that JIMBS and JOMBS, as part

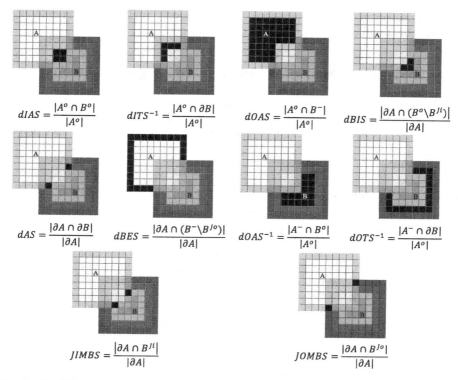

$$dIAS = \frac{|A^o \cap B^o|}{|A^o|}$$

$$dITS^{-1} = \frac{|A^o \cap \partial B|}{|A^o|}$$

$$dOAS = \frac{|A^o \cap B^-|}{|A^o|}$$

$$dBIS = \frac{|\partial A \cap (B^o \setminus B^{Ji})|}{|\partial A|}$$

$$dAS = \frac{|\partial A \cap \partial B|}{|\partial A|}$$

$$dBES = \frac{|\partial A \cap (B^- \setminus B^{Jo})|}{|\partial A|}$$

$$dOAS^{-1} = \frac{|A^- \cap B^o|}{|A^o|}$$

$$dOTS^{-1} = \frac{|A^- \cap \partial B|}{|A^o|}$$

$$JIMBS = \frac{|\partial A \cap B^{Ji}|}{|\partial A|}$$

$$JOMBS = \frac{|\partial A \cap B^{Jo}|}{|\partial A|}$$

Fig. 5. The d^{*+}-metrics including Jordan inner margin boundary splitting (JIMBS), Jordan outer margin boundary splitting (JOMBS), boundary interior splitting (dBIS), boundary exterior splitting (dBES), as well as the six metrics dIAS, dITS^{-1}, dOAS, dAS, dOAS^{-1}, and dOTS^{-1} carried forward from the d*-metrics (Fig. 4).

of a minimal set, can identify all 19 discrete relations. Additionally, we only consider the inner and outer margins as separate components when analyzing A's boundary. As a result, six of the nine d*-metrics remain unchanged (dIAS, sITS^{-1}, dOAS, dAS, dOAS^{-1}, and dOTS^{-1}). Due to the Jordan margins, two d*-metrics are split into a pair of metrics each—dITS splits into dBIS and JIMBS; dOTS splits into dBES and JOMBS. Finally, the d*-metric dES, which was shown to be immaterial when determining DTR$_{Jcb}$, would be also immaterial in the new set and, therefore, is not needed. The set of ten metrics (six carried forward and four from splitting two of the d*-metrics each into a pair of more detailed metrics) is referred to as d^{*+}-*metrics* (Fig. 5).

Since JIMBS and JOMBS absorb area from a region's interior and exterior, respectively, the remaining interiors and exteriors must be large enough to guarantee that a region's boundary is not self-touching. Therefore, the requirements for interior and exterior pixels [19] (Sect. 2) need to hold for $B^o \setminus B^{JI}$ and $B^- \setminus B^{JO}$, rather than for the entire interior and exterior, respectively. As a consequence, an interior of a candidate region must have at minimum 4×4 pixels, which includes the Jordan inner margin. Likewise, the exterior must exhibit the same type of relation with the Jordan outer margin. For purposes of measuring regions, we only apply the Jordan inner and outer margins to object *B*.

7 Analysis of d^{*+}-Metrics for Discrete Topological Relations

Similar to the analysis of the d*-metrics (Table 1), the relevant values for each topological relation can be found for the d*+-metrics (Table 3). Unlike in Table 1, however, each row in Table 3 is unique, which implies that each relation is differentiable with the d*+-metrics.

Table 3. Relevant metric values for each of the d^{*+}-metrics relative to the 19 DTR$_{Jcb}$ (Fig. 2).

Relation	dIAS	dITS^{-1}	dOAS	dBIS	JIMBS	dAS	JOMBS	dBES	dOAS^{-1}	dOTS^{-1}
dDisjoint	0	0	1	0	0	0	0	1	0 < x	0 < x
dDisjointTouch	0	0	1	0	0	0	(0,1)	(0,1)	0 < x	0 < x
dMeet	0	0	1	0	0	(0,1)	(0,1)	(0,1)	0 < x	0 < x
dOverlapMeet	0	(0,1)	(0,1)	0	(0,1)	(0,1)	(0,1)	(0,1)	0 < x	0 < x
dOverlap	(0,1)	(0,1)	(0,1)	(0,1)	(0,1)	(0,1)	(0,1)	(0,1)	0 < x	0 < x
dEqual	1	0	0	0	0	1	0	0	0	0
dOverlapFully	(0,1)	(0,1)	0	0	(0,1)	(0,1)	(0,1)	0	0	0 < x
dOverlapCoveredBy	(0,1)	(0,1)	0	(0,1)	(0,1)	(0,1)	(0,1)	0	0 < x	0 < x
dCoveredBy	1	0	0	(0,1)	(0,1)	(0,1)	0	0	0 < x	0 < x
dCoveredByFully	1	0	0	0	(0,1)	(0,1)	0	0	0	0 < x
dInsideTouch	1	0	0	(0,1)	(0,1)	0	0	0	0 < x	0 < x
dInsideTotal	1	0	0	0	1	0	0	0	0	0 < x
dInside	1	0	0	1	0	0	0	0	0 < x	0 < x
dOverlapCovers	(0,1)	(0,1)	(0,1)	0	(0,1)	(0,1)	(0,1)	(0,1)	0	0 < x
dCovers	(0,1)	(0,1)	(0,1)	0	0	(0,1)	(0,1)	(0,1)	0	0
dCoversFully	(0,1)	(0,1)	0	0	0	(0,1)	(0,1)	0	0	0
dContainsTouch	(0,1)	(0,1)	(0,1)	0	0	0	(0,1)	(0,1)	0	0
dContainsTotal	(0,1)	(0,1)	0	0	0	0	1	0	0	0
dContains	(0,1)	(0,1)	(0,1)	0	0	0	0	1	0	0

Using the same sieve procedure that generated minimal subsets of the d*-metrics (Table 2), we derived the minimal subsets of the d*+-metrics for each of the 19 DTR$_{Jcb}$ (Table 4). Some relations—*dInsideTouch, dEqual, dInside,* and *dContainsTotal*—are easily identifiable in as few as one metric, whereas *dOverlapFully* requires a set of four metrics. This worst-case scenario is the same as for the d*-metrics. Other relations, however, achieve easier differentiability with d*+.

Two different sets of five metrics are enough to determine the explicit relation between two discrete regions: (1) dIAS, JIMBS, dAS, dBES, and dOAS^{-1} or (2) dIAS, dBIS, JIMBS, dAS, and dBES. Since these metrics additionally differentiate *dDisjoint* and *dDisjointTouch* (as well as their *inside* and *contains* cousins), these sets are stronger than the set of five metrics. This difference occurs because of the modification applied to the metrics dITS and dOTS through the insertion of the Jordan inner margin and the Jordan outer margin.

Table 4. Uniquely identifying minimal subsets of the d*$^+$-metrics for the DTR$_{Jcb}$ matrices.

Relation	Minimal subsets
dDisjoint	(1) dOAS, JOMBS; (2) dIAS, JOMBS; (3) dBES, dOTS^{-1}; (4) dIAS, dBES; (5) dITS^{-1}, dBES; (6) dBES, dOAS^{-1}; (7) dOAS, dBES; (8) dBIS, JOMBS, dOAS^{-1}; (9) dBIS, JIMBS, JOMBS, dOTS^{-1}; (10) dITS^{-1}, dBIS, JIMBS, dAS, JOMBS
dDisjointTouch	(1) dAS, dBES, dOTS^{-1}; (2) dITS^{-1}, dAS, dBES; (3) dOAS, dAS, JOMBS; (4) dIAS, dAS, dBES; (5) dIAS, dAS, JOMBS; (6) dITS^{-1}, dAS, JOMBS; (7) dOAS, dAS, dBES; (8) dAS, JOMBS, dOAS^{-1}; (9) dAS, dBES, dOAS^{-1}; (10) dAS, JOMBS, dOTS^{-1}
dMeet	(1) dOAS, dAS; (2) dITS^{-1}, dAS, dBES; (3) dITS^{-1}, dAS, JOMBS; (4) dIAS, JIMBS, dAS; (5) JIMBS, dAS, dOAS^{-1}; (6) dITS^{-1}, JIMBS, dAS; (7) JIMBS, dAS, dOTS^{-1}; (8) dIAS, dITS^{-1}, dAS; (9) dITS^{-1}, dBIS, dAS, dOAS^{-1}
dOverlapMeet	(1) dIAS, JIMBS; (2) dIAS, dOAS; (3) dIAS, dITS^{-1}; (4) dOAS, dBIS, dOAS^{-1}; (5) dBIS, JIMBS, dOAS^{-1}; (6) dITS^{-1}, dBIS, dOAS^{-1}
dOverlap	(1) dBIS, dBES; (2) dOAS, dBIS; (3) dIAS, dBES, dOAS^{-1}; (4) dIAS, dOAS, dOAS^{-1}
dEqual	(1) dAS; (2) dIAS, dOTS^{-1}; (3) dITS^{-1}, dOTS^{-1}; (4) dOAS, JOMBS, dOTS^{-1}; (5) JOMBS, dBES, dOTS^{-1}; (6) dIAS, JIMBS, dOAS^{-1}; (7) dIAS, dBIS, JIMBS; (8) dITS^{-1}, JIMBS, dOAS^{-1}; (9) JIMBS, JOMBS, dBES, dOAS^{-1}; (10) dOAS, JIMBS, JOMBS, dOAS^{-1}; (11) dOAS, dBIS, JIMBS, JOMBS; (12) dITS^{-1}, dBIS, JIMBS, dBES; (13) dBIS, JIMBS, JOMBS, dBES; (14) dITS^{-1}, dOAS, dBIS, JIMBS
dOverlapFully	(1) JIMBS, JOMBS, dBES, dOAS^{-1}; (2) dOAS, JIMBS, JOMBS, dOAS^{-1}; (3) dOAS, dBIS, JIMBS, JOMBS; (4) dIAS, JIMBS, dBES, dOAS^{-1}; (5) dIAS, dBIS, JIMBS, dBES; (6) dOAS, dBIS, JOMBS, dOTS^{-1}; (7) dBIS, JOMBS, dBES, dOTS^{-1}; (8) dITS^{-1}, JIMBS, dBES, dOAS^{-1}; (9) dITS^{-1}, dBIS, JIMBS, dBES; (10) dBIS, JIMBS, JOMBS, dBES; (11) dIAS, dOAS, JIMBS, dOAS^{-1}; (12) dIAS, dOAS, dBIS, JIMBS; (13) dOAS, JOMBS, dOAS^{-1}, dOTS^{-1}; (14) JOMBS, dBES, dOAS^{-1}, dOTS^{-1}; (15) dIAS, dBIS, dBES, dOTS^{-1}; (16) dIAS, dBES, dOAS^{-1}, dOTS^{-1}; (17) dITS^{-1}, dBIS, dBES, dOTS^{-1}; (18) dITS^{-1}, dOAS, JIMBS, dOAS^{-1}; (19) dIAS, dOAS, dBIS, dOTS^{-1}; (20) dITS^{-1}, dOAS, dBIS, JIMBS; (21) dITS^{-1}, dBES, dOAS^{-1}, dOTS^{-1}; (22) dIAS, dOAS, dOAS^{-1}, dOTS^{-1}; (23) dITS^{-1}, dOAS, dBIS, dOTS^{-1}; (24) dITS^{-1}, dOAS, dOAS^{-1}, dOTS^{-1}

(Continued)

Table 4. (*Continued*)

Relation	Minimal subsets
dOverlapCoveredBy	(1) dOAS, dBIS, JOMBS; (2) dIAS, dBIS, dBES; (3) dITS^{-1}, dBIS, dBES; (4) dOAS, JOMBS, dOAS^{-1}; (5) JOMBS, dBES, dOAS^{-1}; (6) dIAS, dBES, dOAS^{-1}; (7) dBIS, JOMBS, dBES; (8) dIAS, dOAS, dBIS; (9) dITS^{-1}, dBES, dOAS^{-1}; (10) dIAS, dOAS, dOAS^{-1}; (11) dITS^{-1}, dOAS, dBIS; (12) dITS^{-1}, dOAS, dOAS^{-1}
dCoveredBy	(1) dAS, JOMBS, dOAS^{-1}; (2) dBIS, dAS, JOMBS; (3) dIAS, dBIS, dAS; (4) dIAS, dAS, dOAS^{-1}; (5) dITS^{-1}, dBIS, dAS; (6) dITS^{-1}, dAS, dBES, dOAS^{-1}; (7) dITS^{-1}, dOAS, dAS, dOAS^{-1}; (8) dITS^{-1}, JIMBS, dAS, dOAS^{-1}
dCoveredByFully	(1) dAS, JOMBS, dOAS^{-1}; (2) dBIS, dAS, JOMBS; (3) dIAS, dBIS, dAS; (4) dIAS, JIMBS, dOAS^{-1}; (5) dIAS, dBIS, JIMBS; (6) dBIS, JIMBS, JOMBS; (7) dIAS, dAS, dOAS^{-1}; (8) JIMBS, JOMBS, dOAS^{-1}; (9) dITS^{-1}, dAS, dOAS^{-1}; (10) dITS^{-1}, JIMBS, dOAS^{-1}; (11) dITS^{-1}, dBIS, JIMBS; (12) dITS^{-1}, dBIS, dAS, dBES; (13) dITS^{-1}, dOAS, dBIS, dAS
dInsideTouch	(1) JIMBS, dAS; (2) dBIS, dAS
dInsideTotal	(1) JIMBS; (2) dIAS, dBIS, dAS; (3) dIAS, dAS, dOAS^{-1}; (4) dITS^{-1}, dAS, dOAS^{-1}; (5) dAS, dOAS^{-1}, dOTS^{-1}; (6) dITS^{-1}, dBIS, dAS, dBES; (7) dOAS, dBIS, dAS, JOMBS; (8) dBIS, dAS, dBES, dOTS^{-1}; (9) dOAS, dAS, JOMBS, dOAS^{-1}; (10) dAS, JOMBS, dBES, dOAS^{-1}; (11) dITS^{-1}, dOAS, dBIS, dAS; (12) dOAS, dBIS, dAS, dOTS^{-1}; (13) dBIS, dAS, JOMBS, dBES
dInside	(1) dBIS; (2) dIAS, JIMBS, dAS; (3) JIMBS, dBES, dOAS^{-1}; (4) dIAS, JIMBS, dOAS^{-1}; (5) JIMBS, dBES, dOTS^{-1}; (6) dIAS, JIMBS, dOTS^{-1}; (7) dOAS, JIMBS, dOAS^{-1}; (8) dOAS, JIMBS, dOTS^{-1}; (9) dITS^{-1}, JIMBS, dAS, dBES; (10) dOAS, JIMBS, dAS, JOMBS; (11) dITS^{-1}, dOAS, JIMBS, dAS; (12) JIMBS, dAS, JOMBS, dBES
dOverlapCovers	(1) JIMBS, dBES, dOAS^{-1}; (2) dBES, dOAS^{-1}, dOTS^{-1}; (3) dOAS, JIMBS, dOAS^{-1}; (4) dOAS, dOAS^{-1}, dOTS^{-1}; (5) dIAS, dBIS, JIMBS, dBES; (6) dIAS, dOAS, dBIS, JIMBS; (7) dIAS, dBIS, dBES, dOTS^{-1}; (8) dIAS, dOAS, dBIS, dOTS^{-1}
dCovers	(1) dAS, dBES, dOTS^{-1}; (2) dOAS, JIMBS, dAS; (3) dOAS, dAS, dOTS^{-1}; (4) dIAS, JIMBS, dAS, dBES; (5) JIMBS, dAS, dBES, dOAS^{-1}; (6) dITS^{-1}, JIMBS, dAS, dBES
dCoversFully	(1) dAS, dBES, dOTS^{-1}; (2) dOAS, JIMBS, dAS; (3) dOAS, JIMBS, JOMBS; (4) dOAS, JOMBS, dOTS^{-1}; (5) JOMBS, dBES, dOTS^{-1}; (6) dOAS, dAS, dOTS^{-1}; (7) JIMBS, dAS, dBES; (8) JIMBS, JOMBS, dBES

(*Continued*)

Table 4. (*Continued*)

Relation	Minimal subsets
dContainsTouch	(1) dAS, dBES, dOTS⁻¹; (2) dITS⁻¹, dAS, dBES; (3) dOAS, dAS, JOMBS; (4) dIAS, dAS, dBES; (5) dIAS, dAS, JOMBS; (6) dITS⁻¹, dAS, JOMBS; (7) dOAS, dAS, dBES; (8) dAS, JOMBS, dOAS⁻¹; (9) dAS, dBES, dOAS⁻¹; (10) dAS, JOMBS, dOTS⁻¹
dContainsTotal	(1) JOMBS; (2) dAS, dBES, dOTS⁻¹; (3) dITS⁻¹, dAS, dBES; (4) dIAS, dAS, dBES; (5) dOAS, dAS, dOTS⁻¹; (6) dIAS, dOAS, dAS; (7) dITS⁻¹, dOAS, dAS; (8) JIMBS, dAS, dBES, dOAS⁻¹; (9) dOAS, JIMBS, dAS, dOAS⁻¹; (10) dOAS, dBIS, JIMBS, dAS; (11) dBIS, JIMBS, dAS, dBES
dContains	(1) dOAS, JOMBS; (2) dIAS, JOMBS; (3) dITS⁻¹, JOMBS; (4) dBES, dOTS⁻¹; (5) dIAS, dBES; (6) dITS⁻¹, dBES; (7) dBES, dOAS⁻¹; (8) dOAS, dBES; (9) dAS, JOMBS, dOTS⁻¹; (10) JIMBS, dAS, JOMBS, dOAS⁻¹

8   Distribution of d*⁺-Metric Values

The relations' A-neighborhood graph offers an opportunity to correlate the metrics' values based on the relations' similarity. The conceptual neighborhood graph of the 19 DTR$_{Jcb}$ [19] (Fig. 6) exposes the same overall structure as the A-neighborhood graph of the 8 region-region relations in $\mathbb{R}^2$ [13]. The center stacks all symmetric relations, while corresponding relations that are mirrored off the verticality are pairs of converse relations. Relations *dDisjoint*, *dInside*, and *dContains* are at the extremes.

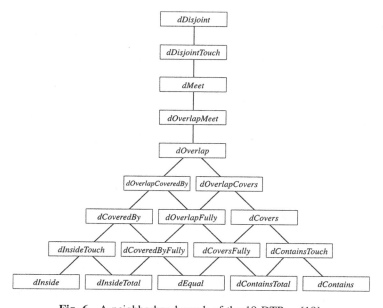

Fig. 6. A-neighborhood graph of the 19 DTR$_{Jcb}$ [19].

The ten d*+-metrics take on the values 0, 1, or between 0 and 1 (0,1). For two metrics—dOAS^{-1} and dOTS^{-1}—a value of greater than 0 (without an upper bound) applies. Mapping these four values for each of the ten metrics over the conceptual neighborhood graph (Fig. 7) reveals that the occurrence of the same values is fairly regularly distributed and follows a pattern of neighborhood.

The same values typically form connected subsets. Only few cases of separations occur (e.g., the two disconnected parts of value 1 in dBES (Fig. 7f), or the two disconnected portions of value 0 in dITS-1 (Fig. 7b) and in dBIS (Fig. 7d)). The two metrics JIMBS and JOMBS have mirror images among those relations that share an interior (Fig. 7i and j). Like dAS, JIMBS and JOMBS have their portions of 0 values split into three separate parts of their graphs.

The distributions also highlight that dAS uniquely identifies *dEqual* (Fig. 7e) as it is the only relation with a dAS value of 1. Likewise, JIMBS is a unique identifier for *dInsideTotal*, as it is the only relation with a JIMBS value of 1. The distributions also confirm visually that the 5-tuples dIAS-dAS-dBES-dOAS^{-1}-JIMBS and dIAS-dBIS-dAS-dBES-JIMBS uniquely identify each of the 19 relations, as for both 5-tuples the value combinations for each relation are unique.

The frequency of the values across the different measures leads to six groups: (1) metrics that result in the value 0 (Fig. 8a), (2) metrics that range between 0 and 1 (Fig. 8b), (3) metrics that result in the value 1 (Fig. 8c), and (4) metrics that result in a value of greater than 0, without an upper bound. While the value 0 occurs only in the

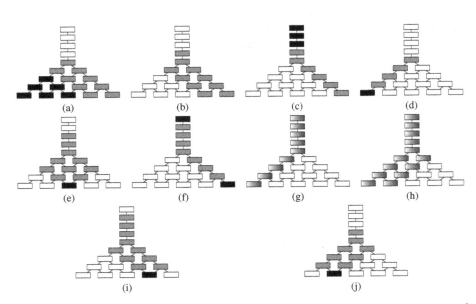

Fig. 7. Metrics displayed with values on the A-neighborhood graph: (a) dIAS, (b) dITS^{-1}, (c) dOAS, (d) dBIS, (e) dAS, (f) dBES, (g) dOAS^{-1}, (h) dOTS^{-1}, (i) JOMBS, and (j) JIMBS. White represents the value 0, grey represents the value $0 < x < 1$, black represents 1, and gradient shading represents values greater than 0 without an upper bound (only applicable for dOAS^{-1} and dOTS^{-1}).

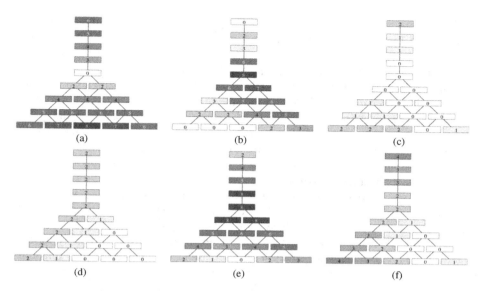

Fig. 8 Frequencies of the values for the d^{*+}-metrics: (a) counts of x=0, (b) counts of 0<x<1, (c) counts of x=1, (d) counts of x>0, (e) combined counts (b)+(d), and (f) combined counts (c)+(d).

first group, the other three groups are not mutually exclusive. These dependencies yields the remaining two value distributions: (5) the combination of $0 < x < 1$ with $x > 0$ (Fig. 8e) and (6) the combination of $x = 1$ with $x > 0$ (Fig. 8f). The distributions of 0 s shows a completely regular pattern from the perspectives of *dDisjoint*, *dInside*, and *dContains*, where the lowest count is at *dOverlap* and the highest count at *dEqual*. The counts of zero gradually decrease from the graph's extremes towards the center. The combined values of $0 < x < 1$ (Fig. 8e) show a regular pattern for the relations between *dDisjoint* and *dInside*, yet not between *dDisjoint* and *dContains*. Their distributions are reverse to those of 0 s, as *dOverlap* carries the highest value, while *dEqual* has the lowest value. The distributions of the combined 1 s (Fig. 8f) shows a similar bias in regularity between *dDisjoint* and *dInside*, while *dContains* is the only relation, across all distributions, that has an insolated count with respect to its neighbor relations. This observation correlates with the behavior of dES (Fig. 7f), which has the value 1 for *dDisjoint* and *dContains*, thereby forming the only separation of neighborhoods of value 1.

9 Conclusions and Future Work

This paper applied pixel counts to determine the explicit topological relation between two discrete regions that are bounded by a digital Jordan curve [19]. Discernment of topological relations in an *ad-hoc* fashion is a fundamental accomplishment for map translations into text or speech [8, 24, 28]. Given the discrete embedding, topological relations need not be stored, using a mere five calculations to compute them when necessary.

The approach used metrics that refine topological relations, a concept that has been demonstrated for continuous objects in multiple fashions [14, 40, 41]. While all previous methods required six metrics to determine all eight binary topological relations in $\mathbb{R}^2$, the discrete embedded resolves the same task with five metrics to determine all 19 relations of Jordan-curve-bounded digital regions. The key to distinguishing more relations with fewer measures is to consider the regions' margins in addition to their boundaries. The reduction of the number of measures represents an opportunity for efficiency gains if the metrics are of comparable computability. The approach presented in this paper is not necessary when considering the hyperraster approach of Winter [44], as the relations are no different than those in $\mathbb{R}^2$, thus making the continuous space metrics applicable [14, 40].

Jordan inner and outer margins are also to distinguish some relations. At the same time, the addition of metrics that consider the Jordan margins reduces the cardinality of the smallest sets needed to determine three relations (*dOverlap*, *dInsideTotal*, and *dContainsTotal*). At the same time the smallest sets maintain their cardinality for ten relations, and for no relation the smallest sets increase in size (Table 5). Other relations may see a similar benefit if each of the 25 9$^+$-intersection intersections were pursued with a specific metric by integrating the Jordan inner and outer margin approach with object A as well.

Table 5. Changes to the cardinality of the smallest set of metrics that fully define each DTR$_{Jcb}$.

Relation	d*	d*$^+$	Net	Relation	d*	d*$^+$	Net
dDisjoint	N/A	2	–	*dInsideTouch*	N/A	2	–
dDisjointTouch	N/A	3	–	*dOverlapFully*	4	4	0
dMeet	2	2	0	*dEqual*	1	1	0
dOverlapMeet	2	2	0	*dOverlapCovers*	3	3	0
dOverlap	3	2	−1	*dCovers*	3	3	0
dOverlapCoveredBy	3	3	0	*dCoversFully*	3	3	0
dCoveredBy	3	3	0	*dContainsTotal*	2	1	−1
dCoveredByFully	3	3	0	*dContains*	N/A	2	–
dInsideTotal	2	1	−1	*dContainsTouch*	N/A	3	–
dInside	N/A	1	–				

As long as co-dimension remains intact with each object relative to one another as well as to the embedding space, the topological relations are typically compatible across embedding spaces [20, 45]. Therefore, metric determinations of pixel space regions equate directly to metric determinations of voxel space volumes. Voxels are an evolving technology [26] that have become more and more necessary with advances in 3D printing [33] and indoor navigation [21]. Similarly, relations from planar spaces have been shown to apply in spherical spaces [12], opening up the digital sphere [25] for study, in the process creating relations like *dAttach*.

This work has several opportunities for refinement and future study. The regions in this work were constrained to only allow margins as digital Jordan curves and furthermore that the margins could not become self-adjacent. When this constraint is lifted, it will produce an environment that might somewhat resemble continuous space, where regions that exhibit the relation *meet* might be confused for *disjoint*, where particular measurements are blinded to 0-dimension intersections. While these measurements would not be blinded, they may have to be adapted to have multiple cases that lead to the same relation, suggesting a different set of metrics within the identifying groups might be favorable. Similarly, the original relations were constrained to not allow small interiors that could be absorbed into a single boundary [19]. Lifting this constraint creates additional relations over thinner regions.

Beyond these types of reasonable constraint relaxations lie further relaxations that go beyond simple regions under digital topological closure. An example of this phenomenon can be found by looking at satellite images of forests. Satellite classification algorithms in land cover scenarios can easily create separated objects or holed objects [22]. While the approach presented can determine that intersections from the 9-intersection perspective are present in the form of a non-zero proportion, further work is necessary to express exact relations, such as the difference between *surrounds* and relations such as *disjoint* or *meet* as discrete features [9].

The Jordan inner and outer margins represent a new approach to considering digital topological relations. While interior and exterior pixels are easily discernable and have enforceable constraints, the Jordan inner and outer margins need future study in an applied manner for the systematic detection of the constraints within the paper.

Acknowledgments. Matthew Dube was partially supported by a Michael J. Eckardt Dissertation Fellowship from the University of Maine. Jordan Barrett was supported by Department of Education grant DOE-P047M130578 and also supported by the Maine Sustainability Solutions Initiative (NSF EPS-0904155). Max Egenhofer's research was partially supported by NSF grant IIS-1016740.

References

1. Ackoff, R.L.: From data to wisdom: presidential address to ISGSR, 1988. J. Appl. Syst. Anal. **16**(1), 3–9 (1989)
2. Alexandroff, P.: Elementary Concepts of Topology. Dover Publishers Inc., New York (1961)
3. Bittner, T., Winter, S.: On ontology in image analysis. In: Agouris, P., Stefanidis, A. (eds.) ISD 1999. LNCS, vol. 1737, pp. 168–191. Springer, Heidelberg (1999)
4. Blaser, A.D., Egenhofer, M.J.: A visual tool for querying geographic databases. In: Proceedings of the Working Conference on Advanced Visual Interfaces, pp. 211–216. ACM (2000)
5. Câmara, G., Egenhofer, M.J., Fonseca, F., Vieira Monteiro, A.M.: What's in an image? In: Montello, D.R. (ed.) COSIT 2001. LNCS, vol. 2205, pp. 474–488. Springer, Heidelberg (2001)
6. Clementini, E., Sharma, J., Egenhofer, M.J.: Modelling topological spatial relations: strategies for query processing. Comput. Graph. **18**(6), 815–822 (1994)
7. Di Sciascio, E., Mongiello, M.: Query by sketch and relevance feedback for content-based image retrieval over the web. J. Vis. Lang. Comput. **10**(6), 565–584 (1999)

8. Dube, M.P., Egenhofer, M.J.: An ordering of convex topological relations. In: Xiao, N., Kwan, M.-P., Goodchild, M.F., Shekhar, S. (eds.) GIScience 2012. LNCS, vol. 7478, pp. 72–86. Springer, Heidelberg (2012)

9. Dube, M.P., Egenhofer, M.J.: Surrounds in partitions. In: Proceedings of the 22nd ACM SIGSPATIAL International Conference on Advances in Geographic Information Systems, pp. 233–242. ACM (2014)

10. Eckhardt, U., Latecki, L.J.: Topologies for the digital spaces $\mathbb{Z}^2$ and $\mathbb{Z}^3$. Comput. Vis. Image Underst. 90(3), 295–312 (2003)

11. Egenhofer, M.J.: Query processing in spatial. J. Vis. Lang. Comput. 8(4), 403–424 (1997)

12. Egenhofer, M.J.: Spherical topological relations. In: Spaccapietra, S., Zimányi, E. (eds.) Journal on Data Semantics III. LNCS, vol. 3534, pp. 25–49. Springer, Heidelberg (2005)

13. Egenhofer, M.J., Al-Taha, K.K.: Reasoning about gradual changes of topological relationships. In: Frank, A.U., Campari, I., Formentini, U. (eds.) Theories and Methods of Spatio-Temporal Reasoning in Geographic Space. LNCS, vol. 639, pp. 196–219. Springer, Heidelberg (1992)

14. Egenhofer, M.J., Dube, M.P.: Topological relations from metric refinements. In: Proceedings of the 17th ACM SIGSPATIAL International Conference on Advances in Geographic Information Systems, pp. 158–167. ACM (2009)

15. Egenhofer, M.J., Franzosa, R.F.: Point-set topological spatial relations. Int. J. Geogr. Inf. Syst. 5(2), 161–174 (1991)

16. Egenhofer, M.J., Herring, J.R.: Categorizing Binary Topological Relations Between Regions, Lines, and Points in Geographic Databases. Technical report, Department of Surveying Engineering, University of Maine (1990)

17. Egenhofer, M.J., Mark, D.M.: Naive Geography. In: Frank, A., Kuhn, W. (eds.) COSIT 1995. LNCS, vol. 988, pp. 1–15. Springer, Heidelberg (1995)

18. Egenhofer, M.J., Shariff, A.R.: Metric details for natural-language spatial relations. ACM Trans. Inf. Syst. 16(4), 295–321 (1998)

19. Egenhofer, M.J., Sharma, J.: Topological relations between regions in $\mathbb{R}^2$ and $\mathbb{Z}^2$. In: Abel, D.J., Ooi, B.C. (eds.) SSD 1993. LNCS, vol. 692, pp. 316–336. Springer, Heidelberg (1993)

20. Egenhofer, M.J., Sharma, J., Mark, D.M.: A critical comparison of the 4-intersection and 9-ntersection models for spatial relations: formal analysis. In: McMaster, R.B., Armstrong, M.P. (eds.) Autocarto 11, pp. 1–11 (1993)

21. Fallah, N., Apostolopoulos, I., Bekris, K., Folmer, E.: Indoor human navigation systems: a survey. Interact. Comput. 25(1), 21–33 (2013)

22. Friedl, M.A., McIver, D.K., Hodges, J.C.F., Zhang, X.Y., Muchoney, D., Strahler, A.H., Woodcock, C.E., Gopal, S., Schneider, A., Cooper, A., Baccini, A., Gao, F., Schaaf, C.: Global land cover mapping from MODIS: algorithms and early results. Remote Sens. Environ. 83(1), 287–302 (2002)

23. Galton, A.: The mereotopology of discrete space. In: Freksa, C., Mark, D.M. (eds.) COSIT 1999. LNCS, vol. 1661, pp. 251–266. Springer, Heidelberg (1999)

24. Giudice, N.A., Bakdash, J.Z., Legge, G.E.: Wayfinding with words: spatial learning and navigation using dynamically updated verbal descriptions. Psychol. Res. 71(3), 347–358 (2007)

25. Huo, M.L.: The basic topology model of spherical surface digital space. In: Proceedings of the 20th International Society for Photogrammetry and Remote Sensing Congress, pp. 1–6 (2004)

26. Kaufman, A., Cohen, D., Yagel, R.: Volume graphics. Computer 26(7), 51–64 (1993)

27. Khalimsky, E., Kopperman, R., Meyer, P.R.: Computer graphics and connected topologies on finite ordered sets. Topology Appl. 36(1), 1–17 (1990)

28. Klippel, A.: Spatial information theory meets spatial thinking: is topology the rosetta stone of spatio-temporal cognition? Ann. Assoc. Am. Geogr. **102**(6), 1310–1328 (2012)

29. Klippel, A., Li, R., Yang, J., Hardisty, F., Xu, S.: The egenhofer-cohn hypothesis or, topological relativity? In: Raubal, M., Mark, D., Frank, A. (eds.) Cognitive and Linguistic Aspects of Geographic Space, pp. 195–215. Springer, Heidelberg (2013)

30. Kong, T.Y., Rosenfeld, A.: Digital topology: a comparison of the graph-based and topological approaches. In: Reed, G.M., Roscoe, A.W., Wachter, R.F. (eds.) Topology and Category Theory in Computer Science, pp. 273–289. Oxford University Press, Oxford (1991)

31. Kurata, Y.: The 9^+-intersection: a universal framework for modeling topological relations. In: Cova, T.J., Miller, H.J., Beard, K., Frank, A.U., Goodchild, M.F. (eds.) GIScience 2008. LNCS, vol. 5266, pp. 181–198. Springer, Heidelberg (2008)

32. Kurata, Y., Egenhofer, M.J.: The 9^+-intersection for topological relations between a directed line segment and a region. In: Gottfried, B. (ed.) Workshop on Behaviour and Monitoring Interpretation, Technical report 42, Technologie-Zentrum Informatik, pp. 62–76. University of Bremen, Germany (2007)

33. Lipson, H., Kurman, M.: Fabricated: The New World of 3D Printing. Wiley, Indianapolis (2013)

34. Mark, D.M., Comas, D., Egenhofer, M.J., Freundschuh, S.M., Gould, M.D., Nunes, J.: Evaluating and refining computational models of spatial relations through cross-linguistic human-subjects testing. In: Frank, A., Kuhn, W. (eds.) COSIT 1995, LNCS, vol. 988, pp. 553–568. Springer, Heidelberg (1995)

35. Mezaris, V., Kompatsiaris, I., Strintzis, M.G.: Region-based retrieval using an object ontology and relevance feedback. Eurosip J. Appl. Sig. Process. **6**, 886–901 (2004)

36. Nedas, K.A., Egenhofer, M.J., Wilmsen, D.: Metric details of topological line-line relations. Int. J. Geogr. Inf. Sci. **21**(1), 21–48 (2007)

37. Nittel, S., Stefanidis, A., Cruz, I., Egenhofer, M., Goldin, D., Howard, A., Labrinidis, A., Madden, S., Voisard, A., Worboys, M.: Report from the first workshop on geo sensor networks. ACM SIGMOD Rec. **33**(1), 141–144 (2004)

38. Randell, D.A., Cui, Z., Cohn, A.G.: A spatial logic based on regions and connection. In: Nebel, B., Rich, C., Swartout, W.R. (eds.) KR 1992, pp. 165–176 (1992)

39. Rosenfeld, A.: Digital topology. Am. Math. Monthly **86**, 621–630 (1979)

40. Shariff, A.R.B., Egenhofer, M.J., Mark, D.M.: Natural-language spatial relations between linear and areal objects: the topology and metric of english-language terms. Int. J. Geogr. Inf. Sci. **12**(3), 215–245 (1998)

41. Sridhar, M., Cohn, A.G., Hogg, D.C.: From video to RCC8: exploiting a distance based semantics to stabilise the interpretation of mereotopological relations. In: Egenhofer, M., Giudice, N., Moratz, R., Worboys, M. (eds.) COSIT 2011. LNCS, vol. 6899, pp. 110–125. Springer, Heidelberg (2011)

42. Sutherland, W.A.: Introduction to Metric and Topological Spaces, 2nd edn. Oxford University Press, Oxford (1975)

43. Vince, A., Little, C.H.: Discrete jordan curve theorems. J. Comb. Theory, Series B **47**(3), 251–261 (1989)

44. Winter, S.: Topological relations between discrete regions. In: Egenhofer, M.J., Herring, J.R. (eds.) SSD 95, LNCS, vol. 951, pp. 310–327. Springer, Heidelberg (1995)

45. Zlatanova, S.: On 3D topological relationships. In: DEXA Workshop 2000, pp. 913–919. IEEE Computer Society (2000)

Language and Space

Where Snow is a Landmark:
Route Direction Elements in Alpine Contexts

Ekaterina Egorova[1,2(✉)], Thora Tenbrink[3], and Ross S. Purves[1]

[1] Department of Geography, University of Zurich, Zurich, Switzerland
{ekaterina.egorova,ross.purves}@geo.uzh.ch
[2] University Priority Research Programme Language and Space (URPP SpuR),
University of Zurich, Zurich, Switzerland
[3] School of Linguistics and English Language, Bangor University, Bangor, UK
t.tenbrink@bangor.ac.uk

Abstract. Route directions research has mostly focused on urban space so far, highlighting human concepts of street networks based on a range of recurring elements such as route segments, decision points, landmarks and actions. We explored the way route directions reflect the features of space and activity in the context of mountaineering. Alpine route directions are only rarely segmented through decision points related to reorientation; instead, segmentation is based on changing topography. Segments are described with various degrees of detail, depending on difficulty. For landmark description, direction givers refer to properties such as type of surface, dimension, colour of landscape features; terrain properties (such as snow) can also serve as landmarks. Action descriptions reflect the geometrical conceptualization of landscape features and dimensionality of space. Further, they are very rich in the semantics of manner of motion.

Keywords: Route directions · Natural environment · Segmentation · Landmarks · Mountaineering

1 Introduction

In 1995, Max Egenhofer and David Mark proposed the notion of Naive Geography for the body of knowledge that lay people have about the surrounding geographic world [15] as a counterpart to the formalizations used by professional geographic community. As well as an underlying scientific motivation, they stressed a real practical need for the incorporation of such naive geographic knowledge into GIS, bridging the gap between an average citizen's needs from a GIS, and the (sometimes abstract) spatial concepts embedded in the latter.

Reaching this aim requires that we also understand how space is perceived and conceptualized, not just by experts involved in the implementation of GIS, but also by a greater cross section of society. One oft cited way of gaining such insights is through the prism of route directions. Locomotion is a major way humans discover, and thus presumably construct mental representations of, environmental space [32], and human

© Springer International Publishing Switzerland 2015
S.I. Fabrikant et al. (Eds.): COSIT 2015, LNCS 9368, pp. 175–195, 2015.
DOI: 10.1007/978-3-319-23374-1_9

concepts and schematizations of space are systematically encoded in language whenever routes are described [48]. Thus, route directions are a readily available external representation of spatial concepts, revealing structures in thinking about and using space. Further, as navigational services become ubiquitous on mobile devices for many modes of locomotion, route direction studies are increasingly relevant in terms of real practical applicability – for example, in choosing which real world features are likely to be salient for a particular application [24].

However, with few exceptions, most research on route directions has focused on urban environments: outdoor (e.g., campus areas [12], neighbourhoods [2], downtown areas [19], cities [13]), indoor (e.g., complex buildings [46], airports [37]), transitional spaces [23]. As observed in [6], one of the few works on non-urban space, extending the range of studies to natural environments remains an important research challenge. Moreover, investigating natural space presents an opportunity to explore the degree to which results from very different urban environments are transferable, and can potentially provide avenues giving new insights into ways in which space may be conceptualized.

To address this gap, we explore alpine route directions and thus discover the features of spatial concepts reflected in this fundamentally different type of environment. Specifically, we address the ways in which the structure of route directions is affected by the properties of the considerably less structured space, and by the more complex activity of mountaineering, as opposed to walking in a city or building. For this purpose, we initially explore the scope of information found to be relevant in alpine route directions, beyond the basic spatial directions. Furthermore, we investigate some major conceptual route elements as known from urban environments: segments and nodes, landmarks and action descriptions.

2 Related Work

2.1 Route Direction Elements

According to [12], the route description process involves three cognitive operations. The first one is the activation of the internal representation of the environment in question by the speaker, who then plans the route by defining a sequence of segments connecting starting and destination points. The result is what [48] describe as route schematization, namely a network of segments and nodes, i.e., decision points involving (potential) changes of direction. The third stage is the formulation of the procedure, resulting in the verbal description of the route.

Despite, or perhaps because of, the large volume of research on route directions, no single analytical framework with clearly defined units of analysis exists. One reason might be the variety of research questions posed within several disciplines, such as linguistics, cognitive psychology, geography, and computer science [13]. [45] identified several essential building blocks that are frequently mentioned in the studies of route directions – starting point and destination, intermediate decision points, route segments, actions and movement directions, reorientations, landmarks, regions and areas, and distances. In the following we examine critically varying definitions of some of these elements, with a view on their transferability to the context of mountaineering.

Segments and Decision Points. Route segments and decision points (or, links and nodes) are key conceptual elements of route schematizations. There are two different interpretations of the way segments are represented in texts. For [48], a segment is a unit containing enough information to go from node to node. It consists of a starting point, reorientation, path/progression and an end point. Essentially, a segment corresponds to the change of direction, as its starting and end points are decision points. However, as the authors note, this is not a necessary condition, since major intersections or landmarks might also separate segments without a direction change. For [29], on the other hand, decision points and their associated reorientation instructions are not integrated into segments. The latter are seen as straightforward parts of the route (as in, "follow the path", "walk along"). Similarly, [2] singles out pathways (nouns referring to actual or potential channels of movement, such as streets, sidewalks, or trails) and choice points (nouns referring to places where options with regard to the further path exist, with intersection as the most typical example).

Independent of their representation in an analytic framework, segments and decision points are critical conceptual elements of the route schematization. Crucially, they reflect the structure of the environment in an urban context, as segments become synonymous with pathways and are associated with linear features, allowing straightforward progression to the next decision point. Decision points, in turn, are often associated with intersections within a structured urban context. In an urban context, therefore, segmentation of a route as such does not pose any major conceptual challenges. By contrast, it is an open question how routes might be segmented in an environment that offers far less structure, such as the natural setting of a mountain.

Landmarks and Action Descriptions. Theoretically, the path from one decision point to the other could be described using metrics (e.g., length of the route segment), as done frequently (and almost exclusively) in automatically generated route directions. However, humans rarely describe routes in this way – typically, references to landmarks are used to demarcate qualitatively the end (or position within) a segment. For Denis [12], landmarks and action descriptions (referred by the author as "prescriptions") are the two essential components of route directions.

Again, definitions vary. According to Denis [12], landmarks can be 3D (building) or 2D (street, square) features of the environment. Within route directions, they can have one of three functions: signalling sites where actions are to be accomplished, helping to locate other landmarks, or confirming the route. In this framework, actions are often prescribed in relation to landmarks, as in "cross X" (X – a street, a bridge, a place) and "take X" (X – a street, a road, a path). Similarly, Montello [33] points out that landmarks are not restricted to point-like features – linear and areal features (e.g., paths, regions) can serve as landmarks just as well. In contrast, Allen [2] regards landmarks as environmental features serving as subgoals on the way from the point of origin to the destination along a specified path of movement. Within his framework, landmarks and pathways (e.g., streets, sidewalks, trails) are two separate elements of route directions. Thus, Allen's pathways would be classified in Denis' framework as landmarks, incorporated as proper parts into the route.

Further addressing the extent to which landmarks are incorporated within a route, [29] differentiate between the functions of the landmarks depending on whether they are on-route or off-route. A special term – *routemark* – has been used for a landmark that represents part of the route and determines the direction of movement (as in, "follow the river") [38].

Since landmarks have been defined in many different ways in the analysis of route descriptions, the question remains as to which features of the environment are essential in serving as landmarks. In the sentence "Walk along the street till the next intersection, where the bakery is, and turn left", the bakery is clearly a landmark, serving to identify the intersection and thereby the decision point. But what about "street" and "intersection", which are not by all authors identified as landmarks? Both are integral parts of the structure of the environment, and serve to segment and structure the route. In urban environments, they represent non-unique features within a network of streets and may not share one of the main characteristics of a more typical landmark, namely saliency. From a more linguistically oriented point of view, streets and intersections appear in descriptions in a similar way to (other) landmarks, reflecting their status as relevant and referable (and thus, arguably, sufficiently recognizable or salient) entities in the speaker's mind. Hence, [6], following [12], annotate all references to geographic objects as landmarks. This approach appears promising for a more natural, non-structured context, especially given the challenge as to understanding "how a continuous land surface, a landscape, becomes cognitive entities" [31].

As already indicated, landmarks are often linguistically related to action description – another important element of route directions. Two major classes of actions are often recognized: changing orientation (as in, "turn right") and proceeding (as in, "walk straight ahead") [12]. These elements are represented by verbs of motion, which fall in the semantic categories of "go" and "turn" [2]. [48] report that the most common actions in their case study were *turn, take a, make a, and go*; specifically, for the verbs expressing progression, the two most frequent ones were *go* and *follow*, used for straight and curved paths respectively. Beyond movement, possible actions are positioning and inspection, such as a check that the current orientation is the intended one (as in, "When you arrive here, you should have the school on your left and the market on your right") [12]. These are related to perceptual experience (as in, "You will see a stop sign") and are therefore often represented by verbs of perception (almost always vision) [2].

Further Descriptive Elements. The spatial elements described so far are typically recognized as a minimum set necessary for successful wayfinding. To capture any remaining elements of route descriptions, [48] differentiate between critical and supplementary information, and [29] note the existence of redundant information in route directions. [12] identifies descriptive components that may specify topological relations between objects and landmark properties, or provide various types of comments and encyclopedic knowledge without direct relevance for the instruction.

However, [16] argue that the type of information included is affected by the purpose of the activity that wayfinding is embedded in. The authors point to several attributes of activities (such as time pressure, effort, focus on destination) that are linguistically indicated by specific markers (such as *quick* or *fast* in the case of time pressure). Hence, it is conceivable that some types of information may be redundant or non-essential in some contexts, while constituting a highly relevant and integral part of a description in other contexts. Mountaineering represents precisely the kind of context where a simple, spatially focused route description is not always sufficient. In the following, we will take a closer look at this kind of context.

2.2 Mountains as Outdoor Natural Space. Mountaineering as Activity

Considering the properties of mountains as a specific type of space, one major distinguishing property pertains to scale. [32] differentiated between four types of psychological spaces on the basis of the projective size of the space relative to the human body and the differing ways in which humans can apprehend them: figural, vista, environmental and geographical. In this framework, a mountain might represent an environmental space, which cannot be apprehended without locomotion – however, it is possible that it may be apprehended by "direct experience" alone [32]. In this respect, a mountain is comparable to an urban space; however, the ways in which the environment can be explored and the kinds of expectations about the environment that can be made on the basis of the information gained from a current position (i.e., within vista space) differ fundamentally.

This is related to another crucial space property – namely, its structure. While built urban space is seemingly structured by objects with more or less bona fide borders (streets, buildings, etc.), natural space represents a (more or less) continuous land surface, raising the question as to how exactly the human mind might structure it into entities [10, 41]. In [23], comparing indoor and outdoor settings, the authors identified further distinguishing structural elements with possible relevance to the mountaineering context. Indoor environments are essentially three-dimensional, while street networks are described in terms of two-dimensional concepts. Landmarks differ structurally – only outdoor environments offer global landmarks such as the sun. Indoor spaces restrict movement in all directions and also fundamentally obscure sight, while outdoor spaces are more flexible and may offer unconstrained lines of sight. Extending these insights, it is fair to say that mountains are likely to be conceived as three-dimensional, they can offer both global and local landmarks as well as an unconstrained line of sight, and they can have restrictions of movement in all directions, depending on the terrain. However, these aspects may vary as mountains are rather heterogeneous and changeable (according to weather conditions, seasons, as well as evergoing natural processes) – and indeed this heterogeneity and changeability is a key distinguishing property of mountains as space.

Unlike navigation in urban space, mountaineering is an activity that requires specific skills in terms of locomotion and navigation. Reaching the summit safely can be a major challenge and thereby constitutes a conceptual goal in itself. While in most contexts wayfinding is a necessity in order to reach a certain destination [16], wayfinding in mountaineering constitutes an essential part of the activity.

2.3 Open Research Questions

To explore human concepts of space in a mountaineering context, it makes sense to start from natural descriptions of using this kind of space. As a genre, alpine literature has a long history. The non-fiction part comes mostly in the form of accounts of ascents published in journals and yearbooks of Alpine Clubs since the 1860s[1], scientific journals [30] and privately [50], offering a rich potential for exploring how this kind of space is conceptualized. For instance, [3, 22] examined the meaning of mountains for the British during the 19th century. Some authors have used the digitized Swiss Alpine Club year-books[2] for quantitative analysis, e.g., to address motivation in mountaineering [7], to investigate how texts change over time [8] and for research in geographic information retrieval, with, for example, [36] investigating the possibilities of automatic route extraction, and [14] linking descriptions to geospatial footprints to examine how landscape descriptions vary across space.

While alpine literature thus provides a rich data source for addressing a wide spectrum of research questions, this has, to our knowledge, not yet been used to investigate human route concepts in mountaineering, as seen in contrast to urban space. A variety of approaches have explored how continuous landscapes are deconstructed into discrete entities and represented in language [10, 41]; in particular, the impact of factors such as experience [34], familiarity with landscape [51], local ecology, culture and language [5, 18, 21, 26] has been addressed. However, few authors have studied route directions in a natural context. [6] explored landmark- and action-based elements in orienteering route directions and identified various constructs from the point of view of geometry. [40] investigated the role of landmarks in summer and winter hiking along a specific route in a national park.

In this paper we aim to shed further light on how humans segment space in an unstructured alpine environment, and, furthermore, how landmarks and action descriptions are referred to in this context. In addition, we address the impact of activity on route directions, pursuing insights by [6, 11] based on case studies on orienteering. In particular, we address the following research questions:

1. What is the *content* and *scope* of mountaineering route directions; to what extent are they focused on spatial information?
2. What constitutes *decision points* and *segments* in an alpine context?
3. In what ways do *landmarks* and *action descriptions* reflect features of alpine space and activities?
4. Finally, we wish to explore whether *generally applicable new insights* into the ways in which space is perceived and conceptualized can be gained by moving from primarily urban, highly structured spaces, to more natural landscapes.

---

[1] http://www.alpinejournal.org.uk/.
[2] http://textberg.ch.

3   Data

Typically, research on route directions draws on controlled data collected from participants in a specific place. The increasing volume of user-generated content found online provides an alternative source of data, which some authors have begun to use for wayfinding research [17]. While such data provide little control over (or insight about) participants and circumstances, they offer a rich diversification of places described in route directions, overcoming some limitations of controlled studies that are necessarily constrained to specific populations and environments [12]. This may facilitate research on the specifics of space structure as reflected in the route directions, and help to uncover systematic patterns in texts of the same kind, independent of place.

For current purposes, our data source consists of 19 texts gathered from www.summitpost.org, a US-based platform for "a collaborative content community focused on climbing, mountaineering, hiking and other outdoor activities"[3]. The site's content is created and maintained by its members, who have profiles with basic personal information (including location, age, gender, date of registration). One section of the website is dedicated to routes, to be selected through an advanced search. Search parameters include location (continent, country), route type (e.g., mountaineering, bouldering, scrambling, mixed, etc.), rock difficulty, and grade. The set structure for route directions consists of the sections "Getting There", "Route Description", "Essential Gear", "Commentary", "When to Climb", and "Images". While some of these sections may be omitted, most of the route descriptions provide content for at least the first three of them.

When collecting texts for our small corpus, we extracted the "Route Description" section only, and chose texts of approximately the same length (350–400 words). We ensured that the same author did not appear twice in the corpus, and that the authors appeared to be native speakers of American English (judging from the location indicated in the users' profiles). By setting the route type (mountaineering) and grade (IV and above) as search parameters, we collected routes running on mixed types of terrain (rock, snow, ice) requiring certain skills and equipment from a mountaineer.

4   Analysis

Following the principles of Cognitive Discourse Analysis (CODA) [44], we started out with a detailed examination of the content of descriptions in our corpus so as to gain an intuitive understanding of the concepts expressed by route givers. These insights were then operationalized towards a systematic analysis procedure, which involved identifying meaningful segments, specifying their content, and identifying linguistic markers associated with the concepts in question. Since our research questions related to different kinds of linguistic features, this procedure was followed for each of them separately, as detailed below.

Generally, the concept of a motion event has proved to be particularly relevant to our analysis. According to [43], a motion event consists of the elements *Figure* (an object

---

[3] https://www.facebook.com/summitpost.org/info?tab=page_info.

moving), *Ground* (object in respect to which the Figure is moving), *Motion* ("presence *per se* of motion"), and *Path* (the course followed by the Figure with respect to the Ground). Further, as [20] notes, *Paths* can contain information about the starting point of the motion event (called FROM paths), the end point (TO paths) and about the path itself, where the Figure moves along the Ground (VIA paths).

Additionally, the *Manner of Motion* can be included in the verbalization of a motion event. Verbs of motion fall into two broad categories: Manner verbs (e.g., *walk, run, crawl*) and Path verbs (e.g., *enter, descend, ascend*), which convey a sense of directionality but remain neutral about manner [35, 43]. The English language is known to have a large variety of Manner verbs, directionality is then usually expressed by additional elements, such as prepositions ("run into the room") [42]. We use these concepts in the analysis below.

Content and discourse analysis procedures typically require iterative loops [25, 44]. Here we aimed at a coding scheme with exhaustive and mutually exclusive categories (wherever possible) that were clearly defined for replication; this could only be achieved as a result of multiple iteration and modification, with continuous double checking by two of the authors to ensure consistency in the coding. In the following, we present the operationalized analysis undertaken together with our results for each research question sequentially.

4.1   General Scope of Alpine Route Directions

To analyse the overall scope of information included in route directions, we identified content categories in the corpus as follows, and counted the words in each category. These categories are mutually exclusive and exhaustively cover all data in our corpus.

Route (3,018 words): This category comprises any information about the route as such: general comments on the route, route segments and decision points, introduction of landmarks used for route confirmation, directions, as well as route options. These can be seen as central (prototypical) elements of route directions, paralleling those found in urban contexts.

Terrain and Difficulty (1,056 words): This comprises information on elevation, gradient of terrain and exposure, type of surface, and technical difficulty. The following markers (and their derivatives) are typical for the aspect of terrain: *elevation, high, steep, flat, angle, vertical, horizontal, exposure, ice, snow, gravel, sand, covered, rock, surface, textured, slippery, loose, rotten, broken, melting, soft, unstable, decomposed, mixed, pure*. Technical difficulty is often expressed quantitatively, and is typically represented by terms such as *crux, class, grade, resistance, challenge, negotiate, rate, attempt, hard, committing, easy, non-trivial, uneventful, climbable, complicated, technical, manageable, advanced*. Both aspects are interlinked and overlapping, as terrain properties are typically made relevant in the context of activity and challenge.

Obstacles (314 words) includes warnings about permanent obstacles such as crevasses, signalled by the following markers: *deal with, beware of, avoid, bypass, watch for, obstacle, detour, hidden, buried*.

Hazards (189 words) contains cautions concerning possible hazards, such as avalanches, rockfalls or strong wind. Typical markers are: *exposed to, hazard, falling rock/rockfall, wind gusts, avalanche, prone to.*

Safe Locomotion (412 words) comprises instructions concerning equipment: introducing protection opportunities and places where certain locomotion techniques should be used. The markers are mostly mountaineering jargon: *crampons, rappel stations, footwear, climbing shoes, helmet, boots, ice axe, rope (up), chain, strap up, belay, simulclimb.*

Past Experience (429 words) accounts for any experience of previous mountaineers (or the author himself) on this particular route. These units are generally marked by verbs in the past tense and first or third person pronouns.

Miscellaneous (588 words) contains less frequent units with various further types of information, such as spots for camping and repose, availability of water, traffic on the route, accounts of the views and references to time.

Altogether, these additional content categories add up to roughly the same amount of words (2,988) as the main Route category (3,018). Thus, we note that the overall scope of alpine route directions is centered on spatial information just as much as on further vital aspects of the mountaineering challenge (Fig. 1). In the following, we pursue the spatial aspects pertaining to alpine route concepts by applying notions known from urban route contexts: segmentation, and the core concepts of landmarks and actions.

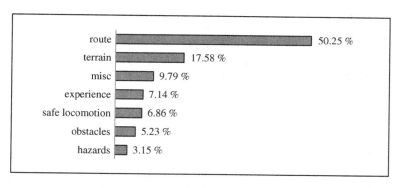

Fig. 1. Information categories in percent (n = 6,006)

4.2   Segments and Decision Points

Route Segments. Conceptually, route segments are links between nodes. In verbal route directions, these are units containing information of how to get from one decision point to another. We identified segments in the corpus based on the idea that each segment should add a minimum of spatial knowledge about a specific portion of the route – using Talmy's terminology, the Path of a motion event [43]. To direct from node to node, a segment should include the FROM, the VIA and the TO elements of the Path [20], as in "Climb the final steep slopes from the top of the ropes to the summit". However, such a complete description of a motion cannot be always expected [48]. One relevant reference can often suffice to describe a segment, as in "Climb up towards the bergschrund where the angle steepens".

For the FROM and TO elements the typical linguistic markers are the prepositions *from, out, of, away* and *to, towards, on(to), in(to)* respectively, as well as Path verbs such as *reach* and *leave* [4, 20, 27]. Upon inspection of our data we further identified the verbs (or verb phrases) *arrive at, encounter, gain access to, obtain, deposit on, lead to* and *take to*. The last three of these were used in the corpus in relation to landscape features (representing fictive motion [43]). Furthermore, subordinate clauses that start with *once* or *after* (and the like), and that contain result-oriented Path verbs (as in, "Once you top out in the Chute", "After cresting the ridge"), indicate the FROM of the next segment with a focus on reaching the TO of the previous segment. Other types of subordinate clauses describe the location of the starting point of a segment; these typically start with *when* (as in, "When it flattens out slightly"). The VIA element was identified as any information on the Path between FROM and TO. It does not necessarily represent a linear landscape feature, can also be area-like, as well as a reference to the terrain properties (as in, "descend the same route", "continue on the close to the north side of the ridge", "deal with the 45 degree crux").

While a prototypical way of encoding a segment could be expected to be a motion event with one of the indicated Path elements, according to our data the presence of a motion verb is not necessary. A segment can be represented by a reference to the VIA element, without a concrete motion action description (as in, "There are 3–4 passes over the range that are negotiable"). As a result, units representing segments were very heterogeneous in terms of linguistic structure. Some contained one or a set of sentences, as in "There's a broad snow covered "pre-summit" ahead of you when you're on the snow. Pass it on the south side or you have to deal with crevasses". Other segments were more simply represented by smaller units such as clauses, as in the following set of segments:

1. From this step a traverse is made left to a small shoulder
2. which is climbed a short ways
3. before traversing left again on to the east face to the second couloir
4. which is climbed for about 25 m.

Altogether, we identified 253 units containing segments according to our definition. For each segment, we annotated the presence of FROM, VIA and TO elements. Out of all segments, 19 (7.51 %) contained all three elements, 96 (37.94 %) contained two elements and as many as 138 (54.55 %) contained only one element. The VIA element was encoded in 74.21 % per cent of segments, the FROM element in 28.57 %, and the TO element in 50.79 %. The high frequency of the inclusion of the VIA element reflects its importance in alpine route direction; also, the end point of a segment is typically more relevant in a route description than the starting point.

Some further peculiarities are worth noting. In some cases, segments were not necessarily ordered, and did not always pertain to the same level of granularity. A straightforward example is: "The trip starts out as a hike along the Heliotrope

Ridge <u>trail</u>. After passing <u>Kulshan Creek</u> the trail curves left and wraps around a small ridge".

Here, the second sentence elaborates the first by specifying the nature of a subsection of the segment. The nature of other parts of this trail remains unspecified. In other cases in our corpus, the same segment could be introduced twice, adding more spatial information the second time: "The other option, and reportedly safer, is to <u>descend into the Hot Rocks area from the Hogsback</u> instead of climbing and traversing under the cliffs, possibly getting pelted by falling ice/rock. You would simply <u>traverse left and down from the Hogsback until below the Chute</u> and then <u>ascend to the ridge</u>". It appears furthermore that the amount of information provided for a specific segment depended on its difficulty; while easy segments were only referred to briefly, more difficult ones were elaborated by prescribing a specific action or describing terrain properties (or resulting difficulty). We leave a more detailed analysis of these interesting granularity phenomena for future work.

Decision Points. We now address the ways in which routes are segmented by decision points. Since decision points are the starting and end points of segments, they can be analyzed through the prism of the FROM and TO elements.

We extracted 198 units of this kind from the annotated segments. Next, we were interested in the conceptual features within these units, so as to gain further insights about the nature of decision points in a mountaineering context. Based on iterative inspection we identified the following mutually exclusive categories, which exhaustively cover all references to decision points in our data:

Intersection (4 cases): This category comprises all units with a lexeme semantically related to an intersection. In our data, we identified the noun *fork* and the verbs *to branch* and *to fork* as markers of intersections. ("At about 4,700 feet the trail <u>forks</u>", "until it <u>branches</u> about 300 foot up").

Landscape feature (129 cases) is comprised of units with nouns that refer to a landscape feature (as in, "From the <u>rock tower</u>", "to the <u>ridge</u>").

Spatial part of a landscape feature (21 cases) includes units referring to the regions of the object on the basis of its inherent orientation [28], marked by nouns such as *base, edge, margin, end, top*. They are more specific in their reference to location than units of the previous category and imply certain geometric properties: landscape features in this category are conceptualized as linear ("From the <u>end</u> of the <u>ledge</u>"), or areal ("Walk across the plateau to it's [sic] northeast <u>edge</u>"), or three-dimensional ("From the <u>base</u> of the <u>rocks</u>").

Accomplishment (25 cases) comprises units where the decision point is not referred to as an identifiable location as such, but rather conceptualized as lying outside the landscape feature that has just been passed. These units are typically marked by prepositions such as *above, below, past* or by result-oriented verbs in the present perfect tense in subordinate clauses starting with *once, after*. Many of these encode a difficulty of the previous segment, as in "<u>Once past</u> the large bergschrund", and "<u>Once you have crossed</u> the tricky crevassed section".

Terrain change (8 cases) comprises units marking a location by referring to a terrain property, usually implying a change, as in "When you arrive at a <u>flatter</u> section" and "When the ridge finally <u>goes vertical</u>".

Miscellaneous (11 cases) comprises the remaining units that did not fall into these categories, such as *after a while* and *from there*.

Figure 2 illustrates the distribution of conceptual decision point categories in our data. Notably, the notion of *intersection* is virtually unknown, and only ever used in the context of an actual trail or a trail-like feature of the environment (here, a gulley). Instead, the descriptions rely heavily on *landscape features*, which are sometimes further specified by references to their *spatial parts* (usually those related to "the end" or "the beginning" of the landform). In other cases, when no specific landscape feature appears to be available to mark a location, the accomplishment of a segment or a change in the terrain serve as reference.

It appears that any change of topography has the potential of a decision point, paralleling intersections in an urban environment. This is clearly visible in the following example, where the traveller is advised to keep going although topography is changing: "The <u>ridge</u> eventually <u>disappears</u> but is <u>trail</u> like [sic] still <u>heading in the same direction</u>".

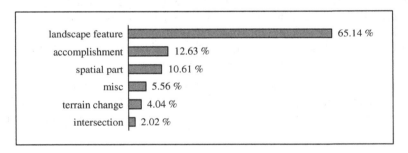

Fig. 2. Types of decision points in percent (n = 198)

4.3   Landmarks and Action Descriptions

Landmarks. We classified all non-quantitative references to locations and geographic objects as landmarks. Four major (mutually exclusive) types emerged from our data (Fig. 3).

Landscape Feature (334 cases): This category constitutes the vast majority of landmarks. These are not only landforms (e.g., *lake, river, gulley*), but also features associated with mountainous landscape (e.g., *moraine, couloir, chute, saddle*), meronyms of a mountain (e.g., *summit, peak, face*), as well as features of a smaller scale (e.g., *step, gendarme, crack*) and non-permanent features (e.g., *bergschrund, crevasse, cornice, snow bridge*). More anthropogenic landmarks include certain areas (e.g., *ski area*), roads and trails, camps and bivouacs, as well as small-scale non-geographic activity-related objects with a fixed location, as in "You will pass <u>one rappel station</u> mid way up this ramp"). Further, this category contains a high number of toponyms, as in "This would take you over <u>Mississippi Head</u> into <u>Zig Zag Canyon</u> (cliffs)".

Spatial Part of a Landscape Feature (50 cases) follows the same definition as in the section on decision points. Some of the most frequent concepts include *side, base, bottom, top* (as in, "the north <u>side</u> of the ridge", "<u>base</u> of the east ridge", "<u>bottom</u> of ramp", "<u>top</u> of the tower"). Also, identifiers such as *upper, lower* are often used (as in, "<u>upper</u> Easton Glacier", "the <u>lower</u> left part of the face").

Terrain (23 cases) comprises references to locations through terrain properties, which are sometimes accompanied by nouns such as *terrain, ground, section* (as in, "scramble up some nasty loose <u>terrain</u>"). Common terrain properties are those related to surface, such as snow, rock and ice (as in, "climb steep exposed <u>snow</u>"), gradient (as in, "where it is almost <u>flat</u>") and difficulty (as in, "ascend <u>easy</u> ground").

Constellation (5 cases) contains units referring to a group of landscape features seen as a whole and is marked by the use of collective nouns (as in, "<u>a series</u> of steep steps") or the plural form of the nouns (as in, "rock <u>islands</u>").

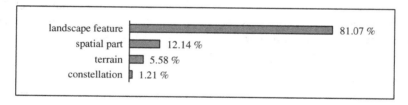

Fig. 3. Types of landmarks in percent (n = 412)

References to landmarks often contain further descriptive information, related to their properties: type of surface, steepness, dimension (size, depth, width), shape, and colour. Also, visual saliency can be addressed, signalled by lexemes such as *distinct, obvious, prominent, main* (as in, "a <u>distinct</u> red-colored sand peak"), as well as the order in which similar features are encountered in space, made possible by the linear progression along the route and signalled by ordinal numbers as well as lexemes such as *next, final, initial* (as in, "as you approach the <u>second</u> rock pillar").

Such information highlights what constitutes "landmarkness" [39] in our context. Within all references to landmarks, we counted the mention of various properties. For the *Terrain* category, this meant annotating additional information about the terrain – for instance, in "climb steep exposed snow", "snow" was annotated as a landmark of the *Terrain* category, whereas "steep" and "exposed" were annotated as further properties (gradient and exposure). While most landmark references did not contain further features (77.18 %, i.e., 318 cases), 74 references (17.96 %) included one feature, 18 (4.37 %) included two features, and 2 (0.49 %) included three features.

Figure 4 highlights their semantic distribution. References to surface, dimension and gradient were most frequent, followed by linear order, difficulty, saliency and colour.

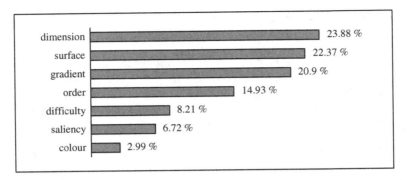

Fig. 4. Types of landmark properties in percent (n = 116)

Actions. In urban contexts, the main actions in route directions pertain to proceeding (e.g., *go, follow*), change of direction (e.g., *turn*), and inspection (e.g., *see*). To address the scope of actions in the mountaineering context, we identified all action-related verbs in our corpus. This excludes, inter alia, verbs related to the description of the terrain and topological relations (e.g., *eases, flattens, drops, joins*). However, we did include fictive verbs of motion [43] that were used with landscape features (as in, "A boulder field leads (N) to the upper Arben Glacier.") since in our context they imply a mountaineer's actions. Altogether, we identified 384 action verbs (tokens). These were further categorized according to the following (exhaustive and mutually exclusive) scheme (see Fig. 5).

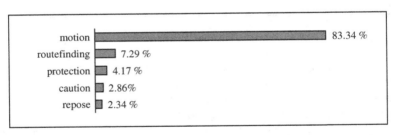

Fig. 5. Action classes in percent (n = 384)

Motion (320 tokens) contains verbs directly related to various types of motion (e.g., *climb, ascend, go, head to).*

Routefinding (28 tokens) includes verbs of vision (e.g., *look, see*), verbs related to locating objects (e.g., *find, locate, notice, recognize, ignore*), verbs and verb phrases related to finding (or missing) the right path (e.g., *miss, make a mistake, check options*) as well as efforts at remembering places (e.g., *make a mental note*).

Protection (16 tokens) includes verbs and expressions referring to safe locomotion, such as *rope up, set belay, strap up (crampons), use (piton, crampons), protect.*

Caution (11 tokens) comprises verbs and expressions such as *beware, exercise caution, be careful, make sure.*

Repose (9 tokens) includes verbs referring to rest, such as *camp, enjoy, rest, pitch (a tent).*

Clearly, Motion verbs are most prominent in our data, as could be expected. On further examination of this category we identified 224 Path verbs and 96 Manner verbs. The Path verbs demonstrate a rich semantic variety reflecting various conceptualizations of the geometrical properties of the Path and the Ground, as well as spatial relations between the Figure and Ground. In relation to the Path, there are verbs reflecting the directionality on the vertical plane (e.g., *ascend, descend*) as well as change of direction on the horizontal plane (e.g., *turn, head, veer*). Also, there are verbs specifying contour [43], also referred to as the global shape of the Path [49] (e.g., *contour, swing, curve, wrap*). In relation to the Ground, a rich variety of spatial relations is encoded: approaching the end point (e.g., *approach, get closer*), reaching the end point or leaving the starting point (e.g., *leave, reach, attain, obtain, arrive, get to, come to*). Further, there are verbs encoding Goal and Source as "containers" (e.g., *enter, exit*), which can be also vertical (as in, "top <u>out</u> in the Chute"). The geometric properties of the Ground element in the motion events are also reflected in verbs that encode one or more dimensions, such as: 1D (e.g., *follow*), 2D (e.g., *cross, traverse*), 3D (e.g., *ascend, descend, drop*). Finally, there is a class of Path verbs and verb phrases related to avoiding the Ground (e.g., *detour, avoid*) or navigating around multiple Grounds (obstacles) on the way to the end point (e.g., *make your way to, mantle your way to, navigate*).

Manner verbs are characterized by a wide spectrum of semantics as well, ranging from relatively general verbs (e.g., *move, go, walk, hike, climb*) to more specific mountaineering jargon (e.g., *downclimb, scramble, glissade, belay, simulclimb, rappel, pitch*).

5   Discussion

We set out to explore how language was used to describe mountains as space and mountaineering as an activity through the prism of route directions. In what follows we discuss our results with respect to our research questions and outline the general insights that we derive.

To explore our first research question concerning the scope of alpine route directions, we categorized content and linguistic indicators in our corpus of mountaineering route directions. We found information going far beyond basic spatial information directly related to wayfinding with, for example, comments on terrain properties and difficulty of the route reflecting the central importance of the *locomotion* aspect in navigation [1]. References to obstacles highlight how the structure of alpine space is characterized by difficult or dangerous places that may necessitate careful avoidance. While following the general route may not be a problem, and indeed it may often be visible given long lines of sight in mountaineering contexts, wayfinding is important at a much more local, small-scale level. Frequent comments on the experiences of other parties on the route may be an indication of the changeability and unpredictability of certain properties of space, such as the type of surface to be found at a particular point of time at a specific location. The data contained also a large amount of further "miscellaneous" information seen as relevant for mountaineers.

Our second research question concerned the ways in which routes were segmented and indicated by decision points in an environment that does not afford obvious intersections such as those structuring urban street networks. Our results suggested that topographical changes can serve these functions in a strikingly similar way. Decision points are places where one landscape feature is conceptualized as adjoining another, or, on a smaller scale, where some terrain property change is perceivable. Thus, a generally continuous landscape is conceptually structured into discrete landscape features (or landforms). This type of categorization process has already been noted in previous literature – for instance, in regard to differences in the way cultures delimit and label landscape aspects [31]. In a mountaineering context, this categorization reflects the necessity of structuring the landscape into manageable and referable segments. Segmentation according to more local terrain changes may be particularly relevant for small-scale locomotion.

Linguistically, segments vary in terms of Path (FROM/VIA/TO) elements encoded. Only a few segments contain all three elements of the Path. 30 % of segments do not encode the VIA element, and 85 % of segments miss either the FROM or the TO element. In an urban setting Tversky and Lee [48] reported 45 % and 75 % for the same phenomena respectively. While the VIA element encoding the Path between the nodes may be more relevant in an alpine context, the distribution is still remarkably similar. This result calls for future experiments using controlled variation, ideally as a within-subjects design with different spatial structures as the independent variable. This would allow for more profound examination and comparison of the conceptual saliency of different elements of a motion event according to context requirements.

A further finding concerns the prevalence of the TO element (present in about half of the segments) over the FROM element (less than a third). A general bias towards referring to the goal of a motion event has been found in the previous linguistic studies [27]. In alpine contexts, this may, in some settings, be supported by more open vistas than in an urban context, where visibility of the end point of a segment as well as the path towards it is potentially more common.

Furthermore, the amount of information provided in route descriptions does not necessarily indicate their effectiveness [2, 12, 13]. In [48], the authors suggest two common rules of inference, namely *forward progression* and *continuity*, i.e., if the starting point is omitted, it coincides with the end point of the previous segment, and vice versa. Our data seemed remarkably similar in this respect, with a potential further rule of inference specific to the mountaineering context: *upward* progression, as in "Proceed to the top of Liberty Cap!" The goal of climbing to the summit is clearly common ground for mountaineers, allowing for inferences in this regard [9].

Our third research question pertained to landmarks and "landmarkness" [39] in our natural context. In line with earlier findings [6], landmarks were overwhelmingly represented by landscape features. The three other types of landmarks found in our data – spatial part of a landscape feature, terrain property and constellation of objects – represent different levels in the hierarchical structure of mental spatial representation. General references to landscape features pertain to navigation on a higher level, leading to the necessity of changing spatial strategies at certain points where the landscape changes – and this sometimes requires more precise information about the landscape features. References relating to terrain properties directly pertain to the lower level of locomotion, which is known to require more detailed small-scale information [47]. A major implication of our findings is the dependence of the granularity of location description on the level of navigation at a specific point of time. Effectiveness of verbal route guidance of mountaineering thus appears to be rooted in flexible switching between granularities. In general, our analysis has reflected some of the findings and current issues discussed in the research on landmarks as summarized and outlined in [39]: the graded membership of the landmark category with better prototypical members (e.g., "distinct red-colored peak") and more uncertain cases (e.g., "when on the snow"), the close interconnection between the properties contributing to saliency (e.g., "when it flattens" has both visual and structural distinctiveness), the role of configurational qualities for landmark perception (e.g., the role of proximity and similarity in the Constellation category), the general high dependency of landmarks on the context (e.g., the case of "rappel station" and "fixed ropes" in our corpus).

Finally, we addressed the ways in which actions were represented in a mountaineering context, and identified an impressive range of variety. In urban contexts, motion-related instructions are typically represented by Path verbs, with some geometric conceptualizations reflected by verbs such as *follow* and *cross*. In contrast, in our data Manner verbs such as *hike* and *simulclimb* frequently attest to the relevance of locomotion in the activity. Furthermore, the wide variety of Path verbs such as *follow*, *traverse*, and *ascend* highlights various geometrical primitives, in line with previous findings by [6]. Verbs such as *exit, top out* further reflect the close interaction with complex space structure by the diverse ways in which landscape features are conceptualized.

A number of insights can be gained from our analysis concerning the diversity of mental representation of space. First, the role of *change* in the natural environment as a structure-imposing factor for segmentation and landmark identification purposes appears to be crucial, and clearly needs further investigation. Second, small-scale spatial relationships appear to be central in mountaineering, reflecting more direct interaction with space and thereby a different mental representation of the environment. This is seen in the frequent description of spatial and topological relations as well as geometric properties of geographic objects on a considerably finer level of detail than usually seen in route directions. Third, from a linguistic point of view, the variety of linguistic structures encoding elements of route directions (in particular, decision points and segments) is intriguing as it surpasses any previous accounts of linguistic features in route descriptions seen in the literature so far. Insights in this area may serve as a contribution to research on automatic itinerary reconstruction from route directions and texts, as well as route generation in navigation systems.

6   Conclusion and Future Work

Alpine route directions, as investigated in our, admittedly small, corpus, are semantically very rich, and thereby provide a strong contrast to urban route directions. This pertains not only to the wide spectrum of information that route providers find relevant, both spatial and non-spatial, but also to the highly diverse and creative ways in which segments, landmarks, and actions are conceived and represented in language. Applying cognitive discourse analysis [44] to such data unveils the spatial conceptualizations that underlie the systematic linguistic choices made by speakers.

In natural environments like mountains, routes can be segmented on the basis of changing topography and are generally conceptualized as a sequence of landscape features. Landmarks range from landscape features to references to changes in terrain, reflecting the role of scale in the activity where locomotion is an important component. Action descriptions are rich in the manner of motion as well as in spatial semantics, which further supports the role of close interaction with space.

Future research is needed to address a range of aspects seen in this paper. These include, for example, the geometric conceptualizations of spatial features as reflecting the way complex spatial environment is abstracted and represented, patterns of granularity switches, the role of the element of uncertainty in both the communication situation (anonymity of the receiver) and the space structure (changeability of space). Given the increasing interest in contextual aspects of wayfinding, the investigation of route directions in an alpine environment contributes to our knowledge of how space properties and activities influence the mental as well as the linguistic representation of space.

Acknowledgements. Ekaterina Egorova is thankful for the support of the University Research Priority Programme "Language and Space" of the University of Zurich.

We would like to thank all four referees of this paper for their constructive and useful pointers which helped improve the final version of this work.

References

1. Allen, G.L.: Spatial abilities, cognitive maps, and wayfinding: Bases for individual differences in spatial cognition and behavior. In: Golledge, R. (ed.) Wayfinding Behavior: Cognitive Maps and Other Spatial Processes, pp. 46–80. Johns Hopkins University Press, Baltimore (1999)
2. Allen, G.L.: Principles and practices for communicating route knowledge. Appl. Cogn. Psychol. **14**(4), 333–359 (2000)
3. Bainbridge, S.: Romantic writers and mountaineering. Romanticism **18**(1), 1–15 (2012)
4. Beavers, J., Levin, B., Tham, S.W.: The typology of motion expressions revisited. J. Linguist. **46**(02), 331–377 (2010)
5. Bromhead, H.: Ethnogeographical categories in English and Pitjantjatjara/Yankunytjatjara. Lang. Sci. **33**(1), 58–75 (2011)
6. Brosset, D., Claramunt, C., Saux, E.: Wayfinding in natural and urban environments: a comparative study. Cartographica: Int. J. Geogr. Inf. Geovisualization **43**(1), 21–30 (2008)
7. Bubenhofer, N., Scheurer, P.: Warum man in die Berge geht. Das kommunikative Muster 'Begründen' in alpinistischen Texten. In: Hauser, S., Kleinberger U., Kersten S.R. (eds.) Musterwandel – Sortewandeln. Aktuelle Tendenzen der diachronen Text(sorten)linguistik, pp. 245–275. Peter Land, Bern (2014)

8. Bubenhofer, N., Schröter, J.: Die Alpen. Sprachgebrauchsgeschichte–Korpuslinguistik–Kulturanalyze. In: Historische Sprachwissenschaft. Erkenntnisinteressen, Grundlagenprobleme, Desiderate. Studia Linguistica Germanica, vol. 110, pp. 63–287 (2012)

9. Clark, H.H.: Using Language. Cambridge University Press, Cambridge (1996)

10. Comber, A.J., Wadsworth, R.A., Fisher, P.F.: Using semantics to clarify the conceptual confusion between land cover and land use: the example of 'forest'. J. Land Use Sci. 3(2–3), 185–198 (2008)

11. Crampton, J.: A cognitive analysis of wayfinding expertise. Cartographica: Int. J. Geog. Inf. Geovisualization 29(3), 46–65 (1992)

12. Denis, M.: The description of routes: a cognitive approach to the production of spatial discourse. Cahiers de psychologie cognitive 16(4), 409–458 (1997)

13. Denis, M., Pazzaglia, F., Cornoldi, C., Bertolo, L.: Spatial discourse and navigation: An analysis of route directions in the city of Venice. Appl. Cogn. Psychol. 13(2), 145–174 (1999)

14. Derungs, C., Purves, R.S.: From text to landscape: locating, identifying and mapping the use of landscape features in a Swiss Alpine corpus. Int. J. Geogr. Inf. Sci. 28(6), 1272–1293 (2014)

15. Egenhofer, M.J., Mark, D.M.: Naive geography. In: Frank, A.U., Kuhn, W. (eds.) COSIT 1995. LNCS, vol. 988, pp. 1–15. Springer, Berlin (1995)

16. Hirtle, S.C., Timpf, S., Tenbrink, T.: The effect of activity on relevance and granularity for navigation. In: Egenhofer, M., Giudice, N., Moratz, R., Worboys, M. (eds.) COSIT 2011. LNCS, vol. 6899, pp. 73–89. Springer, Heidelberg (2011)

17. Hirtle, S., Richter, K.F., Srinivas, S., Firth, R.: This is the tricky part: When directions become difficult. J. Spatial Inf. Sci. 1, 53–73 (2015)

18. Holton, G.: Differing conceptualizations of the same landscape: the Athabaskan and eskimo language boundary in Alaska. In: Mark, D.M., Turk, A.G., Burenhult, N., Stea, D. (eds.) Landscape in Language: Transdisciplinary Perspectives, pp. 225–239. John Benjamins Publishing, Amsterdam (2011)

19. Hölscher, C., Tenbrink, T., Wiener, J.M.: Would you follow your own route description? Cognitive strategies in urban route planning. Cognition 121(2), 228–247 (2011)

20. Jackendoff, R.S.: Semantics and Cognition. MIT Press, Cambridge (1983)

21. Jett, S.C.: Landscape Embedded in Language: the Navajo of Canyon de Chelly, Arizona, and their named places. In: Mark, D.M., Turk, A.G., Burenhult, N., Stea, D. (eds.) Landscape in Language: Transdisciplinary Perspectives, pp. 327–343. John Benjamins Publishing, Amsterdam (2011)

22. Kearns, G.: The imperial subject: geography and travel in the work of Mary Kingsley and Halford Mackinder. Trans. Inst. Br. Geogr. 22(4), 450–472 (1997)

23. Kray, C., Fritze, H., Fechner, T., Schwering, A., Li, R., Anacta, V.J.: Transitional spaces: between indoor and outdoor spaces. In: Tenbrink, T., Stell, J., Galton, A., Wood, Z. (eds.) COSIT 2013. LNCS, vol. 8116, pp. 14–32. Springer, Heidelberg (2013)

24. Klippel, A., Winter, S.: Structural salience of landmarks for route directions. In: Cohn, A.G., Mark, D.M. (eds.) COSIT 2005. LNCS, vol. 3693, pp. 347–362. Springer, Heidelberg (2005)

25. Krippendorff, K.: Content Analysis: An Introduction to its Methodology, 2nd edn. Sage, London (2004)

26. Levinson, S.C.: Landscape, seascape and the ontology of places on Rossel Island, Papua New Guinea. Lang. Sci. 30(2–3), 256–290 (2008)

27. Lakusta, L., Landau, B.: Starting at the end: The importance of goals in spatial language. Cognition 96(1), 1–33 (2005)

28. Landau, B., Jackendoff, R.: "What" and "Where" in spatial language and spatial cognition. Behav. Brain Sci. 16(2), 217–265 (1993)

29. Lovelace, K.L., Hegarty, M., Montello, D.R.: Elements of good route directions in familiar and unfamiliar environments. In: Freksa, C., Mark, D.M. (eds.) COSIT 1999. LNCS, vol. 1661, pp. 65–82. Springer, Heidelberg (1999)

30. Mackinder, H.J.: A journey to the summit of Mount Kenya, British East Africa. Geogr. J. **15**(5), 453–476 (1900)

31. Mark, D.M., Turk, A.G., Burenhult, N., Stea, D. (eds.): Landscape in Language: Transdisciplinary Perspectives. John Benjamins, Amsterdam (2011)

32. Montello, D.R.: Scale and multiple psychologies of space. In: Campari, I., Frank, A.U. (eds.) COSIT 1993. LNCS, vol. 716, pp. 312–321. Springer, Heidelberg (1993)

33. Montello, D.R.: Cognitive geography. In: Kitchin, R., Thrift, N. (eds.) International Encyclopedia of Human Geography, vol. 2, pp. 160–166. Elsevier Science, Oxford (2009)

34. Montello, D.R., Sullivan, C.N., Pick, H.L.: Recall memory for topographic maps and natural terrain: Effects of experience and task performance. Cartographica **31**, 18–36 (1994)

35. Papafragou, A., Massey, C., Gleitman, L.: Shake, rattle, 'n' roll: The representation of motion in language and cognition. Cognition **84**(2), 189–219 (2002)

36. Piotrowski, M., Läubli, S., Volk, M.: Towards mapping of alpine route descriptions. In: 6th Workshop on Geographic Information Retrieval. ACM, New York (2010)

37. Raubal, M., Worboys, M.F.: A formal model of the process of wayfinding in built environments. In: Freksa, C., Mark, D.M. (eds.) COSIT 1999. LNCS, vol. 1661, pp. 381–399. Springer, Heidelberg (1999)

38. Richter, K.-F., Klippel, A.: A model for context-specific route directions. In: Freksa, C., Knauff, M., Krieg-Brückner, B., Nebel, B., Barkowsky, T. (eds.) Spatial Cognition IV. LNCS (LNAI), vol. 3343, pp. 58–78. Springer, Heidelberg (2005)

39. Richter, K.F., Winter, S.: Landmarks. GIScience for Intelligent Services. Springer International Publishing, Switzerland (2014)

40. Sarjakoski, T., Kettunen, P., Halkosaari, H.M., Laakso, M., Rönneberg, M., Stigmar, H., Sarjakoski, T.: Landmarks and a hiking ontology to support wayfinding in a national park during different seasons. In: Raubal, M., Mark, D.M., Frank, A.U. (eds.) Cognitive and Linguistic Aspects of Geographic Space, pp. 99–119. Springer, Heidelberg (2013)

41. Smith, B., Mark, D.M.: Do mountains exist? Towards an ontology of landforms. Environ. Plann. B **30**(3), 411–428 (2003)

42. Slobin, D.I.: The many ways to search for a frog: Linguistic typology and the expression of motion events. In: Strömqvist, S., Verhoeven, L. (eds.) Relating Events in Narrative: Typological and Contextual Perspectives, pp. 219–257. Lawrence Erlbaum Associates Inc., Mahwah (2004)

43. Talmy, L.: Toward a Cognitive Semantics. MIT Press, Cambridge (2000)

44. Tenbrink, T.: Cognitive Discourse Analysis: Accessing cognitive representations and processes through language data. Lang. Cogn. **7**(1), 98–137 (2015)

45. Tenbrink, T.: Relevance in spatial navigation and communication. In: Stachniss, C., Schill, K., Uttal, D. (eds.) Spatial Cognition 2012. LNCS, vol. 7463, pp. 358–377. Springer, Heidelberg (2012)

46. Tenbrink, T., Bergmann, E., Konieczny, L.: Wayfinding and description strategies in an unfamiliar complex building. In: Carlson, L., Hölscher, C., Shipley, T.F. (eds.) Proceedings of the 33rd Annual Conference of the Cognitive Science Society, pp. 1262–1267. Cognitive Science Society, Austin (2011)

47. Timpf, S., Frank, A.U.: Using hierarchical spatial data structures for hierarchical spatial reasoning. In: Hirtle, S.C., Frank, A.U. (eds.) COSIT 1997. LNCS, vol. 1329, pp. 69–83. Springer, Berlin Heidelberg (1997)

48. Tversky, B., Lee, P.U.: How space structures language. In: Freksa, C., Habel, C., Wender, K.F. (eds.) Spatial Cognition 1998. LNCS (LNAI), vol. 1404, pp. 157–175. Springer, Heidelberg (1998)
49. Van der Zee, E., Nikanne, U., Sassenberg, U.: Grain levels in English path curvature descriptions and accompanying iconic gestures. J. Spat. Inf. Sci. **1**, 95–113 (2015)
50. Whymper, E.: Scrambles Amongst the Alps in the Years 1860-69. J. Murray, London (1872)
51. Williams, M., Kuhn, W., Painho, M.: The influence of landscape variation on landform categorization. J. Spat. Inf. **5**, 51–73 (2012)

Spatial Natural Language Generation for Location Description in Photo Captions

Mark M. Hall[1], Christopher B. Jones[2(✉)], and Philip Smart[2]

[1] University of Edgehill, Ormskirk, UK
[2] Cardiff University, Cardiff, UK
jonescb2@cardiff.ac.uk

Abstract. We present a spatial natural language generation system to create captions that describe the geographical context of geo-referenced photos. An analysis of existing photo captions was used to design templates representing typical caption language patterns, while the results of human subject experiments were used to create field-based spatial models of the applicability of some commonly used spatial prepositions. The language templates are instantiated with geo-data retrieved from the vicinity of the photo locations. A human subject evaluation was used to validate and to improve the spatial language generation procedure, examples of the results of which are presented in the paper.

Keywords: Vague spatial language · Natural language processing · Human-subject experiments · Spatial prepositions · Field-based spatial models · Locative expressions

1 Introduction

Spatial locational expressions (or locative expressions) are used in many aspects of written and spoken communication to describe where things or people are located. Typically the expressions involve the use of spatial relational terms, such as *in* or *at*, to associate the located object to another reference or landmark object. Sometimes an expression may involve the composition of several such spatial relationships, such as "I am on the street in front of the house". In the geographical domain there are many different contexts in which the same spatial relation may be used to refer to phenomena of different types and different scales (buildings, rivers, cities etc.). The overloading and contextual dependence of spatial language is well known (e.g. [11,29]) but presents challenges in developing automated methods to interpret spatial language and to generate spatial language. In this paper we consider the task of generating captions for photos taken with location-aware devices and in that context confine ourselves to urban environments and relatively localised (i.e. small regions of) rural environments. In the locational expressions the located object may be either the photo itself, or the imputed subject of the photo, while the reference objects are confined to named buildings (of various types), streets, settlements and containing regions.

© Springer International Publishing Switzerland 2015
S.I. Fabrikant et al. (Eds.): COSIT 2015, LNCS 9368, pp. 196–223, 2015.
DOI: 10.1007/978-3-319-23374-1_10

Keeping track of where and when digital photos were taken can be a challenge and has led to interest in methods for automated tagging and captioning (e.g. [15,16]). It is possible to exploit the time and date stamps that most digital cameras provide and, with GPS-enabled devices, it is also possible to access the camera's geographical coordinates. In order to translate the coordinates to a more human-readable form, automated reverse geo-coding services can be used to generate a place name. If other people have taken a photo at a similar location and uploaded it to a photo sharing site such as Flickr then it is possible to suggest tags from the other photos, as proposed by [15]. For previously well photographed scenes a more automatic approach has been presented in which image matching is used to find similar photos and then to inherit the captions of those images [26]. The latter procedure will only work if many people have already taken a photo at the same location. The work presented here differs from these other approaches in automatically generating a complete natural language caption to describe the geographical context of a photo using locational expressions that are based on analyses of existing captions and on task-specific human-subject experiments. The following is an example of a caption generated fully automatically with our system:

Table 1. Examples of manually generated captions on the Geograph website

Woodland north of Bouverie Avenue, Harnham, Salisbury
The George and Dragon, Castle Street, Salisbury,
Bridge across to Idustrial area at Littlehampton
Cliff Road near Newquay railway station.
Railway bridge over Stratford-on-Avon Canal
Farmland east of Fryern Court Wood
Towards Pendle Hill from York Road, Lanho
Riverside promenade at Brecon
The Monnow above Skenfrith
Postbox on the corner of Linden Road and Gloucester Place
Farmland between Whitsbury & Rockbourne

Rijksmuseum photographed at 2.15 pm at the corner of Stadhouderskade and Museumstraat near Spiegelgracht in Amsterdam, Netherlands.

The objective is to emulate typical locational expressions found in image captions in which some care has been taken to describe the geographical context. Evidence of the structure of such captions was obtained from an analysis of the titles of photos uploaded to the Geograph web site. This site is dedicated to providing "geographically representative photographs and information for every square kilometre of Great Britain and Ireland" (www.geograph.org.uk). It is in some contrast to sites such as Flickr in which most photos have only very short captions if any. Some example Geograph captions are listed in Table 1. Our focus here is on the language structure and the selection of spatial prepositions. The appropriate selection of toponyms including reference to their salience [27] is equally important in producing a useful caption and the system presented here uses methods for selecting and ranking toponyms that are described in [30], but these issues are not considered further in this paper as the emphasis is upon caption language structure.

To understand typical language structures of photo captions and characteristic usage of different prepositions in the context of photo captioning we searched for recurring patterns in language structure, and counted the frequency of occurrence of different spatial prepositions. The pattern analysis led to the creation of a set of language templates of varying levels of complexity. For templates that include a spatial preposition a method is required to determine the most applicable preposition given the configuration of the photo location or subject and the location of candidate toponyms that may serve as referents in a prepositional phrase. Thus decisions need to be made regarding the relative applicability of for example "near <toponym>", "north of <toponym>", "next to <toponym>", "on < street toponym>" and "at the corner of <street toponym A> and <street toponym B>". Knowledge of the applicability of different prepositions was acquired through a set of human subject experiments conducted in a lab and online, in which participants were asked to rate the suitability of a set of prepositions (based on the prior caption analysis) to particular configurations of the located object and a reference location (<toponym>) to which it is related by the preposition. These experiments were similar to those of for example Worboys [32] and Robinson [20,21]. They differ though in that the subjects were told the context of the task was photo captioning, the scale of map data was adapted to the typical scale found in the caption analysis experiments and the subjects were asked to provide ratings of applicability of given prepositions using values on a Likert scale from 1 to 9. The results of these experiments were used to build, for each preposition, a spatial density field model that fitted a smooth surface to the discrete sample values. The density field models in combination with prior evidence of the frequency of use of particular prepositions were used to select prepositions to instantiate the language templates.

The main contribution of the work presented here is the design and implementation of a fully automatic natural language photo caption generation procedure that uses a selected set of spatial relations to create a simple description of the geographic context of the photo location. It is based on an analysis of existing caption language patterns to create language templates; analysis of the frequency of use of spatial prepositions in caption language; selection and salience determination of relevant toponyms; modelling of the applicability of a set of spatial prepositions relative to the reference location for the specific context of photo captioning; and instantiation of the language templates with selected spatial prepositions and toponyms.

In the following section we review related work with regard to photo captioning and acquisition and modelling of knowledge of the use of spatial prepositions in geographic contexts. This is followed in Sect. 3 by a description of the caption language analysis while Sect. 4 explains the approach used to create the density field models of prepositions based on human-subject experiments. Section 5 provides an overview of the functionality of the caption generation system that employs the results of the caption analysis and the studies of applicability of spatial prepositions. The section includes a description of the process of selecting and filtering toponyms and an explanation of the selection and instantiation of caption language templates. An initial evaluation of the results of the approach is described in Sect. 6 along with a discussion of how the results of this evaluation informed various modifications to the final system to take account of insights obtained in the evaluation. The paper concludes in Sect. 7.

2   Related Work

2.1   Photo Captioning

A system with related objectives to our own was the PhotoCompas captioning system [16] which categorises groups of photos according to units of space and time, and can link a photo collection to a neighbouring place using an expression such as "35 km S of Los Angeles, CA". While this functionality appears analogous to that provided in our system, it is not clear what prepositions were implemented as only the example preposition of "S" is provided and there is no explanation or discussion of how the spatial prepositional phrase was generated or chosen. Notably their system operates at a relatively coarse scale with the reference locations being cities that may be tens of kilometres distant. This is in contrast to the captions we analysed, in which the reference place (ground location) was usually within 5 km of the photo location and much smaller in dimension than the cities that were used in PhotoCompas.

Another system for organising photo collections was presented in [28] which exploits GPS data to generate times and locations (from a gazetteer) which are related to the photo in terms of distance and cardinal direction. A single structure for annotation is employed and the issues of application of vague spatial language are not addressed. As indicated above several systems, e.g. [15,26] have been described that attempt to adopt the tags or captions of Flickr photos taken at the same location, but these systems are not concerned with automatic spatial language generation.

2.2   Spatial Language

There have been numerous studies concerned with understanding concepts of spatial language and spatial prepositions (e.g. [1,11]), in particular with regard to the context of use and to distinguishing between frames of reference that may be *relative* to an observer or an object, or based on the properties of an object (*intrinsic*), or *absolute*, such as compass directions, (e.g., [9,12,29,31]). Locative, or locational, expressions are commonly composed of various forms of figure and ground entities that spatially relate a located objected to a reference object ([29]

describes different forms of figure and ground as well as distinguishing between static and dynamic contexts). While photo captions can contain a wide variety of forms of spatial language we are concerned here primarily with locational expressions that are independent of an observer, i.e. non-deictic, the intention being to generate descriptions from knowledge of the locations of spatial objects and regions in the vicinity of the photo obtained from geo-spatial data sources (if the camera direction is known it would be possible to generate deictic spatial relations but that is not considered here). The focus here is upon external relations (in the sense of for example [12,31]) where the photo location or photo subject is the figure (equivalently locatum/located object/trajector), while retrieved places serve as the ground location (relatum/reference location/landmark). We consider ground locations that may be point-based, path-based (roads) or regional. The spatial relations between figure and ground in our system are either independent of a coordinate system, as in linguistic topological relations such as *in, next to, at the corner of,* and *between* (following the terminology of Levinson [12]), or based on an absolute coordinate system that supports the cardinal directions of *north, south, east* and *west.* A further aspect of locational descriptions that is relevant to caption generation is that of geographic hierarchies in which a place description will often encompass multiple levels of granularity [19]. This, in combination with evidence from existing photo captions, motivated the inclusion in the present system of support for geographic containment hierarchies when the relevant toponyms are available.

2.3 Modelling the Applicability of Spatial Prepositions

A notable example of creating density field models (or potential fields) of the applicability of spatial prepositions ("regions of acceptability") with evidence from human subject experiments is provided by [13]. Although that work did not have a geographical context, our methods are clearly related in that we also asked subjects to rate the applicability of spatial prepositions at locations relative to a reference location. Of particular relevance to our task of generating locational expressions in specific geographic (as opposed to "table top") contexts are a number of empirical, human-subject studies of the use of vague spatial language concepts that have been concerned with the possibility of fitting models to the experimental data. For example, Robinson conducted studies to acquire fuzzy membership functions to represent the concept of nearness [20,21] with regard to the relationship between settlements that were mostly tens of kms apart. Using a system that learnt the fuzzy membership function, the subjects were asked to specify the truth or falsehood of nearness for specific instances of pairs of settlements, one of which was the ground location. The latter of these studies emphasised the significant differences between the five participants. Fisher and Orf [5] in a study of the terms *near* and *close* in the context of a university campus found such large variations that they were not able to create a consistent formal model of nearness. Our experiments differed from these latter experiments in using a Likert-like scale to record degrees of applicability of various prepositions and we found that with increasing quantities of data more stable patterns of applicability could be obtained. Gahegan [6] asked subjects to rate the closeness of points to a reference point on a diagram with no absolute scale. In varying the objects in the diagram the study revealed that perceived distance was affected by the presence of neighbouring objects. In our work we do not attempt to consider such "distractor" effects [10], as they are beyond the scope of this study, though it is quite possible that they may have affected the decisions of participants in our human subject studies.

The study in [32] demonstrated the potential value of some alternative approaches to modelling the results of nearness in which human subjects were asked to judge whether it was true that particular landmarks on a university campus were near (or, in other questionnaires, not near) to a specified reference location. The data were modelled in terms of three-valued logic (corresponding to a broad boundary spatial model), fuzzy distance membership functions and four valued logic. The experiment and analysis was extended in [33] to consider leftness and to adopt Dempster-Shafer belief functions. Using about 22 subjects these studies again revealed individual differences between subjects but also found striking regularities. Our human-subject experiments were analogous to these studies with regard to the type of question asked, though in our experiments judgements were made on a numerical scale from 1 to 9 for multiple spatial relationships, as indicated above. Also, importantly the context of our task was specifically designated as that of photo captioning, unlike any of the previously mentioned studies.

2.4   Spatial Natural Language Processing with Density Field Models

An early example of the application of density field models to generate natural language was presented in Schirra [22] to model the applicability of spatial prepositions to the locations of objects and soccer players in the SOCCER system. It generated a natural language description of a soccer game (based on automatically generated data describing the locations of players and of the ball). Superimposing all fields for a given location, multiple prepositions could be invoked if sufficiently applicable. The form of the field models was based on pre-specified functions that adapt to the shape of referent objects (apparently without any reference to psychometric studies). Our approach is analogous to that of Schirra, in using the models to assess the applicability of different prepositions for the purpose of generating natural language, but we generate the field models from human-subject experiments that were specifically designed for the task of photo captioning.

The use of human subject experiments to build density field models, to represent a variety of spatial preposition terms such as *between*, *near*, *far*, and *to the right of*, was described by Mukerjee et al. [14]. Where significant patterns were detected in a particular direction, linear regression methods were used to model the trend of the observations to create ellipsoidal shaped fields. The models were employed to translate from natural language to scene descriptions, and as in Schirra they were superimposed where multiple prepositions applied to the vicinity of a reference object. We have also fitted models to smooth and interpolate our experimental data (with splines and kriging) and we have adopted a similar approach to deciding which is the most applicable of several candidate prepositions (represented by field models).

2.5   Mining Text for Evidence of the Use of Spatial Language

An example of how natural language texts can be exploited to model the applicability of a prepositional phrase for a specific geographic context was provided by Schockaert et al. [23] who analysed text in hotel web sites to produce a fuzzy model of the phrase "within walking distance". Actual distances for instances of the phrase were calculated based on knowledge of a hotel's location in combination with geocoding the location of the named place to which the phrase was applied. An analogous approach was adopted in [7] in which photo captions were mined from the Geograph web site to create spatial field models representing the use of the preposition *near* and the four cardinal directions. Further data on these prepositions was acquired in human subject experiments [8] that employed maps in which participants were asked to rate the applicability of individual prepositions to describe the relationship between multiple named places and a referenced location in a rural setting. Kernel density modelling and kriging [17] interpolation methods, respectively, were used to build the field models for these two studies. The results of [8] are employed in the present paper for captioning in a rural environment.

The study in [8], which was concerned with automated interpretation of the spatial footprints implied by locational expressions in photo captions, revealed three common patterns of caption language. These were a noun phrase giving just a place name, a noun phrase in combination with a prepositional phrase, such as "Windsor Castle near Eton" and comma separated noun phrases that correspond to a geographical hierarchy such as "Buckingham Palace, London, England". The present study is complementary to that work and builds upon these three basic patterns to generate caption sentences containing locational expressions.

2.6   Natural Language Generation Systems

There has been a considerable body of work on generating natural language and aspects of the overall design of our system build on some well established methods [18] based on which we adopt a data-driven approach to content creation, while discourse planning and linguistic realisation are based on evidence from analysis of existing captions and from human-subject studies. There are many examples of spatial language generation in various domains, notably robotics [10,24]. In the geographic domain most such systems have focused on navigational instructions, e.g. [3,4]. While such systems have some generic aspects in common with ours, the approach that we adopt differs with regard to the specific context (which is important given the strongly context-dependent nature of spatial language) and the methods adopted to discover language patterns and to create density models of the applicability of spatial language. An example of language generation that might be regarded as closer in domain to ours is that of a system to create personalised place descriptions for tourists [2], but that work was not concerned with modelling the use of spatial prepositions or the characteristic structure of location descriptions.

3   Analysis of Photo Captions

To gain insight into typical usage of spatial language in photo captions, a set of about 350,000 geocoded photo captions from the Geograph project was analysed. The emphasis of captions in Geograph is upon the geographical context and they are rich in spatial language, in addition to which the photos are usually quite accurately geo-coded. The Geograph dataset was analysed with regard to the frequency of use of different spatial prepositions, the situations in which some prepositions were actually used, with respect to distance and orientation of the photo location relative to a reference location, and the language patterns that were employed in the captions. As with many on-line data sources based on public participation, there is a bias in the data as roughly 90 % of the image captions were produced by about 2 % of the contributors. To avoid this bias affecting the analyses, only one caption per participant was used in caption language structure analysis. Subsequent equivalent analysis of the full caption set (i.e. with multiple captions from the same contributors) found the same patterns identified with the same relative frequencies, indicating no significant effect of participant bias, which then justified using the full set of images for some of the other quantitative analyses.

3.1   Preposition Frequency Analysis in Geograph

The analysis of the frequency of use of spatial prepositions found the top ten in descending order to be *at*, *near*, *to*, *on*, *from*, *in*, *north of*, *west of*, *east of* and *south of*. Their numerical frequencies are listed in Table 2 (which includes captions from the same contributors). The numbers were based on analysis of all words in the captions, which were then filtered manually. Of these top ranked prepositions all are used in our system with the exception of *to* and *from* as these tend to be associated either with information about routes that requires additional information sources or with a view-direction specific description, neither of which we attempt to support in this work. We do however support some other spatial relations such as *next to*, *at the corner of* and *between*, where in the latter cases we use retrieved street names to instantiate the prepositional phrases. The knowledge of preposition frequency is also used in this work to allocate "popularity" weights to prepositions to assist in making decisions about the most appropriate prepositions to employ (in combination with the use of density fields).

Table 2. Most frequent spatial prepositions in Geograph

Preposition	Frequency
At	21754
Near	18589
To	15476
On	13698
From	12886
In	10754
North	5336
West	5230
East	4763
South	4756

3.2 Figure Ground Relationships for Selected Prepositions in Geograph

The preposition *near* and the four cardinal directions were the subject of analysis, in the Geograph dataset, of the distance between photo location and reference location (in the caption) for *near*, and for both distance and angle for cardinal directions relative to a reference location. The measurements of distance and orientation were made following part of speech analysis to detect patterns of the form <subject> <preposition> <toponym> in combination with geocoding the <toponym>, i.e. reference location, and the location of the photo. Details of this form of analysis for the rural use of the cardinal directions were reported in [7], where it was observed that the distances between the photo location and the reference toponym were mostly less than 3 km though ranging up to about 5 km. The same characteristic distance range has been found in the subsequent analysis of the use of *near* from the same data.

This type of analysis could not be performed automatically with the preposition *at* due to the difficulty of geocoding and disambiguating what were often quite obscure geographic features that were not found in the gazetteers that we employed. A quantitative analysis of *at* was however performed in the human subject experiments described below. The latter experiment was also used to investigate the spatial context of the usage of the path-related prepositions of *on*, *to* and *from*. In the study of Geograph photos it was found that *on* is commonly used in association with a reference toponym that may be visible in the photo, while *to* and *from* often refer to locations that could be quite distant from the visible content of the photo (as for example in the Geograph caption "Gloucester to Swindon Railway, near Minety Cross"). *To* and *from* are therefore harder to employ automatically when attempting to provide the geographic context of a location and are not considered further in the current study.

3.3   Caption Language Pattern Analysis

As a foundation for creating language templates for generating photo captions, the language structure of the Geograph captions was analysed using methods similar to those described in [8], where the aim was automated interpretation of photo captions, as opposed to the generation of photo captions that we are concerned with here. A set of 580 captions, all from different contributors, was derived from the initial collection. To detect language patterns, an iterative

Table 3. Result of initial collocation analysis of part of speech tags applied to Geograph captions. NNP - proper noun; NN - noun, IN - preposition; DET - determiner ('a' or 'the'); CC - conjunction ('and'); , - comma (',').

Tag 1	Tag 2	Frequency
NNP	NNP	632
IN	NNP	149
,	NNP	110
NNP	IN	109
NNP	,	109
NNP	NN	72
DET	NNP	68
NN	IN	62
IN	DET	53
NNP	CC	30

Table 4. Examples of phrase construction generalisation rules, prior to subsequent collocation analysis of the phrase tags. The first seven rules are the result of the first round of generalisation. The subsequent examples illustrate some of the generalisations generated at later rounds. NPhr - noun phrase; IPhr - prepositional phrase; CommaPhr - comma phrase.

Tag 1	Tag 2	Tag 3	Generalisation
NNP	NNP		NPhr
NNP	NN		NPhr
NN	NNP		NPhr
NN	NN		NPhr
DET	NPhr		NPhr
IN	NPhr		IPhr
NPhr	,	NPhr	CommaPhr
NPhr	IPhr		FigureGroundPhr
NPhr	,	CommaPhr	ContainPhr
NPhr	CC	NPhr	ConjunctivePhr

process of collocation (bi-gram) analysis was performed on part of speech (POS) tags and subsequently on phrase tags that were substituted for the initially detected tag collocations. The phrase tags were attached to high frequency POS tag collocations that were identified as English grammatical phrase units. Thus considering the initial POS tag collocation analysis (see Table 3), combinations such as NNP NNP, i.e. two proper nouns, corresponding to two-word place names such as "Chipping Campden", were selected as a noun phrase, designated NPhr, and IN NPhr, i.e. a preposition and a noun phrase, was selected as a prepositional phrase (IPhr) corresponding for example to "near Bussage" or "in Chipping Campden" (see Table 4). Note that NNP IN was not selected as a useful phrase for our purposes as it does not represent a typically meaningful phrase in its own right (it could correspond for example to part of a noun phrase that is a relatively unusual proper name, or the first two words of "Bussage in Goucestershire", or be part of a form of a path description such as "Blogton to Brighton" that we are not seeking to generate in this work).

Three iterations of collocation analysis and generalisation phrase creation were performed working from right to left of the sentences (to maintain the structure of noun phrases that may be qualified by a preceding preposition, determiner or adjective). Table 4 illustrates rules generated at the first round (above the separating line) and some of the rules generated at subsequent rounds. The most frequent resulting collocation patterns resulting from the final generalisation process revealed three major caption patterns that account for 70 % of all captions. These are captions that consist of (1) only a noun phrase (NPhr), for example just the toponym "Merthyr Tydfil"; (2) a noun phrase in combination with a prepositional phrase (FigureGroundPhr), for example "Pontsticill Reservoir near Merthyr Tydfil" and (3) captions consisting of a list of comma-separated noun phrases corresponding to a hierarchical toponym such as "Roath Park, Cardiff, Wales" (ContainPhrase). These patterns (Fig. 1) provide the basis of a set of building blocks for the caption generation process described in Sect. 6. It should be noted that while the most common pattern found in Geograph was just the noun phrase representing a single place name, and can therefore be regarded as a typical style of caption, it cannot be regarded as necessarily the most desirable, as it could reflect a minimum effort attitude on the part of the author of the caption. As became apparent in the user evaluation, more complex and hence more informative captions may be preferable to the user. This motivated the creation of captions that combine several templates to create a richer description.

4   Field Based Modelling of Spatial Prepositions from Human Subject Experiments

In the caption generation system spatial prepositions play a key role, in combination with toponyms, in creating natural-sounding locational expressions. Two human-subject surveys were conducted to acquire knowledge of the applicability of selected spatial prepositions with regard to distances and orientations between

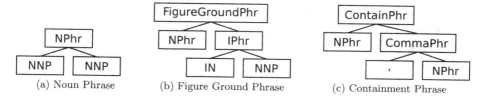

Fig. 1. The three common caption language patterns detected from analysis of Geograph captions. (a) The most common, simple, Noun Phrase structure for a caption, such as 'Merthyr Tydfil'. (b) The Figure Ground Phrase consists of two noun phrases linked by a spatial preposition as in 'Pontsticill reservoir near Merthyr Tydfil'. (c) The Containment Phrase consists of a set of noun phrases separated by commas as in 'Roath Park, Cardiff, Wales'

figure and ground locations, corresponding respectively to photo locations and reference toponyms. The surveys were based separately on rural and urban contexts. The rural experiment, described previously in [8] for purposes of caption language interpretation, involved 24 undergraduates and university staff. Each participant was shown a map with named point-referenced places and was asked to answer questions of the form of the primer phrase "This photo was taken in < toponym> which is <spatial preposition> Cowbridge", where Cowbridge was centrally located on the map. The overall geographic extent of the map was about 25 km which was informed by a prior analysis of the distances between figure and ground locations used in Geograph captioning where it was found that the vast majority of figure and ground locations were within 5 km of each other. The context, of taking a photo, was explained but no photos were shown to the participants, as the intention was to describe the geographic content of the camera location, not the content of a photo. For each question the spatial preposition was fixed, while an alphabetically ordered list of toponyms from the map was provided. For each toponym the user was asked to rate on a scale from 1 ("not at all") to 9 ("perfectly") how well it fitted the phrase. The above primer phrase was used to evaluate the prepositions *near* and *north of, south of, east of* and *west of*. The result of each answer was a set of figure location points each of which was rated with its applicability for use of the respective preposition relative to the single ground location. For each figure location (corresponding to a named place on the map) the median value of the confidence values was computed, resulting in a set of points with respective confidence values distributed in the vicinity of the single ground location. For each of the prepositions *near* and the four cardinal directions a field model was created using Kriging to perform spatial interpolation between the points. To increase the stability of the resulting fields, given the limited number of figure locations (17), the cardinal direction data points were mirrored across their respective directional axis (east-west and north-south) while the *near* observations were mirrored across both horizontal and vertical axes. Examples of the fields for *near* and *north* are illustrated in

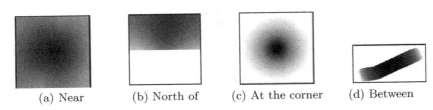

(a) Near   (b) North of   (c) At the corner   (d) Between

Fig. 2. Field models for (a) near (b) north of, (c) at the corner, (d) between. The ground location is at the centre of all fields, except (d) for which it is a line along the centre path of the strip (corresponding to a road)

Figs. 2a and b respectively, with the ground location located at the centre of the square (which has truncated the boundaries of the fields in the diagrams).

The urban experiment (not previously reported) was conducted using a web-based questionnaire that was sent to the same population invited to join the rural experiment. After filtering out participants whose first language was not English, a total of 1042 participants (688 female and 354 male) provided responses.

The usage of the six spatial prepositions *near, north of, next to, at, at the corner* and *between* was investigated. The setup for all core questions was the same. On the left side of the screen a square map of a part of the city of Cardiff was displayed. To avoid the participants treating the questionnaire as a map-reasoning task, only a satellite image was displayed. The primer phrase presented to the participants was of the form "Photo taken <spatial expression>", with the spatial expression constructed using one or more toponyms and one of the spatial prepositions, for example "Photo taken near the Wales Millenium Centre". The toponyms in the primer phrase and the photo location were highlighted on the satellite image. The participants were given a nine-point rating scale to rate how applicable the spatial expression was to the spatial configuration shown in the map. The rating elements 1 and 9 were annotated as "does not fit at all" and "fits perfectly" respectively. The core questions were presented in random order, to minimise memory effects. To provide a rough idea of the distances used in the questions, for each spatial preposition the participants would see the closest and most distant photo points first, before seeing the intermediate points. However, to ensure that they would not treat the experiment as a simple geometric reasoning problem the participants were not informed of the order in which the points would be displayed.

Relative to the rural experiments, the urban questionnaire was subject to considerable constraints on the placement of hypothetical photo locations, due to the presence of buildings which were not regarded as available for placing the locations. This resulted in greater sparseness of measurement points (with the exception of the *near* experiment). In the case of the *near* experiment the data values of some individual points differed considerably from the overall pattern for reasons that are unclear but are expected to relate to the higher density of obstacles and of roads visible in the satellite image, as compared with the relatively simple map employed in the rural experiment. For the cardinal

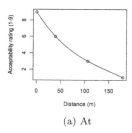
(a) At

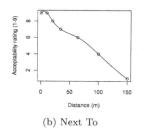
(b) Next To

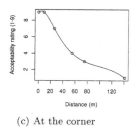
(c) At the corner

Fig. 3. Natural splines fitted through human subject experimental data values obtained for applicability (y-axis; 1 [low] - 9 [high]) of the prepositions 'at', 'next to' and 'at the corner' as a function of distance (x-axis) in the urban setting

direction *north of* a similar pattern to the rural data was observed, but this was the only cardinal direction that was surveyed in the experiment (in order to control the subject effort of the urban experiment). For these reasons it was decided to employ scaled versions of the rural density fields for these prepositions in their urban context. For the urban *near* data, the closest point to have the lowest median value was about 500 m from the ground location, which is one tenth the equivalent distance, i.e. about 5 km, that was found with the rural model (though the majority of points for the latter model were within 3 km). The scaling factor was therefore chosen as a division by 10 for both *near* and cardinal directions.

In the case of *at*, *next to* and *at the corner* there were too few data points to be subject to Kriging interpolation. Instead a spline function was fitted to the data points. These functions are illustrated for *at*, *next to* and *at the corner* in Fig. 3. The function was assumed to be equally applicable in all (radial) directions from a central ground point, producing in the case of the *at the corner* preposition a field of the form illustrated in Fig. 2c, in which the ground location is in the centre of the diagram.

For the path like preposition of *between*, the questionnaire used a primer phrase of the form "on street A between street B and street C", with the candidate locations being positioned at various places along the path of street A. The resulting data values were mirrored lengthwise across the centre point of the street to increase the stability of the fitted spline function. To instantiate the vague field for a particular situation the street was assumed to have a width of 20 m and the spline function was scaled to start and end at the junctions with the streets B and C (wherever they were in practice). The spline function was then used to determine field values for locations on the street according to their distance from the start location, resulting in fields such as that illustrated in Fig. 2d.

It is important to stress that the rural and urban human subject experiments were conducted with a view to creating approximate models of the applicability of a working set of prepositions for the specific context of photo captioning. It is well known (as mentioned earlier) that the use of spatial prepositions is highly context dependent and there is no pretence here of having created models that can be regarded as accurate for all types of rural and urban environment with their respective variety of feature types and scales. The experiments were simply part of a pragmatic approach to demonstrating a proof of concept for automated generation of potentially useful locational expressions for photo captions.

5   The Caption Language Generation System

The previous two sections have described the processes of analysing existing photo caption language structure and conducting human subject experiments on the use of various spatial prepositions. The caption analysis resulted in the creation of three forms of caption template, reflecting the three most commonly occurring language patterns, describing subject, relative and containment toponyms. These templates become instantiated with relevant toponyms and spatial prepositions. The human subject studies resulted in the creation of density field models that can be anchored to a toponym location and used to decide the most applicable spatial preposition for the respective photo location. In this section we provide a brief overview of the locational expression generation system that builds on the results of these prior analyses to create the caption. The system applies the process illustrated in Fig. 4, which consists of the following four main components:

- The *Meta-Data Extraction* component extracts the location and optional direction information from the image's meta-data.
- The *Meta-Gazetteer* uses the location and direction information to retrieve candidate toponyms that will be used to instantiate the subject, relative and containment language templates. Toponyms are retrieved from a number of sources, ranked based on their salience for captioning, and then filtered as described in Sect. 5.1. The resulting set of toponyms is then passed to the *Captioner*.
- The *Captioner* generates a set of natural language captions using the image's location information, the candidate toponyms, density field models and language templates (Sect. 5.2). For the candidate relative toponyms, decisions on instantiating the relative language templates are based on measuring the level of applicability (at the camera location) of all potentially relevant preposition density fields when they are anchored at the toponym location. Templates are merged and, following a linguistic realisation phase, multiple versions of each caption are generated, relating to alternative possible prepositions and toponyms. Each caption is ranked based on a combination of the toponym salience, the applicability of preposition density fields and the popularity of the prepositions.
- The *Meta-Data Embedder* embeds the highest-ranked caption in the image's meta-data.

5.1   Selection and Filtering of Toponyms with a Meta-Gazetteer

The caption language generation system employs a set of caption templates that can be instantiated with spatial prepositions (as explained in the next section) and with toponyms retrieved from the vicinity of the camera location. The toponyms are classified as belonging to one of either subject (S), relative (R) or containment (C) data models. The subject toponyms are ones that occur in a sector in front of the camera, and are only generated if the image's meta-data has orientation direction information (Dir choice in Fig. 4), the relative toponyms occur anywhere in the circular buffer surrounding the camera location as provided by GPS coordinates, while the containment toponyms provide the geographical regional hierarchy of the photo location. The subject and relative toponyms are allocated salience values that can be used in selection and filtering of names using reverse geocoding and salience measurement methods (Fig. 4, *Rank Toponyms*). The toponym reverse geo-coding methods employ a meta-gazetteer that accesses multiple data sources, described in [25,30], while the toponym ranking process uses methods described and evaluated in [30]. In this paper we are concerned primarily with caption language structure rather than the issue of the appropriate selection of toponyms and so in the human subject evaluation experiments described here the toponyms were selected manually for

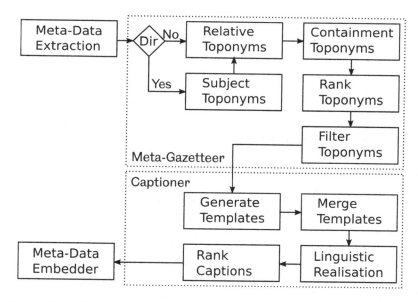

Fig. 4. Overview of components of the caption generation system. The *Meta-Gazetteer* retrieves candidate toponyms that are passed to the *Captioner*, which combines the toponyms with the density field models in order to instantiate the language templates and hence generate the locational expressions. The system follows the same process for each image, except for the *Subject Toponyms* which are only generated if the image's meta-data contains orientation information (Dir choice-point).

input both to the language generation procedures and to the human-subject map annotators and evaluators. There were no subject toponyms and all salience values were equal. In the full implemented system all toponyms and their salience values are obtained automatically.

As the data source for subject, relative and containment toponyms could be the same, the possibility arises of the same toponyms occurring in the respective data models. This could result in captions of the form "Cardiff photographed in Cardiff" and thus the toponyms are filtered (Fig. 4, *Filter Toponyms*) before being passed to the *Captioner*. First the presence of containment toponyms within the relative and subject toponyms data models is checked. If any are found, they are removed from the S and R data models. Similarly, it is also necessary to remove subject toponyms from the relative toponym models, to avoid captions such as "Wales Millenium Centre near the Wales Millenium Centre".

A further form of filtering is performed to reduce what might be regarded as redundant information, which can occur when a relative toponym provides locational content that is of less semantic salience than the subject toponym. A hypothetical example would be "The Eiffel Tower near the Wagamamma Restaurant", in which the subject is clearly the better known and more unique landmark. This is performed using the semantic salience measures that were generated as part of the captioning system, but are not considered further in this paper as explained above.

5.2   Generating Caption Language Templates

Following creation of the data models containing candidate toponyms to be used in instantiation of the individual caption templates, a discourse modelling phase takes places in which language templates are selected as the basis of what may be multiple candidate caption structures that combine different templates containing different candidate toponyms and candidate spatial prepositions (Fig. 4, *Generate Templates*). Which templates are selected depends upon the availability of toponyms and their relative salience. The discourse template models are based on the three major patterns identified in the caption structure analysis (Sect. 3). Thus the single noun phrase and containment (hierarchy) phrase patterns form the Containment template, which may consist of one or more toponyms. The figure-ground phrase pattern results in the Relative template which combines one or more toponyms with a spatial preposition. The noun phrase by itself leads to the Subject template, consisting of one or more toponyms that may be combined with an element representing a conjunction phrase. To reflect the common usage of terms such as *on*, *at the corner of* and *between*, referring to path objects as identified in the data mining experiments (and validated in the evaluation experiment described in Sect. 6) several Road templates are employed (see below). In addition to these, an optional Time template was created with a view to adding additional information within captions, though it was not based on the initial caption structure analysis. The use of the time template is illustrated subsequently in some examples but as it is not a necessary component of the localisation expressions it is not discussed further in this paper.

The templates are combined into the top level discourse model illustrated in Fig. 5. According to availability of toponyms and their salience values, the model is populated from left to right.

Fig. 5. Top level discourse model. All elements are optional, except the containment element.

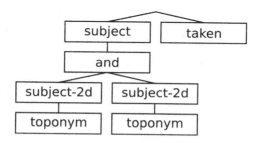

Fig. 6. Template structure for Subject elements.

If subject toponyms are present they are used to instantiate the subject template (Figs. 6, 7). The 'and' element of the template is used if there is more than one subject toponym. The template includes a 'taken' element that serves as padding to provide more well rounded captions. If a subject is present it can be followed by the word "photographed" (as in the example "Solomon's Temple photographed in the afternoon in Buxton, United Kingdom" - see Table 8 for other examples of captions generated by the automated system), while in the absence of subject toponyms the 'taken' element can still be used and is realised linguistically with the words "Photo taken" (as in the example "Photo taken near Chatsworth House in the Peak District National Park, United Kingdom.").

The road templates implement several phrases that refer to road or street objects. The horizontal support element (Fig. 8a), realised by "on <streetname>", can be invoked if the photo location lies within a road as determined by the use of a crisp field model that takes account of road width. All intersections between the road on which the photo is located and other roads are identified and for each intersection that is found an "at the corner" field is instantiated. If the photo location has a field value greater than 0.4 (the cut-off values were determined empirically) then an additional intersection element is created and associated with the two streets, resulting in a phrase of the form "at the corner of <streetname> and <streetname>" and wrapped in a proximal close element denoting a short distance from the intersection point (Fig. 8b). Finally the procedure generates a

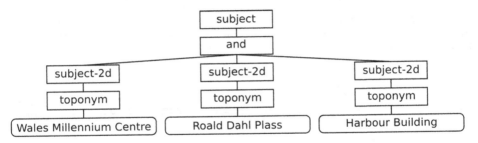

Fig. 7. Template structure for Subject elements with example instantiation. The element would be realised with the phrase "Wales Millennium Centre, Roald Dahl Plass and the Harbour Building".

"between" template for roads that have at least two intersections and for which the between field has a value exceeding 0.4 at the location of the photo. The full "between" template incorporates a horizontal support element (see Fig. 9). It is realised by a phrase of the form "on <streetname> between <streetname> and <streetname>"

The relative templates are constructed by iterating over the list of relative toponyms and for each relative toponym instantiating the vague fields for all supported spatial prepositions. A procedure to distinguish between urban and rural situations (using land cover digital map data) is applied to determine the candidate spatial prepositions and hence which field definitions to use. In the rural context, *near* and the cardinal directions are available, while in the urban context *near*, *at*, *next to* and the cardinal directions are instantiated. For each instantiated field (anchored at the candidate toponym location) the value at

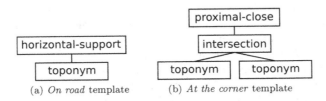

(a) *On road* template   (b) *At the corner* template

Fig. 8. Road-based templates. (a) Basic road-based template realised with a phrase of the form "*on* <streetname>". (b) *At the corner* template corresponding to the intersection of two named roads.

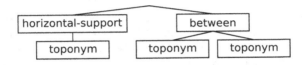

Fig. 9. The *between* road-based template references the names of two roads that intersect the current road.

the photo location is measured and if it is greater than 0.4 a template for the spatial preposition is generated with the toponym as its parameter. In the rural case only fields derived from Kriging are used whereas in the urban context a distinction is made between the Kriging-based field, for the preposition *near* and the cardinal directions, and spline-based fields for the spatial prepositions *at* and *next to*. Each resulting template is realized by phrases such as "near Little Boddington", "north of Blimpsfield" and "next to St Pauls Cathedral".

An additional filtering step is invoked if at least one road template has been generated, which causes all roads to be ignored when instantiating these vague fields. This avoids generating expressions such as "on Princess street near Princess street".

A single containment template is generated representing the containment hierarchy specified as a list from the most specific to the highest-level containment toponym in the data model. This produces a nested structure as in Fig. 10a, in which the containment elements are always interpreted as "the region defined by the left-hand child is contained in the region defined by the right-hand child". The most deeply nested toponym is contained in a special element "world" that is not instantiated to a toponym. This is to ensure that in the structure a containment element always has two child elements, thus simplifying the interpretation, while at the same time providing an explicit end-of-hierarchy marker. The linguistic realisation of the containment template consists in preceding the leftmost toponym (treated as the "root" toponym) with *in* while all subsequent toponyms are concatenated with commas, reflecting the comma phrase pattern that was found in pattern analysis. An example instantiation of the template with three toponyms is illustrated in Fig. 10b realised by the containment phrase "in Roath Park, Cardiff, United Kingdom".

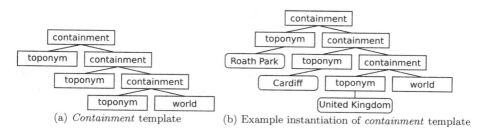

(a) *Containment* template   (b) Example instantiation of *containment* template

Fig. 10. Containment templates. (a) The structure of the containment template. (b) Example instantiation of the containment template, which would be realised with the phrase "in Roath Park, Cardiff, United Kingdom"

The final captions are created by merging the templates (Fig. 4, *Merge Templates*) and then generating the linguistic realisations of the merged templates (Fig. 4, *Linguistic Realisation*). Each caption is given a score based on the sum of individual scores for each subject and relative toponym (Fig. 4, *Rank Captions*). In the case of subject toponyms the scores are based only on the salience value attached to the toponym, while for relative toponyms their scores are calculated by multiplying the toponym salience by a factor representing the preposition density field value at the camera location and by a weight associated with each preposition, which is based on its popularity as measured by the data from Geograph on preposition frequency. By default the highest scored caption is selected, but in the automated system the user can be given the option of selecting from other lower ranked captions.

6   Results and Evaluation

To evaluate the quality of the natural language expression language caption generation system, the automated captions were compared by human evaluators with human-generated captions, where the evaluators were not told which one of the captions was computer-generated. The human-generated captions were created for a set of eight locations, four of which were urban, while the other four were rural. For each of the eight locations, three human annotators were given a labelled map and asked to create a caption describing the location of the photo, using the named places on the map and employing a list of spatial prepositions, corresponding to those available to the automated system (note that the choice of prepositions was itself based on the prior study of spatial natural language employed in photo captions, as described in Sect. 3). In addition to the list of spatial prepositions, the annotators were provided with a list of all toponyms that were on the map. The computer generated captions were created using the same set of toponyms that was available to the human annotators, i.e. all toponyms on their maps.

The same maps used by the caption creators (annotators) were shown to 85 evaluators, with the locations of a notional photo marked on the maps. For each of the human and computer-generated captions (see Tables 5 and 6), which were mixed randomly (i.e. *not* as in these figures), the evaluators were asked to rate on a scale of 1 to 9 how well the caption fitted the given location, where 1 indicated a caption that did not fit at all and 9 represented a perfect caption. The intention was to measure the effectiveness of the caption language in describing the location. For each caption the scores of all evaluators were analysed to calculate the median value and inter-quartile ranges.

It is important to stress that the human generated captions against which the computer generated captions were compared should be regarded as an upper bound. Thus coming close to the manually crafted captions should be regarded as excellent performance.

The first notable outcome of the experiment was that the agreement between evaluators regarding the quality of the human-generated captions was not very high, For almost all human-created captions the inter-quartile range was at least 2 or higher, indicating a large amount of variance in the data.

For the rural urban captions the median values of ratings for the automated captions were 4, 6, 4 and 5 respectively with values ranging between 5 and 8 for the human-generated captions. For the four urban captions the median values

Table 5. Human-generated and computer generated captions for the urban evaluation experiment. Evaluation areas were: CF 1 & CF 2 – Cardiff; EDI 1 & EDI 2 – Edinburgh. Ann. - Human Annotator generated the caption. Algo. - Algorithm generated the caption.

Area	Source	Caption
CF 1	Ann. 1	On Castle Street, between Cardiff Castle and Cathedral Rd.
	Ann. 2	On Castle St West of Cardiff Castle (or East side of Taff River on Castle St.)
	Ann. 3	On Castle Street near the South West corner of Bute Park
	Algo.	On Castle St near Cardiff Castle in Cardiff
CF 2	Ann. 1	East of Mermaid Quay, South-West of Nat. Assemb. of Wales
	Ann. 2	Between NAW & Mermaid Way
	Ann. 3	In Cardiff Bay between the National Assembly + Mermaid Quay
	Algo.	Near the National Assembly of Wales in Cardiff
EDI 1	Ann. 1	Near Scott Monument, West of Scott Monument
	Ann. 2	Next to Scott Monument
	Ann. 3	On Princess Street between Waverly Bridge and the Royal Scottish Academy
	Algo.	On Princess Street next to the Royal Scottish Academy in Edinburgh
EDI 2	Ann. 1	South of Greyfriars Kirk, on Lauriston Pl
	Ann. 2	On Lauriston Pl near George IV Bridge
	Ann. 3	On Lauriston Pl. West of Univ. of Edinburgh
	Algo.	On Lauriston Pl near the University of Edinburgh in Edinburgh

for the computer-generated captions were all 5, while they ranged from 6 to 8 for the human-generated captions. The inter quartile ranges for the computer generated captions tended to be high, six of them having a value of 3 and two with a value of 2, demonstrating low agreement between evaluators of the quality of these computer-generated captions.

To provide a qualitative representation of the evaluation, the automated system's ratings were classified into three categories for each caption and each evaluator (see Table 7). The categories are "as good", if the rating for the automatically generated caption is as high or higher than the rating of at least one of the three manually generated captions; "almost as good" if the rating was at most one level lower than the lowest rating of the three manually generated captions and "not as good" if the rating was more than one level below the rating of the lowest rated human generated caption. This was measured for each manual assessment of each photo location tested. The results are illustrated in Fig. 12. In summary, we see that the four urban captions were rated "as good" by 35 %, 35 %, 29 %, and 27 % of evaluators respectively, with between 45 % and 54 % of evaluators rating them as either "as good" or "almost as good". For the rural

Table 6. Human-generated and computer generated captions for the rural evaluation experiment. Evaluation areas were: BB 1 & BB 2 – Brecon Beacons; PD 1 & PD 2 - Peak District. Ann. - Human Annotator generated the caption. Algo. - Algorithm generated the caption.

Area	Source	Caption
BB 1	Ann. 1	North of Capel Y Ffin
	Ann. 2	Between Velindra and Urishay
	Ann. 3	North of Capel Y Ffin, half-way between Velindra and Urishay
	Algo.	Near Craswall in the Brecon Beacons National Park
BB 2	Ann. 1	North of Coelbren, Near Coelbren
	Ann. 2	Near Coelbren
	Ann. 3	North of Coelbren
	Algo.	Near Coelbren in the Brecon Beacons National Park
PD 1	Ann. 1	Between Wildboardclough and Brandside, North of Quarnford
	Ann. 2	North of Quarnford
	Ann. 3	North of Quarnford half-way between Wildborough and Brandside
	Algo.	Near Dove Head in the Peak District
PD 2	Ann. 1	North of Barbrook Res., West of Totley, Near Owl Bar
	Ann. 2	North of Owl Bar
	Ann. 3	NW of the Owl Bar
	Algo.	Near Owl Bar in the Peak District

captions there was much more variability in the ratings of the four test cases. In one case the computer generated caption was regarded by 71 % of the evaluators as "as good" and by 22 % as "almost as good" (thus 93 % as either "as good" or "almost as good") and for one of the other captions 61 % of evaluators regarded the caption as "as good" and 14 % as "almost as good". The other two computer generated captions were rated by 75 % and 81 % respectively as "not as good" and we discuss the reasons for this shortly.

A further measure of the quality of the automated captions can be found by comparing the spatial prepositional phrases (such as "on Castle Street") and just the selected toponyms of the automated and manually generated captions respectively. With regard to spatial prepositional phrases that combine a preposition and a toponym, in five out of eight of the test locations the automated system generated a prepositional phrase that was the same as that of at least one of the human annotators, examples being "on Princes Street", "on Castle Street" and "near Coelbren" giving a 62.5 % success rate for that measure. In six out of eight of the test locations the automated system selected a toponym that was the same as that of at least one human annotator giving a success rate of 75 %. In three test cases two toponyms were matched, as for example in the Edinburgh 2 location where both "Princes Street" and "University of Edinburgh" were selected.

Table 7. Percentages of evaluators' answers in the categories "as good as human" (AG), "almost as good as human" (AAG), "not as good as human" (NAG) for the urban and rural evaluation experiments.

Urban evaluation				Rural evaluation			
Area	AG	AAG	NAG	Area	AG	AAG	NAG
CF 1	.35	.19	.46	BB 1	.19	.06	.75
CF 2	.29	.24	.47	BB 2	.71	.22	.07
EDI 1	.35	.16	.49	PD 1	.12	.07	.81
EDI 2	.27	.18	.55	PD 2	.61	.14	.24

Table 8. Examples of captions generated by the fully automated system

Pierhead Building and Norwegian Church photographed in the morning near Wales Millennium Centre in Cardiff, United Kingdom.

Photo taken on Queen Street near the Thistle Parc Hotel in Cardiff City Centre, Cardiff, United Kingdom

Solomon's Temple photographed in the afternoon in Buxton, United Kingdom

Photo taken near Chatsworth House in the Peak District National Park, United Kingdom

Ladybower Reservoir photographed in the early afternoon near Snake Pass in the Peak District National Park, United Kingdom

Rijksmuseum photographed at 2.15pm at the corner of Stadhouderskade and Museumstraat near Spiegelgracht in Amsterdam, Netherlands

Photo taken at the corner of Karolinenstraße and Geyerswörthplatz near Schlenkerla in Bamberg, Germany

Close inspection of the cases where the automated system performed poorly in the evaluation revealed some general limitations that were subsequently rectified. One of these was that the automated system was preferring *near* over *north of* due to a distance decay factor value, based on an analysis of Geograph captions, that was zeroing the cardinal direction fields at a distance notably shorter than the total extent of the near field, so that at the greater distances *near* was always applied rather than a cardinal direction. The Geograph effect for cardinal directions was not in fact seen in human subject experiments and it was decided therefore to omit the scaling factor in the modified version of the system.

Another issue was that in general *near* was being preferred to cardinal directions because the "popularity" weights for each of the cardinal directions reflected their frequency of use in Geograph as described in Sect. 3.1. These weights are independent of the applicability weighting at a given location, which is calculated with a density field. Thus each cardinal direction was used with about a quarter of the frequency of *near* and this had resulted in any cardinal direction being used by the program with only a quarter of the frequency of *near*. This was not really appropriate, as when all four directions are considered, a cardinal direction, i.e. any one of the four directions, should be used with similar frequency to *near* (with each one only occurring about 1 in 4). The popularity weighting for use of cardinal direction was therefore modified to be similar to that of *near*. A similar effect resulted in "between" being used less frequently in the automated system, as it is only applicable in quite specific less common situations, for which the initial low weight was acting as a deterrent to its use in those situations. Its popularity weight was therefore also increased.

The evaluation also indicated that the evaluators preferred more detailed captions, when choosing between all captions including the manually-generated captions. This provides scope for further modifications, whereby multiple cardinal direction phrases could be implemented provided each one exceeded a given threshold of applicability. This could then be expected to emulate a human generated caption such as "north of Capel Y Ffin and east of Velindra". Note however that in the automated system that uses a gazetteer, when many toponyms are available for a given location the captioning system can produce relatively detailed captions as illustrated for example in Table 8.

7 Summary and Conclusions

This paper has presented a set of methods to generate natural language photo captions that employ locational expressions to describe the geographic context of the photo. The captions are based on knowledge of the camera location in combination with access to geo-data resources that relate to the location. There is no reference to the image content. Evidence for the typical structure of caption language was obtained from analysis of systematically authored photo captions on the Geograph web site, resulting in a set of caption language templates corresponding to some of the most common caption patterns that were found. One of the main patterns was a prepositional phrase linking a preposition to one or more toponyms. In order to understand how to select appropriate spatial prepositions that could be applied to particular configurations of distance or orientation between the camera location and named reference locations (toponyms), some human-subject experiments were conducted at rural and urban scales. The results were used to build density field-based spatial models of the applicability of the respective prepositions in the vicinity of the reference toponym. Thus on anchoring the density field to a candidate toponym the applicability of the respective preposition (e.g. west of) could be judged by the value of the field at the camera location.

The resulting locational expression generation system presents a significant step forward as it demonstrates that a purely data-driven approach can successfully be used to model and operationalize spatial natural language in order to automatically generate realistic natural language locational expressions. Compared to existing approaches, the data-driven approach makes it easy to implement a wider range of spatial prepositions, making it possible to create a diversity of natural language locational expressions. This in turn enables the generation of an appropriate expression for a given spatial configuration. The data-driven nature also means that the system can easily be extended to use more spatial prepositions or adapted to other use contexts, by simply providing more quantitative models.

The focus of this paper has been on caption language rather than selection of appropriate toponyms and so an evaluation was performed in which the toponyms were selected manually and provided both to the captioning system and to some human annotators who were asked to create captions using a map containing the selected names. For each notional photo location the automatically generated caption was inserted in a set of human-generated captions and a group of evaluators was asked to judge how well each caption fitted the photo location. Overall just over half the auto-generated captions were judged either "as good" or "almost as good" as the manually generated captions. The best two (of eight) computer generated captions were rated by 93 % and 75 % of evaluators as either "as good" or "almost as good" as the manually generated captions. This evaluation revealed some problems in the automated system with regard to an inappropriate constraint on the extent of applicability of directional preposition fields and on the expected frequency of use of cardinal directions relative to use of *near*. These problems were corrected in the automated system and the captions re-generated, showing very clear improvement, though no further human-subject evaluation was conducted. Given adequate toponyms for a given location the system can always create well formed captions with good English locational expressions that can include multiple toponyms and multiple prepositions, as illustrated by the examples.

Future work includes conducting further human subject studies on the modified system in a wider range of contexts and using a wider range of spatial prepositions. This will include the use of the automatically generated toponyms that are retrieved with the multi-source "meta-gazetteer". The system presented here has been designed to support realisation of captions in different languages and future work may be conducted on multi-lingual caption generation. An important aspect of the quality of a caption relates to the appropriateness of the selected toponyms, particularly when they might be local landmarks. This issue was not addressed in the paper but has been the subject of related research and further studies will be conducted to determine automatically the salience of particular toponyms, which may be a function of the interests of the owner or user of the photo.

Acknowledgements. This work was supported by the EC TRIPOD project (FP6 045335).

References

1. Bateman, J.A., Hois, J., Ross, R.J., Tenbrink, T.: A linguistic ontology of space for natural language processing. Artif. Intell. **174**(14), 1027–1071 (2010)
2. Carolis, B.D., Cozzolongo, G., Pizzutilo, S., Silvestri, V.: Mymap: generating personalized tourist descriptions. Appl. Intell. **26**(2), 111–124 (2007)
3. Dale, R., Geldof, S., Prost, J.: Using natural language generation in automatic route. J. Res. Pract. Inf. Technol. **36**(3), 23 (2004)
4. Dethlefs, N., Wu, Y., Kazerani, A., Winter, S.: Generation of adaptive route descriptions in urban environments. Spat. Cogn. Comput. **11**(2), 153–177 (2011)
5. Fisher, P.F., Orf, T.M.: An investigation of the meaning of near and close on a university campus. Comput. Environ. Urban Syst. **15**(1–2), 23–35 (1991)
6. Gahegan, M.: Proximity operators for qualitative spatial reasoning. In: Kuhn, W., Frank, A.U. (eds.) COSIT 1995. LNCS, vol. 988, pp. 31–44. Springer, Heidelberg (1995)
7. Hall, M., Jones, C.: Quantifying spatial prepositions: an experimental study. In: Proceedings of the ACM GIS 2008, pp. 451–454 (2008)
8. Hall, M., Smart, P., Jones, C.: Interpreting spatial language in image captions. Cogn. Process. **12**(1), 67–94 (2011)
9. Herskovits, A.: Semantics and pragmatics of locative expressions. Cogn. Sci. Multi. J. **9**(3), 341–378 (1985)
10. Kelleher, J., Costello, F.: Applying computational models of spatial prepositions to visually situated dialog. Comput. Linguist. **35**(2), 271–306 (2009)
11. Landau, B., Jackendoff, R.: "What" and "where" in spatial language and spatial cognition. Behav. Brain Sci. **16**(2), 217–238 (1993)
12. Levinson, S.: Space in Language and Cognition: Explorations in Cognitive Diversity. CUP, Cambridge (2003)
13. Logan, G., Sadler, D.: A computational analysis of the apprehension of spatial relations. In: Bloom, P., Peterson, M., Garrett, M., Nadel, L. (eds.) Language and Space, pp. 493–529. MIT Press, Cambridge (1996)
14. Mukerjee, A., Gupta, K., Nautiyal, S., Singh, M., Mishra, N.: Conceptual description of visual scenes from linguistic models. Image Vis. Comput. **18**(2), 173–187 (2000)
15. Naaman, M., Nair, R.: Zonetag's collaborative tag suggestions: what is this person doing in my phone? IEEE MultiMedia **15**(3), 34–40 (2008)
16. Naaman, M., Song, Y., Paepcke, A., Molina, H.G.: Automatic organization for digital photographs with geographic coordinates. In: JCDL, pp. 53–62 (2004)
17. Oliver, M., Webster, R.: Kriging: a method of interpolation for geographical information systems. Int. J. Geogr. Inf. Syst. **4**(3), 313–332 (1990)
18. Reiter, E., Dale, R.: Building Natural Language Generation Systems. Cambridge University Press, Cambridge (2000)
19. Richter, D., Vasardani, M., Stirling, L., Richter, K.F., Winter, S.: Zooming in - zooming out: hierarchies in place descriptions. In: Krisp, J.M. (ed.) Progress in Location-Based Services, pp. 339–355. Springer, Heidelberg (2013)
20. Robinson, V.: Interactive machine acquisition of a fuzzy spatial relation. Comput. Geosci. **16**, 857–872 (1990)
21. Robinson, V.: Individual and multipersonal fuzzy spatial relations acquired using human-machine interaction. Fuzzy Sets Syst. **113**(1), 133–145 (2000)

22. Schirra, J.: A contribution to reference semantics of spatial prepositions: the visualization problem and its solution in VITRA. In: Zelinsky-Wibbelt, C. (ed.) The Semantics of Prepositions: From Mental Processing to Natural Language Processing, pp. 471–515. Mouton de Gruyter, Berlin (1993)

23. Schockaert, S., de Cock, M., Kerre, E.: Location approximation for local search services using natural language hints. Int. J. Geogr. Inf. Sci. **22**(3), 315–336 (2008)

24. Skubic, M., Perzanowski, D., Blisard, S., Schultz, A., Adams, W., Bugajska, M., Brock, D.: Spatial language for human-robot dialogs. IEEE Trans. Syst. Man Cyber. Part C Appl. Rev. **34**(2), 154–167 (2004)

25. Smart, P.D., Jones, C.B., Twaroch, F.A.: Multi-source toponym data integration and mediation for a meta-gazetteer service. In: Fabrikant, S.I., Reichenbacher, T., van Kreveld, M., Schlieder, C. (eds.) GIScience 2010. LNCS, vol. 6292, pp. 234–248. Springer, Heidelberg (2010)

26. Snavely, N., Seitz, S., Szeliski, R.: Modeling the world from internet photo collections. Int. J. Comput. Vis. **80**(2), 189–210 (2007)

27. Sorrows, M.E., Hirtle, S.C.: The nature of landmarks for real and electronic spaces. In: Freksa, C., Mark, D.M. (eds.) COSIT 1999. LNCS, vol. 1661, pp. 37–50. Springer, Heidelberg (1999)

28. Spinellis, D.: Position-annotated photographs: a geotemporal web. IEEE Pervasive Comput. **2**(2), 72–79 (2003)

29. Talmy, L.: How language structures space. In: Pick Jr., H.L., Acredolo, L.P. (eds.) Spatial Orientation, pp. 225–282. Plenum, New York (1983)

30. Tanasescu, V., Smart, P., Jones, C.: Reverse geocoding for photo captioning with a meta-gazetteer. In: SIGSPATIAL 2014. ACM Press (2014)

31. Tenbrink, T.: Reference frames of space and time in language. J. Pragmatics **43**, 704–722 (2011)

32. Worboys, M.: Nearness relations in environmental space. Int. J. Geogr. Inf. Sci. **15**(7), 633–651 (2001)

33. Worboys, M., Duckham, M., Kulik, L.: Commonsense notions of proximity and direction in environmental space. Spat. Cogn. Comput. **4**(4), 285–312 (2004)

More Than a List: What Outdoor Free Listings of Landscape Categories Reveal About Commonsense Geographic Concepts and Memory Search Strategies

Flurina M. Wartmann[1(✉)], Ekaterina Egorova[1,2], Curdin Derungs[1,2], David M. Mark[3], and Ross S. Purves[1,2]

[1] Department of Geography, University of Zurich, Zürich, Switzerland
`flurina.wartmann@geo.uzh.ch`
[2] University Priority Research Program Language and Space (URPP SpuR), University of Zurich, Zürich, Switzerland
`{ekaterina.egorova,curdin.derungs,ross.purves}@geo.uzh.ch`
[3] Department of Geography, University at Buffalo, Buffalo, NY, USA
`dmark@buffalo.edu`

Abstract. Categorization is central to abstraction from real world geographic phenomena to computational representations, and as such has been the subject of considerable research. We report on one common approach, free listing, in an outdoor setting and explore terms elicited in response to the question '*What is there for you in a landscape?*'. We collected term lists, and explanations for the strategies used from 89 participants in two mountain and one parkland setting. We analyzed results not only using term frequency, but also by cognitive saliency, exploring list structures, and building aggregated networks visualizing links between terms. We observed memory search strategies, such as exploiting and switching semantic clusters in our data, with participants using for example not only the local setting to start clusters, but also memories of familiar landscapes to switch between clusters. Our results reveal that simple free listing experiments can help us understand how categories are linked, and also highlight ways in which landscapes are conceptualized.

Keywords: Geographic categories · Landscape categorizations · Commonsense geography · GIS · Memory search · Free lists

1 Introduction

Fundamental concepts like categories [1] are important when considering abstractions from real world geographic phenomena to computational representations [2–4]. Knowing how humans parcel up geographic space into objects, and identifying shared categories, is an essential step on the way to supporting non-specialist use of geographic information. However, gaining knowledge about categories is non-trivial, and a wide range of approaches have been used. These range from elicitation tasks based on a variety of stimuli in relatively controlled settings (e.g. [5, 6]), through ethnographic methods [7], to crowd sourcing experiments [8] and, analysis of user generated content [9, 10].

© Springer International Publishing Switzerland 2015
S.I. Fabrikant et al. (Eds.): COSIT 2015, LNCS 9368, pp. 224–243, 2015.
DOI: 10.1007/978-3-319-23374-1_11

Recent work demonstrates empirically that categorization of landscape features varies linguistically and culturally [11, 12]. Therefore, the importance, but also the challenge, of understanding categorization in a practical sense becomes increasingly apparent, since, as succinctly summed up by Smith and Mark, ontologies, and thus categories:

> can help us to understand how different groups of people exchange (or fail to exchange) geographical information, both when communicating with each other and also when communicating with computers. ([13], p. 592).

In this paper we explore free listing as one of the simplest, but also most common approaches to exploring categories. We do so by exploring an explicitly geographic domain (landscape categories), with real societal relevance – for example only defined categories can be considered in managing and quantifying landscapes [14, 15]. We go beyond previous work on free listing of geographic categories, by exploring not only the frequencies of the terms stated, but also their order, their cognitive salience, and compare results in different landscape settings. For this study, based on work in ethnoecology, we defined landscape as an arrangement of culturally recognized biotic, abiotic and cultural or anthropogenic landscape elements that are designated by common rather than proper nouns [16]. In analyzing our results we set out to demonstrate how combining empirical fieldwork with theory from cognitive research can help us not only to better understand our data, but also suggest potential recommendations with practical implications.

In the following, we first give an overview of the different disciplinary perspectives of categorization, providing the theoretical background for this paper, followed by the methods used to elicit and analyze free lists of landscape categories. We then show how the landscape in which outdoor experiments are conducted seems to influence the terms listed, their frequency and saliency, as well as the memory search strategies participants apply.

2 Background

Categorization has been a subject of study in various research fields ranging from anthropology, linguistics and psychology [17–22] to geography and information science [23, 24]. For example, seminal work on prototype theory and the existence and primacy of basic-levels established categorization as a major field of study within cognitive psychology [1]. Following Rosch and Lloyd's definition, a category is usually identified by a name (animal, bird) and consists of a number of members that share some common attributes [1].

One way of examining categorization is through the study of category norms. Free listing tasks (also known as semantic fluency tasks) are a common method of elicitation, in which participants are asked to list examples for named categories such as 'food items' or 'colors', the goal being to define the elements of a cognitive domain for a cultural group [18, 25]. Another purpose is to define norms that serve as a basis for more in-depth psychological experiments [26, 27].

2.1 Research on Categories in the Geographic Domain

Early studies on geographic categories were conducted by psychologists as part of their free listing experiments. Of the 56 categories elicited in the classic category norm study by Battig and Montague one concerned the geographic domain [26]. As examples of 'a natural earth formation' participants most frequently listed mountain, hill, valley and river [26]. Tversky and Hemenway investigated basic-level categories of what they called outdoor 'environmental scenes' [5]. Based on the number of attributes, norms and activities listed for the four most frequently mentioned outdoor scenes (park, city, beach and mountains), these scenes were established as basic-levels. However, the scenes are not a taxonomy of the geographic domain, but rather settings where geographic objects may be situated [23]. Lloyd et al. [28] postulated that categories of administrative units in the United States (e.g. state and city) were basic-level geographic categories. However, members of categories investigated were not sub-categories, but rather instances such as 'Georgia' for the category 'state'.

Smith and colleagues argued that the domain of geography is ontologically distinct from other domains, in that geographic objects are characterized by specific properties that might influence category formation, for instance, a minimal scale, bona fide and fiat boundaries, and structural properties inherited from space [13, 29, 30]. Such considerations, in combination with a more general interest in how lay people conceptualize the geographic domain, triggered studies that specifically investigated commonsense geographic categories.

A series of experiments with US-American university students [13, 23] revealed that results were influenced by the choice of wording in the elicitation task. For instance, the phrase 'something that could be portrayed on a map' produced a higher mean number of terms per participant (8.21) and more anthropogenic elements such as road or city, while the phrase 'a kind of geographic feature' yielded a mean of 7.15 terms per participant, consisting predominantly of physical geographic features such as mountain, river, lake, ocean, valley and hill [13]. Replications of these experiments included a study with Portuguese university students [31], as well as a study with Greek students that further tested for the differences in understanding of geographic concepts between experts and non-experts [32]. For non-experts (Greek high-school students and first year college students) the most frequent categories were the Greek terms for mountain, sea, lake, plain, and river [32]. As these experiments in different language settings produced comparable top ten frequency terms, this gave rise to the argument that geographic category norms may be shared cross-culturally [31]. However, the 'non-experts' in the aforementioned studies were students who more or less recently had gone through geography classes at high school. Thus, whether results reflect similarity in geography curricula in these countries or are in fact generalizable to a broader population remains questionable.

A study in Portugal used videos of landscapes (one familiar and one unfamiliar to participants) as stimuli to elicit landform categorizations from people living in two different villages. The results showed how familiarity with landscapes increased the number of terms listed, as familiar landscapes triggered memories of nearby areas not shown in the videos, for which participants then also listed terms [33].

2.2   Free Listing and Memory Search

Most of the aforementioned studies of geographic categories used free listing, reporting the resulting frequencies of terms and/or the term order, but treating items in lists as independent. However, research in other domains has revealed structure of free lists to contain other interesting information, such as how participants perceive the relationships between categories in a domain [34] or how information is recalled from memory [35].

How humans perform memory search has been the subject of intensive investigations. For instance, in the spreading-activation theory of semantic processing, a semantic network consists of concepts seen as nodes that are linked to other concepts sharing the same properties [36]. The more properties, and thus links, two concepts share, the more closely related they are. In memory search, when the first concept is activated, it activates a semantically similar concept in turn. Semantically related terms are thus often produced together, indicating that people apparently come up with terms by searching in 'semantic fields' or clusters and listing whatever items they discover in these clusters [37]. As the links get weaker and the number of available links decreases, the production of terms slows down during free listing tasks [38].

Hills et al. [39] drew an analogy between heuristic animal foraging strategies (find a resource patch, exploit it, switch to the next patch) that follow the marginal value theorem [40] and memory search strategies applied by humans in free listing tasks (find a semantic cluster, exploit it, switch to the next cluster). In human minds, the two distinct processes of exploiting clusters ('clustering') and moving to the next cluster ('switching') are argued to be linked to specific regions in the brain [35]. Clustering is taken to reflect semantic storage searching in the temporal lobe and switching between clusters to represent frontal lobe executive control mechanisms [41]. The empirically demonstrated link between clustering performance in free listings and mental illnesses such as Alzheimer's have led to practical applications of free listings in clinical diagnostics [41, 42]. The underlying assumption in these diagnostic tests is often that the existence of externally defined clusters indicates how "well" participants organized their memory. A cluster of the categories *orange-lemon-tangerine* is considered "organized", while a cluster of *apple–orange–cherry–blackberry–pear–tangerine–banana–raspberry–lemon–apricot* would be considered "disorganized" [43].

However, assessing clusters on close semantic proximity (defined as a high number of common properties) alone fails to account for the fact that one shared property may be a sufficient link between two concepts. For some people, *penguins* and *pandas* form a cluster because they are both black and white [44]. Furthermore, using pre-defined semantic clusters to analyze free lists fails to account for idiosyncratic clusters formed as a result of experience, for instance "all the animals I saw yesterday in the zoo", making *pandas–gorillas–meerkats-polar bears* an organized cluster. Such clusters can only be explored by complementing free listing exercises with qualitative interviews on how participants came up with the terms in the task [44].

2.3   Research Gaps and Research Questions

Participants. Researchers studying geographic categories through free listing have typically recruited participants among university students. To draw conclusions reflective of commonsense geographic concepts, a broader range of people should be included in the sample, which requires adapting the methodology (e.g. elicitation task) from previous studies to reflect the diverse backgrounds of participants.

Internal Structure of Free Lists. How participants come up with geographic categories and what the resulting lists may reveal about how knowledge on geographic phenomena is stored in, and retrieved from, memory has so far remained uninvestigated. Methods for analyzing the internal structure in free lists exist, but these have not yet been applied to free lists of the geographic domain.

Outdoor Elicitation. Participants in free listing experiments usually complete the task in indoor settings. It remains uninvestigated whether outdoor settings influence the strategies of participants, resulting for example in different frequencies and clusters of geographic categories for different landscapes.

Research Questions. From the identified research gaps, the following research questions emerge:

- RQ1: Does the landscape in which a free list experiment is conducted influence the elicited terms?
- RQ2: What memory search strategies do participants apply in outdoor settings for listing landscape terms and are these strategies reflected in free lists?

Our first hypothesis was that the landscape setting, particulary the visible landscape elements, influences the content of free lists. We expected to find similar terms for free lists elicited in similar landscapes, and different terms for free lists elicited in different landscapes. We selected three study sites, two in similar landscapes in the Swiss mountains, and one in a park on the Swiss central plateau. Based on the finding that the order of terms in free lists is important and reveals memory search strategies [39, 44], our second hypothesis was that landscape influences the memory search strategies and therefore the structure of terms in free lists. We expected to find evidence for clustering in free lists, with clusters consisting of elements that are semantically related, but, as a specific feature of geographic information, we also expected to find clusters of elements that are perceived as belonging together in the landscape based on topological relations or spatial proximity.

3   Methods

3.1   Data Collection Protocol

We conducted free listing tasks on landscape categories with participants in outdoor areas. Between 20 and 30 participants are usually considered sufficient to establish a cognitive domain, with 80 to 90 participants considered a good sample [45]. We therefore aimed for at least 80 participants for our study. Participants were recruited on site by the field researchers, asking people present in the location if they would volunteer to take part in a study. If they agreed, we informed them about the procedure of the survey and handed them an informed consent sheet. The verbatim statement posed to participants in Swiss German was: *'Was hätts für Sie inere Landschaft drin?'* (*'What is there for you in a landscape?'*). In a pre-test, other statements such as *'What geographic categories can you name?'* or *'What landscape elements can you identify?'* were not well understood, and made participants feel uneasy up to the point of saying: *'I am sorry I cannot answer this'*. In comparison, participants felt more comfortable with the revised and relatively open statement used in the experiments. Participants were able to question the experimenter, who then clarified the task by saying: *'Was chönnd Sie mir säge was es für Sie eso inere Landschaft drin hätt? Sie chönnd alles säge, es gitt kei richtig oder falsch.'* (*, What can you tell me that for you there is in a landscape? You can say anything, there is no right or wrong. '*). The field researchers noted down elicited landscape terms by hand on a sheet of paper in the order stated by the person taking the survey. There was no time constraint for the task, and participants often indicated that they had finished the listing task by stating: *'That's all I can think of'*. They were then asked in a short interview to explain how they came up with the terms during the task. To assess site familiarity of participants, we asked where people were from and how many times they visited the interview location. In Val Müstair, the participants were a mix of local people living in the valley and Swiss tourists. In Flims, participants were all tourists skiing in the area, while in Irchelpark, participants were people living in the city of Zurich visiting the Irchelpark. We estimate the age range of participants to be between 30 and 70. We conducted the elicitation in February 2015 at three different sites in Switzerland. To avoid possible influences of weather conditions, we conducted all three experiments on sunny and clear days. However, as the study was carried out in winter, both mountain sites were covered in snow, while Irchel Park was not.

3.2 Study Sites

We selected two study sites in mountainous areas in the Canton of Grisons, and one site on the Swiss Central Plateau in Zurich.

Val Müstair in the south-easternmost part of Switzerland bordering Italy is a mountain valley characterized by forests of Swiss pine (*Pinus cembra*), alpine meadows and mountains. In winter, tourists visit Val Müstair to practice snowshoe-hiking and skiing. The elicitation tasks mainly took place in the small ski area Minschuns in the outside seating area of the restaurant Alp da Munt at an elevation of 2150 m (Fig. 1a.). We collected 19 interviews for Val Müstair, 12 with tourists and 7 with local people. We attempted to collect more data, but the weather conditions drastically changed after the first three days of data collection and when we returned to collect data again 2 weeks later, snow melt had set in and changed the visible aspects of the landscape.

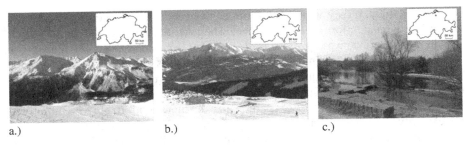

Fig. 1. View from the survey locations in (a.) Val Müstair, (b.) Flims, and (c.) Irchel Park

The second study site in the mountains was Flims, a major winter sports destination in the Canton of Grisons. The landscape is characterized by alpine meadows, pine forests and mountains. We conducted elicitation tasks with 40 participants (all tourists) in the ski area of Flims in the outside seating area of the restaurant at Nagens at an elevation of 2127 m (Fig. 1b.). As a third study site we chose the urban park landscape of Irchel Park to contrast with the two mountain landscapes. Irchel Park is public park in Zurich near a university campus at 479 m above sea level, centered on a pond and surrounded by jogging and pedestrian paths, benches, barbecue places, grass areas, stands of deciduous and coniferous trees as well as shrubs. The park is. The elicitation tasks took place during weekends and on afternoons during the week to avoid sampling university students. The tasks were conducted on the paths around the pond (Fig. 1c.).

3.3   Data Analysis

We analyzed the data collected at the three sites using the combination of quantitative and qualitative methods described below.

Description of Free Listing Data. Descriptive statistics provided a first overview of the free listing data. Furthermore, we identified location specific terms that only occurred in one study site. To exclude idiosyncratic terms, we removed terms only listed by one participant. Although the oral responses by participants were in Swiss German, we present the results in Standard German, the official written language used in the German speaking part of Switzerland. Terms are presented in singular or plural as elicited, since singular and plural forms may represent different categories.

Cognitive Salience Index. For the analysis of free lists, term frequencies are often used (e.g. [45]). Another common measure is mean rank, which often correlates with term frequency [46]. Both measures assess aspects of psychological salience: a tendency to occur at the beginning of lists and to be referenced across participants [18]. However, the two measures generate two different sets of terms. Therefore, in order to combine term frequency and mean rank into a single measure, Sutrop [47] developed a cognitive salience index (S) calculated as:

$$S = F/(NR) \tag{1}$$

where F is term frequency, N is the number of participants, and R the mean rank A term named by all participants, and always in the first rank thus has a maximum cognitive salience of 1. Less salient terms, mentioned by few participants, and towards the end of lists approach a minimum salience value of 0. Based on the cognitive salience index, we compared the top ten salient terms across the study sites. Furthermore, we determined the number of shared terms between sites as an approximation for similarity, using the thirty most salient terms to strike a balance between highly salient and less salient terms that may contain particular terms describing a study site.

Interviews on Participants' Free Listing Strategies. We first applied open coding to derive codes from the actual interview data consisting of: 'senses' (see, smell, taste, touch, and hear), 'personal memories', 'expertise' (job-related or other), 'value-judgement' (positive and negative) as well as the code 'inner picture'. Secondly, we applied structured coding using the codes derived from the data as well as Rosch's [1] criteria: attributes, activities, and parts. In addition, we analyzed the raw free list data for evidence of search strategies not mentioned by participants, but that could be expected for the geographic domain, namely scale (e.g. listing landscape terms ordered from large scale to small scale), partonomy (e.g. listing parts of mountains) and topology (listing topologically related objects in a landscape).

Network Visualizations. Networks are well suited to visualizing internal structure of free lists, for instance, as they allow display of sequentially adjacent terms as nodes connected by edges. The form of the network allows identification of grouped terms, which in our case are candidates for semantic clusters. Only terms listed by more than two participants, and occurring sequentially adjacent in more than one list were included. For example, if in one list the sequence *mountain–river* occurred, and in another list *river–mountain*, we included this pair of terms in the network. We produced the networks using R [48], where edge width represents frequency of a link, node size cognitive salience (calculated as Sutrop's index, [47]), and the node label size the connectivity or betweenness of a node [49]. For the node distribution, we used the Fruchterman-Reingold force directed graph algorithm [50] that applies attractive force to connected nodes and repulsive force to unconnected nodes. Furthermore, chi-values were also calculated between all connected nodes to indicate which connections are overrepresented, given their expected probability based on frequency [51].

4 Results and Interpretation

First, we report on descriptive statistics of the free listing data and the results for the cognitive saliency index, before presenting analysis of participants' explanations for strategies used during the free listing tasks. Finally, the links between sequentially adjacent terms are visualized in network diagrams for each study site as a way to explore the internal structure of free lists. Comparing the resulting networks with qualitative interview data from participants on their strategies during the free list tasks links our results to theories of memory search.

4.1 Description of Free Listing Data

In total, we elicited 89 free lists (Table 1). Flims had the highest mean number of terms per list, and Irchel Park the lowest. For all sites, the number of location specific terms (unique terms) was more than half of all terms. Often, unique terms were low frequency terms, which we can further distinguish. On the one hand, unique terms were listed only once, and in many of these cases we could not establish a semantic link between the term and the study site. On the other hand, unique terms listed by two or more people are candidates for being particular location-specific terms for which instances occur in the landscape of the study site. For example, in Val Müstair, two participants named *Arven* (Swiss pines, *Pinus cembra*), a characteristic tree of the valley and *Kloster* (monastery), with the world heritage monastery of the Convent of St. Johns in Müstair village. In Flims, four participants mentioned *Gletscher* (glacier), probably because of the Vorab glacier accessible from the ski area. In Irchel Park, two or more participants listed *Enten* (ducks) and *Möwen* (gulls), as well as *Haselstrauch* (hazel bush), instances of which occur in Irchel Park, but not in mountain locations. These location-specific terms often occurred towards the end of lists.

Table 1. Descriptive statistics of free listing data from the three different study sites

	Val Müstair	Flims	Irchel Park
N	19	40	30
Mean per participant (± StDev)	11.05 ± 3.17	14.88 ± 7.44	10.77 ± 4.55
Median	12	12.5	10
No. of terms	159	291	179
No. of terms > 1	32	75	41
No. of location specific terms	103	211	116
No. of locations specific terms > 1	8	34	13

4.2 Cognitive Salience Index

For both Val Müstair and Flims, *Berge* (mountains) was the most salient term, while for Irchel Park, *Bäume* (trees) was the most salient term (Table 2). Val Müstair and Flims shared 15 out of the 30 most salient categories, followed by Flims and Irchel Park sharing 13 categories. Val Müstair and Irchel Park shared only 8 categories of 30. The most cognitively salient terms of all three study sites include several highly frequent terms, indicating that frequent categories are often also named first.

Table 2. The top 10 most salient categories ranked according to Sutrop's index [47]

Val Müstair			Flims			Irchel Park		
(S)	German	English gloss	(S)	German	English gloss	(S)	German	English gloss
0.32	*Berge*	mountains	0.38	*Berge*	mountains	0.37	*Bäume*	trees
0.11	*Tal*	valley	0.11	*Bäume*	trees	0.13	*Wiesen*	meadows
0.11	*Wald*	forest	0.05	*Felsen*	rocks, cliffs	0.06	*Wasser*	water
0.07	*Hügel*	hill/hills	0.05	*Wald*	forests	0.05	*Tiere*	animals
0.06	*Wälder*	forests	0.04	*Seen*	lakes	0.05	*Hügel*	hill/hills
0.04	*See*	lake	0.04	*Wiesen*	meadows	0.04	*Berge*	mountains
0.04	*Gipfel*	peak	0.04	*Wälder*	forests	0.04	*Teich*	pond
0.04	*Felder*	fields	0.04	*Hügel*	hill, hills	0.04	*Vögel*	birds
0.04	*Bach*	stream	0.04	*Schnee*	snow	0.04	*Sträucher*	bushes
0.03	*blauer Himmel*	blue sky	0.04	*Wasser*	water	0.04	*See*	lake

4.3   Interviews on Participants' Free Listing Strategies

The following results illustrate what free lists in combination with interviews can reveal about possible memory search strategies, such as clustering and switching. We documented interview data from 63 participants. Each participant described one or more strategies that he or she had used during the free listing.

A total of 25 participants said they used visual stimuli provided by the landscape. For instance: *'I looked around and named what I saw'* or *'I looked at the landscape'*. 22 participants used past memories of landscapes they had visited, while 20 conjured up what they called 'an inner image'. For example: *'I had an image in my mind of different landscapes'*. This strategy was also used to come up with additional terms for the landscape the participants found themselves in, for instance: *'I made myself an inner image of this landscape how it looked like in summer'*. This participant in Val Müstair looked around and first named visible elements of the landscape such as *Hügel–Bach–Bäume–Felsen–Wälder* (hill–stream–trees–rocks–forests). Then, the switch to memory took place and the participant listed non-visible elements such as *Magerwiesen* (rough pastures). Such a combination of visual stimuli and personal memories was mentioned by 7 participants.

A similar switching strategy 21 participants said that they used was recalling geographic locations different to their current position. This was indicated by use of toponyms in explanations, as well as sometimes in free lists themselves. For example, a participant in Irchel Park stated that:

> First I looked around and thought of why I come to the Irchel Park, then my holidays last week in Engelberg next to Titlis and the hike we did there.

The strategy of recalling particular places is reflected in the free list. After naming categories visible in Irchel Park such as trees and walking paths, the participant stated

Titlis. This mountain toponym indicates a switch, followed by a new cluster: *Luftseil-bahn–See–Skifahrer–Skipisten* (cable car–lake–skier–ski slopes). It thus seems that toponyms may be indicators for switches in memory search. However, toponyms are not always named directly in free lists, as sometimes a generic is used:

> *I started with the Irchel Park and then where I was at the weekend, there I was at the lake.*

In this case, 'the lake' is a reference to the Lake of Zurich, the largest instance of its kind around Zurich. Based on the interview data, this switch was visible in the free list. After terms such as *Haselstrauch–Park–Studenten–ältere Leute* (hazel bush–park–students–elderly people), the switch took place and the participant started a new cluster *See–Schilfgürtel* (lake–reedbed). A complete free list from a participant in Irchel Park (Table 3) illustrates how interviews may help identifying potential clusters and switches.

Table 3. Free list and interview data of a participant showing indications for clustering and switching

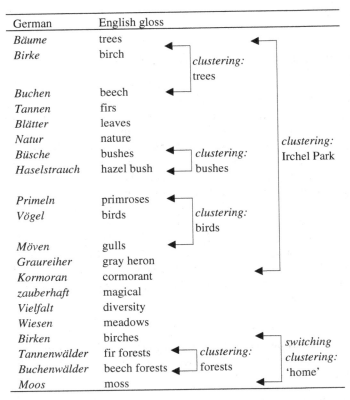

German	English gloss
Bäume	trees
Birke	birch
Buchen	beech
Tannen	firs
Blätter	leaves
Natur	nature
Büsche	bushes
Haselstrauch	hazel bush
Primeln	primroses
Vögel	birds
Möven	gulls
Graureiher	gray heron
Kormoran	cormorant
zauberhaft	magical
Vielfalt	diversity
Wiesen	meadows
Birken	birches
Tannenwälder	fir forests
Buchenwälder	beech forests
Moos	moss

clustering: trees

clustering: bushes

clustering: birds

clustering: forests

clustering: Irchel Park

switching clustering: 'home'

While some semantic clusters in free lists are identifiable from the lists alone, switches are more challenging to identify, unless participants themselves provide additional information. For instance, this participant explained: *'I first thought of the Irchel Park, and then of my home, where there are many birches and birch forests'*. Only with the qualitative data from the interview is this switch from Irchel Park to a landscape remembered as 'home' identifiable. Furthermore, the free list in Table 3 exhibits indications for hierarchical clustering, with several sub-clusters for the landscape in Irchel Park, and a sub-cluster for the landscape 'home'.

4.4 Network Visualizations

The participants stated several strategies for coming up with terms in a free list. We then tested whether these strategies are also visible in aggregated data for all free lists at a study site by visualizing sequentially adjacent terms for all lists at a single location as a network. Because the same criteria for including nodes and links in the networks were applied, but the sample sizes between the study sites differ, the graphs are differently populated. Therefore, rather than comparing the networks, the focus lies on qualitatively assessing what relationships between sequentially adjacent terms emerge for the three sites.

In the network visualization for Irchel Park (Fig. 2) the most salient terms *Bäume* (trees) and *Wiesen* (meadows) are at the center of the network. These two nodes each have several links to other nodes and therefore a high value of betweenness, represented by a large label size. Several clusters are visible, for instance, *Tiere-Vögel* (animals-birds) and a cluster consisting of *Teich-Wasser-Bach* (pond-water-stream).

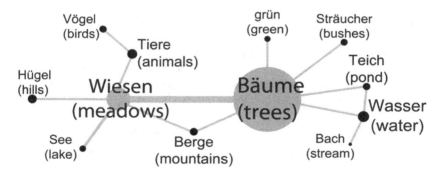

Fig. 2. Network visualization of sequentially adjacent terms in free lists for Irchel Park

For Flims, with a higher sample size, the network is more populated with a total of 33 nodes in one major network and two small unconnected networks (Fig. 3). In the main network, bio-physical elements of landscape such as *Berge* (mountains), *Wald* (forest), *Bäume* (trees), *Hügel* (hill/hills), *Seen* (lakes) form one part of the network, while anthropogenic landscape elements form a semantic cluster of 'human settlements' consisting of terms such as *Städte* (cities), *Dörfer* (villages), *Häuser* (houses), *Weiler* (hamlet/hamlets). The network structure indicates that when listing landscape terms, participants listed natural features separately from anthropogenic features. Several other clusters are identifiable, for instance the combination of *Sonne-Himmel-Wolken* (sun-sky-clouds) forming a small second network. The third network consists of the two nodes *Bach* (stream) and *Fluss* (river) only, which form a small semantic cluster of 'bodies of flowing water'. Less prominent, and thus arguable relations, are identifiable between living things such as *Tiere-Pflanzen* (animals-plants), water related terms such as *Wasser-Fluss-Seen-Flüsse-Bäche* (water-river-lakes-sea-rivers-streams). Partonomic relations include for instance *Wald-Bäume* (forest-trees). Antonymic use of landscape terms may be expressed through the connection between *Berge-Täler* (mountains-valleys). Cognitively salient terms, due to their frequent occurrence, are often strongly linked (edge width) and interlinked, resulting in a spider-web pattern radiating out and around the central salient term *Berge* (mountains).

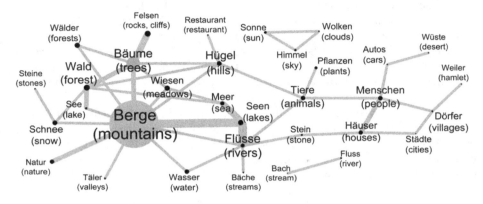

Fig. 3. Network visualization of sequentially adjacent terms in free lists for Flims

In addition to the network, we calculated chi-values to explore term associations which occur more often than expected based on raw frequencies. Table 4 shows pairs of terms for Flims that are overrepresented, forming parts of semantic clusters, such as *Bach-Fluss* (stream-river) and *Städte-Dörfer* (cities-villages).

Table 4. Chi-values for overrepresented pairs of sequentially adjacent terms

Term A	Term B	Observed	Chi-Value
Bach (stream)	*Fluss* (river)	2	12.42
Autos (cars)	*Wüste* (desert)	2	9.81
Städte (cities)	*Dörfer* (villages)	2	9.32
Dörfer (villages)	*Weiler* (hamlet)	2	7.94
Sonne (sun)	*Wolken* (clouds)	2	5.8

The term combination of *Autos-Wüste* (cars-desert) occurred in two lists, but from the sequence *Dörfer-Menschen-Autos-Wüste-Urwald-Insel* (villages-people-cars-desert-jungle-island) in one list and the corresponding interview ('*I thought of my home and then of travels*') we can assess these two terms were rather listed before and after a switch and do not form a semantic cluster such as 'car rally'. High chi-values appear to be indicators for strong semantic relations. The pairs of terms with high chi-values cannot be explained by geographic characteristics such as co-location in a landscape or partonomy, but may be prototypical categories sharing a high number of properties.

For Val Müstair with relatively low sample size (n = 19), the network visualization consists of two unconnected networks with 3 nodes each (Fig. 4). The term *Berge* (mountains) at the center of one network is most cognitively salient for that location.

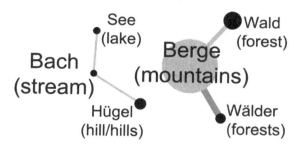

Fig. 4. Network visualization of sequentially adjacent terms in free lists for Val Müstair

In general, the cluster 'water bodies' occurred with variations in the nodes in all three networks, indicating that participants often consecutively list water features while exploiting a cluster in memory. The nodes in all the networks were predominantly terms of which instances occur in all the study sites. The most obvious exception is the node *Berge* (mountains) in Irchel Park that was listed despite the lack of mountains visible from Irchel Park.

5 Discussion

We aimed to study landscape terms as elicited in a free listing task and explore memory search strategies linked to such terms. Using free listing and interviews in outdoor settings in three study sites, we aimed to investigate the influence of landscape on *what* terms participants listed, as well as *why* they listed these terms, that is, their memory search strategies. In the following, we discuss our findings with respect to the research questions set out in Sect. 2.3, before making some more general observations on the wider implications of our results.

5.1 RQ1: Does the Landscape in Which a Free List Experiment is Conducted Influence the Elicited Landscape Terms?

Differences between the study sites are visible in the ranking of the most salient terms. The cognitive salience value for the term *Berge* (mountains) differs considerably from the mountain study sites, where it is the most salient term, to the city park, where it is only the 6th most salient term. Interestingly, in previous studies, students in classroom settings in an urban environment listed mountain as one of the most frequent terms (e.g. [13, 27, 33]). Half of the 30 most salient terms were listed for all three study sites. This high number of shared terms could imply that participants listed category norms, such as forests, lakes, and mountains relatively early in their lists. In addition to the difference in ranking of salient terms, we observed that good candidates for describing particularities of a landscape occur in the long tail distribution of less cognitively salient terms, such as *Kloster* (monastery) and *Arve* (Swiss pine) for Val Müstair, and *Gletscher* (glacier) for Flims. Summing up, to determine the influence of landscape, both ranking of cognitively salient terms, as well as (some) terms in the long tail distribution contain the most useful information, whereas a set of more general terms is shared between different landscapes. Importantly, many previous works have only reported the most frequent terms [13, 33], but we would argue that discarding this information makes replication and detailed comparative analyses difficult.

In interviews, participants stated that they looked at the landscape for coming up with terms, indicating that visual stimuli play a major role for free listing. In our case these visual stimuli take the form of the landscape where the free listing task was carried out and we discuss this in more detail below.

5.2   RQ2: What Memory Search Strategies do Participants Apply in Outdoor Settings for Listing Landscape Terms and are These Strategies Reflected in the Free Lists?

Our methodical approach combining free listing with interviews allowed us to better understand memory search strategies for landscape terms. For instance, the most prominent strategy participants mentioned was to start naming terms by first looking around and using the visual stimuli the landscape provided. This resulted in what we called 'geo-semantic' clusters in free lists consisting of terms for which instances occur in the landscape. Such instances are perceived as spatially related, for example, the forest is on the mountain, the trees are in the park or the lake is near the hill. The second most prominent strategy was to recall memories of particular familiar places.

Participants often combined these two strategies, whereby they first listed visible elements of the landscape (geo-semantic clustering), and then used the memory of a familiar place (switching) to name terms for that landscape (geo-semantic clustering). Additionally to published findings [35, 52], we found that each of the landscape clusters (surroundings and familiar place), was associated with a number of (sub-) clusters, consisting of geographic features found in that particular landscape. This indicates that memory search strategies for landscape terms as part of the geographic domain consist of clusters at multiple hierarchies, and these hierarchies often coincide with differing spatial scales.

Another particularity of the geographic domain was that in this study, toponyms indicated a switch between clusters. The listing of toponyms rather than generic terms in free listings has been documented before for categories such as 'outdoor scenes' [5]. However, in the absence of toponyms in free lists, switches are difficult to identify without the use of additional data from the participants themselves. In our experimental setting, where we directly interacted with study participants, it was possible to elicit qualitative information in interviews about memory search strategies participants were conscious of having used. The combination of free listing tasks followed by interviews has been productively used before to study memory search strategies in clinical patients, documenting a diversity of sometimes idiosyncratic ways of clustering and switching [44]. To go beyond descriptions at the individual level, we used network visualizations for each study site to explore the aggregated internal structure of free lists. When idiosyncratic switching strategies were filtered out, several groups of terms were retained, which often represent semantic clusters, such as 'water bodies' or 'settlement types', which also had a high chi-value, as well as geographic partonomies (forest-trees), and co-occurrence in a landscape (mountains-valleys). The network visualizations therefore provide a means to go beyond the individual level by exploring whether certain clusters are occurring repeatedly in the data.

5.3 Implications

While it is important to recognize the relatively small number of participants and sites at which we performed our research, we nonetheless believe that some more general implications can be drawn from our work.

Firstly, in previous work where participants were limited to 30 s to write their responses [13, 32] the mean number of terms per participant was lower than in our study. Our participants were not time limited, but typically rapidly listed terms within less than a minute. Since we found important information in the long tail, we suggest that imposing an artificial temporal limit may obscure relevant ways in which terms are used. In our context this is particularly important, since we are interested in finding out how landscapes are conceptualized. Our results suggest that lists contain many relevant terms, but not all of these are to be found in the most common terms.

Secondly, it is apparent from our results and previous work [33] that the setting of a free listing task plays an essential role in the responses gained. This points to the importance of not just the ethnopysiographic hypothesis [53], but also the notion that the same people might respond differently in different settings. However, by exploring not only term lists, but also cognitive saliency values and network visualizations, we can learn both about the ways in which individuals addressed this task, and how we might design a task to more exhaustively capture landscape terms. For example, participants recalled memories of previous experiences in the landscape such as journeys by train, or hiking trips, as well as memories of familiar places (identified by toponyms) as effective switching strategies. This suggests that we might make use of activities, experiences or toponyms as prompts to elicit landscape terms (c.f. [33]).

Since the outdoor experimental setting in this study differs from other studies on geographic categories [13, 32], resulting differences may also reflect methodological differences rather than differences in categorizations. The closest experimental setting to our study used videos of landscapes for elicitation [33]. However, substantial filtering was applied to the elicited terms based on notions of what constituted valid answers. For instance, vegetation terms listed by participants in the original free lists [54] were not reported in the final publication [33]. We urge researchers to also report unfiltered results of what people stated to be part of landscape or the geographic domain, rather than what researchers think are participants' correct answers to the elicitation task, since we observe that such terms may also link clusters. For example, in Table 3 birds (gray heron – cormorant) are linked to landscape qualities (magical – diversity) and lead back to landscape terms (meadows). Such linkages are, from a geographic perspective, essential in understanding how landscapes are conceptualized, suggesting, for example, issues of scale [55] and the use of partonomic relationships [29] (for instance in Table 3 the participant firstly positioned her/himself in Irchel Park, identifying elements and qualities of this landscape, before zooming out to another landscape 'at home').

Finally, though toponyms do not appear in our aggregated data, they had an important bridging role in lists, and might provide examples of instances of particular landscapes. This further points to the status of toponyms in language [56], and the rich potential of structured free lists for exploration.

6   Conclusions and Further Work

We believe that our results suggest a number of important methodological and thematic avenues for further research on geographic categories, through both simple methods such as free listing, and more complex qualitative studies:

1. Free lists contain more information than simply frequencies – by considering sequential adjacency of terms and calculating cognitive salience we were able to extract candidates for semantic clusters and build useful aggregating network visualizations.
2. By combining the free listing task with short interviews, it was possible to link theory on memory search strategies directly to our data and thus to identify meaningful structures not possible from the free lists alone.
3. The setting of the elicitation task has clear implications for the terms and linkages used. Nonetheless, some terms appear to be shared across our three landscapes, and may represent more basic categories of landscape terms [5], providing potential insights as to variation in landscape perception.
4. Our lists, and their analysis reveal once again the richness of the geographic domain for such analysis – and importantly that landscapes are conceptualized in a multitude of, equally valid, ways extending far beyond the simple listing of geographic features.

We close this paper by suggesting avenues for further research. When we started this work we simply aimed to replicate some previous studies [13, 26, 31–33] in a new setting. However, we believe that the combination of methodologies applied here and the richness of geographical settings clearly illustrates the potential for further studies exploring the semantics of landscapes, and linking this back to the ways in which we represent these in information systems. We suggest that combining mobile eye tracking [57, 58] with outdoor free listing experiments has the potential to link the visual stimuli provided by the landscape more directly to the clusters produced in free lists. In an era of big data, crowd sourcing and Citizen Science, we note that a simple free listing experiment, with roughly 90 participants who also explained their strategies in a few sentences, combined with hypotheses derived from existing theory in cognitive research, was a very rich source for analysis.

Acknowledgements. The research in this study was funded by the 'Forschungskredit' of the University of Zurich, grant no. FK-13-104 and the University Research Priority Program Language and Space (URPP SpuR) of the University of Zurich. We thank all participants in Val Müstair, Flims and Irchel Park who took part in this study.

References

1. Rosch, E.: Principles of categorization. In: Rosch, E., Lloyd, B. (eds.) Cognition and Categorization. Erlbaum, Hillsdale (1978)

2. Schuurman, N.: Formalization matters: critical GIS and ontology research. Ann. Assoc. Am. Geogr. **96**(4), 726–739 (2006)

3. Gahegan, M., Takatsuka, M., Dai, X.: An exploration into the definition, operationalization and evaluation of geographical categories. In: Pullar, D.V. (ed.) GeoComputation 2001. Brisbane, Australia (2001)

4. Fonseca, F.T., Egenhofer, M.J., Davis, C.A., Borges, K.A.V.: Ontologies and knowledge sharing in urban GIS. Comput. Environ. Urban Syst. **24**(3), 251–272 (2000)

5. Tversky, B., Hemenway, K.: Categories of environmental scenes. Cogn. Psychol. **15**(1), 121–149 (1983)

6. Klippel, A., Weaver, C., Robinson, A.C.: Analyzing cognitive conceptualizations using interactive visual environments. Cartography Geogr. Inf. Sci. **38**(1), 52–68 (2011)

7. Mark, D.M., Turk, A.G.: Landscape categories in yindjibarndi: ontology, environment, and language. In: Kuhn, W., Worboys, M.F., Timpf, S. (eds.) COSIT 2003. LNCS, vol. 2825, pp. 28–45. Springer, Heidelberg (2003)

8. Mark, D., Klippel, A., Wallgrün, J.O.: A crowd-sourced taxonomy for the common-sense geographic domain. In: Stewart, K., Pebesma, E., Navratil, G., Fogliaroni, P., Duckham, M. (eds.) Extended Abstract Proceedings of the GIScience 2014, pp. 358–361. Vienna University of Technology, Vienna (2014)

9. Edwardes, A.J., Purves, R.S.: A theoretical grounding for semantic descriptions of *Place*. In: Ware, J., Taylor, G.E. (eds.) W2GIS 2007. LNCS, vol. 4857, pp. 106–120. Springer, Heidelberg (2007)

10. Rorissa, A.: User-generated descriptions of individual images versus labels of groups of images: a comparison using basic level theory. Inf. Process. Manage. **44**(5), 1741–1753 (2008)

11. Burenhult, N., Levinson, S.C.: Language and landscape: a cross-linguistic perspective. Lang. Sci. **30**(2–3), 135–150 (2008)

12. Mark, D.M., Turk, A.G., Burenhult, N., Stea, D. (eds.): Landscape in Language. Transdisciplinary Perspectives. John Benjamins, Amsterdam (2011)

13. Smith, B., Mark, D.M.: Geographical categories: an ontological investigation. Int. J. Geog. Inf. Sci. **15**(7), 591–612 (2001)

14. Robbins, P.: Fixed categories in a portable landscape: the causes and consequences of land-cover categorization. Environ. Plann. A **33**(1), 161–179 (2001)

15. de Groot, R., Alkemade, R., Braat, L., Hein, L., Willemen, L.: Challenges in integrating the concept of ecosystem services and values in landscape planning. Manage. Decis. Making. Ecol. Complex. **7**(3), 260–272 (2010)

16. Johnson, L.M.: Introduction. In: Johnson, L.M., Hunn, E.S. (eds.) Landscape Ethnoecology: Concepts of Biotic and Physical Space, pp. 1–15. Berghahn Books, New York (2010)

17. Berlin, B.: Ethnobiological Classification: Principles of Categorization of Plants and Animals in Traditional Societies. Princeton University Press, Princeton (1992)

18. Berlin, B., Kay, P.: Basic Color Terms: Their Universality and Evolution. University of California Press, Berkley (1969)
19. Hunn, E.S.: Toward a Perceptual Model of Folk Biological Classification. Am. Ethnologist 3(3), 508–524 (1976)
20. Taylor, J.R.: Linguistic Categorization. Oxford University Press, Oxford (2003)
21. Rosch, E., Lloyd, B. (eds.): Cognition and Categorization. Erlbaum, Hillsdale (1978)
22. Lakoff, G.: Women, fire, and dangerous things. What Categories Reveal about the Mind. University of Chicago Press, Chicago (1987)
23. Mark, D.M., Smith, B., Tversky, B.: Ontology and geographic objects: an empirical study of cognitive categorization. In: Freksa, C., Mark, D.M. (eds.) COSIT 1999. LNCS, vol. 1661, pp. 283–298. Springer, Heidelberg (1999)
24. Shatford, S.: Analyzing the subject of a picture: a theoretical approach. Cataloging Classif. Q. 6(3), 39–62 (1986)
25. Hough, G., Ferraris, D.: Free listing: a method to gain initial insight of a food category. Food Qual. Prefer. 21(3), 295–301 (2010)
26. Battig, W.F., Montague, W.E.: Category norms of verbal items in 56 categories. a replication and extension of the connecticut category norms. J. Exp. Psychol. 80(3), 1–46 (1969)
27. van Overschelde, J.P., Rawson, K.A., Dunlosky, J.: Category norms: an updated and expanded version of the norms. J. Mem. Lang. 50(3), 289–335 (2004)
28. Lloyd, R., Patton, D., Cammack, R.: Basic-level geographic categories. Prof. Geogr. 48(2), 181–194 (1996)
29. Smith, B.: Mereotopology: a theory of parts and boundaries. Data Knowl. Eng. 20(3), 287–303 (1996)
30. Smith, B., Mark, D.M.: Do mountains exist? towards an ontology of landforms. Environ. Plann. B 30(3), 411–427 (2003)
31. Pires, P.: Geospatial conceptualisation: a cross-cultural analysis on portuguese and American geographical categorisations. In: Spaccapietra, S., Zimányi, E. (eds.) Journal on Data Semantics III. LNCS, vol. 3534, pp. 196–212. Springer, Heidelberg (2005)
32. Giannakopoulou, L., Kavouras, M., Kokla, M., Mark, D.: From compasses and maps to mountains and territories: experimental results on geographic cognitive categorization. In: Raubal, M.M., Mark, D.M., Frank, A.U. (eds.) Cognitive and Linguistic Aspects of Geographic Space. LNGC, pp. 63–81. Springer, Heidelberg (2013)
33. Williams, M., Kuhn, W., Painho, M.: The influence of landscape variation on landform categorization. JOSIS 5, 51–73 (2012)
34. Brewer, D.D.: Patterns in the recall of persons in a student community. Soc. Netw. 15(4), 335–359 (1993)
35. Troyer, A.K., Moscovitch, M., Winocur, G.: Clustering and switching as two components of verbal fluency: evidence from younger and older healthy adults. Neuropsychology 11(1), 138–146 (1997)
36. Collins, A.M., Quillian, M.R.: Retrieval time from semantic memory. J. Verbal Learn. Verbal Behav. 8(2), 240–247 (1969)
37. Gruenewald, P.J., Lockhead, G.R.: The free recall of category examples. J. Exp. Psychol.: Hum. Learn. Mem. 6(3), 225 (1980)
38. Bousfield, W.A., Sedgewick, C.H.W.: An analysis of sequences of restricted associative responses. J. Gen. Psychol. 30(2), 149–165 (1944)
39. Hills, T.T., Jones, M.N., Todd, P.M.: Optimal foraging in semantic memory. Psychol. Rev. 119(2), 431–440 (2012)
40. Charnov, E.L.: Optimal foraging, the marginal value theorem. Theor. Popul. Biol. 9(2), 129–136 (1976)

41. Troyer, A.K., Moscovitch, M., Winocur, G., Alexander, M.P., Stuss, D.: Clustering and switching on verbal fluency: the effects of focal frontal- and temporal-lobe lesions. Neuropsychologia **36**(6), 499–504 (1998)
42. Monsch, A.U., Bondi, M.W., Butters, N., Salmon, D.P., Katzman, R., Thal, L.J.: Comparisons of verbal fluency tasks in the detection of dementia of the alzheimer type. Arch. Neurol. **49**(12), 1253–1258 (1992)
43. Reverberi, C., Laiacona, M., Capitani, E.: Qualitative features of semantic fluency performance in mesial and lateral frontal patients. Neuropsychologia **44**(3), 469–478 (2006)
44. Body, R., Muskett, T.: Pandas and Penguins, Monkeys and caterpillars: problems of cluster analysis in semantic verbal fluency. Qual. Res. Psychol. **10**(1), 28–41 (2013)
45. Davies, I., Corbett, G.: The basic color terms of Russian. Linguistics **32**(1), 65–90 (1994)
46. Bousfield, W.A., Barclay, W.D.: The relationship between order and frequency of occurrence of restricted associative responses. J. Exp. Psychol. **40**(5), 643–647 (1950)
47. Sutrop, U.: List task and a cognitive salience index. Field Methods **13**(3), 263–276 (2001)
48. R Development Core Team: R: A Language and Environment for Statistical Computing. R Foundation for Statistical Computing, Vienna (2008)
49. Freeman, L.C.: A set of measures of centrality based on betweenness. Sociometry **40**(1), 35–41 (1977)
50. Fruchterman, T.M.J., Reingold, E.M.: Graph drawing by force-directed placement. Softw.: Pract. Exp. **21**(11), 1129–1164 (1991)
51. Wood, J., Dykes, J., Slingsby, A., Clarke, K.: Interactive visual exploration of a large spatio-temporal dataset: Reflections on a geovisualization mashup. IEEE Trans. Vis. Comput. Graph. **13**(6), 1176–1183 (2007)
52. Troyer, A.K.: Normative data for clustering and switching on verbal fluency tasks. J. Clin. Exp. Neuropsychol. **22**(3), 370–378 (2010)
53. Turk, A.G., Mark, D.M., Stea, D.: Ethnophysiography. In: Mark, D.M., Turk, A.G., Burenhult, N., Stea, D. (eds.) Landscape in Language: Transdisciplinary Perspectives, pp. 25–45. John Benjamins Publishing, Amsterdam (2011)
54. Williams, M.C.: Contribution towards Understanding the Categorisation of Landforms. MSc Thesis Instituto Superior de Estatística e Gestão de Informação ISEGI, Universidade Nova, Lissabon (2011)
55. Fisher, P., Wood, J., Cheng, T.: Where is helvellyn? fuzziness of multi-scale landscape morphometry. Trans. Inst. Br. Geogr. **29**(1), 106–128 (2004)
56. Levinson, S.C.: Foreword. In: Mark, D.M., Turk, A.G., Burenhult, N., Stea, D. (eds.) Landscape in Language: Transdisciplinary Perspectives, pp. ix–x. John Benjamins Publishing, Amsterdam (2011)
57. Kiefer, P., Giannopoulos, I., Kremer, D., Schlieder, C., Raubal, M.: An outdoor eye tracking study of tourists exploring a city Panorama. In: Qvarfordt, P., Witzner Hansen, D. (eds.) Proceedings of the Symposium on Eye Tracking Research and Applications - ETRA 2014, pp. 315–318. ACM, New York (2014)
58. Henderson, J.M.: Human gaze control during real-world scene perception. Trends Cogn. Sci. **7**(11), 498–504 (2003)

Signs, Images, Maps, and other Representations of Space

Identifying the Geographical Scope
of Prohibition Signs

Konstantin Hopf$^{(\boxtimes)}$, Florian Dageförde, and Diedrich Wolter$^{(\boxtimes)}$

Faculty of Information Systems and Applied Computer Science,
University of Bamberg, Bamberg, Germany
{konstantin.hopf,diedrich.wolter}@uni-bamberg.de,
dagefoerde.florian@gmail.com

Abstract. Prohibition signs warn of actions considered dangerous or annoying. Typically, these signs are located near the beginning of their scope, but knowledge about applicable prohibitions is important at any place within the scope. We developed an automated method to determine the scope of signs, aiming to support volunteered geographic information (VGI) applications that wish to capture prohibitions. In this paper we investigate the problem of computing the scope of geo-referenced signs that refer to human outdoor activities using OpenStreetMap (OSM) data. We analyze the problem and discuss the specific challenges faced. From the analysis we derive a symbolic representation that links activities with (OSM) map features, enabling semantic assessment of map features with respect to a prohibition and reasoning to infer its scope. In a comparative evaluation we demonstrate that our spatial-semantic approach significantly outperforms a previous method based on proximity.

Keywords: Spatial semantics · Semantic assessment · Prohibition signs · Volunteered geographic information (VGI) · OpenStreetMap (OSM)

1 Introduction

Signs are widely applied to convey that some action may or may not be performed in a certain area, for example to keep a dog on the leash in a park, not to camp on the beach, etc. While some of the prohibition signs address neighborliness of daily social life, others are meant to prevent committing something dangerous, for example taking a swim in areas where a hazardous but hidden riptide is present. Moreover, not knowing about a prohibition can imply breaking an important law unintentionally. For mobile computing applications it would be desirable to obtain any prohibition applicable to a specific location, allowing assistance systems to present important prohibitions in an electronic map display. While maps may already register prohibition signs *and* their geographical scope, there exists so far – to the best of our knowledge – no reliable service provider that combines this information. However, the popular volunteered geographic information (VGI) service OpenStreetMap (OSM)[1] already includes

---

[1] http://www.openstreetmap.org/about, last accessed 06.04.2015.

S.I. Fabrikant et al. (Eds.): COSIT 2015, LNCS 9368, pp. 247–267, 2015.
DOI: 10.1007/978-3-319-23374-1_12

prohibition signs related to road traffic, with their location, orientation, and designation. For users, it would be valuable to highlight the scope of any sign in a map visualization, in particular if the sign itself is not located within the part of the displayed map. The OSM community already develops a specification for registering prohibition signs jointly with their scope, but at the time of writing this paper no such information was found for the local environments in Germany, that we used to examine the evaluation of our approach. In general, the annotation of entries regarding prohibitions in the whole world OSM database is at an early stage: from over 3 billion of objects in OSM, only 5 % that are classified as "building" in combination with "amenity" (indicating public facilities provided) are related to an allowance or prohibition of smoking and, only 5'789 objects are annotated with the key "dog" or "dogs"[2] and provide therefore information to dog owners, such as whether dogs are allowed or have to be kept on a leash. The scarcity of prohibition information makes the database unreliable for the time being and shows the need for an automated method for determining the scope of a prohibition. Such a system would support contributors of a VGI system, by presenting them a plausible interpretation of the scope for any geo-localized prohibition sign, thereby easing the registration process. An automated method solves a challenging data integration task and contributes to the state-of-the-art in semantic map processing.

Automatically computing the scope of a sign turns out to be no easy task, it requires considering the semantics of regions and prohibitions in order to counter-act ambiguous placement of signs and uncertainty in input data. The contribution of this paper is to show how the necessary semantics can be captured and how an automated method can be realized. A knowledge-based approach allows us to integrate the taxonomy of OSM entities captured by their key-value coded tags associated with objects in the database. We describe a rule-based system that implicitly captures a small spatial ontology that focuses on reasoning in human outdoor activities. The architecture of the developed system is shown in Fig. 1. A fundamental issue of presenting knowledge to users which has been obtained by automatic inference from uncertain input is that sometimes it is not the objectively correct answer, i.e., the true scope of a sign. Thus there is a high risk for users that rely on an algorithmic decision. We aim to counter-act this problem by not aiming to compute *the* scope, but to evaluate *potential alternative* interpretations too. In an experimental evaluation we demonstrate that the new method significantly outperforms an existing approach based on proximity.

The remainder of this paper is organized as follows. In Sect. 2 we first put our approach in context of related work involved with the semantics of regions and maps. In Sect. 3 we analyze the problem in detail and derive individual measures to counter-act its challenges. Section 4 presents our knowledge-based approach and Sect. 5 details the implementation. In Sect. 6 a comparative evaluation is presented. The paper concludes in Sect. 7 by a discussion of the achievements and open issues.

---

[2] http://taginfo.openstreetmap.org/keys, last accessed 15.06.2015.

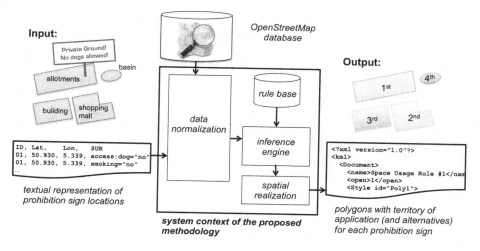

Fig. 1. Determining plausible scopes of prohibition signs – system overview.

2   Related Work

To the best of our knowledge, the first approach to automated interpretation of prohibition signs with respect to OSM has been presented by Samsonov et al. [11]. It indicates that the respective scope of the sign is – in almost all cases – given by the geographic entity closest to the sign. In this work we demonstrate that choosing the nearest geographic entity gives poor results when handling a variety of prohibition signs. We give a detailed analysis of the problem from which we derive an approach exploiting spatial and conceptual semantics.

From a more general perspective, several research directions relate to interpreting signs automatically. In the area of visual sign recognition various approaches have been pursued. While Samsonov et al. [11] apply machine learning for recognizing still images, spatio-temporal reasoning enables robust tracking and detection of signs in video streams from moving vehicles [14]. The principle applicability of computer vision to source a VGI system database has recently been demonstrated by the technology company Bosch who publishes the app *MyDriveAssistant*[3] that detects traffic signs in video streams and shares the information among users. Due to the clear network structure of roads, determining the scope of a traffic sign is often easy.

With signs referring to human activities, we are however involved with open spaces. Penn and Turner [10] show that routes taken by visitors in indoor environments can be determined using a set of space syntax rules. This generates a network structure, but a set of rules for general outdoor activity has not been proposed yet and it seems more difficult to design than to solve our original problem. Spatial and ontological knowledge similar to space syntax rules

---

[3] http://iphone.bosch.com/mydriveassist/, last accessed 27.03.2015.

becomes necessary to interpret a prohibition robustly. Formal languages to capture those semantics of everyday geographic entities have been investigated by Bennett [2] who defined them with spatial and conceptual properties. This line of foundational ontology research also enables abstract query languages grounded in geographic databases [3,6]. As we will see, determining the scope of a sign goes beyond querying for geographic objects but requires us to exploit the semantics to compute an hypothetic entity not registered in the database but which likely exists in the real world. Using a highly expressive logic language applied in formal ontology, this task is computationally infeasible. In context of visualization of virtual 3D cities, Ulmer et al. [13] present an efficient approach to deriving ecological uses of geographic regions by representing the requirements of vegetation and animal habitats, using manually designed rules. Since this approach is intended to populate free space in city models, the application to find the scope of prohibition signs is limited. Tailored to declarative programming, van Hage et al. present the Prolog *Space Package* to provide an infrastructure for dealing with space and semantics [15]. This package exceeds the requirements of developing our research prototype by far since we only process OSM data that can be easily accessed in any programming environment due to its XML organization. In our implementation we opt for a lightweight declarative representation tailored to efficient reasoning that is accompanied by external utility programs.

Reasoning involves assessing the suitability of a region to perform an action, which is closely related to determining similarity on a semantical level. This task has been analyzed by Janowicz et al. [7] and existing approaches have been surveyed by Schwering [12]. In context of OSM tags to present semantics, Ballatore et al. [1] propose an automated similarity measure based on analyzing the link structure of the web pages documenting the OSM tags. In our approach, we derive similarity from a representation of actions and OSM entities, but we additionally consider uncertain interpretation of OSM concepts.

Finally, noteworthy methods of machine learning in spatial environments have been proposed [8]. In general, those approaches are not applicable to our problem for two reasons: first, the effort of collecting a sufficient set of training examples for each prohibition type in several spatial settings is high and second, machine learning algorithms may solve the classification problem of identifying a suitable entity, they cannot be applied to generate a missing entity.

3   Problem Analysis

The following problem analysis surveys key difficulties in interpreting prohibitions with respect to a spatial database. Along the analysis we derive ingredients that allow us to overcome the individual difficulties while assembling all pieces in a knowledge-based computational approach.

The starting point of our work is the knowledge about a prohibition sign, its semantics and geographical location, as well as knowledge about the surrounding environment, given by data presented in OSM. The legal background on signage with prohibition signs in the area of our field studies, Germany, gives a first decisive hint to computing the scope of signs: prohibition signs have to be erected in the immediate surrounding of the area they refer to [4,5, Sects. 41 and 42]. Therefore, a first approach could be to assign a prohibition sign to the closest geographical entity represented. While first experiments documented about 97.6 % correctness with this approach for no-smoking signs [11], the results do not generalize to signs in general.

In the experimental evaluation with 77 prohibition sign instances of several prohibition types that we present in this paper, the closest OSM entity conforms to ground truth only in fewer than 30 % of the cases. We reviewed the incongruent examples and identified four difficulties that are discussed consecutively.

Spatial Misalignment. In reality, signs are not put immediately in front of the region they refer to, but at a suitable position close by. The location of a sign may thus be closer to entities not in its scope than to the entity it refers to. Moreover, we are confronted with data in VGI that is subject to considerable limits in accuracy and we have to accept errors in global positioning system (GPS) localization of partly more than 10m, especially in cities with high-rise buildings. Figure 2 shows two examples of the spatial misalignment problem in the case of GPS uncertainty. All pictures have been taken on the street near the entrance doors, but the recorded GPS location differs significantly from the photographer's position. Besides that, further uncertainty is added by considering the geo-reference of photographs as the position of the sign they depict. These two types of inaccuracy lead to an ambiguous interpretation of the applicability, that we can aim to overcome by considering the sign's and geographic entities's semantics, e.g., that dogs are not prohibited in public spaces but possibly in retail stores.

Semantic Misalignment. Prohibition signs refer to activities or actions that can only be performed in certain areas. In our work we capture the semantics of signs in the sense that we will not only represent a label like "no swimming" but develop a representation that explains important characteristics of swimming. OSM entities are widely tagged with semantic knowledge too, using an open informal ontology. Technically, we find tags like "building=yes" or "amenity=restaurant".

However, semantics captured in OSM have not been designed for the purpose of reflecting prohibition signs and so misalignment can occur. First, several OSM concepts – possibly organized across different primary map features such as natural, land use, or amenity – may suit an activity to be performed (e.g., (no) swimming in natural water entities ("natural=water"), basins (e.g., "landuse=basin"), or fountains ("amentiy=fountain"). Second, only some instances of an OSM concept may allow an activity to be performed. For example, swimming is only possible in fountains of considerable size, the possibility of driving a car onto an industrial premises depends on the surface conditions and absence of buildings, and so on. If only the semantic knowledge provided by the OSM taxonomy is available, it may thus not be possible to decide applicability of a sign – residual uncertainty must be considered with any decision taken. The

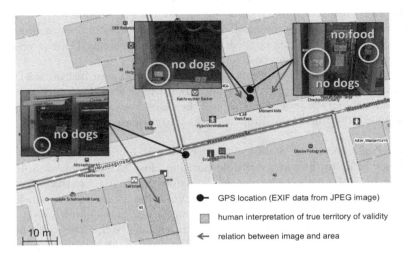

Fig. 2. Example for spatial misalignment: the sign locations recorded via a camera-internal GPS sensor cannot be interpreted unambiguously.

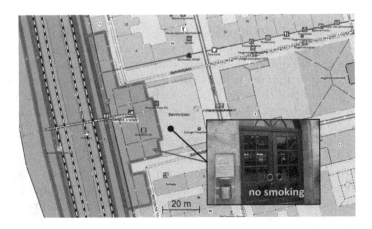

Fig. 3. Semantic misalignment example: The sign is located on the forecourt of a railway station; smoking is only prohibited within the building (red) (Color figure online).

local context may however provide valuable information. For example, if a fountain is the only water-related entity close to a no-swimming sign, the sign likely refers to the fountain. In Fig. 3 we present an example for semantic misalignment: the GPS position pins the sign in the forecourt, which is a semantically possible interpretation – in some outdoor areas smoking is not permitted. One can however infer from a side-effect of smoking, namely annoyance by smoke, that smoking is more annoying indoors than outdoors and so the station building presents a viable alternative interpretation of the sign. Indeed, the sign is meant to apply to the station building only.

Conceptual Misalignment. Although prohibition signs apply to a specific spatial area, this area may not be represented in OSM because its underlying concept is not a geographic entity. Understanding the meaning of such a prohibition requires us to regard how an environment will be conceived by a visitor, and which functionality it can fulfill. A typical example is shown in Fig. 4: The no-smoking prohibition applies to the whole train station area, but only isolated parts of the train station like the entrance hall, tracks and platforms are present in OSM. In order to interpret the sign correctly we have to recognize that the sign is erected at the entry of one conceptual region – the train station.

Lack of Data. Last but not least, OSM may simply miss a specific entity. As an example we mention a "do not lean bicycle against"-sign on a bridge that refers to the railing of the bridge – the railing of type "barrier=cable_barrier" is however not registered. In this situation one may still be able to infer a likely interpretation of the sign, considering its typical use case and the background knowledge that bridges have a railing. From the shape of the bridge we can thus determine the area of applicability quite precisely. As a further example, Fig. 5

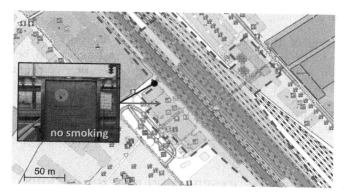

Fig. 4. Example for conceptual misalignment: Smoking is prohibited in the whole area of the train station (marked green), but the objects represented in OSM cover only isolated parts of this area (tracks, parts of buildings and platforms of the train station) (Color figure online).

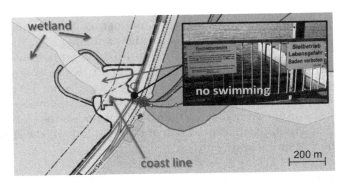

Fig. 5. Further example for conceptual misalignment: swimming is prohibited in the small harbor due hazardous water conditions. The intended territory of validity cannot be retrieved from OSM easily since the water area is uncharted.

illustrates the problem of missing information in context of semantic misalignment. The example shows a no-swimming sign that is meant to apply to a small harbor which is not registered in OSM, but whose layout may be derivable from a water stream going through and the adjacent tidal wetlands, enclosed by man-made quay-like groynes. Making interpretation even more difficult, the coast line is registered to cross the harbor right through. Handling conflicting information is therefore necessary too.

Summing up, we identify several aspects that need to be considered in order to understand the applicability of a prohibition sign. These aspects are:

- proximity
- semantics
- local context
- background knowledge about typical use cases.

Interpretation of the individual aspects may be conflicting and thus all aspects have to be balanced carefully and uncertainty has to be taken into account. In order to compute a reasonable interpretation of the applicability of a prohibition sign we also have to balance between two different approaches:

- select a region from the OSM database (e.g., by selecting the outline of a way that encloses an entity)
- compute a new region, possibly re-using parts from the OSM database.

4   A Rule-Based Representation of Prohibitions

We aim to tackle the challenges described above by constructing a knowledge-based system that infers the plausibility of a region representing the scope of a geo-referenced prohibition sign. The key component of our approach is a mapping from the OSM object taxonomy to the semantic concepts underlying the understanding of prohibition signs. In reminiscence of AI's classic expert systems, we construct rules in the form $\alpha \to \beta$, representing that a precondition α leads to a conclusion β. Since symbolic rules are employed, an explanation of a result can be obtained by considering the sequence of rules applied. If no suitable region of applicability can be found in the geographic database, the rules are exploited to constrain a suitable but unregistered region – which then is computed.

4.1 Inferences Under Uncertainty

Applicability may not be assessable by Boolean truth values due to various aspects of uncertainty discussed in the previous section. Any interpretation is thus subject to uncertainty and it becomes necessary to consider the degree of confidence associated with any decision in order to choose between competing options. To this end, any rule has to accommodate for uncertainty introduced by applying the rule, restricting the confidence of a decision eventually reached. It is not our aim to quantify this uncertainty as precise as possible, e.g., by developing a Bayesian network – as a matter of fact, we conjecture that even an extensive empirical evaluation of the characteristics of OSM knowledge will not substantially contribute to quantifying uncertainty.

For modeling our knowledge base we opt for a qualitative assessment of confidence levels of statements using 5 classes that range from -2 (very confident to not hold) to $+2$ (very confident to hold), using 0 to mark an indecision. During computation, in-between values will be handled too. All inference rules are augmented by this confidence measure. For example, we may say that a water area implies the possibility to swim with high confidence whereas we only assign low confidence to inferring food handling is happening at a retail store. In notation, we write $\texttt{waterArea} \overset{+2}{\to} \texttt{maySwim}$ and $\texttt{store} \overset{+1}{\to} \texttt{foodHandling}$ to represent these cases.

Confidence values need to be propagated along inference chains in order to obtain the confidence of conclusion β with respect to confidence of rule $\alpha \overset{k}{\to} \beta$ and initial confidence of α. Since we are not involved with deep logic inference, we do not need to worry about the many challenges faced in defining multivalued truth semantics. Among the various options considered in multivalued logics, we opt for maximizing confidence over disjunctions and averaging over conjunctions. Negation is determined by complement, i.e., the confidence value of a negated fact is $-c(\alpha)$, if $c(\alpha)$ is the confidence of the original fact. For the binary operations with facts α_1, α_2 we thus have:

$$c(\alpha_1 \wedge \alpha_2) := 0.5\left(c(\alpha_1) + c(\alpha_2)\right) \tag{1}$$
$$c(\alpha_1 \vee \alpha_2) := \max\left\{c(\alpha_1), c(\alpha_2)\right\} \tag{2}$$

In contrast to the predominant approach in Fuzzy logic to minimize confidence over conjunctions (called membership degree μ in Fuzzy logic), we choose to average. We do so to respond to our observation that in context of deciding applicability there is often some contradicting fact which leads to a very low confidence for some influence factor. Taking the minimum of all confidence values would then discard confidences for all factors other than the very low one caused by the contradicting fact. By averaging over all confidences, the effect of a single contradictory fact among many others is leveled out. Of course, this approach may cause a bias when modeling interdependent facts, but there does not seem to be a problem in the small rule base and the shallow reasoning necessary in our application.

4.2 Modeling Prohibitions by Actions and Side-Effects

The key idea underlying our modeling is to represent the possibility of performing the prohibited activity and also its potential side-effects in a specific region. Side-effects of activities are important since they are often the cause of a prohibition being invoked. For example, one possible side-effect of letting off-leash dogs run around freely is that they will hunt potential prey. Regions likely populated with prey (say parks, woods, meadows etc.) thus provide a good explanation why off-leash dogs may be prohibited there. In what follows we will refer to any effect of an activity including side-effects shortly as requirement. Let us assume, OSM objects tagged with key-value pairs have been interpreted in terms of primitives `isPark`, `isWood`, and `isMeadow`. Then our example can be formalized by two rules, the first making the connection between the region and the requirement to hunt prey:

$$\texttt{isPark} \lor \texttt{isMeadow} \lor \texttt{isWoods} \xrightarrow{+2} \texttt{mayHuntPrey}. \qquad \text{(rule type I)}$$

In the same fashion we would state that a park, among other entities, meets a requirement of walking. As second rule we specify the requirements of walking a dog off-leash to involve hunting prey and the ability to walk within the region:

$$\texttt{mayHuntPrey} \land \texttt{walk} \xrightarrow{+2} \texttt{offLeash} \qquad \text{(rule type II)}$$

For realizing the mapping from OSM entities and their various key-value pairs to symbolic primitives (e.g., `isPark`) and further to the requirement (e.g., `mayHuntPrey`), we simply write down individual rules by going through the OSM tag documentation. The rules modeled in our system are only little more complex than the example shown above, but they precisely follow the two-step rule schema just shown. In total, we consider the following requirements that we derived manually from the considered prohibition signs:

outdoors, indoors the requirement to perform an activity outdoors or indoors
land area, water area the requirement to perform an activity in/on a water area or land area
walking access whether pedestrian access is necessary for an activity to be performed
may annoy the requirement to affect other people unpleasantly (e.g., by smoking) is met by confined, highly populated places

may hunt prey whether a dog can encounter animals to hunt

stain/foul food the requirement to stain objects (e.g., in a fancy retail shop) or affect hygiene of food (e.g., by a dog at places where food gets handled)

may ignite the requirement to accidentally set fire to objects (e.g., fuel vapors at a gas station).

For the signs encountered in our field study, we mapped these to disjunctions of conjunctions of action effects – disjunctions serve to collect the different action effects that may arise. We present the mapping of actions to requirements in Table 1 where we list all prohibition signs considered and mark by "×" the requirements associated with a sign, e.g., $\texttt{outdoors} \wedge \texttt{waterArea} \xrightarrow{+2} \texttt{noSwimming}$. Disjunctions are shown as multiple rows in the table. Modeling this part of the knowledge base we do not differentiate confidence values but employ implications $\xrightarrow{+2}$ with equally high confidence only. Here, disjunctions provide sufficient means to represent alternatives. Representing whether a specific region meets a requirement, differentiation of confidence values is however necessary, e.g., to differentiate how likely the different water-related OSM entities allow for swimming (see discussion on semantic misalignment). Due to the variety of tags defined in OSM the mapping of tags to requirements satisfied by the respective entity is rather lengthy, but straightforward to achieve. We therefore only present an example of our model for the requirement 'stain/foul food', using $t(o)$ to refer to the key-value pairs tagged to an object o:

$$\texttt{stain}(o) = \begin{cases} +2 & (\texttt{shop}, x) \in t(o), x \in \underbrace{\{\texttt{bakery}, \texttt{butcher}, \ldots, \texttt{supermarket}\}}_{:=F} \\ +2 & (\texttt{amenity}, x) \in t(o), x \in \{\texttt{bar}, \texttt{biergarten}, \ldots, \texttt{restaurant}\} \\ +1 & (\texttt{shop}, x) \in t(o), x \notin F \\ -2 & \text{otherwise} \end{cases}$$

(3)

In this definition, symbols on the right-hand side stand for the respective keys and values defined in OSM.

4.3 Assessing Plausibility of a Region

In order to assess the suitability of a region for performing an action with all its effects, we apply simple forms of logic reasoning. First, having fixed a specific entity for evaluation we can infer by means of deduction using modus ponens, whether a region meets a requirement (rule type I). This is just a matter of computing the resulting confidence as explained before. Doing so we obtain the set of requirements met by a specific region.

Since we are confronted with a concrete prohibition sign, the conclusion of the second type of rules is known to hold. Second, we thus have to apply abduction to determine a suitable explanation, i.e., a conjunction of requirements met, leading to the fixed conclusion (rule type II). This can be realized by searching for the rule yielding the desired conclusion with highest confidence value. Due

Table 1. Interpreting prohibition signs in terms of requirements and side-effects.

sign	outdoors	indoors	land area	water area	walking access	may annoy	may hunt prey	stain/foul food	may ignite
no dogs	×				×		×		
		×			×			×	
no smoking		×							×
			×			×			
no skating			×			×			
dogs on leash	×		×		×		×		
	×		×		×			×	
no food			×		×			×	
no camping	×		×						
no alcohol			×			×			
don't lean against	×		×		×				
no open fire	×		×		×				×
no littering	×		×		×				
no access			×		×				
no swimming	×			×					
stay off the ice	×			×					
no mobile phones			×			×	×		
			×			×			×
no horse riding	×		×						
no fishing	×			×					

to the simple two-step modeling, this process is non-recursive and can be run in linear time with respect to the size of the knowledge base.

For determining the region with the highest plausibility of being the scope of a prohibition sign, we simply iterate over all candidates from a local context, assessing their plausibility.

4.4   Determining Scope by Instantiating a Region

If an area with non-negative plausibility has been identified, its geographical extent is fixed as the scope of the sign at hand. However, one of the challenges faced is that no geographic entity with non-negative plausibility may exist in the database, for example due to conceptual misalignment or lack of data. As discussed before, a suitable region may in some cases still be inferred. We approach this challenge by instantiating a new object that we then assess as if it was an object already present in the database. The instantiation process is realized as

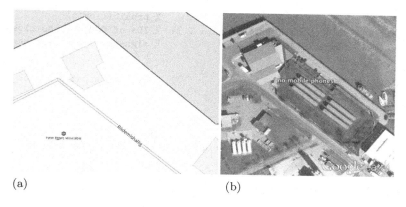

(a)   (b)

Fig. 6. Instantiating a new region to respond to lack of data sensibly. (a) Local map according to OSM not registering a gas storage (b) aerial image from Google Earth with automatically generated scope and position of the prohibition sign

a mixture of conceptual and spatial reasoning and works in opposite direction to assessing suitability.

Given an activity mentioned by a sign and the knowledge base described before, we identify all regions in the surrounding which do not meet the action's requirements in the sense that their confidence level of meeting the requirements is less than zero. These objects are regarded to be incompatible with the prohibition. For example, a sign prohibiting swimming is not agreeable with land areas. We then create a local low-resolution raster map in which all non-applicable areas are marked. Starting at the unmarked position in the map that is closest to the position of the sign, we grow a region using an area-fill algorithm for raster images to create a region. By contour walking the filled region and contour simplification we obtain a polygon. The size of the region created is bounded by the size of the raster map which we set to ±50m around the location of the sign. In order to assess this region as described above, all attributes of OSM objects whose outline overlaps the region are collected. An example of this procedure is shown in Fig. 6 where a sign prohibiting the use of mobile phones at an unregistered gas storage (the six white objects in the aerial image). As an alternative interpretation to the building close by registered in OSM the region highlighted in pink is created that fits between the registered street, quay wall and registered buildings. Here, we obtain a result very close to ground truth.

To respond to situations in which the available knowledge is insufficient to instantiate a new region by reasoning, we additionally generate a second area around the immediate surrounding of a sign. This region is not assessed but tagged with a zero confidence, i.e., it will only be considered if all alternatives available do not agree with a sign in the sense that their confidence value is less than zero. A typical case in which this approach becomes necessary is a no-swimming sign at a location where no water area is registered close-by.

```
restriction([no_dogs,no_camping]).          % prohibition(s)
entities([144523898,51707356,26534312, ...]).  % IDs of OSM objects close by
distance(144523898, 28.42).                  % sign-to-object distances
    ⋮
has(144523898, type, way).                   % key-value pairs for objects
has(144523898, waterway, riverbank).
    ⋮
```

Fig. 7. Excerpt of local environment description input to the Prolog inference engine

5   System Realization

To evaluate our approach we implemented the method using the architecture shown in Fig. 1. The implementation comprises some utility components for data handling and the rule-based inference system written in Prolog.

First, the location of a prohibition sign and surrounding spatial objects from OSM are converted into a uniform internal representation within the *data normalization* component. Based on background knowledge – stored in a *rule base* – the *inference engine* determines the semantic plausibility of a region serving as scope of the sign. Finally, the *spatial realization* component generates polygons representing the geographical scope of the most plausible regions. This step comprises mapping OSM objects (points, lines, etc.) to areas as well as the instantiation of new regions as described above.

5.1   Data Normalization

In this component, the geo-referenced prohibition sign is read first. Our implementation supports a set of individual prohibitions associated with a single sign, these are treated to conjunctively hold.

The OSM database is then queried for nearby objects, retrieving all objects closer than 30 m to the sign location. The parameter value of 30 m is motivated by two considerations: (a) German safety regulations that request warning signs in working areas to be closer than 30 m to the objects they refer to [4] and (b) the GPS error of 10 m, or more in urban areas, that we consider double-fold in this OSM query to retrieve all objects in the surrounding. To ease spatial computations we map all coordinates to a local Euclidean coordinate system using the UTM projection.

The result contains the three types of OSM objects – nodes, ways and relations – each associated with one or more OSM map feature tags. We now merge nodes with closed way objects whenever a node is contained within the way object or, in case of open ways, if the node is closer than 1 m to the way. The rationale behind this step is that nodes represent presence of certain features in the local surroundings (e.g., a cafe or shop as in Fig. 2), but way objects demarcate their geographic extent. During merging we also unite the tags associated with the objects which present the semantic attachment in a key-value paired manner, for example if the object is a building, an area of water, or a cafe. A full list of possible information is maintained within the OSM-Wiki[4] and is continuously extended.

Finally, the distance between entities and location of the sign is determined and the scene description is output as Prolog code, sanitizing symbols to conform with Prolog syntax. See Fig. 7 for an example showing the four types of statements generated.

5.2 Inference Engine

As the core component, the inference engine realizes the search process described in Sect. 4. Realization of the rules is straightforward in Prolog, which is also the reason why we opted for Prolog. Confidence levels have been introduced manually, i.e., we simply pass on confidence levels as an extra argument. For realizing abduction when searching for the most plausible interpretation we make use of Prolog's `findall` primitive, maximizing over the results.

Finally, a list of objects is returned, sorted by an assessment of semantic congruence and distance. Whenever two entities have the same semantic congruence, closer distance to the sign position is used for disambiguation.

5.3 Spatial Realization

The final result of the inference engine is an ordered set of candidate entities that are plausible interpretations of the scope of the given sign. Now, for each object the spatial region is computed as polygon and stored in the Keyhole Markup Language (KML) format. We distinguish between three cases in determining the outputs:

1. The object is a polygon,
2. the object is a point (node), including the special location 'here',
3. a suitable object is missing.

In case (1) the polygon is fixed as a territory of scope. For handling the case (2) we construct a circle with 10 m radius centered at the position of the node. Recall that if plausibilities of all objects are less than zero, a region around the sign is chosen as explained in Sect. 4.4. This is realized by inserting a dummy object into the result list with plausibility zero of type node that is located at the sign position. For handling case (3) as explained in Sect. 4.4 we construct a bitmap with cell size of 1 m and pixel values free, blocked and marked. All pixels are initialized to free, then setting those to blocked that coincide with regions that cannot overlap with the scope of the sign at hand. To this end, we select all objects o in the surrounding, which feature a negative semantic $c(o)$ confidence as determined by the inference engine. Free cells represent areas in space potentially contained in the scope of the sign. We start by marking the free cell closest to the sign position and iteratively mark all free cells adjacent to a cell

---

[4] http://wiki.openstreetmap.org/wiki/Map_Features, last accessed 10.06.2015.

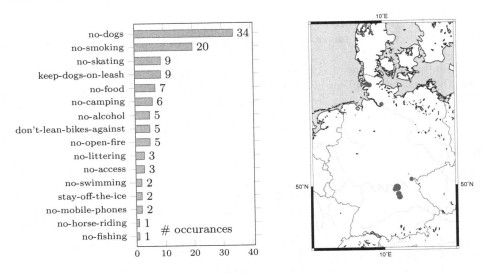

Fig. 8. Overview of the 77 instances of test data collected for evaluation (note: multiple prohibitions per sign can occur) and their geographic distribution (dot size demarcates relative distribution)

marked. Finally we apply an erosion filter in order to remove spuriously marked cells (e.g., due to uncertainty in object locations and discretization effects) and determine the outline of the area using a contour walking algorithm.

6   Experimental Evaluation

We perform an experimental evaluation with test data collected throughout Germany. An overview of the test data is presented in Fig. 8. We started to collect prohibition signs for human outdoor activities systematically in the surrounding of our university campus and extended this collection to local areas. For each identified prohibition sign we obtain a geo-tagged photo of the sign, assigned the corresponding subset of prohibitions in OSM fashioned key-value pairs and we manually determine ground truth by surveying the local context of a sign. The area of applicability we identify is stored as polygon in a KML file or, if OSM already presents the corresponding object, we store the respective OSM unique object identity number. KML files for specifying ground truth are only necessary in situations where the corresponding OSM object is missing – these cases are particularly hard to solve since they require the system to generate a plausible region of applicability. In evaluation, we thus differentiate between instances specified by KML and OSM objects. To quantify the contribution of our approach, we compare our results with the performance of an algorithm that tags the nearest geographical object as described in [11].

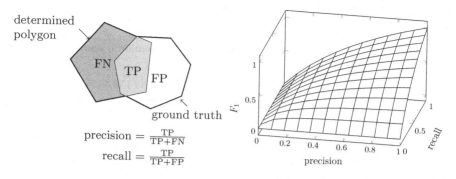

precision $= \dfrac{\text{TP}}{\text{TP+FN}}$

recall $= \dfrac{\text{TP}}{\text{TP+FP}}$

Fig. 9. Illustration of F_1 quality measure (Eq. 4) of determined polygon with respect to ground truth. TP, FN, and FP represent the size of the respective subregions.

6.1 Method

Our implementation computes the five most plausible regions of applicability for each geo-tagged prohibition sign. Then, we determine the quality of the highest-ranked region as well as the highest quality within all five regions. Quality of a region determined is computed by comparison with ground truth, using the F_1 performance measure that determines the harmonic average of precision and recall (see also Fig. 9):

$$F_1 = 2 \cdot \frac{\text{precision} \cdot \text{recall}}{\text{precision} + \text{recall}} \tag{4}$$

This leads to the following measures considered:

- average of F_1 scores of top-1 reply
- percentage of top-1 replies with non-zero score (i.e., not completely wrong)
- average of best F_1 score among top-5 replies
- percentage of top-5 replies with non-zero best score.

The four measures are then further subdivided for problem sub-classes providing ground truth as object in the database (ID instances) or as new polygons not registered in OSM (KML instances). For the comparative evaluation, the nearest object approach described in [11] is adapted to return the 5 nearest objects.

6.2 Discussion

The results of the comparative analysis are presented in Table 2. As can be clearly seen, the presented semantic search for suitable object outperforms the choose-the-nearest approach in all cases. While the overall percentage in which the top-5 entities determines at least overlaps with ground truth ($F_1 > 0$) is slightly higher and the F_1 measure is significantly higher for semantic search.

Table 2. Results of the comparative evaluation of semantic search against the nearest object assignment proposed by [11]

Test condition	Nearest	Semantic search
All 77 instances:		
Avg. F_1 for top-1	0.15 (29.87 % > 0)	0.43 (58.44 % > 0)
Avg. F_1 for best in top-5	0.54 (80.52 % > 0)	0.72 (88.31 % > 0)
51 ID instances:		
Avg. F_1 for top-1	0.20 (29.41 % > 0)	0.55 (62.75 % > 0)
Avg. F_1 for best in top-5	0.68 (78.43 % > 0)	0.92 (92.16 % > 0)
Correct ID as top-1 result	19.61 %	52.94 %
Correct ID in top-5 result	66.67 %	92.16 %
26 KML instances:		
Avg. F_1 for top-1	0.04 (30.77 % > 0)	0.19 (50.00 % > 0)
Avg. F_1 for best in top-5	0.26 (84.62 % > 0)	0.33 (80.77 % > 0)

In our study using a wide range of prohibition signs, the approach of choosing the nearest objects falls short of the 97.6 % accuracy reported in [11] for no-smoking signs. One reason for this difference is that no-smoking signs are often attached to the entrance of restaurants which is easier to interpret than, for example, the correct area in which dogs are to be kept on a leash. Considering the ID instances which "only" require identifying the correct object, the high uncertainty in the data leads to an accuracy for the top-1 result of 52.94 % using our method as compared to 19.61 % using choose-the-nearest. This demonstrates that presentation of alternative interpretations as shown in Fig. 10 is necessary. Here, five plausible interpretations of a no-smoking sign are shown. The second and third alternative are selected since they include a node representing a retail shop, the other two buildings are (erroneously) registered as residential homes.

With semantic search the top-5 objects include the correct entity in 92.16 % of the cases, whereas the accuracy of choose-the-nearest only achieves 66.67 %. The performance with respect to KML instances is as expected lower than for ID instances since no suitable region is contained in the OSM database. This leads to a very poor performance of the choosing the nearest object approach. By contrast, our approach to construct a new region still achieves the F_1 score of 0.19 for the top-1 answer. We observe that by selecting the 5 nearest objects close by and constructing regions around point-objects (nodes), there is a high chance of obtaining at least one region that overlaps with the desired scope, leading to a positive F_1 score. The percentage of having $F_1 > 0$ does thus not provide insight on its own but reveals how many instances contribute to the overall F_1 score achieved.

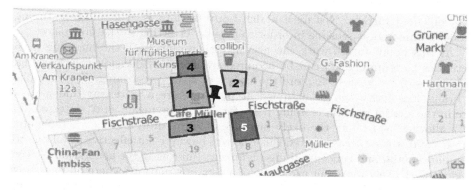

Fig. 10. Correctly classified example with presentation of uncertainty. The pin indicates the "no smoking" sign location that refers to the cafe outlined in green, but only represented as node in OSM. Alternative interpretations of the scope are marked with relative plausibility, color-coded from green to red (Color figure online).

7   Conclusion and Outlook

In this paper we present an approach to identify the geographical scope of prohibition signs related to human outdoor activities that is foremost based on semantics. By comparison with a previously studied choose-nearest-entity approach we demonstrate the utility of considering semantics of prohibition signs and spatial entities. With the performance achieved we believe that VGI applications that wish to capture the scope of prohibitions already can benefit from this method. From the comparison we can also conclude that introducing a semantic level is a crucial step towards automating the interpretation of prohibition signs. We expect that this observation generalizes to similar information retrieval tasks with geo-spatial databases that rely on information that is only implicitly represented by the database. Introducing a semantic level enables reasoning to sensibly combine all factors that contribute to coming up with a good answer. In our application these factors are spatial proximity, type of entities as well ass the activities they allow for, and local context. We propose a rule-based approach to spatial-semantic reasoning that explicitly considers uncertainty.

A specific challenged faced in our application is that OSM in 33 % of the test cases does not include an entity that represents the desired scope, i.e., we are not involved with a classical query task. Sensibly responding to such cases requires us, firstly, to recognize the lack of an appropriate entity within OSM and, secondly, to instantiate a plausible region within the local context of the sign. We tackle this challenge by a declarative rule-based approach to reasoning which allows us flexibly re-use parts of the rule-base for instantiating new regions.

Since one cannot claim to have identified the best-performing approach so far without an excessive empirical evaluation, all experimental data and our implementation is published online[5] and can be used by other researchers and

---

[5] https://github.com/hopfkons/prohibitionSigns.

supplemented further. Not until a rich body of data is available, fine-tuning model parameters is sensible in order to avoid over-fitting in the first place.

At this point, future work can also investigate how well the mapping from activities to requirements can be acquired by means of decision tree learning. Additionally, we wish to explore the utility of spatial reasoning. For example, a bridge-like entity can be assumed at places where a waterway and a road are in the spatial relation crossing (e.g., formalized by means of 9^+-intersection [9]), allowing us to infer the likely presence of a guardrail and its whereabouts. With the test instances collected this rule was however not effective at all due to complex spatial and conceptual misalignment. Improving the generation of new regions that likely represent the scope of a sign in situations where no matching object is registered in the database presents itself as a very difficult but fruitful subject of further studies.

Acknowledgements. Financial support of Technnogieallianz Oberfranken (TAO) is gratefully acknowledged. The approach presented in this paper is a significantly revised method previously submitted to a student's programming competition organized by the German Informatics Society. We thank Alexander Baumgärtner for his contribution to implementing the system.

References

1. Ballatore, A., Bertolotto, M., Wilson, D.: Geographic knowledge extraction and semantic similarity in OpenStreetMap. Knowl. Inf. Syst. **37**(1), 61–81 (2013)
2. Bennett, B.: Space, time, matter and things. In: Welty, C., Smith, B. (eds.) Proceedings of the 2nd International Conference on Formal Ontology in Information Systems (FOIS 2001), pp. 105–116. ACM, Ogunquit (2001)
3. Bennett, B., Mallenby, D., Third, A.: An ontology for grounding vague geographic terms. In: Eschenbach, C., Gruninger, M. (eds.) Proceedings of the 5th International Conference on Formal Ontology in Information Systems (FOIS-08). IOS Press, Amsterdam (2008)
4. Bundesanstalt für Arbeitsschutz und Arbeitsmedizin (BAuAs): Technische Regel für Arbeitsstätten ASR A1.3. Gemeinsames Ministaralblatt (GMBl) 2013(16), 334–347 (2013) (In German)
5. Bundesministerium für Verkehr, Bau und Stadtentwicklung (BMVI), Bundesministerium für Umwelt, Naturschutz und Reaktorsicherheit (BMU): Verordnung zur Neufassung der Straßenverkehrs-Ordnung (StVO) vom 6. März 2013. Bundesgesetzblatt 2013(12), 367–427 (2013) (In German)
6. Fonseca, F.T., Egenhofer, M.J.: Ontology-driven geographic information systems. In: Proceedings of the 7th ACM International Symposium on Advances in Geographic Information Systems, pp. 14–19. ACM (1999)
7. Janowicz, K., Raubal, M., Kuhn, W.: The semantics of similarity in geographic information retrieval. J. Spat. Inf. Sci. **2011**(2), 29–57 (2011)
8. Kanevski, M., Pozdnoukhov, A., Timonin, V.: Machine learning for spatial environmental data: theory, applications, and software. EPFL press, Lausanne (2009)
9. Kurata, Y.: The 9^+-Intersection: a universal framework for modeling topological relations. In: Cova, T.J., Miller, H.J., Beard, K., Frank, A.U., Goodchild, M.F. (eds.) GIScience 2008. LNCS, vol. 5266, pp. 181–198. Springer, Heidelberg (2008)

10. Penn, A., Turner, A.: Space syntax based agent simulation. In: 1st International Conference on Pedestrian and Evacuation Dynamics (2001)
11. Samsonov, P., Tang, X., Schöning, J., Kuhn, W., Hecht, B.: You cant smoke here: Towards support for space usage rules in locationaware technologies. Technical report 14–022, University of Minnesota (2014)
12. Schwering, A.: Approaches to semantic similarity measurement for geo-spatial data: a survey. Trans. GIS **12**(1), 5–29 (2008)
13. Ulmer, A., Halatsch, J., Kunze, A., Müller, P., Van Gool, L.: Procedural design of urban open spaces. Proceed. eCAADe **25**, 351–358 (2007)
14. Šegvič, S., Brkić, K., Kalafatić, Z., Pinz, A.: Exploiting temporal and spatial constraints in traffic sign detection from a moving vehicle. Mach. Vis. Appl. **25**(3), 649–665 (2014)
15. Van Hage, W.R., Wielemaker, J., Schreiber, G.: The space package: Tight integration between space and semantics. Trans. GIS **14**(2), 131–146 (2010)

Conceptualizing Landscapes

A Comparative Study of Landscape Categories with Navajo and English-Speaking Participants

Alexander Klippel[1]([⊠]), David Mark[2], Jan Oliver Wallgrün[1], and David Stea[3]

[1] Department of Geography, The Pennsylvania State University, University Park, USA
{klippel,wallgrun}@psu.edu
[2] Department of Geography, University at Buffalo, Buffalo, USA
dmark@buffalo.edu
[3] Center for Global Justice, San Miguel de Allende, Mexico
david.stea@gmail.com

Abstract. Understanding human concepts, spatial and other, is not only one of the most prominent topics in the cognitive and spatial sciences; it is also one of the most challenging. While it is possible to focus on specific aspects of our spatial environment and abstract away complexities for experimental purposes, it is important to understand how cognition in the wild or at least with complex stimuli works, too. The research presented in this paper addresses emerging topics in the area of landscape conceptualization and explicitly uses a diversity fostering approach to uncover potentials, challenges, complexities, and patterns in human landscape concepts. Based on a representation of different landscapes (images) responses from two different populations were elicited: Navajo and the (US) crowd. Our data provides support for the idea of conceptual pluralism; we can confirm that participant responses are far from random and that, also diverse, patterns exist that allow for advancing our understanding of human spatial cognition with complex stimuli.

Keywords: Landscape · Category construction · Cultural differences

1   Introduction

The way humans understand their natural environments—landscapes—either as individuals or as a collective frames prominent research topics in several disciplines. From a geographic perspective, it can be argued that the "man-land tradition" or in more modern terms "human-environment relation" is one of the four intellectual cores of geography (Pattison, 1964). However, geography is by no means the only discipline interested in landscapes and human-environment relations prominently are featured in anthropology, philosophy, psychology, linguistics, or landscape architecture.

As pointed out by Mark and collaborators (Mark et al., 2011b), there is a surprising lack of scholarly research on how landscapes are conceptualized; in Mark's words "… how a continuous land surface, a landscape, becomes cognitive entities, and how those

© Springer International Publishing Switzerland 2015
S.I. Fabrikant et al. (Eds.): COSIT 2015, LNCS 9368, pp. 268–288, 2015.
DOI: 10.1007/978-3-319-23374-1_13

entities are classified and represented in language and in thought." (p. 1). While geography and related disciplines have a long history of studying toponyms (Jones and Purves, 2008), less focus has been placed on the relation between general geographic/landscape categories and how they are linguistically referred to (for exceptions see, for example, (Mark et al., 2011a)).

The lack of research that Mark and colleagues identified can be partially explained with the associated challenges of cognition in the wild (Hutchins, 1995). From a theoretical perspective, researchers may argue that there is an essentially infinite number of ways to make sense of the world (Foucault, 1994). And even if we are not using the mathematical concept of infinity, people might still argue that the number of ways to conceptualize a part of reality is simply too large and context dependent (Keßler, 2010), to be addressed in lab-style experiments. Others may argue, however, that there are distinct characteristics in our spatial environments that favor a (small) subset of category structures that humans would intuitively focus on. An example of the latter is work by Medin et al. (1987) who make the point that humans normally use and create only a tiny subset of the many ways that information could be partitioned, and that a central question in the cognitive sciences is to reveal principles that underlie category construction behavior (see also [Malt, 1995]).

Along these discussions, we find a parallel debate on the importance of similarity for the construction of categories in that prominent researchers argue for a similarity grounded understanding of human categorization processes (Goldstone, 1994; Goldstone and Barsalou, 1998) but that other researchers make a strong case against similarity arguing that in order to be relevant, similarity needs to be constrained by an explicit understanding of "similar in relations to what" (Rips, 1989).

Our own understanding of the matter can best be described with *conceptual pluralism*, a term we use in the sense of Wrisley (2008). Conceptual pluralism seeks a middle ground between a one-size-fits-all approach and the assumption of an infinite number of intuitive ways the world can be made sense of. We are subscribed to the notion that intuitive conceptual structures exist and that there is not an infinite number that humans use naturally. However, we also acknowledge that the real world (in contrast to most lab experiments) is complex and that there may or may not be a single intuitive understanding/category structure of the world that fits every human being or even the majority of people.

We demonstrate an experimental approach to conceptual pluralism by combining field studies with crowd-based experiments addressing the conceptualization of landscapes. This research has intimate ties to landscape perception (Habron, 1998), image retrieval and image similarity research (Ul-Qayyum et al., 2010), fundamentals of categorical perception (Harnad, 2005), ethnophysiography (Mark et al., 2007), and the role of perceptual similarity (Malt et al., 1999). Given the exploratory nature of our research, to combine field studies with crowdsourcing, we made a conscious decision to skip detailed background discussions that we have provided elsewhere, in favor of more analyses.

2   Experiments

In the following we detail two experiments that provide insights into the challenging question of how humans intuitively conceptualize landscapes and landscape features as seen in photographs. In order to gain an understanding of how diverse landscape concepts can be on the level of individuals as well as for culturally and/or linguistically distinct groups, we discuss and compare two data sets which resulted from an experiment conducted as a field study with Navajo and a crowdsourced study through Amazon Mechanical Turk (AMT), which used the same stimuli but a somewhat different experimental protocol.

2.1   Preliminaries: Selection of Landscape Image Stimuli

In collaboration with Andrew Turk, Mark and Stea have been conducting research on Navajo landscape conceptualization and classification since 2003 (Turk et al., 2011). Most of the work so far has been qualitative, using ethnographic methods. In the early phases of this research, Mark and Turk took many hundreds of landscape photographs in and near the Navajo reservation. In 2004 they chose a stratified sample of 106 photographs to use in sorting and other photo-response experiments. Initially, the 106 photographs included 17 landscape images from northwestern Australia. The remaining 89 images were selected so as to have a relatively uniform number of images across landform types and also across regions of the Navajo Reservation. In the sampling design, they used 7 arbitrary geographic regions of the Reservation, and 8 researcher-defined types of landforms. They hypothesized that participants might group the images by kinds of landforms, by regions, or by traditional (origin) stories. The types of landforms used in the design were: mountains/hills, buttes/monoliths, valleys/canyons, cliffs, flats, watercourses, standing waterbodies, and 'other' (caves, arches, dunes). Of course, these are etic categories defined by English speaking researchers. These 89 landscape images have been used in several experiments and data collection protocols, mostly with unpublished results. The corpus was augmented by a few additional photographs representing landscape categories that were not included among the original 89. For the current experiments, Mark and Stea resampled to select a subset of 54 photographs to use in a tabletop photo sorting experiment in 2010. Those same photographs were used in Experiment 2.

2.2   Experiment 1 - Navajo

Participants. Eight Navajo participants worked in pairs (i.e., four groups) and grouped hard-copy images of the stimuli (described above, see various Figures below and companion website; www.cognitiveGIScience.psu.edu/landscape2210.html). All Navajo participants were volunteers recruited by Carmelita Topaha (the project consultant), and were paid $20 per hour.

We did not ask Navajo participants for their ages. They were all fluent speakers of the Navajo language, and all registered members of the Navajo Nation. The first pair were two male Navajo elders who sorted the photographs in October 2010; both of them had participated earlier in other aspects of the project, and were familiar with some of the photographs; we estimated that they were in their 70 s, and we did not record the duration of this session. The other six participants who performed the sorting were all female Navajos, tested in June 2011. The first pair of female participants were two sisters, estimated to be in their 70 s; this session took 30 min. The second pair were probably in their 50 s or 60 s, and took only 10 min to sort the photographs and describe the groups. The third pair were a mother and daughter, with the mother probably in her 60 s, and their session lasted 18 min. None of these latter six had participated previously in Mark and Stea's landscape research.

Procedure. The rationale for having people work in pairs was to generate discussion preferably in the Navajo language, and to have a consensus process within each session. All sessions were audio-recorded, with the informed consent of the participants.

Mark and Stea had a research permit from the Navajo Nation, but all testing was conducted off the Reservation at San Juan College in Farmington, New Mexico. The 54 photographs were printed as glossy (4 by 6 inch) prints on photographic paper. The photographs were shuffled and given to the participants, who were asked to divide the photographs into as many groups as they wanted. They laid out the images on a large table in rows and columns, sorting as they laid them out (see Fig. 1). The participants then rearranged the photographs as needed until they were satisfied with their groups. After this, they were asked to give a brief description of each group, and to select one photograph from each group that best represented the group (a prototype).

Fig. 1. Two Navajos part way through the sorting of the landscape images; note that the "buttes and monoliths" pictures already form a cluster along the side of the table.

Some Results. The four pairs of Navajo participants (NAV) created on average 13.5 groups (SD 2.65) with a minimum of 11 and a maximum of 17. No two pairs of participants created the same number of groups. The relatively small number of participants does not allow for a purely quantitative analysis but focusing on some qualitative characteristics (supported by cautiously using quantitative support) provides valuable insights. One approach was to label the groups for each pair of participants, place those group labels in a table, and sort, to see which subsets of images were always put together. For example, two views of Shiprock and one other rocky butte were clustered by all four pairs of Navajo participants (Fig. 2, upper row left). Several other upstanding rock outcrops were put with these for 3 of the 4 Navajo groups. A cluster of upstanding rock outcrops was nominated by all four Navajo group-participants as most representative of landscapes in Navajo country. Otherwise, while there was considerable similarity in the image-grouping behavior of the Navajo participants, this was not reflected in unanimous consensus core groups of images.

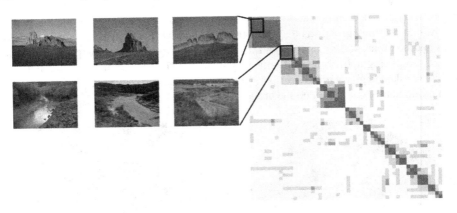

Fig. 2. Heatmap reordered based on cluster analysis (Ward's method). The cluster analysis is not the focus here but allows for a clearer picture. There are five distinct color groups according to the five different similarity ratings any pair of images could receive (0 to 4). Darker (red) shades indicates high similarity. Images (left) are those grouped together by all participants (Color figure online).

We visualized the category construction behavior of the four NAV paired-participants by using a heatmap that was ordered by a cluster analysis (Ward's method, which is not the focus but allows for making the heatmap readable), see Fig. 2. Certain groups of landscape images form relatively strong categories while other are more varied; this of course is based in part on the selection of images for the test set, if similar pictures were chosen or not. What we did not see were clusters based on geographical areas of the Reservation, or groups based on connections in origin stories. Instead, groups seemed to be based on similar landforms, or similar land cover, or mere visual resemblance of the photographs (colors, patterns, textures). Such a response is similar to what we have found for English-speaking participants from the general US population (see Experiment 2).

One of the largest, most consistent groups is composed of visually salient convex landscape features. A sub-group is displayed in Fig. 2. This group corresponds to an image pre-classification by Mark and collaborators: buttes and monolith. All other groups are essentially smaller and more diverse.

Figure 2 shows a second example of three images that all participants placed into the same category. While this is only a small piece of the grouping behavior, it demonstrates an important component of landscape perception: water features are often used to ground conceptualization processes as they are relatively easy to identify (Sparks et al., 2015b).

Prototypes. After participants performed the category construction experiment they were asked to select the most prototypical image for each category they created. Figure 3 shows the four most frequently selected images by the NAV participants. We only present the first four prototypes as many images were tied with a rating of two.

Fig. 3. The four most frequently selected prototypical images by NAV participants. Numbers below the images indicate frequency.

Linguistic Descriptions. Figure 4 shows a Wordle based on the short descriptions for each category that the NAV participants provided. We will revisit the linguistic descriptions when we compare them to descriptions provided by AMT participants.

Fig. 4. Wordle of the short linguistic labels from NAV participants.

2.3 Experiment 2: American English Speakers Recruited through AMT

To contrast the landscape concepts elicited form Navajos with a more general perspective on landscape concepts, we conducted a second experiment with the general US public through the crowdsourcing platform Amazon Mechanical Turk.

Participants. We recruited 40 participants through AMT. Three of these participants were excluded from the analysis for the following reasons: one focused on only the sky in the images, and two participants provided identical descriptions for two different categories. Average age of the remaining 37 participants was 33.55, 26 participants were male. Participants received $1.20 plus $.25 bonus for their participation.

A Note on AMT. There is a substantial amount of discussion regarding the validity of AMT for scientific research. While this article is not focusing on evaluating AMT, we will include here some general comments on the validity of using AMT for our experiments. We are calculating the payment such that on average participants receive at least the minimum pay in the US. That means we do not run a sweatshop and our experiments are popular and provide an incentive to actually spend some time with them. There are numerous articles that generally attest to the validity of experiments run through AMT (Crump et al., 2013; Rand, 2012) and point out the greater variety of people who participate, that is, the sample reflects the general public better than classic student samples. For an article that addresses conceptual pluralism such as the one presented here, this is a positive feature. We would also like to point out that our experiments are very different from standard/most AMT HITs (Human Intelligence Tasks). The recently discussed fatigue syndrome of people doing hundreds of highly similar HITs (Marder, 2015) does not apply to our experiments as they are rather unique. Finally, and we hope that this will convince critics the most, we have gone through each response individually. While there is the occasional inconsistency in the categorization of images, all responses make sense. In the spirit of making science transparent, we have placed the anonymous data collected in this experiment onto the companion website such that, if a reader is in doubt, he or she may convince herself that the provided responses are thoughtful and consistent.

Materials. The images were the same as in Experiment 1. The photographs were reduced in size and displayed as 160×120 pixel images on the screen as part of the CatScan software package (Klippel et al., 2008).

Procedure. Individual experiments were posted to AMT's website as HITs, Once a HIT was accepted by a worker, she was instructed to download the standalone Java version of CatScan and work on the experiment with a unique participant number assigned to her/him. At the beginning of the experiment, participants were required to enter their demographic information such as age, gender, native language, and educational background. After that, participants were asked to read the experiment instructions (see below), which introduced the basics of the experiment. Participants were only allowed to proceed after a certain time and had to enter text into a box to ensure that they read and understood the instructions. A warm-up task was set up to acquaint participants with the interface and the idea of category construction by sorting animals into groups. In the main experiment, all 54 icons were initially displayed on the left panel of the screen. Participants were asked to sort icons into categories they had to create on the right panel of the screen. Once all icons were sorted into categories, they were able to proceed to the second part of the experiment. Here they were presented with the categories they created, one category at a time, and asked to provide a short label (no more than five words) and a detailed description to articulate the rationale(s) of their category construction behavior. They were also asked to select one image in each group as a prototypical representation of that category.

Results. We will review AMT specific results but also include a direct comparison to results from Experiment 1. Participants created on average five groups with a standard deviation of 2.31. The minimum number of groups created was three, the maximum 16. This means that AMT participants created significantly fewer groups than NAV participants ($t = 6.91$, $df = 39$, $p < .001$). Average grouping time was 6.5 min (SD 233 s) for AMT, and 19.3 min for NAV.

Cluster Analysis/Cluster Validation. The larger number of participants allows for statistical approaches to understand the category construction behavior. The category construction behavior of each participant was recorded by CatScan in an individual similarity matrix (ISM). An ISM is a 54×54 binary matrix that encodes the similarity rating between all pairs of images (54 is the total number of images used in the experiment). For each pair of icons in the experiment, the corresponding similarity rating is 1 if they were placed into the same group, and 0 if not. By summing up all 37 ISMs in the experiment, an overall similarity matrix (OSM) is obtained. In the OSM, the similarity rating for a pair of icons ranges from 0 (lowest similarity possible) to 37 (highest similarity possible based on 37 participants in AMT). The corresponding similarity values for NAV are 0 to 4. For the quantitative analysis we focus on AMT data.

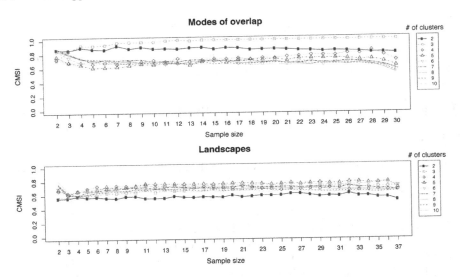

Fig. 5. Top: CMSI plots from an experiment on overlap relations showing that the three-cluster solution stands out with optimal CMSI values even for small numbers of participants. Bottom: CMSI plot for Experiment 2 showing that CMSI values remain comparatively low for all cluster sizes and all sample sizes.

To reveal the dominant category construction behavior of all participants, we performed cluster analyses and cluster validation based on the OSM. To this end, we have developed a cluster validation approach called Cross-Method-Similarity-Index, short CMSI, that incorporates a variety of best practice suggestions from the literature (for details see (Wallgrün et al., 2014)). In a nutshell, the CMSI captures the similarity of the groupings we get from applying three different hierarchical clustering methods: Ward's method, average linkage and complete linkage. CMSI values are computed for different number of clusters/categories. We are computing average CMSI values for samples of varying size from the participant pool, essentially a bootstrapping approach (Boos and Stefanski, 2010). A CMSI value of 1 means that the resulting groupings are identical for all three methods, while, if the groupings strongly differ, the CMSI value will be significantly lower. We then produce and analyze plots that show the average CMSI values for different sample sizes.

Without going into too much detail, the main result is that there is no unique (one-size-fits-all) clustering solution for the landscape images of AMT participants. This is an indication of conceptual pluralism and is in strong contrast to experiments on fundamental spatial concepts such as topology (Klippel et al., 2013) that produce, in all of our previous experiments, a unique category structure. For illustration purposes, Fig. 5 displays reanalyzed results from an experiment on overlap relations (Klippel et al., 2013) against the results from the current landscape experiment. The important thing to note here is that in the analysis of the overlap data it is sufficient to have 11 participants to reach a perfect score, which indicates that, abstracting from individual differences, a three category solution (non-overlapping, overlapping, proper part relations) is strong and consistent and a viable generalized view on how many spatial relations humans intuitively distinguish with respect to overlapping spatially extended entities. In contrast, the landscape data shows conceptual pluralism that does not allow a single best solution to surface as the most valid interpretation of the category structure participants have applied to these stimuli.

Despite the existence of conceptual pluralism, we strongly argue against the possibility of an infinite (essentially random) approach by participants, or data that is simply too diverse to be analyzed statistically. While there are clearly competing perspectives possible on the stimulus the participants were presented with, it is not the case that every participant created his or her own category structure. Salient environment features are selected to anchor the category construction behavior and there is not an infinite number of salient environmental features from a cognitive perspective. What makes the analysis, however, challenging is that there are not only different perspectives on the stimuli but that the perceptual and conceptual salience of individual images varies, too. We will use this case study to systematically look into these issues that are relevant for cognitive and information scientists alike. To this end we have developed a number of analysis methods specifically for this paper, that is, they have not been reported elsewhere.

Statistical Significance of the Grouping Behavior. In order to exclude the possibility that there is an infinite number of possibilities present in the participants grouping behavior, we tested the OSM for statistical significance. We compared the results from the experiment to Monte Carlo simulations of an experiment with the same number of participants and icons. Computing the z-scores for the variances of the rows in the OSM (each row corresponds to a particular icon, see also Figs. 10 and 11) resulted in highly significant z-scores (associated p values all < .001), clearly showing that the grouping behavior is far from being random. The z-scores are consistent with the results in Section image-variance analysis in that images identified as conceptually more ambiguous have lower z-scores.

Given the inconclusive results of the CMSI analysis we chose to visualize the participants grouping behavior using multi-dimensional scaling (MDS) as it attempts to make sense of the category construction behavior in a continuous rather than categorical sense. Figure 6 shows the results of this analysis for both the AMT participants and the Monte Carlo simulation. The AMT MDS plot has three clearly distinguishable axes along which the images overall are positioned. We labeled these axes (their intersections) as: flatlands, protrusions, water features. To better understand the MDS results, we visualize the three corners/intersection of the axes individually in Figs. 7, 8 and 9, replacing the dots in the MDS plots with actual landscape images.

Identifying Images Grouped Together Frequently (Grouping Frequency). This analysis looks at the dendrogram resulting from a hierarchical cluster analysis of the overall similarity matrix (OSM). It starts at the root and from there moves through the split points in the dendrogram until it reaches the last split point at the bottom where the first two icons are merged. For each split point, it makes the cut through the dendrogram to get the groups existing at that level. For each of these groups, it checks how many (and which) participants have put the icons from that group together into one group. This gives an idea of how frequently the theoretical groups from the hierarchical cluster analysis actually appeared in the grouping results of the

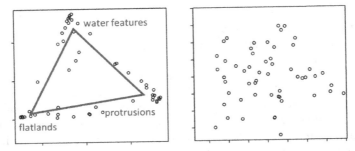

Fig. 6. An MDS plot of the grouping behavior of AMT participants (left) and an MDS plot of the randomized grouping behavior (right). Despite individual differences, the MDS analysis reveals a clear tripartite organization of the grouping data: flatlands, protrusions, and water features. The three parts are shown in detailed figures below.

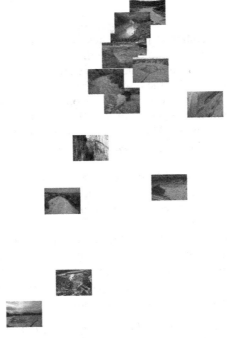

Fig. 7. Water features. The upper left corner of the MDS plot for AMT participants includes mainly washes (dry stream beds), flowing watercourses, and pools.

participants. Results from applying this method together with the method described in the following will be given below.

Image-Variance-Analysis. In order to deepen our understanding of conceptualization processes as an interplay of participants' individual perspective but also effects of the stimuli, we developed a method referred to as *image-variance-analysis*. It allows for

Fig. 8. Protrusions. The lower right corner of the MDS plot for AMT participants includes mainly buttes, mesas, and cliffs.

Fig. 9. Flatlands. The lower left corner of the MDS plot for AMT participants includes flatlands, and other unobstructed views of the horizon.

assessing how distinct the grouping behavior for a particular image is. The analysis takes a row of the OSM, which codes for each image how often it has been placed into the same category with any other image across all participants, and calculates the variance for this category construction behavior. Higher variances mean more distinct grouping behavior; lower variances mean that an icon has similarities across many or potentially all other images.

To demonstrate the effectiveness of these two analyses, we selected two groups of images, one that was consistently placed into the same category by the majority of participants and the group of the four images with the lowest variance values. The analysis nicely reveals the high (Fig. 10) and low (Fig. 11) conceptual distinctness of images. Images of high distinctness show a combination of very high and very low similarity values in the plots of the corresponding row; images of low distinctness are more balanced.

Fig. 10. Identification of some of the least ambiguous landscape images. Shown are images from highly consistent group together with the image mean/variance analysis for each image.

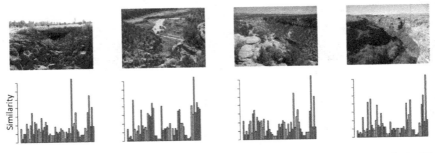

Fig. 11. Conceptually most ambiguous images as identified by image-variance-analysis. These are all longitudinal depressions.

Participant Similarity Analysis. The mutual similarity of the grouping behavior of the participants can also be computed from the image grouping results. The clustering decisions of the Navajo participants were entered into CatScan by the authors, to facilitate comparison of the Navajo and AMT participants. Participant similarity analysis measures the similarity between participants based on individual similarity matrices (ISMs). To this end, a 41-by-41 between-participant similarity matrix (BSM) was constructed to encode the similarity of category construction behavior for each pair of participants. In the BSM, the similarity between a pair of participant is determined by computing the Hamming distance between the ISMs of two participants. Cluster analysis using Ward's method performed on the BSM allowed us to identify participants who employed similar category construction strategies.

All four Navajo participants (numbers ending 101, 102, 103, and 104) fell into a distinct group of six participants. Evidently, the image clustering decisions of the Navajo participants were distinctly different from most of the AMT participants. We could see nothing in the background information about the two AMT participants who grouped

Fig. 12. Wordle of word frequencies, short labels, AMT participants.

with the Navajos (035, 013) to account for their results being more similar to those of the Navajos. Participant 013 created the largest number of groups of any AMT participant (16), participant 035 created seven groups.

Word Frequency Analysis and Comparison. All participants in both experiments provided verbal descriptions of criteria for each of their groups, except for the first pair of Navajos. For the other Navajo participants, we have one description per pair of participants per group. However, the Navajos formed more groups and supplied longer descriptions. We eliminated 10 check words such as articles and conjunctions from the descriptions before the analysis. The Navajo descriptions from three pairs of participants included 295 non-check words, and 161 distinct words, of which 41 were used more than once. The 37 AMT participants used a total of 365 non-check words, with 50 words used more than once (see Fig. 12).

The most obvious difference is "landscape" (see Table 1), which was mentioned by about half of the AMT participants, but by none of the Navajos. "Desert" is also missing from the Navajo descriptions. The low frequency of "mountain" among the Navajos is interesting. In unpublished results of photo description experiments, Mark and Stea noted that Navajos often paid more attention to the foregrounds of the images, especially the vegetation, than English-speaking Americans, who tended to focus attention on larger more distant landforms in the background.

Comparing Linguistic and Conceptual Variation. Adding to the analysis of linguistic descriptions of landscape images, we compared the labels that have been provided for the images with the highest and lowest conceptual variability as revealed by the image-variance-analysis.

Table 1. Nine terms with the largest excess of occurrence for the AMT/Anglo participants, compared to the Navajos, adjusted for the total number of words.

Term	AMT	Navajo	Total
Landscape	18	0	18
Water	23	5	28
Mountain	17	4	21
Desert	12	0	12
Formation	12	2	14
Rocks	30	17	47
Road	8	0	8
Rivers	7	0	7
Flat	6	0	6

The results of analyzing the linguistic labels, as varied as they are, shows that in case of Fig. 13 all labels indicate concepts potentially found in the lower left corner of the MDS plot (see Fig. 9), hence, as diverse as the linguistic labels are, the conceptual distinctness of this image is high. In contrast, the image in Fig. 14, which has the lowest variability of all images, has relations to all three main axes identified in the MDS analysis (see Fig. 6).

Fig. 13. Short labels associated with the image with the highest conceptual variance (most distinct grouping).

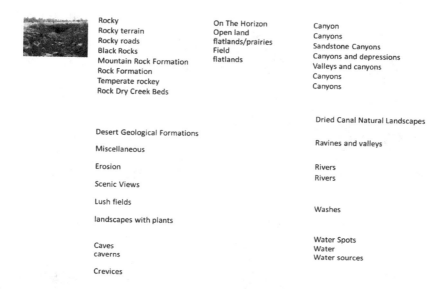

Rocky	On The Horizon	Canyon
Rocky terrain	Open land	Canyons
Rocky roads	flatlands/prairies	Sandstone Canyons
Black Rocks	Field	Canyons and depressions
Mountain Rock Formation	flatlands	Valleys and canyons
Rock Formation		Canyons
Temperate rockey		Canyons
Rock Dry Creek Beds		
		Dried Canal Natural Landscapes
Desert Geological Formations		
		Ravines and valleys
Miscellaneous		
Erosion		Rivers
		Rivers
Scenic Views		
Lush fields		
		Washes
landscapes with plants		
		Water Spots
Caves		Water
caverns		Water sources
Crevices		

Fig. 14. Short labels associated with the image with the lowest conceptual variance (least distinct grouping).

Prototypes. Figure 15 shows the nine images most frequently selected as prototypes by AMT participants. Not surprisingly, the image with the greatest conceptual distinctness (Fig. 13) is the third most frequently selected image (14 times); in contrast, the image with the least conceptual distinctness (Fig. 14) was selected four times. This is, however, above the median (three).

An examination of the prototypes provides some interesting insights into the image sorting procedure. An important thing to keep in mind is that a prototype is a prototype for some particular set of images or for some abstract classification procedure. But almost every group of images is unique to one participant. So, when we count how often a particular image was selected as a prototype, we must keep in mind that it was being a prototype for different groups. Each participant categorized 54 images. The more categories a participant created, the smaller the average size of the group. With one prototype per group, the smaller group (fewer instances), the more chance of each member of the group being selected as a prototype.

In addition, there is a feedback effect: selection of prototypes may occur during the division of the images into groups and then similarity to a candidate for prototype might influence the construction of the groups themselves. If all N images in a group are equally likely to be selected as the prototype, then the chance of being a prototype is 1/N. This principle could be a null hypothesis and used to assess the significance of being selected as a prototype. If participants create a small number of image groups, then on average the groups must be large, and particular images will have a low a priori chance of being selected as prototypes. Until a statistical test is developed that incorporates this principle, our interpretation of prototype frequency will remain qualitative.

Fig. 15. Shown are the nine images that have been selected most frequently as prototypes by the AMT participants. Below the image is the raw number. The black boxes indicate images that are also in the top four of the NAV participants.

2.4 Discussion

Although most of our analyses are exploratory, that is, we did not test a specific hypothesis, there are a number of important insights. Certain images are perceptually and in turn conceptually more salient than others. Perceptual salience can be induced by a single feature (e.g., a monolith) or by a prototypical landscape (e.g., flatland). The focus on perceptual salience is an indication for perceptually grounded, similarity based categorization processes, especially as it occurred in both experiments.

Despite the lack of a statistically identifiable clustering structure (CMSI analysis) that fits all AMT participants, the MDS analysis revealed a high-level category structure that intuitively makes sense (water features, flatlands, protrusions) and is also reflected in a qualitative analysis of category construction behavior of many participants. What we find is that participants with finer distinctions often partition within this larger tripartition. The reason that the tripartition does not surface statistically as the overarching structure is that there are alternative views (e.g., a focus on vegetation, a focus on soil color, a separation of near or distant features) and the existence of perceptually and conceptually ambiguous landscapes (images).

The possibility of cultural or linguistic differences in perception or cognition is a 'Holy Grail' or an anathema for cognitive researchers, depending on their beliefs about the Sapir-Whorf hypothesis of linguistic relativism (Boroditsky, 2001; Gumperz and Levinson, 1996). Due in part to differences in the experimental protocols, the results of this study do not demonstrate language-based differences between Navajo and English-language conceptualization of landscape images. Ethnophysiography (Turk et al., 2011), which uses qualitative ethnographic research methods to elicit categories for landscape entities, has shown cultural/linguistic differences in basic landform categories. We found statistical differences in the number of categories the two groups of participants created and different ways of naming categories. Despite these differences, both groups placed a strong focus on perceptually identifiable characteristics.

It is interesting that the image that is conceptually confusing for AMT participants is the image most frequently selected as a prototype by NAV. It seems to be visually distinctive for both groups, but familiar and meaningful to Navajos (connected to an origin story) but unfamiliar and strange to the AMT participants.

These results, as a combination of qualitative and quantitative analysis together with the finding of grouping principles being (highly) significantly different from being random is interpreted by us as conceptual pluralism: A number of distinct perspectives exist on the stimuli used in these experiments. These perspectives are grounded in perceptually salient characteristics. There is not an infinite number of salient character-istics such that participants grouping behavior is significantly different from being random. While a larger number of potential perspectives on the stimuli exist, they often center on the most salient distinctions which, as a result, are the dimensions in the MDS plot.

3   Conclusions and Outlook

Our research takes place at the intersection of different research efforts on landscape conceptualization that we will briefly address in distinct paragraphs.

The crowd as a contributor to scientific endeavors is an omnipresent phenomenon in current research. To quote a new European funded project (Cobweb: cobwebpro-ject.eu): "The Citizen OBservatory WEB project seeks to increase the value and inter-operability of crowdsourcing technology to policy makers by enabling the fusion of citizen-sourced data with reference data from a range of sources including data published by public authorities." Other project such as GeoWiki echo the support for crowd-based earth observation in a similar direction: "Crowdsourcing is just starting for environ-mental monitoring [...] There have been lots of crowdsourcing projects in astronomy, archaeology, and biology, for example, but there hasn't been much on land use, and there is huge potential there." (Fritz, 2014). A crucial missing piece in order to under-stand the potential and limitations of citizens as sensors is an understanding of how the human conceptual and perceptual systems operates in interaction with real world infor-mation, whether in the field or in interaction with representations of spatial environments (see also Sparks et al., 2015a). By developing a solid basis through methodological inventions we have started to contribute to these efforts from a more basic science oriented perspective focusing on cognitive categories. Through methodological advancements (see below) and a focus on basic research questions we aim for making crowdsourced research a reliable source for earth observations.

Many of the analyses we introduced in this paper are still at the experimental stage, that is, while they are aided computationally through various software implementations, the challenge of understanding real world intuitive conceptualization processes (i.e., the associated complexity, or one might say messiness, of the data) requires substantial human interpretation and extensive hours of human analysis. Our goal is to scale up this process by refining methodological aspects. We strongly believe that the research efforts we placed into methodologically advancing category construction experiments will result in scientific revenue as the methods allows for truly connecting human concepts, human language, and environmental characteristics in real world settings.

We briefly discussed our current approach to assess the similarity of participants based on Hamming distance. This approach has limitations as it compares participants based on a single, undifferentiated value, that is, the overall similarity of two matrices. This method does not identify local characteristics of the matrices, which would be desirable, comparable to global and local spatial analysis methods (Anselin, 1995). In order to obtain a more precise assessment of participants' category construction behavior we are working on a local matrix similarity assessment. To this end, we are also in the process of using one potential advantage of crowdsourcing, that is, we are planning to scale up the number of participants to capture human conceptualization processes in its diversity.

The research presented here provides insights into participants' category construction behavior but also allows for formulating specific hypotheses that can and will be tested within our framework. Examples are: While there are individual differences among the general US public regarding their conceptualization of landscape images, the overall tendency of a tripartition (flatlands, protrusions, water features) will emerge in experiments with different stimuli. And, using crowdsourcing platforms that allow for easier cross linguistic distribution, we can test whether or not the tripartition into flatlands, protrusions, water features is universal across languages/cultures.

Finally, the image retrieval community (Kwaśnicka and Jain, 2011) has developed a number of approaches that allow for a more objective characterization of images with respect to their salience but also with respect to their content. One of the next steps in our analysis will be to relate behavioral data to more objective characterizations of images.

References

Anselin, L.: Local indicators of spatial association - lisa. Geogr. Anal. **27**(2), 93–115 (1995)

Boos, D., Stefanski, L.: Efron's bootstrap. Significance **7**(4), 186–188 (2010). doi:10.1111/j.1740-9713.2010.00463.x

Boroditsky, L.: Does language shape thought?: Mandarin and English speakers' conceptions of time. Cogn. Psychol. **43**, 1–22 (2001)

Crump, M.J.C., McDonnell, J.V., Gureckis, T.M., Gilbert, S.: Evaluating amazon's mechanical turk as a tool for experimental behavioral research. PLoS ONE **8**(3), e57410 (2013). doi:10.1371/journal.pone.0057410

Foucault, M.: The Order of Things: An Archaeology of the Human Sciences, Vintage Books edn. Vintage Books, New York (1994)

Fritz, S. (2014). *ERC funds IIASA crowdsourcing project - 2014 - IIASA*. Accessed on http://www.iiasa.ac.at/web/home/about/news/20140114-ERC-Crowdland.html

Goldstone, R.: The role of similarity in categorization: providing a groundwork. Cognition **52**(2), 125–157 (1994)

Goldstone, R.L., Barsalou, L.W.: Reuniting perception and conception. Cognition **65**, 231–262 (1998)

Gumperz, J.J., Levinson, S.C. (eds.): Rethinking Linguistic Relativity. Cambridge University Press, Cambridge (1996)

Habron, D.: Visual perception of wild land in Scotland. Landscape Urban Plann. **42**(1), 45–56 (1998). doi:10.1016/S0169-2046(98)00069-3

Harnad, S.: To cognize is to categorize: Cognition is categorization. In: Cohen, H., Lefebvre, C. (eds.) Handbook of Categorization in Cognitive Science, 1st edn. Elsevier, Amsterdam (2005)

Hutchins, E.: Cognition in the Wild. MIT Press, Cambridge (1995)

Jones, C.B., Purves, R.S.: Geographical information retrieval. Int. J. Geogr. Inf. Sci. **22**(3), 219–228 (2008). doi:10.1080/13658810701626343

Keßler, C.: Context-aware semantics-based information retrieval. Univ., Diss.–Münster (2010). Dissertations in geographic information science, vol. 3. Amsterdam, Heidelberg: IOS-Press; AKA Akadem. Verlagsges

Klippel, A., Wallgrün, J.O., Yang, J., Mason, J.S., Kim, E.-K., Mark, D.M.: Fundamental cognitive concepts of space (and time): using cross-linguistic, crowdsourced data to cognitively calibrate modes of overlap. In: Tenbrink, T., Stell, J., Galton, A., Wood, Z. (eds.) COSIT 2013. LNCS, vol. 8116, pp. 377–396. Springer, Heidelberg (2013)

Klippel, A., Worboys, M., Duckham, M.: Identifying factors of geographic event conceptualisation. Int. J. Geogr. Inf. Sci. **22**(2), 183–204 (2008)

Kwaśnicka, H., Jain, L.C. (eds.): Innovations in Intelligent Image Analysis. Springer, Heidelberg (2011)

Marder, J. (2015). The Internet's hidden science factory. http://www.pbs.org/newshour/updates/inside-amazons-hidden-science-factory/. Accessed 21 March 2015

Malt, B.C., Sloman, S.A., Gennari, S.P., Shi, M., Wang, Y.: Knowing versus naming: Similarity and the linguistic categorization of artifacts. J. Mem. Lang. **40**, 230–262 (1999)

Malt, B.C.: Category coherence in cross-cultural perspective. Cogn. Psychol. **29**, 85–148 (1995)

PBS. (2015). The Internet's hidden science factory. Accessed on http://www.pbs.org/newshour/updates/inside-amazons-hidden-science-factory/

Mark, D.M., Turk, A.G., Burenhult, N., Stea, D. (eds.): Landscape in Language. John Benjamins Publishing Company, Amsterdam (2011a)

Mark, D.M., Turk, A.G., Burenhult, N., Stea, D.: Landscape in language: an introduction. In: Mark, D.M., Turk, A.G., Burenhult, N., Stea, D. (eds.) Landscape in Language, vol. 4, pp. 1–24. John Benjamins Publishing Company, Amsterdam (2011b)

Mark, D.M., Turk, A.G., Stea, D.: Progress on yindjibarndi ethnophysiography. In: Winter, S., Duckham, M., Kulik, L., Kuipers, B. (eds.) COSIT 2007. LNCS, vol. 4736, pp. 1–19. Springer, Heidelberg (2007)

Medin, D.L., Wattenmaker, W.D., Hampson, S.E.: Family resemblance, conceptual cohesiveness, and category construction. Cogn. Psychol. **19**(2), 242–279 (1987)

Pattison, W.D.: The four traditions of geography. J. Geogr. **63**(5), 211–216 (1964)

Rand, D.G.: The promise of Mechanical Turk: How online labor markets can help theorists run behavioral experiments: evolution of Cooperation. J. Theor. Biol. **299**, 172–179 (2012). doi:10.1016/j.jtbi.2011.03.004

Rips, L.J.: Similarity, typicality and categorisation. In: Vosniadou, S., Ortony, A. (eds.) Similarity and Analogical Reasoning, pp. 21–59. Cambridge University Press, Cambridge (1989)

Sparks, K., Klippel, A., Wallgrün, J.O., Mark, D.M.: Citizen science land cover classification based on ground and aerial imagery. In: Fabrikant, S.I., Raubal, M., Bertolotto, M., Davies, C., Freundschuh, S.M., Bell, S. (eds.) COSIT 2015. LNCS, vol. 9368, pp. xx–yy. Springer, Heidelberg

Sparks, K., Klippel, A., Wallgrün, J.O., Mark, D.M.: Crowdsourcing landscape perceptions to validate land cover classifications. In: Ahlqvist, O., Janowicz, K., Varanka, D., Fritz, S. (eds.) Land Use and Land Cover Semantics. Principles, Best Practices and Prospects, pp. 296–314. CRC Press, Boca Raton (2015b)

Turk, A.G., Mark, D.M., Stea, D.: Enthnophysiography. In: Mark, D.M., Turk, A.G., Burenhult, N., Stea, D. (eds.) Landscape in Language, vol. 4, pp. 25–45. John Benjamins Publishing Company, Amsterdam (2011)

Ul-Qayyum, Z., Cohn, A.G., Klippel, A.: Psychophysical evaluation for a qualitative semantic image categorisation and retrieval approach. In: García-Pedrajas, N., Herrera, F., Fyfe, C., Benítez, J.M., Ali, M. (eds.) IEA/AIE 2010, Part III. LNCS, vol. 6098, pp. 321–330. Springer, Heidelberg (2010)

Wallgrün, J.O., Klippel, A., Mark, D.M.: A new approach to cluster validation in human studies on (geo)spatial concepts. In: Stewart, K., Pebesma, E., Navratil, G., Fogliaroni, P., Duckham, M. (eds.) Extended Abstract Proceedings of the GIScience 2014, pp. 450–453. Hochschülerschaft, TU Vienna, Vienna (2014)

Wrisley III, G.A.: Realism and Conceptual Relativity (Doctor of Philosophy). The University of Iowa (2008)

Citizen Science Land Cover Classification Based on Ground and Aerial Imagery

Kevin Sparks[1]([⊠]), Alexander Klippel[1], Jan Oliver Wallgrün[1],
and David Mark[2]

[1] The Pennsylvania State University,
University Park, State College, PA 16801, USA
{kas5822,klippel,wallgrun}@psu.edu
[2] NCGIA and Department of Geography,
University at Buffalo, Buffalo, NY 14228, USA
dmark@buffalo.edu

Abstract. If citizen science is to be used in the context of environmental research, there needs to be a rigorous evaluation of humans' cognitive ability to interpret and classify environmental features. This research, with a focus on land cover, explores the extent to which citizen science can be used to sense and measure the environment and contribute to the creation and validation of environmental data. We examine methodological differences and humans' ability to classify land cover given different information sources: a ground-based photo of a landscape versus a ground and aerial based photo of the same location. Participants are solicited from the online crowdsourcing platform Amazon Mechanical Turk. Results suggest that across methods and in both ground-based, and ground and aerial based experiments, there are similar patterns of agreement and disagreement among participants across land cover classes. Understanding these patterns is critical to form a solid basis for using humans as sensors in earth observation.

Keywords: Land cover · Citizen science · Classification

1 Introduction

Land cover data is often a critical parameter in geophysical research. Climate modeling, food security, and biodiversity monitoring are a few areas that have recently increased in importance, all of which land cover are central to. With the recent availability of different types of earth observation data (e.g., high resolution aerial photos, ground-based photos), and the growth of citizen science and crowdsourcing, new opportunities have opened up for environmental monitoring and data creation from non-authoritative sources (i.e. novice citizens). One of the main advantages of using citizen science in any area of research is its efficiency. It is critical however that we ensure this new source of monitoring and data creation is as reliable and consistent as possible. This paper follows up results found in Sparks et al. [1], and will examine how citizen science can best be used for the purposes of environmental monitoring and land cover classification.

© Springer International Publishing Switzerland 2015
S.I. Fabrikant et al. (Eds.): COSIT 2015, LNCS 9368, pp. 289–305, 2015.
DOI: 10.1007/978-3-319-23374-1_14

There are multiple global land cover datasets, many of which are created for a specific purpose or bias in mind. These specialized purposes often lead to disagreements between datasets and their classification schemes. This variation is illustrated by the Geo-Wiki project [2] showing the differences between GLC-2000, MODIS land cover products, and GlobCover. Variation in the classification schemes can occur via differing land cover classes (e.g., the inclusion or exclusion of a Grassland class), changed meanings of shared terminology (e.g., defining a Grassland class in different ways), and differences in the interpretation and perception of the land cover classes among users. Non-standardized data is a challenge to be faced in all fields of research. In our specific example of land cover, Ahlqvist [3] and Comber et al. [4] discuss the need for a more standard interpretation in light of the subjectivity in the dataset creation process, and in the interpretation of that data by the user. Comber et al. [4] specifically discusses the point how the perception of geographic terms in a land cover classification scheme will differ depending on the purpose of why that dataset was created. Likewise, the users' perception is often influenced by cultural and individual differences. If citizen science is to be incorporated in the environmental dataset/classification creation process, than there is a need for standardizing class definitions [5]. This need is proven through Robbins [6] discussion on land cover and land use classification choices by foresters and herders in a local Indian community. He shows how their classification choices for a surrounding area in a local community are influenced by these people's cultural and political roles in that community. This example further illustrates the point of class interpretation variation between producers and users of the data.

In order to solve this challenge of interpretability of land cover classification schemes, we must gain a deeper understanding of how humans perceive land cover classes [7]. This means exploring users' natural concepts of the environment, and being concerned with cognitive models about the geographic world. Coeterier [8] discusses that even when asking citizens to compare landscapes that are vastly different from each other, there is agreement among the importance of higher-level attributes of those landscapes. These high-level attributes are not necessarily features or objects within the landscape (i.e., trees, grass, water), but instead the landscape's use, naturalness, and spaciousness. These types of high-level attributes can nonetheless be quantified and used in a classification scheme. Continuing with these higher-level attributes of landscapes, Habron [9] analyzes the perceptual variation of what is considered Wild Land in Scotland. He concludes that while there is variation among sections of the population, there is a general agreement on the core definition of what Wild Land is. Also, by identifying that human impact has a large effect on what is considered Wild Land, he implicitly notes that citizens can consistently identify and distinguish non-natural environmental features from natural environmental features. These types of cognitive model processes discussed by Coeterier [8] and Habron [9] are the issues we must be aware of when attempting to increase interpretability on land cover classes.

In addition to interpretation variation and accuracy related issues, unknown variation gets introduced when a given dataset changes its methodology for the creation process. Comber et al. [10] uses the example of the Great Britain Datasets LCM1990 and LCM2000 to illustrate how a new methodology in the dataset creation process can create uncertainty between either observing land cover change, or simply observing a change in how the land cover is represented semantically. Furthermore, Foody [11] reiterates that land cover is dynamic. The earth's surface will change in the time it takes to update datasets especially for developed regions. This alone encourages a method of collecting data that is quick and reliable in order to keep up with the changing earth surface.

While there is a lot of interest surrounding the opportunities of the crowd, there is a high demand for systematic evaluations of how much improvement in environmental monitoring can be achieved using crowd-based assessments. In response to these challenges, this paper follows up on experiments described in Sparks et al. [1], and analyzes the consistency and reliability of humans' classification of land cover given ground-based and aerial-based photos of landscapes. If citizen science is to be incorporated into the evaluation of land cover data, there needs to be a more rigorous understanding of how humans perceive and conceptualize land cover types and a more detailed assessment of how well humans perform in recognizing predefined land cover classes. We are reporting on two experiments that provide insights on the relationships between human conceptualizations of land cover and land cover classifications using novices. Our findings suggest inter-participant agreements are not random but rather systematic to unique land cover stimuli and unique land cover classes, but are not greatly influenced by additional information such as aerial photos.

2 Background

Recent advancements in engineering and technology have created an opportunity for citizen science to have a significant impact on scientific research. We have seen examples of this impact across many research fields through the discovery of protein structures [12], the identification of galaxies [13], and the validation of land cover classes [2]. This last reference [2], referring to the Geo-Wiki project, is the most recent example of a crowdsourcing effort to assist in environmental monitoring. The Geo-Wiki project identifies locations where global land cover datasets disagree on a given land cover classification. It then solicits crowdsourced participants, provides them with aerial imagery, and asks them to make a classification choice for that location of disagreement. This data shows a lot of promise in validating land cover datasets, but like most sources of citizen science and crowdsourced data, it fails to assure reliability and consistency.

In order to ensure reliability and consistency, most attempts to gather data come from more authoritative sources. The Land Use/Cover Area Frame Survey (LUCAS) [14] is an example of a more authoritative source that attempts to capture land use/cover data. LUCAS, commissioned by Eurostat, uses trained surveyors to collect and create land cover data rather than relying on novice citizens. These land surveyors personally visit many locations, recording land transects, taking photos, and determining land use/cover at a given location. In this example coming from a more authoritative source, efficiency is sacrificed for reliability and consistency.

Citizen science needs to be able to guarantee a relatively high amount of reliability and consistency, along with being efficient. In the context of using citizen science for environmental monitoring, the Citizen Observatory Web (COBWEB) [15] uses citizens living in biosphere reserves across Europe to collect environmental data using mobile devices. Like LUCAS, COBWEB uses humans to collect data in the field, with the difference coming from trained (LUCAS) versus novice (COBWEB) sensors. The project's aim is to gain a deeper understanding of environmentally crowdsourced data by working with citizens throughout the process of data creation. By quality controlling this information, COBWEB hopes to impact environmental policy formation and more general societal and commercial benefits through the use of citizen science.

Data from Geo-Wiki project [2] has been analyzed to measure the quality of humans' classification of land cover given aerial photos [16–20]. See et al. [17] and Comber et al. [19] focus on the differences between expert Geo-Wiki participants and non-expert Geo-Wiki participants when classifying land cover given aerial imagery. See et al. [17] reports averaged agreement rates between participants of 66 %–76 % agreement when classifying land cover, noting experts generally having a higher maximum agreement than non-experts. Comber et al. [19] concludes with a similar result of experts being different than non-experts, but still calls for "…further investigation into formal structures to allow such differences to be modeled and reasoned with" ([19], pg 257). And while expertise has a general influence, that influence is varied across land cover classes, with expertise playing a larger role in certain types of classes [20].

While aerial photos have been available for some time, access to quality datasets of ground-based photos have recently emerged. The Geo-Wiki campaigns offer insight on humans' land cover classification using aerial photos. Others have attempted to test the effectiveness of using ground-based photos for humans' land cover classification [21, 22]. While no research has tested these ground-based photos on a large number of crowdsourced participants, research has concluded that ground-based photos are a valid data source when attempting to classify land cover [21, 22].

In combination with the success of the Geo-Wiki project in contributing to the growth of land cover datasets, OpenStreetMap [23] has also shown success in the contribution of environmental information from citizen science. OpenStreetMap is an open source dataset that is built from citizens volunteering and creating geographic information. Arsanjani et al. [24] analyzed OpenStreetMap land use/cover contributions to measure the accuracy of participants. He concludes that OpenStreetMap, and in general other forms of crowdsourced geographic data, can be reliable and consistent sources for mapping land use.

To summarize, citizen science and crowdsourced geographic data, while being efficient, are largely critiqued for being unreliable and inconsistent. In the context of land cover data, there has been preliminary research that suggests citizen science is promising for monitoring, validating, and creating environmental data. However, as projects like COBWEB show, there is a need to further understand how humans perceive and classify environmental features in order to determine reliable practices. Furthermore, various environmental information channels (ground-based versus aerial-based photos) and methods of classification need to be tested to determine best practices for citizen science involvement in environmental monitoring.

3   Experiments

To further advance our understanding of the potentials and limits of the human sensory and conceptual system in contributing to earth observations, we systematically extend our previous experiments [1] in two ways: first, we replicate an experiment on land cover classes but use a different methodology; second, participants received multiple perspectives on the same environment, that is, ground-based photos were complemented by aerial photos. The rational for these changes are explained in more detail below. Experiments in previous work [1] tested varying levels of participant expertise when classifying ground-based photos into land cover classes. The overall task in the experiments reported here remains the same: Participants are asked to classify photos into land cover classes.

3.1   Experiment 1 – Ground-Based Photos

The first question we address is a methodological one: Do we change the results of previous studies when the experimental setup is changed. Specifically, instead of using CatScan [25] the experimental setup was switched to Qualtrics [26]. CatScan is a card sorting tool that presents participants with stimuli/icons on the left half of the screen, and empty groups on the right half of the screen. Participants are asked to click and drag these stimuli/icons from the left half of the screen into groups on the right half of the screen based on their similarity. Qualtrics is an online survey platform with a lot of customizability. The driving force behind this change is the greater flexibility for non-free classification of land cover photos that Qualtrics offers (see Fig. 1): (a) Images can be presented individually allowing for higher resolution, (b) additional information for individual photos can be obtained such as how certain or uncertain a selected land cover class is, finally, (c) in preparation for experiment 2, the display real estate can be used to provide additional information for solving the classification task.

The first experiment asks participants to choose a land cover class, based on the National Land Cover Dataset (NLCD) [27] classification scheme, for ground-based photos of land cover. The experiment is a replication of the experiment of Sparks et al. [1] with the methodological change mentioned above.

Materials. Two datasets were used in experiment 1: First, ground-based photos of landscapes provided by the Degree Confluence Project (DCP) (confluence.org). Second, the National Land Cover Dataset (NLCD) 2006 provided by the United States Geological Survey (USGS) Land Cover Institute [27].

The DCP is a website which provides a platform for collecting crowdsourced photos of the environment at confluence points across the world. The word confluence as defined for the purposes of the DCP is the location where two integer latitude and longitude coordinate lines meet. A total of 799 photos were collected across the continental United States, which we sampled from for experiment 1 and experiment 2.

The NLCD 2006 is thematic land cover data for the United States prepared from Landsat 7 Enhanced Thematic Mapper Plus and Landsat 5 Thematic Mapper imagery collected between 2001 and 2006. We use NLCD as an authoritative dataset to measure participants' classification against. NLCD data however is not being used as ground truth, or being used to determine accuracy of participant classification. The data is only being used to see how much participants agree with an authoritative dataset. NLCD also provides the scheme from which participants can choose land cover classes in the experiments.

Latitude and longitude coordinates from the DCP data were used to spatially join their corresponding land cover class from the level II NLCD 2006. The level II NLCD classification scheme has a total of 16 land cover classes, but after spatially joining the 799 DCP photos, only 11 were returned. We aggregated Deciduous Forest, Evergreen Forest, and Mixed Forest into one Forest class (removing 2 options from the 16), and Developed Medium Intensity, Developed High Intensity, and Perennial Ice/Snow did not return enough photos (removing 3 options from the 16). The DCP images, now each assigned to 1 of 11 land cover classes, were sorted into bins based on their land cover class. 7 photos were randomly selected within each class, totaling 77 images. These 11 land cover classes make up the categorical choices for each question in the experiment.

Participants. 20 lay participants (non-experts, 11 female) were recruited through the crowdsourcing platform Amazon Mechanical Turk (AMT); average age 32.4 years; reimbursement: $1.80. Eight participants have postsecondary degrees. Participants were asked to provide the type of landscape they live in, given the options of Rural, Sub-urban, and Urban. Participants were not provided with definitions of Rural, Sub-urban, and Urban, and we did not verify their response. Of the three options for the currently lived in landscape, 3 participants live in Rural, 9 in Sub-urban, and 8 in Urban.

We believe that it is important to note that crowd science does not necessarily mean large samples. We have looked into calculating effect size but this does not seem to be straightforward for classification tasks. While it is possible to pick up smaller effects with larger samples, the goal is not to show that there is a difference even if it takes 10,000 participants in each experiment. Given that the patterns that we observed are present not only across methods but also across different participant groups (experts versus novices), we believe that it is not pertinent for this paper to increase the sample size.

While this paper is not a review on the validity of using AMT to solicit participants for academic studies, we will provide the following comments on our use of AMT for this research. As seen in Sparks et al. [1], AMT participants and expert participants solicited personally from a university campus performed the classification task with very similar results (statistically speaking, significantly not different). This would suggest the lack of any influence from AMT "super workers" that perform similar tasks multiple times on AMT. To reinforce this, our experimental surveys, being environmental classification tasks, are relatively unique from other AMT HITS.

Procedure. Qualtrics is an online survey service that we used to build our surveys and record data. Participants are solicited through AMT and directed to the Qualtrics survey via a link. Once participants begin the survey, they are asked basic demographic questions (age, gender, level of education). After providing this personal information, participants are given the definitions of each of the 11 land cover classes. The definitions of each class are taken directly from the NLCD classification scheme. Participants must confirm that they have read and understood each definition before they can progress to the main experiment. They are also given prototypical photos of what each land cover type looks like. Each participant has access to these definitions and prototypical photos at any point throughout the experiment. The participant is then shown each of the 77 ground-based photos, one at a time, and asked to make a classification decision given the 11 land cover class options. The participant is also asked to give their level of confidence about their choice of land cover: Sure (most confident), Quite sure, Less sure, and Unsure (least confident). The directions were explicit in informing the participants that Sure meant most confident, and Unsure meant least confident. Once a selection had been made, the participants could not go back and revisit previous questions. The participants had to finish the experiment in one sitting (i.e. they could not stop, exit the survey, and revisit the survey at a later time to finish). An example of what a question in the survey looks like can be seen in Fig. 1.

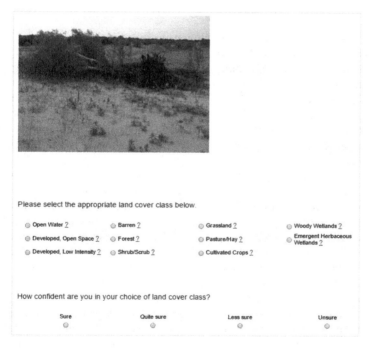

Fig. 1. Qualtrics interface showing one land cover photo participants are asked to classify (the image in the actual experiment is in color). Each photo is shown with the 11 potential land cover classes and a scale for the confidence of the selected classification. The blue question marks next to the land cover class options are links to the definition and prototypical photo of that land cover class.

Results. Results are reported as the level of agreement between participants' classification choice and the classification from the NLCD for a given photo. Once again, we are not claiming this NLCD classification as accuracy or ground truth, rather we are using it to measure percent agreement between an authoritative dataset and participants. Overall agreement with NLCD is 45.97 %. The average length of an experiment was 21 min 13 s. Time data for each individual question was not recorded.

After performing a Chi Square analysis, the following land cover classes significantly agreed with NLCD more frequently than expected by having a standardized residual value greater than 1.96 (Table 1): Developed, Low Intensity (dL), Forest (FO), Open Water (OW), and Shrub Scrub (SS). The following land cover classes significantly disagreed with NLCD more frequently than expected by having a standardized residual value less than -1.96: Barren (BA), Emergent Herbaceous Wetlands (EW), Pasture/Hay (PH), and Woody Wetlands (WW).

Table 1. Standardized residuals for experiment 1.

	BA	CC	dL	dO	EW	FO	GS	OW	PH	SS	WW
Correct	−6.11	−0.06	4.20	−0.95	−7.17	9.18	−0.59	13.09	−6.46	2.06	−7.17

Confusion matrices were created to allow us to see participant agreement across the 11 land cover classes (see Fig. 2). The confusion matrices' classes (columns and rows) come from the NLCD classification scheme, and illustrate the grouping behavior between participants across classes. For example, looking across the first row (BA), 21.43 % of the photos that NLCD classified as Barren (BA), participants also agreed were Barren (BA). However, 55 % of the photos NLCD classified as Barren (BA), participants thought were Shrub/Scrub (SS). Most importantly, these matrices allow us to visualize participants' classification patterns. As previously mentioned, this is the most important metric.

Along with the relatively low overall agreement with NLCD (45.97 %), we can see from the confusion matrix in Fig. 2 that overall, the level of participant agreement varies across land cover classes. Classes like Open Water (OW), Forest (FO), and the Developed classes (dL, dO) show a relatively large amount of agreement among participants. Otherwise, participants show a relatively large amount of disagreement with each other among the rest of the land cover classes. Specifically, participants are classifying a variety of photos into classes like Emergent Herbaceous Wetlands (EW) and Grasslands (GS), suggesting these classes are more heterogeneous compared to classes like Open Water (OW).

We also created a confusion matrix for photos for which participant indicated a high level of certainty (Sure) in their responses (Fig. 3). The Total Confident column at the end of the matrix shows how many Sure confidence responses that particular land cover class received across the 20 participants out of a total of 140.

Overall participant agreement with NLCD for Sure confidence responses is 71.19 %. Participants who indicated to be Sure of classifying Forest (FO) and Open Water (OW) were in 100 % agreement with each other, and NLCD. No one was confident in

	BA	CC	dL	dO	EW	FO	GS	OW	PH	SS	WW
BA	21.4	0.71	0	0	14.3	5	1.43	0	2.14	55	0
CC	6.43	45.7	0	0	4.29	0	22.9	0	16.4	4.29	0
dL	0	0	62.9	30	0	0	5	0	2.14	0	0
dO	0	0	45.7	42.1	0	0	2.86	0	9.29	0	0
EW	0.71	2.14	0	0	17.1	30.7	5	0	15	22.1	7.14
FO	0	0	0	0	2.86	82.9	0.71	0	0	9.29	4.29
GS	7.14	13.6	0	0	2.86	0	43.6	0	15.7	16.4	0.71
OW	0	0	0	0	0.71	0	0	98.6	0	0	0.71
PH	2.86	4.29	0	2.14	2.14	14.3	34.3	0	20	17.9	2.14
SS	22.9	0	0	0	2.86	1.43	15	0	3.57	54.3	0
WW	0	0	0	0	7.86	72.1	0	0	0	2.86	17.1

Fig. 2. Confusion matrix for experiment 1 (ground-based photos) showing percentages of participant agreement. Agreement between 5 % and 25 % is indicated by light grey, agreement between 25 % and 50 % is grey, and agreement above 50 % is dark grey.

	BA	CC	dL	dO	EW	FO	GS	OW	PH	SS	WW	Total Confident
BA	33.33	0	0	0	16.66	0	0	0	0	50	0	30
CC	0	68.08	0	0	0	0	17.02	0	10.63	4.25	0	47
dL	0	0	67.18	28.12	0	0	3.12	0	1.56	0	0	64
dO	0	0	46.42	41.07	0	0	1.78	0	10.71	0	0	56
EW	0	2.56	0	0	10.25	53.84	0	0	12.82	17.94	2.56	39
FO	0	0	0	0	0	100	0	0	0	0	0	39
GS	3.33	26.66	0	0	0	0	50	0	3.33	16.66	0	30
OW	0	0	0	0	0	0	0	100	0	0	0	114
PH	8.69	4.34	0	0	0	0	43.47	0	17.39	26.08	0	23
SS	26.66	0	0	0	0	0	6.66	0	0	66.66	0	30
WW	0	0	0	0	0	0	0	0	0	0	0	0

Fig. 3. Confusion matrix for only *Sure* confidence responses.

classifying any Woody Wetlands (WW) photos. When comparing this Sure confusion matrix (Fig. 3) with the previous confusion matrix (Fig. 2), the low percentages of agreement (light pink squares) disappear for the Sure confusion matrix, and the high percentages of agreement intensify. Higher percentage of agreement can be seen specifically for the Pasture/Hay (PH) row and Emergent Herbaceous Wetlands (EW) row. Simply looking at the last column, Total Confident, provides insight into the most interpretable land cover classes. As perhaps expected, participants were more frequently confident when classifying a photo as Open Water (OW). Both Developed classes showed frequent confidence too. Conversely, no participants were confident when classifying a photo as Woody Wetlands (WW), and were not frequently confident when classifying a photo as Barren (BA), Grassland (GS), Pasture/Hay (PH), or Shrub/Scrub (SS).

One of the motivations to switch from CatScan to Qualtrics was the possibility to obtain additional information such as the confidence a participant has making a classification. As mentioned previously, we repeated the analysis above considering only Sure responses. The agreement between NLCD and participants increases significantly ($x2 = 24.1902$, df = 1, p-value < 0.001). Overall agreement with NLCD (71.91 %) starts to approach the overall accuracy of Level II NLCD (78 %) [28]. A similar pattern can be seen between the two confusion matrices, only with a less amount of low disagreement, and more high agreement in the Sure confusion matrix. Participants are Sure most often when classifying Open Water (OW) and, to a lesser extent, the Developed land cover classes. This is perhaps expected as water and developed features are more easily distinguishable from natural features.

Discussion. The change in the experimental setup has allowed us to confirm results obtained previously but also add to our understanding of how humans might aid in earth observations. Overall percent agreement of Sure responses with NLCD (71.19 %) is comparable to the overall range of accuracy reported in Geo-Wiki campaigns of 64–84 % and 66–76 % accuracy. The overall results, not considering the certainty of the classification, are very similar to the results seen in previous experiments that use the same ground-based photos, but a different experimental interface (CatScan). The overall agreement with NLCD differs only 1.62 % from Sparks et al. to the experiment above. More so, the pattern in the confusion matrices from Sparks et al. and the experiment above are very similar. Barren (BA) and Shrub/Scrub (SS) frequently are confused with one another, participants are generally in high agreement on what is Developed but vary between what is Open Space and what is Low Intensity, and participants are in high agreement on what is a Forest (FO) and Open Water (OW) photo. This suggests that participants' classification choices are not being influenced differently from the CatScan interface to the Qualtrics interface.

3.2   Experiment 2 – Ground and Aerial Based Photos

To further advance our understanding of the human potential as an earth observer, experiment 2 extends experiment 1 by allowing participants access to not only ground-based photos but also corresponding aerial photos showing the area in question. The aerial photos contribute additional context information that might aid in achieving consistent classifications.

Materials. The National Agricultural Imagery Program (NAIP) imagery provided by the United States Department of Agriculture (USDA) is used in combination with the materials described in experiment 1. Using the latitude and longitude coordinates from each of the 77 ground-based photos, NAIP imagery of those same locations were downloaded. The NAIP imagery has a spatial resolution of 1 meter and is taken during the agricultural growing season across the continental United States. The NAIP images are shown to the participants at a 1:2000 scale, and cover an extent of 300 meters ($\sim$ 984 feet) by 300 meters. We added a white square centered on the image that is 30 meters ($\sim$ 98 feet) by 30 meters. This white square represents where the corresponding ground-based photo should be located. Participants are asked to only make their classification choice based on both information sources, the ground-based photo and what is inside the white square (see Fig. 4). Everything outside of the white square is there to provide context of the surrounding area. Participants are encouraged to consider this surrounding area when making their classification choice.

Participants. 20 lay participants (non-experts, 6 female) were recruited through AMT; average age 32.9 years; reimbursement: $1.80. Seven participants have postsecondary degrees. Of the three options for currently lived in landscape, 3 participants live in Rural, 9 Sub-urban, and 8 Urban. These 20 participants for experiment 2 were a separate group from the previous 20 participants in experiment 1.

Procedure. The NAIP photos were added to each question in the survey. When considering the NAIP photo, participants are asked to make their decision based off of the region inside the white square in the center of the photo. Otherwise the procedure is the same as experiment 1. An example of what a question in the survey looks like can be seen below (Fig. 4).

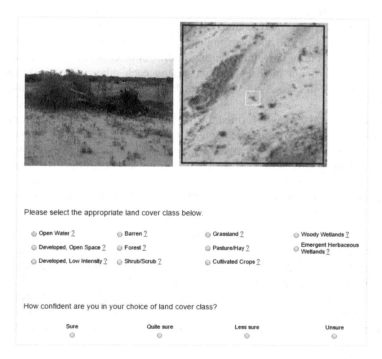

Please select the appropriate land cover class below.

- Open Water ?
- Barren ?
- Grassland ?
- Woody Wetlands ?
- Developed, Open Space ?
- Forest ?
- Pasture/Hay ?
- Emergent Herbaceous Wetlands ?
- Developed, Low Intensity ?
- Shrub/Scrub ?
- Cultivated Crops ?

How confident are you in your choice of land cover class?

Sure	Quite sure	Less sure	Unsure
○	○	○	○

Fig. 4. Qualtrics interface showing one land cover photo participants are asked to classify (the images in the actual experiment are in color).

Results. Overall agreement with NLCD is 42.79 %. The average length of an experiment was 21 min 16 s. The drop in agreement with NLCD from experiment 1 (45.97 %) to experiment 2 (42.79 %) is not statistically significant ($x2 = 3.0305$, df = 1, p-value = 0.081). Agreement for Barren (BA) increased significantly from experiment 1 to experiment 2 ($x2 = 13.7708$, df = 1, p-value < 0.001), and agreement for Shrub/Scrub (SS) and Woody Wetlands (WW) significantly decreased from experiment 1 to experiment 2 ($x2 = 4.1362$, df = 1, p-value = 0.04, and $x2 = 4.702$, df = 1, p-value = 0.03).

After performing a Chi Square analysis, the following land cover classes significantly agreed with NLCD more frequently than expected by having a standardized residual value greater than 1.96 (Table 2): developed, Low Intensity (dL), Forest (FO), and Open Water (OW). The following land cover classes significantly disagreed with NLCD more frequently than expected by having a standardized residual value less than −1.96: Emergent Herbaceous Wetlands (EW), Pasture/Hay (PH), and Woody Wetlands (WW).

Table 2. Standardized residuals for experiment 2.

	BA	CC	dL	dO	EW	FO	GS	OW	PH	SS	WW
Correct	0.01	−0.70	2.70	0.37	−7.50	8.97	−0.87	13.45	−7.32	−0.34	−8.76

Similar to experiment 1, we can see from the confusion matrix (Fig. 5) that participant agreement varies across land cover classes. Once again, classes like Open Water (OW), Forest (FO), and Developed (dL, dO) are the exception, showing a relatively large amount of agreement among participants. Otherwise, participants show a relatively large amount of disagreement among each other across rest land cover classes, specifically in classes like Emergent Herbaceous Wetlands (EW) and Grasslands (GS).

Like in experiment 1, we performed the analysis again using only those images that participants indicated they were Sure about (Fig. 6). Overall participant agreement with NLCD for Sure confidence responses is 59.55 %. No land cover class had full agreement among participants who were Sure. Similar to experiment 1, no one was confident in classifying any Woody Wetlands (WW) photos. Participants classified more images as Barren (BA) and Grassland (GS) land cover classes in experiment 2 (for both overall and Sure responses).

Discussion. The introduction of corresponding aerial photos significantly increased Barren (BA) agreement with NLCD from experiment 1, but otherwise agreement with NLCD slightly decreased with the introduction of aerial photos, with Shrub Scrub (SS) and Woody Wetlands (WW) dropping significantly.

These drops in agreement, along with a relatively unchanged number of Sure confidence responses from experiment 1 to experiment 2 across each class, suggests that aerial photos do not provide any more clarity when perceiving and classifying land cover types.

	BA	CC	dL	dO	EW	FO	GS	OW	PH	SS	WW
BA	42.86	0	0	0	11.43	2.86	1.43	0	0.71	40	0.71
CC	5.71	40	0	2.86	1.43	0	32.86	0.71	11.43	4.29	0.71
dL	0.71	0	53.57	35.71	0	0	7.14	0	2.86	0	0
dO	0	4.29	38.57	44.29	0	0.71	4.29	0	7.14	0.71	0
EW	2.86	2.14	0	0	12.86	37.14	8.57	0	12.86	20.71	2.86
FO	0.71	0	0	0	0.71	78.57	1.43	0	2.14	14.29	2.14
GS	12.14	11.43	0.71	0	3.57	0	39.29	0	17.14	15.71	0
OW	0	0	0	0	0.71	0	0	96.43	0	0	2.86
PH	7.14	5.71	0	5.71	3.57	2.14	42.86	0	13.57	17.86	1.43
SS	32.86	0	0	0	1.43	0.71	15.71	0	7.86	41.43	0
WW	0	0	2.14	0.71	0.71	85	0.71	0	0	2.86	7.86

Fig. 5. Confusion matrix for experiment 2 (ground and aerial-based photos).

	BA	CC	dL	dO	EW	FO	GS	OW	PH	SS	WW	Total Confident
BA	64.51	0	0	0	6.45	0	0	0	0	29.03	0	31
CC	0	59.52	0	2.38	0	0	33.33	0	4.76	0	0	42
dL	0	0	58.2	40.29	0	0	1.49	0	0	0	0	67
dO	0	0	55.55	40.74	0	0	1.85	0	1.85	0	0	54
EW	7.31	0	0	0	2.43	68.29	7.31	0	4.87	9.75	0	41
FO	0	0	0	0	0	95.65	0	0	0	4.34	0	46
GS	17.24	27.58	0	0	0	0	44.82	0	3.44	6.89	0	29
OW	0	0	0	0	0	0	0	99.14	0	0	0.85	117
PH	6.45	6.45	0	3.22	0	0	70.96	0	0	12.9	0	31
SS	44.11	0	0	0	2.94	0	14.7	0	0	38.23	0	34
WW	0	0	0	0	0	0	0	0	0	0	0	0

Fig. 6. Confusion matrix for only *Sure* confidence responses.

The introduction of aerial photos could possibly be contradicting what its corresponding ground-based photo is showing. For example, if a given ground-based photo portraying something a participant might classify as Shrub/Scrub (SS) is contradicted by its corresponding aerial photo that portrays something a participant might classify as Barren (BA), perhaps the aerial photo takes precedence for the participant when determining a land cover. The participant might also be considering the surrounding area too much outside of the location of interest in the aerial photo, i.e. the white square. The introduction of a larger region could be leading to more heterogeneous surfaces, making the classification of the location more ambiguous.

Barren environments perhaps benefit from a larger context of the surrounding area. In experiment 1, participants may have been influenced more by the presence of one or two shrubs in the ground-based photo in a mostly barren environment. Whereas in experiment 2, when provided with more of the surrounding area, the participant can see how much of the land cover is truly barren, with sparse shrubs potentially having less of an impact with a larger area. Barren (BA) is also generally a more homogenized land cover class, as is Grassland (GS). This homogenization is perhaps intensified and considered more when an aerial perspective is included. This possibly explains why participants chose Barren (BA) and Grassland (GS) more frequently.

The significant decrease in Woody Wetlands (WW) agreement could possibly be explained by the influence of, in most cases, a homogenized looking canopy from an aerial perspective in the Woody Wetlands (WW) photos. In experiment 1, participants did not have this homogenized canopy influence, and are perhaps more focused on the potential source of water seen from ground-based photos.

Similarly to the reasoning for the increase in participant agreement for Barren (BA), Shrub/Scrub (SS) may have experienced a significant decrease in agreement when considering the larger surrounding area. Participants' classification choice may have been more heavily influenced from the aerial photos rather than the ground-based photo.

4 General Discussion/Outlook

Looking at the results from experiment 1 and experiment 2, as well as the three experiments outlined in Sparks et al. [1], a consistent parameter in all these experiments has been the categorical 11-class classification method. The land cover class semantics are potentially the largest influence on participant agreement. These categorical land cover classification tasks are difficult. A possible explanation for why, is that this categorical classification scheme is generalizing land covers too much, and these classes are at too high of a level that subjectivity overrides objectivity. As discussed in Sect. 2, the earth's surface is often complex and heterogeneous. Forcing this complexity into relatively high-level categorical classes is prone to errors and disagreements.

We have now run experiments using different interfaces (CatScan vs Qualtrics), different users (Amazon Mechanical Turk vs live Experts), and different stimuli (ground vs ground and aerial). Yet with all these changes, a similar pattern seen in the confusion matrices persists (Fig. 7).

Future research will begin to explore non-categorical classification methods, such as free classification of landscape images [29] as well as mimicking a decision tree process of classification. In the case of the latter, participants will make an initial decision regarding the presence of environmental features in the image (e.g., is this image primarily vegetated or primarily non-vegetated) that will then lead to another unique set of choices, and so on. Humans will more likely agree on the presence of environmental features versus higher-level categories shown in the experiments above.

Confusion matrices were also created for each subcategory of landscape each participant said they currently lived in (Rural, Sub-urban, Urban) for both the

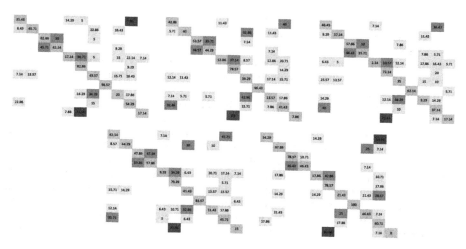

Fig. 7. All 5 confusion matrices.

ground-based, and ground and aerial-based experiments. Data collected is preliminary as each sub-class of participant included less than 10 participants. Generally, rural dwelling participants showed relatively low agreement on the Developed classes and relatively higher agreement on Woody Wetlands (WW) and Cultivated Crops (CC). Urban dwelling participants showed generally higher agreement on Developed classes. While this is preliminary, we intend to explore the influence of lived in landscape for land cover classification in future research.

In the context of projects like Geo-Wiki and COBWEB, we need to be aware of the classification task design and try to make it as objective as possible. As discussed in Sect. 2, the earth's surface is complex and often heterogeneous. In order to use citizen science for environmental monitoring and be as reliable and consistent as possible, we need to continue to explore how humans perceive land cover classes and environmental features, and make sure that the classification task is designed in a way to encourage objectivity and discourage subjectivity.

Acknowledgments. This research is funded by the National Science Foundation (#0924534). We would like to thank the Degree Confluence Project for permission to use photos from the confluence.org website for our research. We would like to thank the members of the Human Factors in GIScience laboratory.

References

1. Sparks, K., Klippel, A., Wallgrün, J.O., Mark, D.: Crowdsourcing landscape perceptions to validate land cover classifications. In: Ahlqvist, O., Janowicz, K., Varanka, D., Fritz, S. (eds.) Land Use and Land Cover Semantics - Principals, Best Practices and Prospects. CRC Press/Taylor & Francis, Boca Raton (2015)

2. Fritz, S., McCallum, I., Schill, C., Perger, C., Grillmayer, R., Achard, F., Kraxner, F., Obersteiner, M.: Geo-Wiki.Org: the use of crowdsourcing to improve global land cover. Remote Sens. **1**(3), 345–354 (2009)

3. Ahlqvist, O.: In search of classification that supports the dynamics of science: the FAO land cover classification system and proposed modifications. Environ. Plan. B Plan. Des. **35**(1), 169–186 (2008)

4. Comber, A., Fisher, P., Wadsworth, R.: What is land cover? Environ. Plan. B **32**, 199–209 (2005)

5. Jepsen, M.R., Levin, G.: Semantically based reclassification of Danish land-use and land-cover information. Int. J. Geogr. Inf. Sci. **27**(12), 2375–2390 (2013)

6. Robbins, P.: Beyond ground truth: GIS and the environmental knowledge of herders, professional foresters, and other traditional communities. Hum. Ecol. **31**(2), 233–253 (2003)

7. Ahlqvist, O.: Semantic issues in land use and land cover studies – foundations, application and future directions. In: Giri, C. (ed.) Remote Sensing of Land Use and Land Cover: Principles and Applications, pp. 25–36. Taylor and Francis, Boca Raton (2012)

8. Coeterier, J.F.: Dominant attributes in the perception and evaluation of the Dutch landscape. Landsc. Urban Plan. **34**(1), 27–44 (1996)

9. Habron, D.: Visual perception of wild land in Scotland. Landsc. Urban Plan. **42**(1), 45–56 (1998)

10. Comber, A., Fisher, P., Wadsworth, R.: Integrating land-cover data with different ontologies: identifying change from inconsistency. Int. J. Geogr. Inf. Sci. **18**(7), 691–708 (2004)

11. Foody, G.M.: Status of land cover classification accuracy assessment. Remote Sens. Environ. **80**, 185–201 (2002)

12. Khatib, F., DiMaio, F., Cooper, S., Kazmierczyk, M., Gilski, M., Krzywda, S., Zabranska, H., Pichova, I., Thompson, J., Popović, Z., Jaskolski, M., Baker, D.: Crystal structure of a monomeric retroviral protease solved by protein folding game players. Nat. Struct. Mol. Biol. **18**(10), 1175–1177 (2011)

13. Clery, D.: Galaxy zoo volunteers share pain and glory of research. Science **333**, 173–175 (2011)

14. LUCAS: Land use/cover area frame statistical survey. http://www.lucas-europa.info/

15. COBWEB: Citizen observatory web. https://cobwebproject.eu

16. Perger, C., Fritz, S., See, L., Schill, C., van der Velde, M., McCallum, I., Obersteiner, M.: A campaign to collect volunteered geographic information on land cover and human impact. In: GI_Forum 2012: Geovizualisation, Society and Learning (2012)

17. See, L., Comber, A., Salk, C., Fritz, S., van der Velde, M., Perger, C., Schill, C., McCallum, I., Kraxner, F., Obersteiner, M., Preis, T.: Comparing the quality of crowdsourced data contributed by expert and non-experts. PLoS ONE **8**(7), e69958 (2013)

18. Foody, G.M., See, L., Fritz, S., Van der Velde, M., Perger, C., Schill, C., Boyd, D.S.: Assessing the accuracy of volunteered geographic information arising from multiple contributors to an internet based collaborative project. Trans. GIS **17**(6), 847–860 (2013)

19. Comber, A., Brunsdon, C., See, L., Fritz, S., McCallum, I.: Comparing expert and non-expert conceptualisations of the land: an analysis of crowdsourced land cover data. In: Tenbrink, T., Stell, J., Galton, A., Wood, Z. (eds.) COSIT 2013. LNCS, vol. 8116, pp. 243–260. Springer, Heidelberg (2013)

20. Comber, A., See, L., Fritz, S.: The impact of contributor confidence, expertise and distance on the crowdsourced land cover data quality. In: GI_Forum 2014 - Geospatial Innovation for Society (2014)

21. Iwao, K., Nishida, K., Kinoshita, T., Yamagata, Y.: Validating land cover maps with degree confluence project information. Geophys. Res. Lett. **33**, L23404 (2006)

22. Foody, G.M., Boyd, D.S.: Using volunteered data in land cover map validation: mapping West African forests. IEEE J. Sel. Top. Appl. Earth Obs. Remote Sens. **6**(3), 1305–1312 (2013)
23. OpenStreetMap. https://www.openstreetmap.org
24. Arsanjani, J.J., Helbich, M., Bakillah, M., Hagenauer, J., Zipf, A.: Toward mapping land-use patterns from volunteered geographic information. Int. J. Geogr. Inf. Sci. **27**(12), 2264–2278 (2013)
25. Klippel, A.: Spatial information theory meets spatial thinking - is topology the Rosetta Stone of spatio-temporal cognition? Ann. Assoc. Am. Geogr. **102**(6), 1310–1328 (2012)
26. Qualtrics: online survey software & insight platform. http://www.qualtrics.com/
27. Fry, J., Xian, G., Jin, S., Dewitz, J., Homer, C., Yang, L., Barnes, C., Herold, N., Wickham, J.: Completion of the 2006 national land cover database for the conterminous United States. Photogram. Eng. Remote Sens. **77**(9), 858–864 (2011)
28. Wickham, J.D., Stehman, S.V., Gass, L., Dewitz, J., Fry, J.A., Wade, T.G.: Accuracy assessment of NLCD 2006 land cover and impervious surface. Remote Sens. Environ. **130**, 294–304 (2013)
29. Klippel, A., Mark, D., Wallgrün, J.O., Stea, D.: Conceptualizing landscapes: a comparative study of landscape categories with Navajo and English-speaking participants. In: Fabrikant, S.I., Raubal, M., Bertolotto, M., Davies, C., Freundschuh, S.M., Bell, S. (eds.) Proceedings, Conference on Spatial Information Theory (COSIT 2015), Santa Fe, NM, USA. Springer, Berlin, 12–16 Oct 2015

Qualitative Spatio-Temporal Reasoning and Representation II

Swiss Canton Regions: A Model for Complex Objects in Geographic Partitions

Matthew P. Dube[✉], Max J. Egenhofer, Joshua A. Lewis,
Shirly Stephen, and Mark A. Plummer

School of Computing and Information Science, University of Maine,
5711 Boardman Hall, Orono, ME 04469-5711, USA
{matthew.dube, joshua.lewis,
shirly.stephen}@umit.maine.edu,
max@spatial.maine.edu, mark.a.plummer@maine.edu

Abstract. Spatial regions are a fundamental abstraction of geographic phenomena. While simple regions—disk-like and simply connected—prevail, in partitions complex configurations with holes and/or separations occur often as well. Swiss cantons are one highlighting example of these, bringing in addition variations of holes and separations with point contacts. This paper develops a formalism to construct topologically distinct configurations based on simple regions. Using an extension to the compound object model, this paper contributes a method for explicitly constructing a complex region, called a *canton region*, and also provides a mechanism to determine the corresponding complement of such a region.

Keywords: Spatial reasoning · Topological variability · Complex regions

1 Introduction

The world is full of boundaries, but not all boundaries are made equal. Some boundaries are rigid formations that are physically built within our world (such as the boundary of an island), yet others are lines developed by human intervention to demarcate control. Smith [35] refers to these boundaries as *bona fide* and *fiat* respectively. While *bona fide* boundaries are conceptually rigid, *fiat* boundaries are flexible, and often referred to as "lines in the sand" [30], an expression that suggests both permanence as well as a potentially temporary status, capable of being redefined, a reality that history shows is ever present [2, 18, 30, 41].

Fiat boundaries seemingly abound in an administrative context, yet their properties around the world are different both politically [30] and topologically [11, 23, 24]. While most political subdivisions result in *simple regions* [15, 34], a variety of causes —alliances, physical disasters, railroads, religions—have led to the formation of more complex partitions. For instance, the Republic of San Marino remained independent despite the unification of the other Italian microstates around it in the 19th century, resulting in a hole within Italy [2]. Tiptonville, Kentucky represents a phenomenon of separation in the United States, cut off from the rest of its state by the New Madrid Earthquake of 1811 that rerouted the Mississippi River [33]. The Swiss cantons of

© Springer International Publishing Switzerland 2015
S.I. Fabrikant et al. (Eds.): COSIT 2015, LNCS 9368, pp. 309–330, 2015.
DOI: 10.1007/978-3-319-23374-1_15

Appenzell Innerrhoden and Appenzell Ausserrhoden were partitioned for religious reasons, forcing holes and separations within both [18]. The German railroad Vennbahn has created an exclave scenario within Belgium [41]. The number of holes, separations, and the complexity of their combinations may vary: Switzerland, for instance, has two holes (one filled by a German exclave, the other filled by an Italian exclave); Azerbaijan is separated into two parts; many island nations, such as Japan and Indonesia, are archipelagos, forming a plethora of separations; Austria has a point-connected separation that is otherwise fully surrounded by Germany; in return, Germany has, among its cavities, a fringed hole (i.e., a hole that connects at one point to Germany's boundary); Italy has both holes and separations; the Cooch Behar district in India has a complex combination of holes and separations, yielding a piece of India that is inside a piece of Bangladesh, which in turn is inside another piece of India, which is again inside another piece of Bangladesh. The quest is to develop a generic model that captures completely the topology of such potentially highly complex partitions of space.

While regions with holes and separations are parts of contemporary GIS models—for instance, OGC's simple-features specification includes the type MultiGon [29] — they lack the details needed to distinguish topologically distinct configurations as well as to support qualitative reasoning about their spatial relations. Although the 9-intersection relations [16] or RCC-8 relations [31] apply to these types of objects when specifying spatial relations to one another, these spatial-relation models hide the variability present in the complex regions themselves. The relations displayed in Fig. 1 are all topologically different, yet conventional models do not distinguish them.

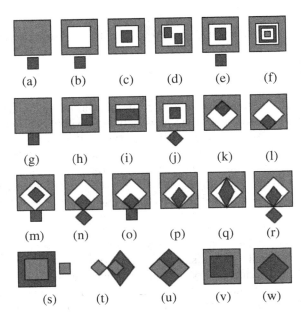

Fig. 1. Different topological relations, due to complex regions, for which 9-intersection [34] or RCC-8 [31] make no distinctions: (a)-(f) map onto *disjoint* (DC in RCC-8); (g)-(u) map onto *meet* (EC in RCC-8); (v) and (w) map in RCC-8 onto EC, while the 9-intersection maps them onto a matrix that is distinct from *meet*, yet it cannot distinguish between the two configurations.

As an homage to the first-level political subdivisions of Switzerland, we name these potentially complex regions *canton regions*, because together they expose a wide range of these oddities (Fig. 2). The model developed for canton regions, however, will not be limited to the current occurrences of the geometries of Swiss cantons. Rather, canton regions stand for potentially arbitrarily complex regions in partitions, which may yield any number and kind of holes and separations. The structure of such a canton region is analogous to that of a multi-part polygon, however this approach also provides a means to reason about such features.

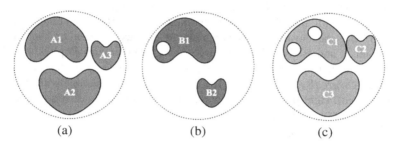

(a)   (b)   (c)

Fig. 2. A generalized representation of various Swiss cantons, each topologically distinct from the others: (a) Schaffhausen with three separations, (b) Thurgau with two separations, one of which having a hole, and (c) Bern with three separations, two coincident at a single point, and two holes within one of those separations.

The remainder of this paper is structured as follows. Section 2 highlights the literature concerning the diversity of spatial regions and formalisms designed to account for such diversity. Section 3 provides the fundamental mathematical definitions for describing regions concisely. Section 4 develops canton regions using the compound object model by further classifying the simple topological relations on the sphere where boundary interplay is possible, utilizing the concept of boundary sequencing from the *o*-notation [24] and *i*-notation [25]. Section 5 addresses how to compute the direct complement of a canton region. Section 6 compares this formalism with that of the *surrounds* relations [8], creating a larger set of *surrounds* constructions than previously realized. Section 7 constructs a representative subset of the Swiss cantons using this method. Section 8 provides conclusions and opportunities for future study.

2 Diversity of Spatial Regions

Independent of human subdivisions of space, spatial regions are inherently diverse. Regions with holes [17, 34, 38–40] and separations [11, 26, 34] quickly extend consideration beyond simple regions. Such spatial diversity is not just a hindrance, but rather an opportunity for designing more comprehensive models that can account for more real-world applications than what current formalisms can achieve. This section reviews the related approaches to modeling complex spatial regions based on relations (Sect. 2.1), boundary sequences (Sect. 2.2), and graphs (Sect. 2.3).

2.1   Relation-Based Approaches

The gold standards of topological/mereotopological relations are the 9-intersection [16] and the region connection calculus [31], forming the backbone of spatial query operations [3] and qualitative spatial reasoning [6]. The eleven 9-intersection relations between two simple regions on the sphere $\mathbb{S}^2$ (Fig. 3) expose some critical algebraic properties: The pairs *inside-contains* and *coveredBy-covers* are pairs of converse relations. The relation *equal* is the identity relation, while *attach* captures the relation between a region and the closure of its complement. The closure of a region's complement yields further dependencies. Replacing in the relation *contains* between A and B the region A with the closure of its complement implies the relation *disjoint*; replacing B with the closure of its complement implies the relation *embrace*, and replacing both A and B with the closures of their complements implies the relation *inside*. The same analysis applies when starting with the relation *covers*, yielding *meet*, *entwined*, and *coveredBy*, respectively. The dependencies are referred to as the left dual (A's complement), right dual (B's complement), and double dual (A's and B's complement) [12, 17]. The operation of complementation requires the usage of $\mathbb{S}^2$ relations, as opposed to their $\mathbb{R}^2$ counterparts.

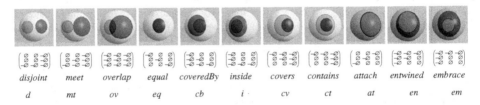

Fig. 3.   The eleven 9-intersection relations between simple regions on the sphere [12].

While these formalisms extend beyond simple regions, they inherently lack specificity when they capture relations involving non-simple regions [26, 34]. For instance, neither the 9-intersection nor RCC *per se* can distinguish whether a holed region contains another object in its hole or in its outer exterior [8, 17].

To accommodate such distinctions, holed regions have been modeled as the closure of the set difference of a simple-region host, the object that contains the hole, and a second simple region *inside* the host [16, 39] (e.g., R = $\overline{(R0 \setminus R1)}|$(R0 *contains* R1)). For regions with multiple holes, the same approach adds constraints about the pairs of regions that form holes, for instance that they must be *disjoint* [40] (e.g., R = $\overline{(R0 \setminus R1 \setminus R2)}|$(R0 *contains* R1 ∧ R0 *contains* R2 ∧ R1 *disjoint* R2)).   Each   holed region then has a *host* and one or more holes within it. Holes are seen as objects that are taken out of the host, then host and hole are related by the relations *contains* or *covers*. No other relation is possible between a host and its holes. Conversely, the hole must be related to its host by either *inside* or *coveredBy*. This approach extends to forming arbitrarily complex objects [11] when simple objects (e.g., a point, line segment, or simple region), constrained by specific topological relations, are combined through set operations (union or set difference). Depending on whether a region is subject to union or set difference it becomes respectively an *additive* or *subtractive* element. One of the additive elements is chosen as the *principal* element. With every set difference, closure is applied to the resulting region. Figure 4 gives examples of how complex regions are formed with this *compound object model* (COM) [11].

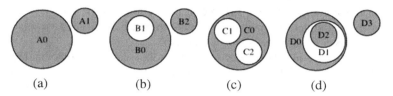

Fig. 4. Construction of complex regions with the compound object model [11]: (a) a separated region A = A0∪A1|(A0 *disjoint* A1), (b) a single-holed region with a separation in the host's exterior B = $\overline{(B0\backslash B1)}$B2|(B0 *contains* B1 ∧ B0 *disjoint* B2), (c) a region with two holes C = $\overline{(C0\backslash C1 \backslash C2)}$|(C0 *contains* C1 ∧ C0 *contains* C2 ∧ C1 *disjoint* C2), and (d) a holed region with two separations, one in the hole, the other in the host's exterior D = $\overline{(D0 \backslash D1)}$∪ D2∪D3c|(D0 *contains* D1 ∧ D1 *contains* D2 ∧ D0 *disjoint* D3).

2.2 Sequence-Based Approaches

While the 9-intersection and RCC-8 specify for two simple regions the topological relations without any boundary interactions (i.e., *disjoint*, *inside*, and *contains* in $\mathbb{R}^2$) up to the level of homeomorphism, for configurations with boundary interactions, these topological relations group similar (yet also topologically distinct cases) together. For instance, any two simple regions with partial boundary interaction and mutually exclusive interiors are categorized by *meet*, independent of the number of separate components in the boundary-boundary intersection between the two regions, and independent of the dimension of each component. Such differences are captured by the components' sequences [14]. When considering *spatial scenes*—entire layouts of spatial objects—not only the sequences of the boundary intersections of two objects need to be recorded, but also the sequence in which boundary segments of multiple objects coincide [24, 25] (Fig. 5).

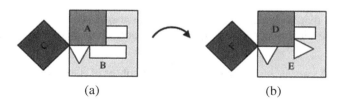

Fig. 5. Two scenes that differ by the boundary sequences of their dimensions, captured by the o-notation [24] (only shown A's and D's boundary): (a) $\partial A : o_{\{BC\}}(1, B, \emptyset)o_{\{BC\}}(1, B, \emptyset)o_{\{BC\}}$ $(0, B, \emptyset)o_{\{BC\}}(0, \{BC\}, \emptyset)$ and   (b) $\partial D : o_{\{EF\}}(1, E, \emptyset)o_{\{EF\}}$ $(0, E, \emptyset)o_{\{EF\}}(1, E, \emptyset)o_{\{EF\}}$ $(0, \{EF\}, \emptyset)$.

2.3   Graph-Based Approaches

Graphs are an alternative model to capture the complexity of a spatial region. A canonical model for areal objects constructs regions with holes and separations as a tree [44]. MapTree [42, 43] models a space based on boundary lines and isolated spaces. Since it is a combinatorial map, it distinguishes topological configurations up to the level of homeomorphism [10, 37]. Yet it only considers the objects' boundaries, rather than their complete geometry. A related approach models space as a bigraph [28], which offers a hierarchical view of a space).

Graph models have also been used to capture topological relations that are beyond what the 9-intersection or RCC-8 distinguishes, particularly configurations with surrounding areas. Although three 9-intersection matrices can map a partitioned *surrounds* configuration [26, 34], two of them are identical to those for a standard *disjoint* relation and a standard *meet* relation. An algebraic model is needed to define *surrounds* relations [8]. Detecting such relations, however, requires partition analysis based on graph-theoretic representations [8].

3   An Inventory of Basic Canton Regions

Let A be a subset of a topological space T. The *interior* of A, denoted by A°, is the union of all open sets contained in A. A's *closure,* denoted by $\bar{A}$, is the intersection of all closed sets containing A. A's *boundary,* denoted by ∂A, is the set difference between A's closure and its interior (i.e., $\partial A = \bar{A} \setminus A^\circ$). A's *exterior*, denoted by A^-, is the difference between the embedding space and A's closure (i.e., $A^- = \mathbb{R}^2 \setminus \bar{A}$ in the plane or $A^- = \mathbb{S}^2 \setminus \bar{A}$ on the sphere).

These basic topological definitions underlie the point-set topological relations constructed under the 4-intersection [15] and the 9-intersection [16]. As such, they are the smallest building blocks of the compound object model [11].

Definition 1: Let X be a closed subset of a topological space T. X is ***properly connected*** if its interior is path-connected within T. This property is also called ***interior-connected***.

Definition 2: Let X be a closed subset of a topological space T. X is ***closure-connected*** if its closure is path-connected within T.

Any region that is properly connected must be also closure-connected, but not all regions that are closure-connected are also properly connected within the standard topology [1].

Definition 3: A *canton region* R is a regularized subset of a topological space T (i.e., $\overline{R^\circ} = \bar{R}$) with x non-empty interior components, y non-empty boundary components, and z non-empty exterior components. A *basic canton region* R is a canton region such that $x + y + z \leq 5$.

Any canton region can be constructed by union or intersection of finitely many basic canton regions. Definitions 4 through 9 represent their basic classes.

Definition 4: Let R be a *basic canton region*. R is a *simple region* if R° is properly connected and R^- is also properly connected (Fig. 6b). Otherwise, it is a *non-simple region* (i.e., either R° or R^- is not properly connected) (Fig. 6c).

Definition 5: A non-simple region R whose interior is not properly connected is an *interior-disconnected region* (Fig. 6d). Otherwise, it is an *exterior-disconnected region* (i.e., R^- is not properly connected) (Fig. 6i).

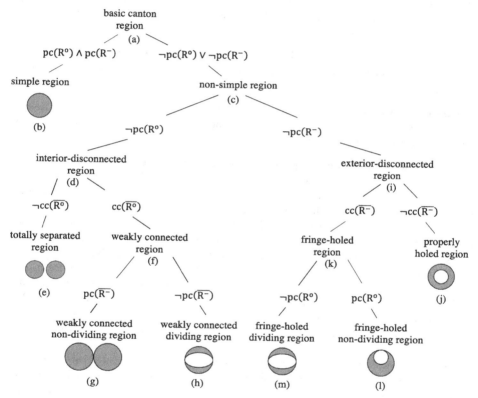

Fig. 6. Basic canton regions distinguished by closure-connected (cc) and properly connected (pc) interiors and exterior. The structure shows the symmetry between interior-disconnected (d) and exterior-disconnected (i) regions, as such regions are closed complements of each other in $\mathbb{S}^2$. A weakly connected dividing region (h) and a fringe-holed dividing region are two different perspectives over the same configurations. Their closed complements also map onto themselves. This structure illustrates the difference between a holed and non-holed region, and is not intended to be a disjunctive tree. It illustrates that there are multiple ways to form a single configuration, such as (h) and (m). These types of objects have been addressed in other manners previously [4, 45].

Definition 6: An interior-disconnected region R is a *totally separated region* if $\overline{R^\circ}$ is not closure-connected (Fig. 6e). Otherwise, it is a *weakly connected region* (i.e., $\overline{R^\circ}$ is closure-connected) (Fig. 6f).

Definition 7: Let R be a weakly connected region. R is a *weakly connected non-dividing region* if R^- is *properly connected* (Fig. 6g). Otherwise, it is a *weakly connected dividing region* (i.e., R^- is not properly connected) (Fig. 6h).

Exterior-disconnected regions are refined analogously to the refinement of interior-disconnected regions.

Definition 8: Let R be an exterior-disconnected region. R is a *properly holed region* if $\overline{R^-}$ is not closure-connected (Fig. 6j). Otherwise, it is a *fringe-holed region* (i.e., $\overline{R^-}$ is closure-connected) (Fig. 6k).

Definition 9: Let R be a fringe-holed region. R is a *fringe-holed non-dividing region* if R° is properly connected (Fig. 6l). Otherwise, it is a *fringe-holed dividing region* (i.e., R° is not properly connected) (Fig. 6m).

Theorem 1: A weakly connected, dividing region, is homeomorphic to a fringe holed, dividing region. Such regions are homeomorphic as long as they have an equivalent number of distinct boundary intersections.

Proof: Consider a region R comprised of two parts, R1 and R2, such that R1 and R2 come together at two points. Consider the smallest closed disk D that is a superset of both R1 and R2. One of the two separated portions of R's exterior must be a superset of D, as its boundary is a cycle that connects the pair of coinciding points and uses non-incident portions of both R1 and R2's boundary. There are two such cycles: one is the boundary of D, whereas the other is not a subset of the boundary, but rather a subset of the closure of D. Call that exterior bounded by that cycle RH. Clearly, RH is a subset of D. Consider D\RH. This object represents a holed region via set difference in the compound object model with two points of incidence with the boundary, eliminating proper connection of D\RH. Therefore, D\RH is a fringe-holed, dividing region. We started however with R1 and R2 as weakly connected, dividing regions. D\H and R1∪R2 by construction are the same object, therefore both types are equivalent.

Now consider a region D with a fringed hole RH, connecting at a pair of points. Consider the boundary of D\RH. Pick a point on the boundary of D and traverse it until encountering the boundary of RH. Rather than following D's boundary, follow the boundary of RH, keeping the interior of D\RH on the same side of the boundary as it had been previously, until encountering the boundary of D again. Now, traverse the boundary of D again, keeping D's interior on the same side of the boundary as before. Follow the boundary until returning to the original point. Define the region comprised only of D's interior bounded by this line as R1. Pick another boundary point of D that is not part of the previous path. Follow the same procedure. Define that region as R2. R1 and R2 meet at two points. By definition, this is a weakly connected, dividing region. Therefore, the fringe-hole dividing region D\RH and the weakly connected, dividing region R1∪R2 are equivalent. ∎

Basic canton regions can be combined under union and intersection to yield more complex canton regions. Yet the mere union and intersection is insufficient to fully specify the complex region. For instance, for the union of a holed region and a fringe-holed region different results emerge as different exteriors can be selected for the placement of two hosts (Fig. 7a). These scenarios discount the added possibilities of the two hosts having boundary contact. The intersection of two regions shows variations due to boundary contact (Fig. 7b).

In order to generate arbitrarily complex canton regions that are homeomorphic it is necessary to consider the exterior separations as well as the boundary interactions.

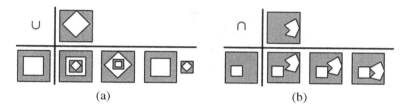

Fig. 7. Combination of basic canton regions under (a) union and (b) intersection, demonstrating that the mere set operations are underspecified, leading to ambiguous results.

4 Developing Canton Regions with COM⁺

The compound object model [11] enables the construction of objects within a planar or a spherical embedding space. It has five components: (1) the cardinality of touching boundary segments (Sect. 4.1), (2) implied external separations (Sect. 4.2), (3) the relation-constrained set operations (Sect. 4.3), (4) the sequence of intersections along boundaries (Sect. 4.4), and (5) the sequence around boundary-boundary intersections (Sect. 4.5). The five components are orthogonal (i.e., none of the five can be replaced by a combination of the other four). While not all five components are needed for the construction of all canton regions, some canton regions cannot be constructed without the entire toolset. The compound object model that uses cardinality-enhanced relations, together with the relation-constrained set operations and the two sequences is referred to as COM⁺.

4.1 Cardinality of Touching Boundary Segments

In order to form single regions, the relations between the constituent regions cannot encompass configurations with 1-dimensional common boundaries, because otherwise the interiors on both sides of the 1-dimensional boundary would merge into a single entity. As such, only 0-dimensional (i.e., point-connected) common boundaries may occur. Such connections, however, are not limited to a single instance between two regions. If the relation between two constituent regions has multiple 0-dimensional boundaries, the resulting canton region will have multiple separations, of exterior or interior. In order to capture the cardinality of the 0-dimensional boundary intersections, the topological relations *meet*, *covers*, and *coveredBy* need to be refined.

Definition 10: Let simple regions x and y have *relation*$(x, y) = meet$ consisting only of 0-dimensional boundary-boundary intersections. The relation $|n|$-*meet* refers to the number of points n in the boundary-boundary intersection of x and y (Fig. 8a).

The relations $|n|$-*coveredBy* (Fig. 8b) and $|n|$-*covers* (Fig. 8c) are defined in an analogous way. While other relations with boundary-boundary intersections could be defined in a similar way, they are not applicable to forming canton regions.

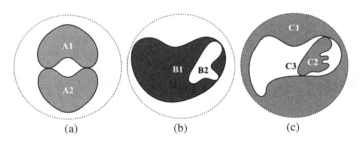

(a)   (b)   (c)

Fig. 8. The four $|n|$-relations for the case of $n = 2$: (a) A1 $|2|$-*meet* A2, (b) B1 $|2|$-*covers* B2, and (c) C2 $|2|$-*coveredBy* C3.

4.2 Implied Separations

Each of the three relations $|n|$-*meet*, $|n|$-*covers*, and $|n|$-*coveredBy* divides the region or its complement into weakly connected entities. When space is divided, one region becomes multiple regions conceptually as each component can be represented by distinct simple disks. Conveniently, $|n|$-relations divide their host region into n separate disks. It is possible, however, for combinations of holes or separations to split their hosts. Therefore, one can only assert that $|n|$-relations divide their host region into n separate disks, but together with other discs they may yield more than n separate disks.

4.3 Relation-Constrained Set Operations

The relation-constrained set operations of the compound object model allow for the combination of the constituent regions to form a canton region. Set union forms separations, while set difference with respect to a host forms a holed region. The relations between each pair of the m constituent regions would require m^2 entries, yet given the converse property of the region-region relations [12] and the trivial diagonal relations *equal*, at most $1/2(m^2 - m)$ relations are necessary to specify a canton region, representing the diagonal symmetry of the table. Relations implied by composition could be further eliminated, but such optimization [9, 32] is not yet pursued here. Whenever two regions form a separation of the exterior (Sect. 4.2), this separation must be added to the constituent regions and its relations with respect to other regions needs to be specified as well.

Figure 9 shows a canton region (Fig. 9a) constructed under the compound object model with its set operations (Fig. 9c), the equation forming the exterior separation (Fig. 9d), and the upper diagonal of table with the constraining relations between the constituent regions (Fig. 9b).

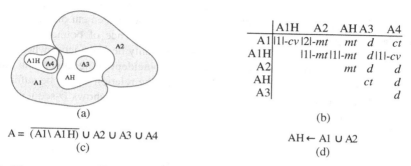

	A1H	A2	AH A3	A4
A1	I1I-*cv*	I2I-*mt*	*mt* *d*	*ct*
A1H		I1I-*mt*	I1I-*mt* *d*	I1I-*cv*
A2			*mt* *d*	*d*
AH			*ct*	*d*
A3				*d*

(a)   (b)

$$A = \overline{(A1 \setminus \overline{A1\,H})} \cup A2 \cup A3 \cup A4$$

(c)

$$AH \leftarrow A1 \cup A2$$

(d)

Fig. 9. The compound object model for (a) a canton region consisting of (b) its relation between the object's components, (c) the set operations upon the components, and (d) the implied external separation.

4.4 Sequences Along Boundaries

While the compound object model manages the containment hierarchy of disks, it does not account for sequences around the boundary, which must be maintained to ensure homeomorphism [1, 14, 24, 25]. Figure 10 displays two non-homeomorphic scenarios that have the same compound object model representation.

Fig. 10. Two non-homeomorphic canton regions that cannot be differentiated solely with the compound object model. The sequence of holes containing satellites is not immediately transformable due to the boundary intersections involved within the canton region's structure.

To differentiate such cases, one must record the sequence in which such intersections occur along each component's boundary, always traversing the boundary in a consistent orientation. While such boundary sequences have been known from binary topological relations [14] and the modeling of entire spatial scenes [24, 25], they have not yet been considered for the modeling of complex spatial objects *per se*.

4.5 Sequences Around Boundary-Boundary Intersections

Sequences along all boundaries, however, are not yet enough to fully specify all canton regions, as differences in the immediate vicinity of a point intersections may occur. All point intersections represent the intersection between a pair (or more) of boundary lines in a touch configuration. Similarly, this cycle around the point intersection must also be recorded, with a consistent orientation, to capture the exact placement of multiple disks sharing the same point. This addendum, called the sequence of boundary-boundary intersections (also referred to as $\partial\cap\partial$ sequence), need only be considered when at least two disks of the same type (interior or exterior) are coincident at the same point and on the same side of the boundary line itself. Otherwise, the relation would be sufficiently specified by the table and the boundary sequence. Figure 11 shows ∂-sequences and $\partial\cap\partial$-sequences for a canton region, highlighting its three boundary-boundary intersections.

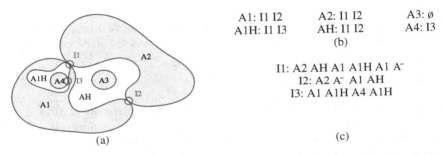

$$A1: I1\ I2 \qquad A2: I1\ I2 \qquad A3: \emptyset$$
$$A1H: I1\ I3 \qquad AH: I1\ I2 \qquad A4: I3$$
(b)

$$I1: A2\ AH\ A1\ A1H\ A1\ A^-$$
$$I2: A2\ A^-\ A1\ AH$$
$$I3: A1\ A1H\ A4\ A1H$$

(a)   (c)

Fig. 11. For (a) a canton region (Fig. 9) with three touching boundary intersections (I1-I3) its (b) boundary sequences and (c) its $\partial\cap\partial$ sequences, all recorded in a clockwise fashion.

Two canton regions that have different constructions under this formalism must be topologically distinct (modulo rotational symmetry and naming conventions). This representation ensures not only the topological relationship between region components, but also two different sequences of boundary interactions, effectively preserving order within the object itself [14].

5 Complement Canton Regions

An object's complement provides critical information for the dependencies of topological relations. For a simple region embedded in $\mathbb{S}^2$, the closure of its complement gives the relation's dual [12, 27] (Sect. 2.1). For the basic canton regions (Fig. 3), the closure of a complement links holed regions with corresponding separated regions. In order to enable the detection of such dependencies for arbitrary canton regions, we develop a method to determine from a COM$^+$ specification its complement. In so doing, the topological concept of *attach* (Fig. 2) can be considered for an arbitrary region, a relation only possible in $\mathbb{S}^2$ between complementary regions.

From a simple region's perspective, its complementary region is given by the relation *attach*. Under composition or duality [12, 27], we can consider the complementary relations between pairs of disks in a straightforward manner. For an arbitrary canton region, modeled with COM$^+$, three additional issues need to be addressed when determining its complement canton region:

1. A complement canton region switches holes and separations, therefore, complementation must exchange union and set difference in the relation-constraint set operations.
2. Holes *meet* components within a host disk, just as separations *meet* components outside of a host disk. The same goes for *disjoint*. Therefore, hole-component relations in a host as well as separation-component relations in the exterior must be maintained in the complement canton region.
3. Just as a hole is *inside* or *coveredBy* a host disk, a separation is *inside* or *coveredBy* a host's exterior disk. Therefore, relations between host and hole, as well as relations between exterior and separation must be maintained in the complement canton region.

5.1   Relation-Constrained Set Operations for a Complement Canton Region

The disk complementary to a canton region's principal region is the starting point for complementing that canton region's relation-constrained set operations. Through complementation, the relations with respect to the principal region migrate to relations with respect to the complement of the principal region. This migration implies that the principal region's relations convert to their left duals, that is, any *meet* relation (including $|n|$-*meets*) converts to its left dual (i.e., *covers* ($|n|$-*covers*), and vice-versa; likewise, any *disjoint* relation converts to a *contains* relation).

All other objects swap types in complementation—additive components become subtractive components, and vice versa. So a satellite becomes a hole of the complement, while a hole in the original canton region becomes a satellite. Since all other pairwise relations involve the swapping of types (interior/exterior) for both disks, they retain their topological relation. For the principal disk, however, exchanging it for the complementary disk does not change its type, namely an interior disk is exchanged with a second interior disk fulfilling relation *attach*.

The set operations also reflect this exchange between additive and subtractive components. A union in the original disk becomes a set difference, and vice-versa. Closure still needs to be applied after each set difference.

The choice of principal object within the canton region is immaterial, insofar as relations between objects remain the same, independent of which one is the principal object. The difference arising from this choice is on the order of the perspective of the general exterior under complementation or the new principal object under complementation.

5.2   Sequences of a Complement Canton Region

The last step in complementation is to account for sequences. Since the principal region is mapped onto its complement, two perspectives exist. On the one hand one may consider that the consistent orientation applies to the embedding space, which implies that all boundary sequences in the complement canton region need to be reversed. On the other hand, the consistent orientation may be considered the property of the principal object, which implies that all complementary boundary sequences remain unchanged. Both perspectives preserve the intersections' neighborhood, which is the critical aspect of the sequences.

5.3 Example of the Complementation of a Canton Region

Figure 12 shows a canton region B, which is the complement for the canton region specified in Figs. 9 and 11. A's components are mapped as follows onto the components of its complement B: $B1 = \widehat{A1}$, $B2 = \widehat{A1H}$, $B1H = \widehat{A2}$, $B3 = \widehat{AH}$, $B3H = \widehat{A3}$, $B2H = \widehat{A4}$. The top row of the constraining relations (Fig. 12b) is defined by the left dual the top row of A's relations (Fig. 9b). All other relations remain unchanged. The set operations (Fig. 9c and d) adapt the mappings and exchange unions for set intersections (Fig. 12c and d), and the sequences (Fig. 11 b and c) apply the mappings (Fig. 12e and f).

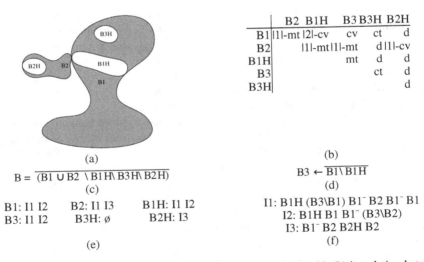

(a)

	B2	B1H	B3	B3H	B2H
B1	l1l-mt	l2l-cv	cv	ct	d
B2		l1l-mtl1l-mt		dl1l-cv	
B1H			mt	d	d
B3				ct	d
B3H					d

(b)

$B = \overline{(B1 \cup B2 \setminus B1H \setminus B3H \setminus B2H)}$

(c)

$B3 \leftarrow \overline{B1 \setminus B1H}$

(d)

B1: I1 I2   B2: I1 I3   B1H: I1 I2
B3: I1 I2   B3H: ∅   B2H: I3

(e)

I1: B1H (B3\B1) B1⁻ B2 B1⁻ B1
I2: B1H B1 B1⁻ (B3\B2)
I3: B1⁻ B2 B2H B2

(f)

Fig. 12. The complement of the canton region displayed in Fig. 9 with (b) its relation between the object's components, (c) the set operations upon the components, and (d) the implied separation.

6 Self-Surrounding Canton Regions

Canton regions with separations of the exterior have the potential to contain satellites in each separation. When these satellites are located within a canton region's topological hull [24], the region partially surrounds itself. Such configurations resemble the *surrounds* relations for holed regions or of collections of regions that form a separation of the exterior, but they are not with respect to another object, but with respect to a part of the same object. While the binary relation *surrounds* comes in four basic forms— *surroundsEmpty*, *surroundsDisjoint*, *surroundsMeet*, and *surroundsAttach*— self-surrounding canton regions expose more detail about the first three relations (Sects. 6.1, 6.2 and 6.3). The fourth relation, *surroundsAttach*, occurs when a region fills an entire hole, which requires a 1-dimensional boundary-boundary intersection. Since canton regions are defined to have only point intersections with other parts of the same canton region, this type of relation is impossible between the parts of a canton region. Similarly, *surroundsAttachHole*, a refinement of *surroundsAttach*, cannot be realized in canton region form.

6.1   SelfSurroundsEmpty

The relation *surroundsEmpty* corresponds to a holed region that has no other parts in its hole(s). Such a canton region is referred to as *selfSurroundsEmpty*. It has three different versions: *selfSurroundsProperHole*, *selfSurroundsFringedHole*, and *selfSurroundsN-FringedHole* (Fig. 13).

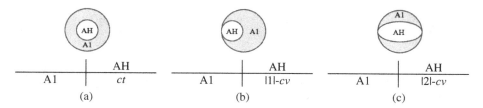

Fig. 13. A *selfSurroundsEmpty* canton region splits into three forms of canton regions: (a) *selfSurroundsProperHole*, (b) *selfSurroundsFringedHole*, and (c) *selfSurroundsNFringed-Hole,* each with A = $\overline{\text{A1} \setminus \text{AH}}$.

The region *selfSurroundsNFringedHole* is an instance of the special case of weakly connected regions. For any value of *n* in |*n*|-*covers*, *n* defines the number of separations of the disk A. Each of the three configurations in Fig. 13 corresponds to a basic canton region: *selfSurroundsProperHole* is a properly holed region (Fig. 6j), *selfSurrounds-FringedHole* is a fringe-holed dividing region (Fig. 6l), and *selfSurroundsNFringed-Hole* is, for N = 2, a fringe-holed non-dividing region (Fig. 6m) as well as a weakly connected dividing region (Fig. 6h).

6.2   SelfSurroundsDisjoint

Adding for each of the three configurations of *selfSurroundsEmpty* (Fig. 13) a satellite into its hole such that the satellite has no contact with the hole's boundary yields three versions of *selfSurroundsDisjoint*: *selfSurroundsProperHoleDisjoint* (Fig. 14a), *self-SurroundsFringedHoleDisjoint* (Fig. 14b), and *selfSurroundsNFringedHole- Disjoint* (Fig. 14a).

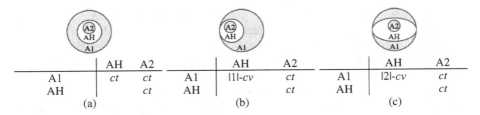

	AH	A2
A1	ct	ct
AH		ct

(a)

	AH	A2
A1	I1I-cv	ct
AH		ct

(b)

	AH	A2
A1	I2I-cv	ct
AH		ct

(c)

Fig. 14. The relation *surroundsDisjoint* splits into three forms of canton regions: (a) selfSurroundsProperHoleDisjoint, (b) selfSurroundsFringedHoleDisjoint, and (c) selfSurroundsNFringedHoleDisjoint, each with A = $\overline{A1 \setminus AH} \cup A2$.

6.3   SelfSurroundsMeet

The relation *surroundsMeet* as a canton has five distinct cases, unlike its predecessors *surroundsEmpty* and *surroundsDisjoint*. The difference between the relations is that the internal disk can exhibit its *1-meet* relation with the disk at one of two types of points: intersection with the outer boundary or away from the outer boundary (Fig. 15).

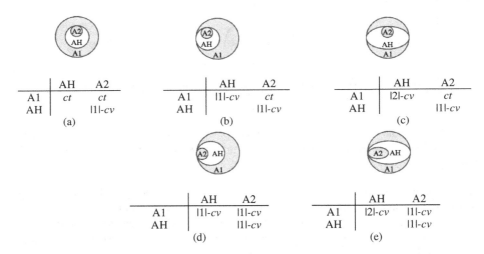

	AH	A2
A1	ct	ct
AH		I1I-cv

(a)

	AH	A2
A1	I1I-cv	ct
AH		I1I-cv

(b)

	AH	A2
A1	I2I-cv	ct
AH		I1I-cv

(c)

	AH	A2
A1	I1I-cv	I1I-cv
AH		I1I-cv

(d)

	AH	A2
A1	I2I-cv	I1I-cv
AH		I1I-cv

(e)

Fig. 15. The relation *surroundsMeet* and its five canton regions: (a) selfSurroundsProperHoleMeet, (b) selfSurroundsFringedHoleMeetInternal, (c) selfSurroundsFringedHoleMeetFringe, (d) selfSurroundsNFringedHoleMeetInternal, and (e) selfSurroundsNFringedHoleMeetFringe, each with A = $\overline{A1 \setminus AH} \cup A2$.

7   Swiss Cantons Defined with the Canton Region Model

The Swiss cantons and similar regions around the world motivated the development of a formal model for complexly structured regions in partitions. Among the 26 cantons in Switzerland are eleven that are simple regions. The remaining 15 Swiss cantons have holes, separations, or both, including a canton with a weekly separated region. Still no canton features a fringed hole, as the weak connection always tucked in between the union of two cantons. One proper hole in Switzerland[1] is not filled by satellites of other cantons (the Italian exclave Campione d'Italia fills the hole in the canton Ticino), therefore, Ticino forms a *selfSurroundProperHole* region. No other self-surround scenarios occur among individual Swiss cantons. We demonstrate for four complexly structured Swiss cantons how COM + models them:

- Schaffhausen is hole-free and totally separated (Fig. 16a).

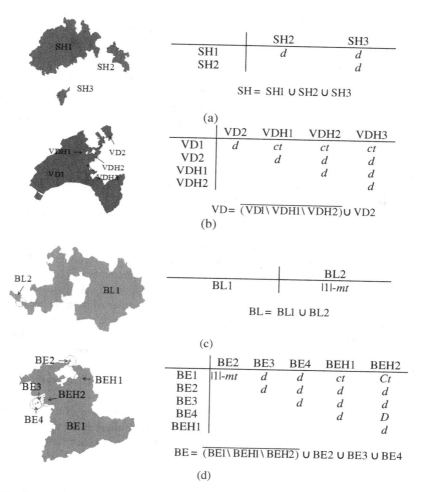

	SH2	SH3
SH1	d	d
SH2		d

$$SH = SH1 \cup SH2 \cup SH3$$

(a)

	VD2	VDH1	VDH2	VDH3
VD1	d	ct	ct	ct
VD2		d	d	d
VDH1			d	d
VDH2				d

$$VD = \overline{(VD1 \setminus VDH1 \setminus VDH2)} \cup VD2$$

(b)

	BL2
BL1	l1l-*mt*

$$BL = BL1 \cup BL2$$

(c)

	BE2	BE3	BE4	BEH1	BEH2
BE1	l1l-*mt*	d	d	ct	Ct
BE2		d	d	d	d
BE3			d	d	d
BE4				d	D
BEH1					d

$$BE = \overline{(BE1 \setminus BEH1 \setminus BEH2)} \cup BE2 \cup BE3 \cup BE4$$

(d)

Fig. 16. Four swiss cantons modeled with COM⁺: (a) Schaffhausen (SH), (b) Vaude (VD), (c) Basel-Landschaft (BL), and (d) Bern (BE).

---

[1] The other hole in Switzerland, filled by the German exclave of Büsingen, is a hole formed by the union of the cantons Schaffhausen, Thurgau, and Zürich so that no single Swiss canton has a hole filled by the German exclave.

- Vaud is totally separated with proper holes[2] (Fig. 16b).
- Basel-Landschaft is a weakly-connected non-dividing region (Fig. 16c).
- Berne is a weakly connected, non-dividing region with two proper holes and two totally separated regions (Fig. 16d).

8 Conclusions and Future Work

Models for spatial relations that can account for properties that only complexly structured regions yield require a solid foundation for constructing arbitrarily complex regions that will support spatial-relation modeling and reasoning. Many regions in the world fail to adhere to simple-region constructs [2, 18, 30, 41]. Motivated by the diversity in the structure of Swiss cantons, this paper developed a theory for regions with potential holes and separations, where the holes and separations may be weakly connected to each other. While these canton regions originate in an administrative structure as fiat objects, they may equally occur in processed geosensor network data. Therefore, any model for canton regions must not be limited by the constraints of current political subdivisions, but needs to be powerful enough to address arbitrarily structured regions in partitions.

The paper first identified seven basic types of canton regions distinguished by closure-connected and properly connected interiors and exterior. While this set of basic canton regions explains holed and separated regions, it also showed that a region with a weakly connected dividing region vs. a fringe-holed dividing region are two different perspectives over the same configurations. The seven prototypical complexly structured regions can, however, not be combined through union or intersection without providing additional detail about the relations between regions' components.

While the compound object model with relation-constrained set operations allows for the construction of canton regions with an arbitrary number of holes or regions, the mere use of the set operations is insufficient in order to create topologically distinct regions. Two additional properties need to be recorded in order: (1) the sequence of intersections around each component's boundary and (2) the sequence of neighbors around each 0-dimensional boundary-boundary intersection. These sequences, together with the relation-constrained set operations, form the extended Compound Object Model (COM$^+$).

The paper also demonstrated the ease by which the complement of a canton region can be derived for a COM$^+$ modeled region. Complementation requires the left-dual of all relations between the canton region's principal object and its other components. All other components swap from hole to satellite, and vice versa. Their relations in COM$^+$ map onto themselves for the complement. The two sequences are either maintained or

---

[2] Although most maps account only two holes in Vaud, a third hole is occupied by a monastery in Avenches, which is an exclave of the canton Fribourg. The swisstopo vector map swissBOUNDARIES3D captures the hole correctly. The 2000 Swiss Census also accounts for this Fribourgeois exclave.

reverted, depending on whether the consistent orientation is seen as a property innate to the embedding space or associated with the principal object.

Another insight is that the topological hull [24, 25] is not needed. It is a detector of holes, rather than a constructor of holes. Its purpose is to fill all holes within an object. Holes can, however, be removed by removing the labeled disk, with all objects contained by it, from the canton region. Similarly they can be filled by computing the union between the canton region and a region that is exactly equal to the hole itself.

The paper also demonstrated the proliferation of *surrounds* cantons, a generalization of *surrounds* relations [8]. Using the framework of *surrounds* not only creates a plethora of pre-defined constructions, but also allows for future streamlining of the overall compound object model by treating a defined *surrounds* canton configuration as a disk type that can be used in concert with other disk unions or differentials. The neighbors approach, used in the *surrounds* detection mechanism [7], however, could provide a fruitful approach for trying to isolate particular types of canton regions within a combinatorial map, or potentially to define the relationship between a pair of cantons [20].

The key area of future work within this domain is the creation of the relational semantics between two such regions so that spatial query languages as well as natural-language constructs as well as qualitative spatial reasoning would be supported. The interplay between different canton regions yields such scenarios as *fully surrounding* regions and *partially surrounding* regions, nested surrounding regions, and combinations of surrounding and externally located regions. Not only do these relations need to be constructed mathematically, but also the human semantics of such associations [7, 19, 21, 22, 35, 43] must be addressed when such complex constructions necessitate the need for an individual or system make correspondingly complex decisions [46]. While topological reasoning systems are often constructed in an *ad hoc* manner from geometry [5, 13, 36], the topological language is ultimately important in establishing proper cognitive semantics.

Future work can also be pursued in a complete representation of an object from the very disks that construct it. In such a model, each disk (and its complement) is stored in a tabular format, and then choices and operations are made from that table, effectively generalizing the COM + model.

Acknowledgments. Matthew Dube is partially supported by a Michael J. Eckardt Dissertation Fellowship from the University of Maine. Max Egenhofer's research was partially supported by NSF grant IIS-1016740. Joshua Lewis is supported by a teaching assistantship at the University of Maine.

References

1. Adams, C.C., Franzosa, R.D.: Introduction to Topology: Pure and Applied. Pearson Prentice Hall, Upper Saddle River (2008)
2. Beales, D.E.D., Biagini, E.F.: The Risorgimento and the Unification of Italy. Pearson Education, Harlow (2002)

3. Clementini, E., Sharma, J., Egenhofer, M.J.: Modelling topological spatial relations: strategies for query processing. Comput. Graph. **18**(6), 815–822 (1994)

4. Cohn, A.G., Renz, J.: Qualitative spatial representation and reasoning. In: van Hermelen, F., Lifschitz, V., Porter, B. (eds.) Handbook of Knowledge Representation, pp. 551–596. Elsevier, Amsterdam (2008)

5. Cohn, A.G., Varzi, A.: Mereotopological connection. J. Philos. Logic **32**, 357–390 (2003)

6. Dube, M.P., Barrett, J.V., Egenhofer, M.J.: From metric to topology: determining relations in discrete space. In: Fabrikant, S.I., et al. (eds.) COSIT 2015. LNCS. vol. 9368, pp. xx–yy. Springer, Heidelberg (2015)

7. Dube, M.P., Egenhofer, M.J.: An Ordering of Convex Topological Relations. In: Xiao, N., Kwan, M.-P., Goodchild, M.F., Shekhar, S. (eds.) GIScience 2012. LNCS, vol. 7478, pp. 72–86. Springer, Heidelberg (2012)

8. Dube, M.P., Egenhofer, M.J.: Surrounds in Partitions. In: Huang, Y., Schneider, M., Gertz, M., Krumm, J., Sankaranarayanan, J. (eds.) ACM SIGSPATIAL 2014, pp. 233–242. ACM Press, New York (2014)

9. Duckham, M., Li, S., Liu, W., Long, Z.: On redundant topological constraints. In: Baral, C., De Giacomo, G., Eiter, T. (eds.) KR 2014. AAAI Press, Menlo Park (2014)

10. Edmonds, J.: A combinatorial representation of polyhedral surfaces. Not. Am. Math. Soc. **7**, 646 (1960)

11. Egenhofer, M.J.: A reference system for topological relations between compound spatial objects. In: Heuser, C.A., Pernul, G. (eds.) ER 2009. LNCS, vol. 5833, pp. 307–316. Springer, Heidelberg (2009)

12. Egenhofer, M.: Spherical topological relations. In: Spaccapietra, S., Zimányi, E. (eds.) Journal on Data Semantics III. LNCS, vol. 3534, pp. 25–49. Springer, Heidelberg (2005)

13. Egenhofer, M.J., Dube, M.P.: Topological relations from metric refinements. In: Agrawal, D., Arefw, W., Lu, C., Mokbel, M., Scheurmann, P., Shahabi, C., Wolfson, O. (eds.) ACM SIGSPATIAL 2009, pp. 158–167. ACM Press, New York (2009)

14. Egenhofer, M.J., Franzosa, R.D.: On the equivalence of topological relations. Int. J. Geogr. Inf. Syst. **9**(2), 133–152 (1995)

15. Egenhofer, M.J., Franzosa, R.D.: Point-set topological spatial relations. Int. J. Geogr. Inf. Syst. **5**(2), 161–174 (1991)

16. Egenhofer, M.J., Herring, J.R.: Categorizing Binary Topological Relations Between Regions, Lines, and Points in Geographic Databases. Technical report, Department of Surveying Engineering, University of Maine (1990)

17. Egenhofer, M., Vasardani, M.: Spatial reasoning with a hole. In: Winter, S., Duckham, M., Kulik, L., Kuipers, B. (eds.) COSIT 2007. LNCS. vol. 4736, pp. 303–320. Springer, Heidelberg (2007)

18. Glass, H.E.: Ethnic diversity, elite accommodation and federalism in Switzerland. Publius **7** (4), 31–48 (1977)

19. Hampe, B., Grady, J.E. (eds.): From Perception to Meaning: Image Schemas in Cognitive linguistics, vol. 29. Walter de Gruyter, Berlin (2005)

20. Hu, Y., Ravada, S., Anderson, R., Bamba, B.: Supporting Topological Relationship Queries for Complex Regions in Oracle Spatial. In: Cruz, I.F., Knoblock, C.A., Kröger, P., Krumm, J., Tanin, E., Widmayer, P. (eds.) SIGSPATIAL 2012, pp. 3–12. ACM Press, New York (2013)

21. Klippel, A.: Spatial information theory meets spatial thinking: is topology the Rosetta Stone of spatio-temporal cognition? Ann. Assoc. Am. Geogr. **102**(6), 1310–1328 (2012)

22. Klippel, A., Li, R., Yang, J., Hardisty, F., Xu, S.: The Egenhofer-Cohn–Hypothesis or, topological relativity? In: Mark, D.M., Frank, A.U., Raubal, M. (eds.) Cognitive and Linguistic Aspects of Geographic Space, pp. 195–215. Springer, Heidelberg (2013)

23. Kurata, Y.: The 9+-intersection: a universal framework for modeling topological relations. In: Cova, T.J., Miller, H.J., Beard, K., Frank, A.U., Goodchild, M.F. (eds.) GIScience 2008. LNCS, vol. 5266, pp. 181–198. Springer, Heidelberg (2008)

24. Lewis, J.A., Dube, M.P., Egenhofer, M.J.: The topology of spatial scenes in $\mathbb{R}^2$. In: Tenbrink, T., Stell, J., Galton, A., Wood, Z. (eds.) COSIT 2013. LNCS, vol. 8116, pp. 495–515. Springer, Heidelberg (2013)

25. Lewis, J.A., Egenhofer, M.J.: Oriented regions for linearly conceptualized features. In: Duckham, M., Pebesma, E., Stewart, K., Frank, A.U. (eds.) GIScience 2014. LNCS, vol. 8728, pp. 333–348. Springer, Heidelberg (2014)

26. Li, S.: A complete classification of topological relations using the 9-intersection method. Int. J. Geogr. Inf. Sci. **20**(6), 589–610 (2006)

27. Li, S., Li, Y.: On the complemented disk algebra. The J. Logic Algebraic Program. **66**(2), 195–211 (2006)

28. Milner, R.: pure bigraphs: structure and dynamics. Inf. Comput. **204**(1), 60–122 (2006)

29. Open GIS Consortium, Inc.: OpenGIS® Simple Feature Specification for SQL. OpenGIS Project Document 99–049 (1999)

30. Parker, N., Vaughan-Williams, N.: Lines in the sand? Towards an agenda for critical border studies. Geopolitics **14**(3), 582–587 (2009)

31. Randell, D.A., Cui, Z., Cohn, A.G.: A Spatial logic based on regions and connection. In: Nebel, B., Rich, C., Swartout, W.R. (eds.) KR 92, pp. 165–176. Morgan Kaufmann, San Francisco (1992)

32. Rodríguez, M.A., Egenhofer, M.J., Blaser, A.D.: Query pre-processing of topological constraints: comparing a composition-based with neighborhood-based approach. In: Hadzilacos, T., Manolopoulos, Y., Roddick, J.F., Theodoridis, Y. (eds.) SSTD 2003. LNCS, vol. 275, pp. 362–379. Springer, Heidelberg (2003)

33. Saucier, R.T.: Evidence for episodic sand-blow activity during the 1811–1812 New Madrid (Missouri) earthquake series. Geology **17**(2), 103–106 (1989)

34. Schneider, M., Behr, T.: Topological relationships between complex spatial objects. ACM Trans. Database Syst. **31**(1), 39–81 (2006)

35. Smith, B.: Fiat Objects. Topoi **20**(2), 131–148 (2001)

36. Sridhar, M., Cohn, A.G., Hogg, D.C.: From video to RCC8: exploiting a distance based semantics to stabilise the interpretation of mereotopological relations. In: Egenhofer, M., Giudice, N., Moratz, R., Worboys, M. (eds.) COSIT 2011. LNCS, vol. 6899, pp. 110–125. Springer, Heidelberg (2011)

37. Tutte, W.T.: What is a map? In: Harary, F. (ed.) New Directions in the Theory of Graphs, pp. 309–325. Academic Press, New York (1973)

38. Tyler, A., Evans, V.: The Semantics of English Prepositions. CUP, Cambridge (2003)

39. Varzi, A.C.: Spatial reasoning in a holey world. In: Torasso, P. (ed.) AI*IA 93, pp. 326–336. Springer, Heidelberg (1993)

40. Vasardani, M., Egenhofer, M.J.: Comparing Relations with a Multi-holed Region. In: Hornsby, K.S., Claramunt, C., Denis, M., Ligozat, G. (eds.) COSIT 2009. LNCS, vol. 5756, pp. 159–176. Springer, Heidelberg (2009)

41. Whyte, B.R.: En Territoire Belge et à Quarante Centimètres de la Frontière: An Historical and Documentary Study of the Belgian and Dutch Enclaves of Baarle-Hertog and Baarle-Nassau. University of Melbourne (2004)

42. Worboys, M.: The maptree: a fine-grained formal representation of space. In: Xiao, N., Kwan, M.-P., Goodchild, M.F., Shekhar, S. (eds.) GIScience 2012. LNCS, vol. 7478, pp. 298–310. Springer, Heidelberg (2012)

43. Worboys, M.: Using maptrees to characterize topological change. In: Tenbrink, T., Stell, J., Galton, A., Wood, Z. (eds.) COSIT 2013. LNCS, vol. 8116, pp. 74–90. Springer, Heidelberg (2013)

44. Worboys, M.F., Bofakos, P.: A canonical model for a class of areal spatial objects. In: Abel, D.J., Ooi, B.C., (eds.) SSD 1993. LNCS, vol. 692, pp. 36–52. Springer, Heidelberg (1993)

45. Worboys, M.F., Duckham, M.: Monitoring qualitative spatial change for geosensor networks. Int. J. Geogr. Inf. Sci. **20**(10), 1087–1108 (2006)

46. Zlatev, J.: Spatial semantics. In: Geeraerts, D., Cuyckens, H. (eds.) The Oxford Handbook of Cognitive Linguistics, pp. 318–350. Oxford University Press Inc., New York (2007)

Spatial Symmetry Driven Pruning Strategies for Efficient Declarative Spatial Reasoning

Carl Schultz[1,3]([✉]) and Mehul Bhatt[2,3]

[1] Institute for Geoinformatics, University of Münster, Münster, Germany
schultzc@uni-muenster.de
[2] Department of Computer Science, University of Bremen, Bremen, Germany
[3] The DesignSpace Group, Bremen, Germany

Abstract. Declarative spatial reasoning denotes the ability to (declaratively) specify and solve real-world problems related to geometric and qualitative spatial representation and reasoning within standard knowledge representation and reasoning (KR) based methods (e.g., logic programming and derivatives). One approach for encoding the semantics of spatial relations within a declarative programming framework is by systems of polynomial constraints. However, solving such constraints is computationally intractable in general (i.e. the theory of real-closed fields).

We present a new algorithm, implemented within the declarative spatial reasoning system CLP(QS), that drastically improves the performance of deciding the consistency of spatial constraint graphs over conventional polynomial encodings. We develop pruning strategies founded on spatial symmetries that form equivalence classes (based on affine transformations) at the qualitative spatial level. Moreover, pruning strategies are themselves formalised as knowledge about the properties of space and spatial symmetries. We evaluate our algorithm using a range of benchmarks in the class of contact problems, and proofs in mereology and geometry. The empirical results show that CLP(QS) with knowledge-based spatial pruning outperforms conventional polynomial encodings by orders of magnitude, and can thus be applied to problems that are otherwise unsolvable in practice.

Keywords: Declarative spatial reasoning · Geometric reasoning · Logic programming · Knowledge representation and reasoning

1 Introduction

Knowledge representation and reasoning (KR) about *space* may be formally interpreted within diverse frameworks such as: (a) analytically founded geometric reasoning & constructive (solid) geometry [21,27,29]; (b) relational algebraic semantics of 'qualitative spatial calculi' [24]; and (c) by axiomatically constructed formal systems of mereotopology and mereogeometry [1]. Independent of formal semantics, commonsense spatio-linguistic abstractions offer a human-centred and cognitively adequate mechanism for logic-based automated reasoning about spatio-temporal information [5].

© Springer International Publishing Switzerland 2015
S.I. Fabrikant et al. (Eds.): COSIT 2015, LNCS 9368, pp. 331–353, 2015.
DOI: 10.1007/978-3-319-23374-1_16

▷ **Declarative Spatial Reasoning.** In the recent years, *declarative spatial reasoning* has been developed as a high-level commonsense spatial reasoning paradigm aimed at (declaratively) specifying and solving real-world problems related to geometric and qualitative spatial representation and reasoning [4]. A particular manifestation of this paradigm is the constraint logic programming based CLP(QS) spatial reasoning system [4,33,34] (Sect. 2).

▷ **Relational Algebraic Qualitative Spatial Reasoning.** The state of the art in qualitative spatial reasoning using relational algebraic methods [24] has resulted in prototypical algorithms and black-box systems that do not integrate with KR languages, such as those dealing with semantics and conceptual knowledge necessary for handling background knowledge, action &change, relational learning, rule-based systems etc. Furthermore, relation algebraic qualitative spatial reasoning (e.g. LR [25]), while efficient, is incomplete in general [22–24].[1] Alternatively, constraint logic programming based systems such as CLP(QS) [4] and others (see [9,10,18,20,28,29]) adopt an analytic geometry approach where spatial relations are encoded as systems of polynomial constraints;[2] while these methods are sound and complete (see Sect. 2.2), they have prohibitive computational complexity, $O(c_1^{c_2^n})$ in the number of polynomial variables n, meaning that even relatively simple problems are not solved within a practical amount of time via "naive" or direct encodings as polynomial constraints, i.e. encodings that lack common-sense knowledge about spatial objects and relations. On the other hand, highly efficient and specialised geometric theorem provers (e.g. [12]) and geometric constraint solvers (e.g. [17,27]) exist. However, these provers exhibit highly specialised and restricted spatial languages[3] and lack (a) the direct integration with more general AI methods and (b) the capacity for incorporating modular common-sense rules about space in an extensible domain- and context-specific manner.

The aims and contributions of the research presented in this paper are two-fold:

1. to further develop a KR-centered declarative spatial reasoning paradigm such that spatial reasoning capabilities are available and accessible within AI

---

[1] *Incompleteness* refers to the inability of a spatial reasoning method to determine whether a given network of qualitative spatial constraints is consistent or inconsistent in general. Relation-algebraic spatial reasoning (i.e. using algebraic closure based on weak composition) has been shown to be incomplete for a number of spatial languages and cannot guarantee *consistency* in general, e.g. relative directions [23] and containment relations between linearly ordered intervals [22], Theorem 5.9.

[2] We emphasise that this analytic geometry approach that we also adopt is not *qualitative spatial reasoning* in the relation algebraic sense; the foundations are similar (i.e. employing a finite language of spatial relations that are interpreted as infinite sets of configurations, determining consistency in the complete absence of numeric information, and so on) but the methods for determining consistency etc. come from different branches of spatial reasoning.

[3] Standard geometric constraint languages of approaches including [12,17,27] consist of points, lines, circles, ellipses, and coincidence, tangency, perpendicularity, parallelism, and numerical dimension constraints; note the absence of e.g. mereotopology and "common-sense" relative orientation relations [35].

programs and applications areas, and may be seamlessly integrated with other AI methods dealing with representation, reasoning, and learning about non-spatial aspects

2. to demonstrate that in spite of high computational complexity in a general and domain-independent case, the power of analytic geometric—in particular polynomial systems for encoding the semantics of spatial relations – can be exploited by systematically utilising commonsense knowledge about spatial object and relationships at the qualitative level.

We present a new algorithm that drastically improves analytic spatial reasoning performance within KR-based declarative spatial reasoning approaches by identifying and pruning spatial symmetries that form equivalence classes (based on affine transformations) at the qualitative spatial level. By exploiting symmetries our approach utilises powerful underlying, but computationally expensive, polynomial solvers in a significantly more effective manner. Our algorithm is simple to implement, and enables spatial reasoners to solve problems that are otherwise unsolvable using analytic or relation algebraic methods. We emphasise that our approach is independent of any particular polynomial constraint solver; it can be similarly applied over a range of solvers such as CLP(R), SMTs, and specialised geometric constraint solvers that have been integrated into a KR framework.

In addition to AI/commonsense reasoning applications areas such as design, GIS, vision, robotics [3, 5–7], we also address application into automating support for proving the validity of theorems in mereotopology, orientation, shape, etc. (e.g. [8, 36]). Building on such foundational capabilities, another outreach is in the area of computer-aided learning systems in mathematics (e.g. at a high-school level). For instance, consider Proposition 9, Book I of Euclid's *Elements*, where the task is to bisect an angle using only an unmarked ruler and collapsable compass. Once a student has developed what they believe to be a constructive proof, they can employ declarative spatial reasoners to formally verify that their construction applies to all possible inputs (i.e. all possible angles) and manipulate an interactive sketch that maintains the specified spatial relations (i.e. dynamic geometry [17]). A further area of interest is verifying the entries of composition tables that are used in relation algebraic qualitative spatial reasoning [30]: given spatial objects $a, b, c \in U$, composition "look up" tables are indexed by pairs of (base) relations $R_{1_{ab}}$, $R_{2_{bc}}$ and return disjunctions of possible (base) relations $R_{3_{ac}}$. For each entry, the task is to prove $\exists a, b, c \in U (R_{1_{ab}} \wedge R_{2_{bc}} \wedge R_{3_{ac}})$ for only those base relations R_3 in the entry's disjunction.

2   Declarative Spatial Reasoning with CLP(QS)

Declarative spatial reasoning denotes the ability of declarative programming frameworks in AI to handle *spatial objects* and the *spatial relationships* amongst them as *native* entities, e.g., as is possible with concrete domains of Integers, Reals and Inequality relationships. The objective is to enable points, oriented points, directed line-segments, regions, and topological and orientation relationships amongst them as *first-class* entities within declarative frameworks in AI [4].

2.1   Examples of Declarative Spatial Reasoning with CLP(QS)

With a focus on spatial question-answering, the CLP(QS) spatial reasoning system [4,33,34] provides a practical manifestation of certain aspects of the declarative spatial reasoning paradigm in the context of constraint logic programming (CLP).[4] CLP(QS) utilises a high-level language of spatial relations and commonsense knowledge about how various spatial domains behave. Such relations describe sets of object configurations, i.e. qualitative spatial relations such as *coincident, left of*, or *partially overlapping*. Through this deep integration of spatial reasoning with KR-based frameworks, the long-term research agenda is to seamlessly provide spatial reasoning in other AI tasks such as planning, non-monotonic reasoning, and ontological reasoning [4]. What follows is a brief illustration of the spatial Q/A capability supported by CLP(QS).

EXAMPLE A. *Massachusetts Comprehensive Assessment System (MCAS).*

Grade 3 Mathematics (2009), Question 12. *Put a square and two right-angled triangles together to make a rectangle. (1) Put the shapes T_1, T_2, S illustrated in Fig. 1(d) together to make a rectangle. (2) Put the shapes T_1, T_2, S in Fig. 1(d) together to make a quadrilateral that is not a rectangle.*
    CLP(QS) represents right-angle triangles as illustrated in Fig. 1(b). Figure 1(a) and (c) present the CLP(QS) solutions.

Grade 3 Mathematics (2013), Question 17. *(1) How many copies of T_1 illustrated in Fig. 1(d) are needed to completely fill the region R illustrated in Fig. 2(a) without any of them overlapping?*
    As presented in Fig. 2, CLP(QS) solves both the geometric definition and a variation where the dimensions of the rectangle and triangles are not given.

EXAMPLE B. *Qualitative Spatial Reasoning with Complete Unknowns.*
    In this example CLP(QS) reasons about spatial objects based solely on given qualitative spatial relations, i.e. without any geometric information.
    Define three cubes A, B, C. Put B inside A, and make B disconnected from C. What spatial relations can possibly hold between A and C?
    CLP(QS) determines that A must be *disconnected* from C and provides the inferred corresponding geometric constraints, as illustrated in Fig. 3.

2.2   Analytical Geometry Foundations for Declarative Spatial Reasoning

Analytic geometry methods parameterise classes of objects and encode spatial relations as systems of polynomial equations and inequalities [12]. For example, we can define a sphere as having a 3D centroid point (x, y, z) and a radius r, where x, y, z, r are reals. Two spheres s_1, s_2 *externally connect* or *touch* if

$$(x_{s_1} - x_{s_2})^2 + (y_{s_1} - y_{s_2})^2 + (z_{s_1} - z_{s_2})^2 = (r_{s_1} + r_{s_2})^2 \tag{1}$$

---

[4] Spatial Reasoning (CLP(QS)). www.spatial-reasoning.com.

```
?- T1  = right_triangle(_,_,W,W),
|    T2  = right_triangle(_,_,W,W),
|    S = rectangle(_,W,W),
|    Solution = rectangle(_,_,_),
|    topology(rcc8(eq), Solution, tile_union([T1,T2,S])),
|    ground_object(Solution).

T1 = right_triangle(orientation(0), point(0, 0), 5, 5),
W = 5,
T2 = right_triangle(orientation(180), point(5, 5), 5, 5),
S = rectangle(point(5, 0), 5, 5),
Solution = rectangle(point(0, 0), 10, 5) .
```

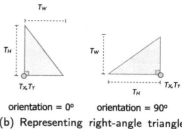

(a) CLP(QS) solution to arranging T_1, T_2, S to form rectangles.

(b) Representing right-angle triangle T in CLP(QS).

```
?- T1  = right_triangle(_,_,W,W),
|    T2  = right_triangle(_,_,W,W),
|    S = rectangle(_,W,W),
|    topology(rcc8(eq), Solution, tile_union([T1,T2,S])),
|    quadrilateral(Solution),
|    not(Solution = rectangle(_,_,_)),
|    ground_object(Solution).

T1 = right_triangle(orientation(90), point(5, 0), 5, 5),
W = 5,
T2 = right_triangle(orientation(270), point(10, 5), 5, 5),
S = rectangle(point(5, 0), 5, 5),
Solution = parallelogram(orientation(0), point(0,0), 10, 5);
...
T1 = right_triangle(orientation(90), point(5, 0), 5, 5),
W = 5,
T2 = right_triangle(orientation(0), point(10, 0), 5, 5),
S = rectangle(point(5, 0), 5, 5),
Solution = trapezoid(orientation(0), point(0, 0), 15, 5, 5);
...
false.
```

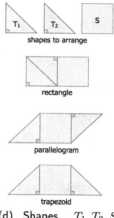

(c) CLP(QS) solution to arranging T_1, T_2, S to form non-rectangular quadrilaterals.

(d) Shapes T_1, T_2, S to be arranged and CLP(QS) solutions.

Fig. 1. Using CLP(QS) to solve MCAS Grade 3 Mathematics Test questions (2009).

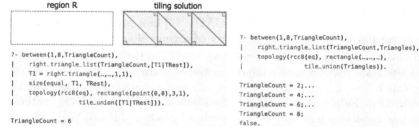

```
?- between(1,8,TriangleCount),
|    right_triangle_list(TriangleCount,[T1|TRest]),
|    T1 = right_triangle(_,_,1,1),
|    size(equal, T1, TRest),
|    topology(rcc8(eq), rectangle(point(0,0),3,1),
|             tile_union([T1|TRest])).

TriangleCount = 6
```

```
?- between(1,8,TriangleCount),
|    right_triangle_list(TriangleCount,Triangles),
|    topology(rcc8(eq), rectangle(_,_,_),
|             tile_union(Triangles)).

TriangleCount = 2;...
TriangleCount = 4;...
TriangleCount = 6;...
TriangleCount = 8;
false.
```

(a) CLP(QS) solution to tiling a rectangular region with triangle T_1.

(b) When no geometric information is given, CLP(QS) determines that the number of right-angle triangles must be even.

Fig. 2. Using CLP(QS) to solve MCAS Grade 3 Mathematics Test questions (2013).

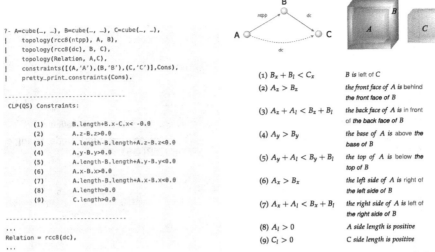

(a) CLP(QS) solution and its corresponding inferred geometric constraints.

(b) Geometric inequalities provided by CLP(QS) that correspond to the qualitative constraints.

Fig. 3. Spatial reasoning about cubes A, B, C with complete geometric unknowns.

If the system of polynomial constraints is satisfiable then the spatial constraint graph is consistent. Specifically, the system of polynomial (in)equalities over variables X is satisfiable if there exists a real number assignment for each $x \in X$ such that the (in)equalities are *true*. Partial geometric information (i.e. a combination of numerical and qualitative spatial information) is utilised by assigning the given real numerical values to the corresponding object parameters. Thus, we can integrate spatial reasoning and logic programming using *Constraint Logic Programming* (CLP) [19]; this system is called CLP over qualitative spatial domains. CLP($\mathcal{QS}$), provides a suitable framework for expressing and proving first-order spatial theorems.

Cylindrical Algebraic Decomposition (CAD) [13] is a prominent sound and complete algorithm for deciding satisfiability of a general system of polynomial constraints over reals and has time complexity $O(c_1^{c_2^n})$ in the number of free variables [2]. Thus, a key focus within analytic spatial reasoning has been methods for managing this inherent intractibility.[5] More efficient refinements of the original CAD algorithm include partial CAD [14]. Symbolic methods for solving systems of multivariate *equations* include the Gröbner basis method [11] and Wu's characteristic set method [40]. In the QUAD-CLP(R) system, the authors improve solving performance by using linear approximations of quadratic constraints and by

---

[5] Important factors in determining the applicability of various analytic approaches are the degree of the polynomials (particularly the distinction between linear and non-linear) and whether both equality and inequalities are permitted in the constraints.

identifying geometric equivalence classes [28]. Ratschan employs pruning methods, also at the polynomial level, in the *rsolve* system [31,32].

Constructive and iterative (i.e. Newton and Quasi-Newton iteration) methods solve spatial reasoning problems by "building" a solution, i.e. by finding a configuration that satisfies the given constraints [27]. If a solution is found, then the solution itself is the proof that the system is consistent – but what if a solution is not found within a given time frame? In general these methods are incomplete for spatial reasoning problems encoded as nonlinear equations and inequalities of arbitrary degree.[6]

3   Spatial Symmetries

Information about *objects* and their *spatial relations* is formally expressed as a constraint graph $G = (N, E)$, where the nodes N of the graph are spatial objects and the edges between nodes specify the relations between the objects. Objects belong to a *domain*, e.g. points, lines, squares, and circles in 2D Euclidean space, and cuboids, vertically-extruded polygons, spheres, and cylinders in 3D Euclidean space. We denote the object domain of node i as U_i (spatial domains are typically infinite). A node may refer to a partially ground, or completely geometrically ground object, such that U_i can be a proper subset of the full domain of that object type. Each element $i' \in U_i$ is called an *instance* of that object domain. A *configuration* of objects is a set of instances $\{i'_1, \ldots, i'_n\}$ of nodes $i_1, \ldots, i_n$ respectively.

A binary relation R_{ij} between nodes i, j distinguishes a set of relative configurations of i, j; relation R is said to *hold* for those configurations, $R_{ij} \subseteq U_i \times U_j$. In general, an n-ary relation for $n \geq 1$ distinguishes a set of configurations between n objects: $R_{i_1, \ldots, i_n} \subseteq U_{i_1} \times \cdots \times U_{i_n}$.

An edge between nodes i, j is assigned a logical formula over relation symbols $R_1, \ldots, R_m$ and logical operators $\vee, \wedge, \neg$. Given an interpretation i', j', the formula for edge e is interpreted in the standard way, denoted $e(i', j')$:

- $R_1 \equiv (i', j') \in R_{1_{ij}}$
- $(R_1 \vee R_2) \equiv (i', j') \in R_{1_{ij}} \cup R_{2_{ij}}$
- $(R_1 \wedge R_2) \equiv (i', j') \in R_{1_{ij}} \cap R_{2_{ij}}$
- $(\neg R_1) \equiv (i', j') \in (U_i \times U_j) \setminus R_{1_{ij}}$.

An edge between i, j is *satisfied* by a configuration i', j' if $e(i', j')$ is *true* (this is generalised to n-ary relations). A spatial constraint graph $G = (N, E)$ is *consistent* or *satisfiable* if there exists a configuration s of N that satisfies all edges in E, denoted $G(s)$; this is referred to as the *consistency task*. Graph G' is a *consequence* of, or *implied* by, G if every spatial configuration that satisfies G also satisfies G'. This is the *sufficiency task* (or *entailment*) that we commonly apply to constructive proofs, where the task is to prove that objects and relations in G are sufficient for ensuring that particular properties hold in G'.

---

[6] That is, constructive methods may fail in building a consistent solution, and iterative root finding methods may fail to converge.

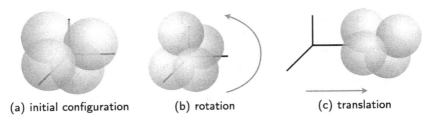

(a) initial configuration   (b) rotation   (c) translation

Fig. 4. Topological relations between four spheres maintained after various affine transformations.

Given graph G, two key questions are (1) how to give meaning, or interpret, the spatial relations in G, and (2) how to efficiently determine consistency and produce instantiations of G. That is, we need to adopt a method for *spatial reasoning*.

3.1   An Example of the Basic Concept

A key insight is that spatial configurations form equivalence classes over qualitative relationships based on certain affine transformations. For example, consider the spatial task of determining whether five same-sized spheres can be mutually touching. Suppose we are given a specific numerically defined configuration of four mutually touching spheres as illustrated in Fig. 4(a), and we prove that it is impossible to add an additional mutually touching sphere to this configuration. That is, let $s_1, \ldots, s_4$ be unit spheres (radius 1), centred on points $p_1 = (0, 0, 0)$, $p_2 = (2, 0, 0)$, $p_3 = (1, \sqrt{3}, 0)$, $p_4 = (1, \sqrt{\frac{1}{3}}, \sqrt{\frac{8}{3}})$, respectively. According to Eq. 1, $s_1, \ldots, s_4$ are mutually touching. We prove that a fifth same-sized, mutually touching sphere cannot be added to this configuration by determining that the corresponding system of polynomial constraints is unsatisfiable (the system consists of four constraints with three free variables $x_{s_5}, y_{s_5}, z_{s_5}$, by reapplying Eq. 1 between s_5 and each other sphere, e.g. s_1 *touches* s_5 is $x_{s_5}^2 + y_{s_5}^2 + z_{s_5}^2 = 4$).

Now consider that we apply an affine transformation to the original configuration such as rotation, translation, scaling, or reflection, as illustrated in Fig. 4(b) and (c). After having applied the transformation, it is still impossible to add a fifth mutually touching sphere, because the relevant qualitative (topological) relations are *preserved* under these transformations. Thus, when we proved that it was impossible to add a fifth same-sized mutually touching sphere to the original given configuration, in fact we proved it for a *class* of configurations, specifically, the class of configurations that can be reached by applying an affine transformation to the original configuration. Now, when determining consistency of graphs of qualitative spatial relations, we are not given any specific spatial configurations to work with (i.e. complete absence of numerical information), and instead need to prove consistency over all possible configurations.

The key is that, each time we ground and constrain variables, we are eliminating a *spatial symmetry* from our partially defined configuration. If we maintain knowledge about symmetries that certain object types have (e.g. spheres have complete rotational symmetry) then we can judiciously "trade" symmetries for unbound variables in our polynomial encoding at a purely symbolic level. Importantly, rather than having to compute symmetries or undertake any complex symmetry detection procedure, we are instead *building knowledge about space and spatial properties of objects into the spatial solver at a declarative level*. Thus, we are able to efficiently reason over an infinite set of possible configurations by incrementally pruning spatial symmetries based on commonsense knowledge about space, and this pruning is exploited by eliminating and constraining variables in the underlying polynomial encoding.

3.2   Theoretical Foundations for Symmetries

Due to the parameterisation of objects, spatial configurations are embedded in n-dimensional Euclidean space $\mathbb{R}^n$ ($1 \leq n \leq 3$) with a fixed origin point. Let V, W be Euclidean spaces in $\mathbb{R}^n$, each with an origin. Given vectors x, y and constant k, a linear transformation f is a mapping $V \rightarrow W$ such that

$$f(x + y) = f(x) + f(y) \quad (additive)$$
$$f(kx) = kf(x) \qquad\quad (homogeneous)$$

An affine transformation f' is a linear transformation composed with a translation. It is convenient to represent a linear transformation on vector x as a left multiplication of a $d \times d$ real matrix Q, and translation as an addition of vector t, $f'(x) = Qx + t$. We denote a transformation T applied to a spatial configuration of objects s as Ts.

We distinguish particular classes of transformations with respect to the qualitative spatial relationships that are preserved, for example, in $\mathbb{R}^2$ the following matrices represent rotation by θ, uniform scaling by $k > 0$, and horizontal reflection, respectively:

$$\begin{pmatrix} cos(\theta) & -sin(\theta) \\ sin(\theta) & cos(\theta) \end{pmatrix}, \begin{pmatrix} k & 0 \\ 0 & k \end{pmatrix}, \begin{pmatrix} -1 & 0 \\ 0 & 1 \end{pmatrix}.$$

Given transformation T we annotate it with its type $c \in C$, e.g. $C = \{translate, rotate, scale, reflect\}$ as T^c. Each spatial relation R belongs to a class of relations in Rel, such as *topology, mereology, coincidence, relative orientation, distance*. Let Sym be a function Sym : Rel $\rightarrow 2^C$ that represents the classes of transformations that preserve a given class of spatial relations. The Sym function is our mechanism for building knowledge about spatial symmetries into the spatial reasoning system. Let Rel_G be the set of classes of the spatial relations that are used in the spatial constraint graph G, and let $\text{Sym}_G = \bigcap_{R \in \text{Rel}_G} \text{Sym}(R)$.

The following formal Condition on Sym_G states that transformations (applied to the embedding space) define equivalence classes of configurations with respect to the consistency of spatial constraint graphs. When satisfied, this condition provides a theoretically sound foundation for symmetry pruning.

Condition 1. *Given spatial constraint graph G, configuration s, and affine transformation T^c with $c \in \text{Sym}_G$ then $G(s)$ is true if and only if $G(T^c s)$.*

Table 1. Polynomial encodings of qualitative spatial relations.

Relation	Polynomial Encoding
Left of (point p, segment s_{ab})	$(x_b - x_a)(y_p - y_a) > (y_b - y_a)(x_p - x_a)$
Collinear (point p, segment s_{ab})	$(x_b - x_a)(y_p - y_a) = (y_b - y_a)(x_p - x_a)$
Right or collinear (point p, segment s_{ab})	$(x_b - x_a)(y_p - y_a) \leq (y_b - y_a)(x_p - x_a)$
Parallel (segments s_{ab}, s_{cd})	$(y_b - y_a)(x_d - x_c) = (y_d - y_c)(x_b - x_a)$
Coincident (point p, segment s_{ab})	collinear$(p, s_{ab}) \wedge x_p \in [x_a, x_b] \wedge y_p \in [y_a, y_b]$
Coincident (point p, circle c)	$(x_c - x_p)^2 + (y_c - y_p)^2 = r_c^2$
Inside (point p, rectangle a)	$(0 < (p - p_{1_a}) \cdot v_a < w_a) \wedge$ $(0 < (p - p_{1_a}) \cdot v_a' < h_a)$
Intersects (point p, rectangle a)	$(0 \leq (p - p_{1_a}) \cdot v_a \leq w_a) \wedge$ $(0 \leq (p - p_{1_a}) \cdot v_a' \leq h_a)$
Boundary (point p, rectangle a)	intersects$(p, a) \wedge \neg$ inside(p, a)
Outside (point p, rectangle a)	$\neg$ intersects(p, a)
Concentric (rectangles a, b)	$\frac{1}{2}(p_{3_a} - p_{1_a}) + p_{1_a} = \frac{1}{2}(p_{3_b} - p_{1_b}) + p_{1_b}$
Part of (rectangles a, b)	$\bigwedge_{i=1\ldots4}$ intersects(p_{i_a}, b)
Proper part (rectangles a, b)	$\neg$ equals$(a, b) \wedge$ part_of(a, b)
Boundary part of (rectangles a, b)	$\bigwedge_{i=1\ldots4}$ boundary(p_{i_a}, b)
Discrete from (rectangles a, b)	$\bigvee_{i=1\ldots4} \big($ right_or_collinear$(a, (p_{i_b}, p_{(i+1)_b})) \vee$ right_or_collinear$(b, (p_{i_a}, p_{(i+1)_a})))$
Partially overlaps (rectangles a, b)	$\exists p_i \in \mathbb{R}^2 \big($ inside$(p_i, a) \wedge$ inside$(p_i, b)) \wedge$ $\exists p_j \in \mathbb{R}^2 \big($ inside$(p_j, a) \wedge$ outside$(p_j, b)) \wedge$ $\exists p_k \in \mathbb{R}^2 \big($ outside$(p_k, a) \wedge$ inside$(p_k, b))$

3.3 Polynomial Encodings for Spatial Relations

In this section we define a range of spatial domains and spatial relations with the corresponding polynomial encodings. While our method is applicable to a wide range of 2D and 3D spatial objects and qualitative relations, for brevity and clarity we primarily focus on a 2D spatial domain. Our method is readily applicable to other 2D and 3D spatial domains and qualitative relations, for example, as defined in [4,9,10,28,29,33,34].

- a *point* is a pair of reals x, y
- a *line segment* is a pair of end points p_1, p_2 ($p_1 \neq p_2$)
- a *rectangle* is a point p representing the bottom left corner, a unit direction vector v defining the orientation of the base of the rectangle, and a real width and height w, h ($0 < w, 0 < h$); we can refer to the vertices of the rectangle: let $v' = (-y_v, x_v)$ be v rotated $90°$ counter-clockwise, then $p_1 = p = p_5, p_2 = wv + p_1, p_3 = wv + hv' + p_1, p_4 = hv' + p_1$
- a *circle* is a centre point p and a real radius r ($0 < r$).

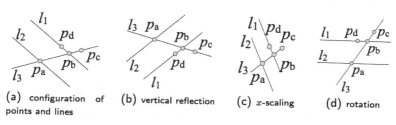

Fig. 5. Affine transformations preserve point coincidence, parallelism, and ratios of distances along parallel lines.

We consider the following spatial relations:

Relative Orientation. Left, right, collinear orientation relations between points and segments, and parallel, perpendicular relations between segments [23].

Coincidence. Intersection between a point and a line, and a point and the boundary of a circle. Also whether the point is in the interior, outside or on the boundary of a region.

Mereology. Part-whole relations between regions [37].

Table 1 presents the corresponding polynomial encodings. Given three real variables v, i, j, let:

$$v \in [i, j] \equiv i \le v \le j \vee j \le v \le i.$$

Determining whether a point is inside a rectangle is based on vector projection. Point p is projected onto vector v by taking the dot product,

$$(x_p, y_p) \cdot (x_v, y_v) = x_p x_v + y_p y_v$$

Given point a and rectangle b, we translate the point such that the bottom left corner of b is at the origin, project p_a on the base and side vectors of b, and check whether the projection lies within the width and height of the rectangle,

$$0 < (p_a - p_{1_b}) \cdot v_b < w_b$$

$$0 < (p_a - p_{1_b}) \cdot v_b' < h_b$$

Convex regions a, b are disconnected iff there is a hyperplane of separation, i.e. there exists a line l such that a and b lie on different sides of l. This is the basis for determining the *discrete from* relation between rectangles.

3.4 Formalising Knowledge About Symmetries

In this section we formally determine the qualitative spatial relations that are preserved by various affine transformations. A fundamental property of affine transformations is that they preserve (a) point coincidence (e.g. line intersections), (b) parallelism between straight lines, and (c) proportions of distances between points on parallel lines [26]. For example, consider the configuration of points p_a, p_b, p_c, p_d and lines l_1, l_2, l_3 in Fig. 5: (a) we cannot introduce new points of coincidence between lines by applying transformations such as translation, scaling, reflection, and rotation. Conversely, if two lines intersect, then they will still intersect after these transformations; (b) lines l_1, l_2 are parallel before and after the transformations; lines l_1, l_3 are always non-parallel; (c) the ratio of distances between collinear points p_a, p_b, p_c is maintained; formally, let s_{ij} be the segment between points p_i, p_j and let $|s_{ij}|$ be the length of the segment. Then the ratio $\frac{|s_{ab}|}{|s_{bc}|}$ in Fig. 5 is the same before and after the transformations.

Based on these properties we can determine the transformations that preserve various qualitative spatial relations.[7]

Theorem 1. *The following qualitative spatial relations are preserved under translation, scale, rotation, and reflection (applied to the embedding space): topology, mereology, coincidence, collinearity, line parallelism.*

Proof. By definition, affine transformations preserve *parallelism* with respect to qualitative line orientation, and point *coincidence*. Due to preservation of point coincidence and proportions of collinear distances by affine transformations, it follows that mereological *part of* and topological *contact* relations between regions are preserved, i.e. if a mereological or topological relation changes between regions a, b, then by definition there exists a point p coincident with a such that the coincidence relation between p and b has changed; but this cannot occur as point coincidence is maintained with affine transformations by definition, therefore mereological and topological relations are also maintained.

The interaction between spatial relations and transformations is richer than we have space to elaborate on here, i.e. not all qualitative spatial relations are preserved under all affine transformations; orientation is not preserved under reflection (e.g. Fig. 5(b) gives a counter example), distance is not preserved under non-uniform scaling. To summarise, we formalise the following knowledge as modular commonsense rules in CLP(QS): point-coincidence, line parallelism, topological and mereological relations are preserved with all affine transformations. Relative orientation changes with reflection, and qualitative distances and perpendicularity change with non-uniform scaling. Spheres, circles, and rectangles are not preserved with non-uniform scaling, with the exception of axis-aligned bounding boxes.

"Trading" Transformations. Symmetries are used to eliminate object variables. As a metaphor, unbound variables are replaced by constant values in

---

[7] The properties of affine transformations and the geometric objects that they preserve are well understood; further information is readily available in introductory texts such as [26]. Our key contribution is formalising and exploiting this spatial knowledge as modular and extensible common-sense rules in intelligent knowledge-based spatial assistance systems.

"exchange" for transformations. We start with a set of transformations that can be applied to a configuration: translation, scaling, arbitrary rotation, and horizontal and vertical reflection. We can then "trade" each transformation for an elimination of variables. Each transformation can only be "spent" once. Theorem 2 presents an instance of such a pruning case.

Table 2 presents a variety of different pruning cases for position variables and the associated combination of transformations, as illustrated in Fig. 6.[8] Some cases require more than one distinct set of parameter restrictions to cover the set of all position variables due to point coincidence being preserved by affine transformations. For example, consider case (f): all pairs of points p_1, p_2 can be transformed into any other pair of points p_i, p_j by translation, rotation, and scaling, iff $p_1 = p_2 \leftrightarrow p_i = p_j$. Thus, to cover all pairs of possible points, we need to consider two distinct parameter restrictions: $p_i = p_j$ and $p_i \neq p_j$; we refer to these as *subcases*.

Many further pruning cases are identifiable. For example, a version of case (i) can be defined without reflection by requiring more sub-cases where $c_4 > c_2$ and $c_4 < c_2$. Case (i) can be extended so that all six coordinates of three points are grounded if we also "exchange" the *skew* transformation (e.g. applicable to object domains like triangles or points).

Theorem 2. *Any pair of object position variables $(x_1, y_1), (x_2, y_2)$ can be transformed into any given position constants $(c_1, c_2), (c_3, c_2)$ such that $(c_1 = c_3 \leftrightarrow (x_1 = x_2 \wedge y_1 = y_2))$ by applying: an xy-translation, a rotation about the origin in the range $(0, 2\pi)$, and an x-scale.*

Proof. The corresponding expression has been verified using the *Reduce* system (*Redlog* quantifier elimination) [15]; all variables are quantified over reals.

$$\forall c_1 \forall c_2 \forall c_3 \forall x_1 \forall y_1 \forall x_2 \forall y_2$$
$$(c_1 = c_3 \leftrightarrow (x_1 = x_2 \wedge y_1 = y_2)) \leftrightarrow \exists t_x \exists t_y \exists d_x \exists d_y \exists s_x ($$
$$(0 < s_x) \wedge (d_x^2 + d_y^2 = 1) \wedge$$
$$let S = \begin{pmatrix} s_x & 0 \\ 0 & 1 \end{pmatrix} \wedge let \ R = \begin{pmatrix} d_x & -d_y \\ d_y & d_x \end{pmatrix} \wedge let T = \begin{pmatrix} t_x \\ t_y \end{pmatrix} \wedge$$
$$\begin{pmatrix} c_1 \\ c_2 \end{pmatrix} = SR \begin{pmatrix} x_1 \\ y_1 \end{pmatrix} + T \wedge \begin{pmatrix} c_3 \\ c_2 \end{pmatrix} = SR \begin{pmatrix} x_2 \\ y_2 \end{pmatrix} + T) \equiv \top$$

We can use this pruning case on any spatial constraint graph G where the graph's spatial relations are preserved by translation, rotation, and scaling. Given graph $G = (N, E)$, the following Algorithm applies the pruning case, with selected constants $c_1 = 0$, $c_2 = 0$, and $c_3 = 1$ or $c_3 = 0$:

1. select object position variables p_1, p_2 from nodes in N
2. copy G to create G_1, G_2
3. in G_1 set $p_1 = (0,0), p_2 = (1,0)$ (*case $c_1 \neq c_3$*)
4. in G_2 set $p_1 = (0,0), p_2 = (0,0)$ (*case $c_1 = c_3$*)
5. if the task is:

---

[8] All cases have been verified using Reduce as presented in Theorem 2.

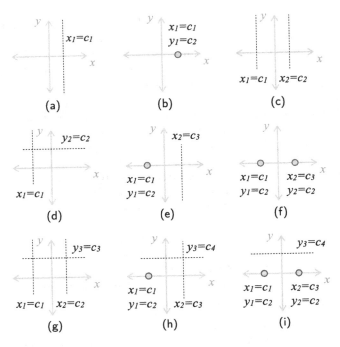

Fig. 6. Cases for pruning position parameters.

(a) *consistency* of G then solve $\bigvee_{i=1}^{2} \exists s \; G_i(s)$

(b) *sufficiency*, $G \rightarrow G'$, then solve
$\bigwedge_{i=1}^{2} \neg \exists s (G_i(s) \wedge \neg G'(s))$

In Step 1 any pair of objects can be selected for which their position variables will be grounded; we also employ policies that target computationally costly subgraphs (for example, pairs of non-equal circles that share a boundary point are often good candidates for this pruning case). Having eliminated free variables from the system of polynomial constraints, the constraints are significantly more simple to solve. Due to the double exponential complexity $O(c_1^{c_2^n})$ reducing n has a significant impact on performance; the system may even collapse from nonlinear constraints to linear (solvable in $O(c^n)$) or constants.

3.5   Combining Symmetry Pruning with Graph Decomposition

In certain cases, spatial constraint graphs can be decomposed into subgraphs that can be solved independently. For example, subgraphs G_1, G_2 can be independently solved if all objects in subgraph G_1 are either:

- *disconnected* from all objects in subgraph G_2;
- a *proper part* of some object in G_2;
- *left of* some segment in G_2;
- only related by *relative size* to some object in G_2, and so on.

In such cases we can reapply spatial symmetry pruning in each independent sub-graph; this commonsense spatial knowledge is modularly formalised within CLP(QS). For example, consider Proposition 22 of Book I of Euclid's Elements (Fig. 7):

Constructing a triangle from three segments. Given three line segments l_{ab}, l_{cd}, l_{ef}, draw a line through four collinear points $p_1, \ldots, p_4$ such that $|(p_1, p_2)| = |l_{ab}|$, $|(p_2, p_3)| = |l_{cd}|$, $|(p_3, p_4)| = |l_{ef}|$. Draw circle c_a centred on p_2, coincident with p_1. Draw circle c_b centred on p_3 coincident with p_4. Draw p_5 coincident with c_a and c_b. The triangle p_2, p_3, p_5 has side lengths such that $|(p_2, p_3)| = |l_{cd}|$, $|(p_3, p_5)| = |l_{ef}|$, $|(p_5, p_2)| = |l_{ab}|$.

In this example, the three segments l_{ab}, l_{cd}, l_{ef} and the remaining objects are only related by the distances between their end points. That is, the relative position and orientation of l_{ab}, l_{cd}, l_{ef} is not relevant to the consistency of the spatial graph; we only need to explore all combinations of segment *lengths*. Thus the solver decomposes the graph into four sub-graphs: (1) l_{ab} (2) l_{cd} (3) l_{ef}, and (4) $p_1, \ldots, p_5, c_a, c_b$. In subgraphs (1),(2),(3) it "trades" translation and rotation to ground $p_a = p_c = p_e = (0, 0)$, and $y_b = y_d = y_f = 0$ and keeps x-scale to cover all possible combinations of segment lengths, i.e. x_b, x_d, x_f are free variables. In subgraph (4) CLP(QS) applies the pruning case of Theorem 2 by grounding p_1, p_4.

Table 2. Cases for pruning parameters for one position point (a,b), two position points (c-f), three position points (g-i). Cases marked with $*$ require arbitrary scaling (i.e. both uniform and non-uniform).

Case	Parameter restrictions	Traded transformations
a	$x_1 = c_1$	x-translate
b	$x_1 = c_1, y_1 = c_2$	xy-translate
c	$x_1 = c_1, x_2 = c_2$, (i) $c_1 \neq c_2$ (ii) $c_1 = c_2$	x-translate, rotate π, x-scale
d	$x_1 = c_1, y_2 = c_2$	xy-translate
e	$x_1 = c_1, y_1 = c_2, x_2 = c_3$, (i) $c_1 \neq c_3$ (ii) $c_1 = c_3$	xy-translate, rotate π, x-scale
f	$x_1 = c_1, y_1 = c_2, x_2 = c_3, y_2 = c_2$, (i) $c_1 \neq c_3$ (ii) $c_1 = c_3$	xy-translate, rotate $(0, 2\pi)$, x-scale
g	$x_1 = c_1, x_2 = c_2, y_3 = c_3$ (i) $c_1 \neq c_2$ (ii) $c_1 = c_2$	xy-translate, rotate π, x-scale
h $*$	$x_1 = c_1, y_1 = c_2, x_2 = c_3, y_3 = c_4$ (i) $c_1 \neq c_3 \wedge c_2 \neq c_4$ (ii) $c_1 = c_3 \wedge c2 \neq c_4$ (iii) $c_1 \neq c_3 \wedge c_2 = c_4$ (iv) $c_1 = c_3 \wedge c2 = c_4$	xy-translate, rotate π, xy-scale, y-reflect
i $*$	$x_1 = c_1, y_1 = c_2, x_2 = c_3, y_2 = c_2, y_3 = c_4$ (i) $c_1 \neq c_3 \wedge c_2 \neq c_4$ (ii) $c_1 \neq c_3 \wedge c_2 = c_4$ (iii) $c_1 = c_3 \wedge c_2 = c_4$	xy-translate, rotate $(0, 2\pi)$, xy-scale, y-reflect

```
?- Sab = segment(Pa,Pb), Scd = segment(Pc,Pd), Sef = segment(Pe,Pf),
|  S12 = segment(P1,P2), S23 = segment(P2,P3), S34 = segment(P3,P4),
|  S14 = segment(P1,P4),
|  coincident(P2,S14), coincident(P3,S14),
|
|  length_dim(equal, S12, Sab),
|  length_dim(equal, S23, Scd),
|  length_dim(equal, S34, Sef),
|
|  Ca=circle(P2,_),     coincident(P1,Ca),
|  Cb=circle(P3,_),     coincident(P4,Cb),
|  coincident(P5,Ca), coincident(P5,Cb),
|  S35 = segment(P3,P5), S52 = segment(P5,P2),
|
|  %% triangle
|  length_dim(equal, S23, Scd),
|  length_dim(equal, S35, Sef),
|  length_dim(equal, S52, Sab).
true.
...
|   (length_dim(not_equal, S23, Scd);
|    length_dim(not_equal, S35, Sef);
|    length_dim(not_equal, S52, Sab)).
false.
```

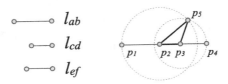

Fig. 7. Constructing a triangle by decomposing the spatial constraint graph.

4   Application-Driven Use Cases

We present problem instances in the classes of *mereology, ruler and compass*, and *contact*. Table 3 presents the experiment time results of CLP(QS) using symmetry pruning compared with existing systems: *z3* SMT solver, *Redlog* real quantifier elimination (in the *Reduce* computer algebra system) [15], and the relation algebraic qualitative spatial reasoners GQR [16] and SparQ [39]. CLP(QS) uses *z3* to solve polynomial constraints (after our pruning), thus z3 is the most direct comparison. Experiments were run on a MacBookPro, OS X 10.8.5, 2.6 GHz Intel Core i7. The empirical results show that no other spatial reasoning system exists (to the best of our knowledge) that can solve the range of problems presented in this section, and in cases where solvers are applicable, CLP(QS) with spatial pruning solves those problems significantly faster than other systems.

4.1   Spatial Theorem Proving: Geometry of Solids

Tarski [36] shows that a geometric point can be defined by a language of mereological relations over spheres. The idea is to distinguish when spheres are concentric, and to define a geometric point as the point of convergence. Borgo [8] shows

Table 3. Time (in seconds) to solve benchmark problems using CLP(QS) with pruning compared to z3 SMT solver, Redlog (Reduce) quantifier elimination, and GQR and SparQ relation algebraic solvers. *Time out* was issued after a running time of 10 min. Failure (*fail*) indicates that the incorrect result was given. Not applicable (*n/a*) indicates that the problem could not be expressed using the given system.

Problem	CLP(QS)	z3	Redlog	GQR	SparQ
Aligned Concentric	6.831	47.651	*time out*	*n/a*	*n/a*
Boundary Concentric	2.036	*time out*	*time out*	*n/a*	*n/a*
Mereologically Concentric	0.105	0.373	*time out*	*n/a*	*n/a*
Angle Bisector	0.931	*time out*	*time out*	*n/a*	*n/a*
Sphere Contact	0.004	*time out*	*time out*	*fail*	*fail*

that this can be accomplished with a language of mereology over *hypercubes*. We will use CLP(QS) to prove that the definitions are sound for rotatable squares.

As a preliminary we need to determine whether the intersection of two squares is non-square (Fig. 8) [34]. Given two squares a, b, the intersection is non-square if a *partially overlaps* b (Table 1) and either (a) a and b are not aligned, $x_{v_a} \neq x_{v_b}$ or (b) the width and height of the intersection are not equal, $w_I \neq h_I$, such that

$$w_I = \min(v \cdot p_{2_a}, v \cdot p_{2_b}) - \max(v \cdot p_{1_a}, v \cdot p_{1_b})$$

$$h_I = \min(v' \cdot p_{4_a}, v' \cdot p_{4_b}) - \max(v' \cdot p_{1_a}, v' \cdot p_{1_b})$$

Aligned Concentric. Two squares A, B are aligned and concentric if: A is *part of* B and there does not exist a square P such that (a) P is covertex with B, and (b) the intersection of P and A is not a square (Fig. 9).

```
| aligned_concentric(A,B) :-
|   mereology(part_of, A,B),
|   not((
|     P=square(_,_,_),
|     covertex(P,B),
|     intersection(non_square,A,P).
|   )).
```

We use CLP(QS) to prove that the definition is sufficient, by contradiction; two squares are covertex if they are aligned and share a vertex, and the relation *concentric* is the geometric definition of concentricity in Table 1 that is used to evaluate the mereological definition of concentricity:

```
?- A=square(_,_,_),
|  B=square(_,_,_),
|  aligned(A,B),
|  not_concentric(A,B),
|  aligned_concentric(A,B).
false.
```

Boundary Concentric. Square A is boundary concentric with square B if: A is *proper part* of B and there does not exist a square Z such that (a) Z is *proper part* of B (b) A is *part* of Z, and (c) Z is not *part* of A (Fig. 9).

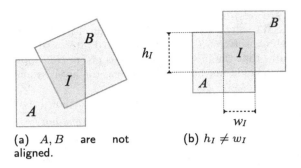

(a) A, B are not aligned.

(b) $h_I \neq w_I$

Fig. 8. Intersection I of squares A, B is non-square.

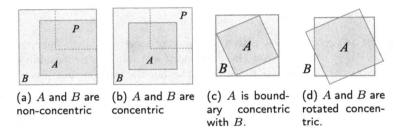

(a) A and B are non-concentric

(b) A and B are concentric

(c) A is boundary concentric with B.

(d) A and B are rotated concentric.

Fig. 9. Characterising aligned (a), (b) and rotated (c), (d) concentric squares using mereology (reproduced from [Borgo, 2013])

```
|  boundary_concentric(A,B) :-
|    mereology(proper_part,A,B),
|    not((
|    Z = square(_,_,_)
|    mereology(proper_part, Z,B),
|    mereology(part_of, A,Z),
|    mereology(not_part_of, Z,A),
|    )).

?- A=square(_,_,_),
|  B=square(_,_,_),
|  not_aligned(A,B),
|  not_concentric(A,B),
|  boundary_concentric(A,B).
false.
```

Mereologically Concentric. Squares A, B are mereologically concentric if: A, B are *aligned concentric* or there exists Q such that (a) Q is *boundary concentric* with B and Q is *aligned concentric* with A or (b) Q is *boundary concentric* with A and Q is *aligned concentric* with B.

Having proved the mereological definitions of *aligned* and *boundary* concentricity, we can replace these with more efficient geometric definitions from Table 1 when proving mereological concentricity.

```
| g_aligned_concentric(A,B) :-
|   aligned(A,B), concentric(A,B).
|
| g_boundary_concentric(A,B) :-
|   mereology(boundary_part, A,B).
|
| m_concentric(A,B) :-
|   g_aligned_concentric(A,B)
|   ;
|   Q = square(_,_,_),
|   g_boundary_concentric(Q,B),
|   g_aligned_concentric(Q,A)
|   ;
|   Q = square(_,_,_),
|   g_boundary_concentric(Q,A),
|   g_aligned_concentric(Q,B).

?- A=square(_,_,_),
|  B=square(_,_,_),
|  not_concentric(A,B),
|  m_concentric(A,B).
false.
```

4.2   Didactics: Ruler–Compass and Contact Problems

Angle Bisector. Let l_a, l_b be line segments that share an endpoint at p. Draw circle c at p. Circle c intersects l_a at p_a and l_b at p_b. Draw circles c_a at p_a and circle c_b at p_b such that p is coincident with both c_a, c_b. Circles c_a and c_b intersect at p and p_c. The line segment from p to p_c bisects the angle between l_a and l_b (Fig. 10).

We use CLP(QS) to prove that the definition is sufficient. The relation *bisects* is used for evaluation by checking if the midpoint of (p_a, p_b) is collinear with l_c (i.e. idealised rulers cannot directly measure the midpoint of a line). An interactive diagram is then automatically generated that encodes the specified program (using the FreeCAD system); see Fig. 11.

```
?- La = segment(P,_),Lb = segment(P,_),
|  orientation(not_parallel, La, Lb),
|  C = circle(P,_),
|  coincident(Pa,La),coincident(Pa,C),
|  coincident(Pb,Lb),coincident(Pb,C),
|  Ca=circle(Pa,_), Cb=circle(Pb,_),
|  coincident(Pa,Ca),coincident(Pa,Cb),
|  coincident(Pc,Ca),coincident(Pc,Cb),
|  not_equal(P,Pc),
|  Lc=segment(P,Pc),
|  bisects(segment(Pa,Pb),Lc).
true.
...
|    not_bisects(segment(Pa,Pb),Lc).
false.
```

Sphere Contact. Determine the maximum number of same-sized mutually touching spheres (Fig. 4 - note that no numeric information about the spheres is given in this benchmark problem).

```
?- sphere_list(5, Spheres),
|  size(equal, group(Spheres),
|  topology(rcc8(ec), group(Spheres) ).
false.
```

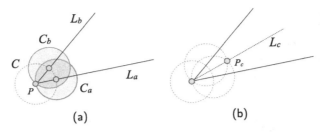

Fig. 10. Ruler and compass method for angle bisection. Line L_c bisects the angle between lines L_a, L_b.

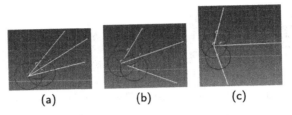

Fig. 11. Interactive diagram encoding the student's constructive proof of Euclid's angle bisector theorem; as the student manipulates figures in the diagram, the other geometries are automatically updated to maintain the specified qualitative constraints.

5   Conclusions

Affine transformations provide an effective and interesting class of symmetries that can be used for pruning across a range of qualitative spatial relations. To summarise, we formalise the following knowledge as modular commonsense rules in CLP(QS): point-coincidence, line parallelism, topological and mereological relations are preserved with all affine transformations. Relative orientation changes with reflection, and qualitative distances and perpendicularity change with non-uniform scaling. Spheres, circles, and rectangles are not preserved with non-uniform scaling, with the exception of axis-aligned bounding boxes. Our algorithm is simple to implement, and is easily extended to handle more pruning cases.

Theoretical and empirical results show that our method of pruning yields an improvement in performance by orders of magnitude over standard polynomial encodings without loss of soundness, thus increasing the horizon of spatial problems solvable with *any* polynomial constraint solver. Furthermore, the declaratively formalised knowledge about pruning strategies is available to be utilised in a modular manner within other knowledge representation and reasoning frameworks that rely on specialised SMT solvers etc., e.g., in the manner demonstrated in ASPMT(QS) [38], which is a specialised non-monotonic spatial reasoning system built on top of answer set programming modulo theories.

References

1. Aiello, M., Pratt-Hartmann, I.E., van Benthem, J.F.: Handbook of Spatial Logics. Springer-Verlag New York Inc., Secaucus (2007). ISBN 978-1-4020-5586-7
2. Arnon, D.S., Collins, G.E., McCallum, S.: Cylindrical algebraic decomposition I: the basic algorithm. SIAM J. Comput. **13**(4), 865–877 (1984)
3. Bhatt, M., Wallgrün, J.O.: Geospatial narratives and their spatio-temporal dynamics: Commonsense reasoning for high-level analyses in geographic information systems. ISPRS Int. J. Geo-Information **3**(1), 166–205 (2014). doi:10.3390/ijgi3010166. http://dx.doi.org/10.3390/ijgi3010166
4. Bhatt, M., Lee, J.H., Schultz, C.: CLP(QS): a declarative spatial reasoning framework. In: Egenhofer, M., Giudice, N., Moratz, R., Worboys, M. (eds.) COSIT 2011. LNCS, vol. 6899, pp. 210–230. Springer, Heidelberg (2011)
5. Bhatt, M., Schultz, C., Freksa, C.: The 'Space' in spatial assistance systems: conception, formalisation and computation. In: Tenbrink, T., Wiener, J., Claramunt, C. (eds.) Representing space in cognition: Interrelations of behavior, language, and formal models. Series: Explorations in Language and Space. Oxford University Press (2013). 978-0-19-967991-1
6. Bhatt, M., Suchan, J., Schultz, C.: Cognitive interpretation of everyday activities - toward perceptual narrative based visuo-spatial scene interpretation. In: Finlayson, M., Fisseni, B., Loewe, B., Meister, J.C. (eds.) Computational Models of Narrative (CMN) 2013, a satellite workshop of CogSci 2013: The 35th meeting of the Cognitive Science Society., Dagstuhl, Germany, OpenAccess Series in Informatics (OASIcs) (2013)
7. Bhatt, M., Schultz, C.P.L., Thosar, M.: Computing narratives of cognitive user experience for building design analysis: KR for industry scale computer-aided architecture design. In: Baral, C., Giacomo, G.D., Eiter, T. (eds.) Principles of Knowledge Representation and Reasoning: Proceedings of the Fourteenth International Conference, KR 2014, Vienna, Austria, 20–24 July, 2014. AAAI Press (2014). ISBN 978-1-57735-657-8
8. Borgo, S.: Spheres, cubes and simplexes in mereogeometry. Logic Logical Philos. **22**(3), 255–293 (2013)
9. Bouhineau, D.: Solving geometrical constraint systems using CLP based on linear constraint solver. In: Pfalzgraf, J., Calmet, J., Campbell, J. (eds.) AISMC 1996. LNCS, vol. 1138, pp. 274–288. Springer, Heidelberg (1996)
10. Bouhineau, D., Trilling, L., Cohen, J.: An application of CLP: Checking the correctness of theorems in geometry. Constraints **4**(4), 383–405 (1999)
11. Buchberger, B.: Bruno Buchberger's PhD thesis 1965: an algorithm for finding the basis elements of the residue class ring of a zero dimensional polynomial ideal (English translation). J. Symbolic Comput. **41**(3), 475–511 (2006)
12. Chou, S.-C.: Mechanical Geometry Theorem Proving, vol. 41. Springer Science and Business Media, Dordrecht (1988)
13. Collins, G.E.: Quantifier elimination for real closed fields by cylindrical algebraic decompostion. In: Brakhage, H. (ed.) GI-Fachtagung 1975. LNCS, vol. 33, pp. 134–183. Springer, Heidelberg (1975)
14. Collins, G.E., Hong, H.: Partial cylindrical algebraic decomposition for quantifier elimination. J. Symbolic Comput. **12**(3), 299–328 (1991). ISSN 0747–7171
15. Dolzmann, A., Seidl, A., Sturm, T.: REDLOG User Manual, Edition 3.0, Apr 2004
16. Gantner, Z., Westphal, M., Wölfl, S.: GQR-A fast reasoner for binary qualitative constraint calculi. In: Proceedings of AAAI, vol. 8 (2008)

17. Hadas, N., Hershkowitz, R., Schwarz, B.B.: The role of contradiction and uncertainty in promoting the need to prove in dynamic geometry environments. Educ. Stud. Mathe. **44**(1–2), 127–150 (2000)
18. Haunold, P., Grumbach, S., Kuper, G., Lacroix, Z.: Linear constraints: Geometric objects represented by inequalitiesl. In: Frank, A.U. (ed.) COSIT 1997. LNCS, vol. 1329, pp. 429–440. Springer, Heidelberg (1997)
19. Jaffar, J., Maher, M.J.: Constraint logic programming: A survey. J. Logic Prog. **19**, 503–581 (1994)
20. Kanellakis, P.C., Kuper, G.M., Revesz, P.Z.: Constraint query languages. In: Rosenkrantz, D.J., Sagiv, Y. (eds.) Proceedings of the Ninth ACM SIGACT-SIGMOD-SIGART Symposium on Principles of Database Systems, Nashville, Tennessee, USA, 2–4 April, 1990, pp. 299–313. ACM Press (1990). ISBN 0-89791-352-3
21. Kapur, D., Mundy, J.L. (eds.): Geometric Reasoning. MIT Press, Cambridge (1988). ISBN 0-262-61058-2
22. Ladkin, P.B., Maddux, R.D.: On binary constraint problems. J. ACM (JACM) **41**(3), 435–469 (1994)
23. Lee, J.H.: The complexity of reasoning with relative directions. In: 21st European Conference on Artificial Intelligence (ECAI 2014) (2014)
24. Ligozat, G.: Qualitative Spatial and Temporal Reasoning. Wiley-ISTE, Hoboken (2011)
25. Ligozat, G.F.: Qualitative triangulation for spatial reasoning. In: Campari, I., Frank, A.U. (eds.) COSIT 1993. LNCS, vol. 716, pp. 54–68. Springer, Heidelberg (1993)
26. Martin, G.E.: Transformation geometry: An introduction to symmetry. Springer, New York (1982)
27. Owen, J.C.: Algebraic solution for geometry from dimensional constraints. In: Proceedings of the First ACM Symposium on Solid Modeling Foundations and CAD/CAM Applications, pp. 397–407. ACM (1991)
28. Pesant, G., Boyer, M.: QUAD-CLP (R): Adding the power of quadratic constraints. In: Borning, A. (ed.) PPCP 1994. LNCS, vol. 874, pp. 95–108. Springer, Heidelberg (1994)
29. Pesant, G., Boyer, M.: Reasoning about solids using constraint logic programming. J. Automated Reasoning **22**(3), 241–262 (1999)
30. Randell, D.A., Cohn, A.G., Cui Z.: Computing transitivity tables: A challenge for automated theorem provers. In 11th International Conference on Automated Deduction (CADE-11), pp. 786–790 (1992)
31. Ratschan, S.: Approximate quantified constraint solving by cylindrical box decomposition. Reliable Comput. **8**(1), 21–42 (2002)
32. Ratschan, S.: Efficient solving of quantified inequality constraints over the real numbers. ACM Trans. Comput. Logic (TOCL) **7**(4), 723–748 (2006)
33. Schultz, C., Bhatt, M.: Towards a declarative spatial reasoning system. In: 20th European Conference on Artificial Intelligence (ECAI 2012) (2012)
34. Schultz, C., Bhatt, M.: Declarative spatial reasoning with boolean combinations of axis-aligned rectangular polytopes. In: ECAI 2014–21st European Conference on Artificial Intelligence, pp. 795–800 (2014)
35. Schultz, C., Bhatt, M., Borrmann, A.: Bridging qualitative spatial constraints and parametric design - a use case with visibility constraints. In: EG-ICE: 21st International Workshop - Intelligent Computing in Engineering 2014 (2014)
36. Tarski, A.: A general theorem concerning primitive notions of Euclidean geometry. Indagationes Mathematicae **18**(468), 74 (1956)

37. Varzi, A.C.: Parts, wholes, and part-whole relations: The prospects of mereotopology. Data Knowl. Eng. **20**(3), 259–286 (1996)
38. Walega, P., Bhatt, M., Schultz, C.: ASPMT(QS): non-monotonic spatial reasoning with answer set programming modulo theories. In: LPNMR: Logic Programming and Nonmonotonic Reasoning - 13th International Conference (2015). http://lpnmr2015.mat.unical.it
39. Wallgrün, J.O., Frommberger, L., Wolter, D., Dylla, F., Freksa, C.: Qualitative spatial representation and reasoning in the SparQ-Toolbox. In: Barkowsky, T., Knauff, M., Ligozat, G., Montello, D.R. (eds.) Spatial Cognition 2007. LNCS (LNAI), vol. 4387, pp. 39–58. Springer, Heidelberg (2007)
40. Wenjun, W.: Basic principles of mechanical theorem proving in elementary geometries. J. Syst. Sci. Math. Sci. **4**(3), 207–235 (1984)

On Distributive Subalgebras of Qualitative Spatial and Temporal Calculi

Zhiguo Long and Sanjiang Li[✉]

Faculty of Engineering and Information Technology,
Centre for Quantum Computation and Intelligent Systems,
University of Technology Sydney, Sydney, Australia
Sanjiang.Li@uts.edu.au

Abstract. Qualitative calculi play a central role in representing and reasoning about qualitative spatial and temporal knowledge. This paper studies distributive subalgebras of qualitative calculi, which are subalgebras in which (weak) composition distributives over nonempty intersections. The well-known subclass of convex interval relations is an example of distributive subalgebras. It has been proven for RCC5 and RCC8 that path consistent constraint network over a distributive subalgebra is always minimal and strongly n-consistent in a qualitative sense (weakly globally consistent). We show that the result also holds for the four popular qualitative calculi, i.e. Point Algebra, Interval Algebra, Cardinal Relation Algebra, and Rectangle Algebra. Moreover, this paper gives a characterisation of distributive subalgebras, which states that the intersection of a set of $m \geq 3$ relations in the subalgebra is nonempty if and only if the intersection of every two of these relations is nonempty. We further compute and generate all maximal distributive subalgebras for those four qualitative calculi mentioned above. Lastly, we establish two nice properties which will play an important role in efficient reasoning with constraint networks involving a large number of variables.

Keywords: Qualitative calculi · Qualitative spatial and temporal reasoning · Distributive subalgebra · Region connection calculus · Rectangle algebra

1 Introduction

A dominant part of qualitative spatial and temporal reasoning (QSTR) research focuses on the study of individual or multiple qualitative calculi. Roughly speaking, a qualitative calculus $\mathcal{M}$ is simply a finite class of relations over a universe $\mathcal{U}$ of spatial or temporal entities which form a Boolean algebra. Usually, we assume that the identity relation is an atomic relation in $\mathcal{M}$ and relations in $\mathcal{M}$ are closed under converse [21]. Well-known qualitative calculi include Point Algebra (PA) [4,29] and Interval Algebra (IA) [1] for representing temporal relations

Work supported by the Australian Research Council under DP120104159 and FT0990811.

© Springer International Publishing Switzerland 2015
S.I. Fabrikant et al. (Eds.): COSIT 2015, LNCS 9368, pp. 354–374, 2015.
DOI: 10.1007/978-3-319-23374-1_17

and Region Connection Calculus RCC5 and RCC8 [25], Cardinal Relation Algebra (CRA) [14,20], and Rectangle Algebra (RA) [3,16] for representing spatial relations.

For convenience, we write RCC5/8 for either RCC5 or RCC8. Since the composition of two RCC5/8 relations R, S is not necessarily a relation in RCC5/8 [13,18], we write $R \diamond S$ for the smallest relation in RCC5/8 which contains $R \circ S$, the usual composition of R and S, and call $R \diamond S$ the *weak composition* of R, S [13,18]. Unlike RCC5/8, the calculi PA, IA, CRA and RA are closed under composition and are all relation algebras. Replacing composition with weak composition, RCC5/8 is also a relation algebra.

Using a qualitative calculus $\mathcal{M}$, we represent spatial or temporal information in terms of relations in $\mathcal{M}$, and formulate a spatial or temporal problem as a set of qualitative constraints (called a *qualitative constraint network* or QCN). A qualitative constraint has the form (xRy), which specifies that two variables x, y are related by the relation R in $\mathcal{M}$. A QCN $\mathcal{N}$ is *consistent* if there exists an assignment of values in $\mathcal{U}$ to variables in $\mathcal{N}$ such that all constraints in $\mathcal{N}$ are satisfied simultaneously. If this is the case, we call this assignment a *solution* of $\mathcal{N}$. We say $\mathcal{M}$ is *minimal* if, for each constraint (xRy) in $\mathcal{N}$, R is the minimal (or *strongest*) relation between x and y that is entailed by $\mathcal{N}$. We say $\mathcal{N}$ is *globally consistent* if every partial solution (i.e. a partial assignment that satisfies all constraints in a restriction of $\mathcal{N}$) can be extended to a solution of $\mathcal{N}$.

The *consistency problem* and the *minimal labelling problem* (MLP) are two major reasoning tasks of QSTR research. The consistency problem is to decide whether a QCN has a solution and the MLP is to decide if it is minimal. These problems have been investigated in depth in the past three decades for many qualitative calculi in the literature, see e.g. [1–4,7,17,20,24,26].

Both problems are in general NP-hard for IA, CRA, RCC5/8, and RA. Local consistency algorithms like path consistency algorithm (PCA) are designed for solving these problems approximately [1]. A QCN $\mathcal{N} = \{v_i R_{ij} v_j : 1 \leq i, j \leq n\}$ is *path consistent* (PC) if each R_{ij} is non-empty and contained in the (weak) composition of R_{ik} and R_{kj} for any k. Applying PCA will either find an inconsistency in $\mathcal{N}$ in case $\mathcal{N}$ is not path consistent, or return a path consistent network that is equivalent to $\mathcal{N}$, which is also known as the *algebraic closure* or *a-closure* of $\mathcal{N}$ [21].

In this paper, we study distributive subalgebras of qualitative calculi. A subalgebra of $\mathcal{M}$ is a subclass of $\mathcal{M}$ that contains all atomic relations and is closed under (weak) composition, intersection, and converse. A subalgebra $\mathcal{S}$ is *distributive* if (weak) composition distributives over nonempty intersection, i.e. $R \diamond (S \cap T) = (R \diamond S) \cap (R \diamond T)$ and $(S \cap T) \diamond R = (S \diamond R) \cap (T \diamond R)$ for any $R, S, T \in \mathcal{S}$ with $S \cap T \neq \varnothing$.

Although distributive subalgebra is a new concept proposed recently in [11, 17], several examples of distributive subalgebras have been studied before. The first such a subalgebra, the subclass of convex IA relations $\mathcal{C}_{IA}$, was found in [19], where Ligozat also proved that path consistent networks over $\mathcal{C}_{IA}$ is globally consistent. As every globally consistent network is minimal, this shows that path consistent networks over $\mathcal{C}_{\text{IA}}$ is also minimal. Later, Chandra and Pujari [7] defined a class of convex RCC8 relations (written D_{41}^8 in [17] and this paper) and proved that every path consistent network over D_{41}^8 is minimal. More recently, Amaneddine and Condotta [2] found another subclass of IA, written as $\mathcal{S}_{\text{IA}}$, and proved that $\mathcal{C}_{\text{IA}}$ and $\mathcal{S}_{\text{IA}}$ are the only maximal subalgebras of IA such that path consistent networks over which are globally consistent. It turns out that these subalgebras are all the maximal distributive subalgebras of IA or RCC8 [17].

The important concept of distributive subalgebra was also found very useful in identifying a subnetwork that is equivalent to a given one but has no redundant constraints. Such a subnetwork is called a *prime* subnetwork in [11,17]. It was proven there that every constraint network over a distributive subalgebra of RCC5/8 has a unique prime subnetwork, which can be found in cubic time; and, in contrast, it is in general NP-hard to decide if a constraint is non-redundant in an arbitrary RCC5/8 constraint network. The cubic time algorithm for finding the prime subnetwork is very useful in applications such as computing, storing, and compressing the relationships between spatial objects and hence saving space for storage and communication. We refer the reader to [17] for a real-world application example and detailed discussions.

As the focus of [11,17] is redundancy in RCC5/8 constraint networks, there are several interesting topics left untouched, which are the subject of this paper. We first give a characterisation of distributive subalgebras in terms of intersections of relations, and then compute and find all maximal distributive subalgebras for every qualitative calculus mentioned before. Lastly, we establish two nice properties regarding partial path consistency [6] and variable elimination [31] of constraint networks over a distributive subalgebra. These properties will play an important role in efficient reasoning with sparse constraint networks involving a large number of variables.

The remainder of this paper is organised as follows. In Sect. 2, we first give a short introduction of the qualitative calculi mentioned above and recall basic notions including weak composition, path and global consistency. Section 3 then presents a characterisation of distributive subalgebras and shows that path consistent networks over a distributive subalgebra are globally consistent in a qualitative sense. Section 4 shows how we compute and find all maximal distributive subalgebras of these calculi. We then prove in Sect. 5 two important properties of distributive subalgebras that will be used in efficient reasoning with large sparse constraint networks. In Sect. 6 we discuss the connection between distributive subalgebras and conceptual neighbourhood graphs, and relation with classical CSPs. The last section then concludes the paper.

2   Qualitative Calculi

In this section, we first recall the qualitative calculi PA, IA, CRA, RCC5/8, and RA, and then, recall some relevant notions and results of these constraint languages.

Suppose $\mathcal{U}$ is a domain of spatial or temporal entities. Write $\mathbf{Rel}(\mathcal{U})$ for the Boolean algebra of binary relations on $\mathcal{U}$. A *qualitative calculus* [21] $\mathcal{M}$ on $\mathcal{U}$ is defined as a finite Boolean subalgebra of $\mathbf{Rel}(\mathcal{U})$ which has an atom that is the identity relation $id_\mathcal{U}$ on $\mathcal{U}$ and is closed under converse, i.e., R is in $\mathcal{M}$ iff its converse

$$R^{-1} = \{(a, b) \in \mathcal{U} \times \mathcal{U} : (b, a) \in R\}$$

is in $\mathcal{M}$ [21]. A relation α in a qualitative calculus $\mathcal{M}$ is *atomic* or *basic* if it is an atom in $\mathcal{M}$. Note that the set of basic relations of a qualitative calculus is *jointly exhaustive and pairwise disjoint* (JEPD). Well-known qualitative calculi include, among others, PA [4,29], IA [1], CRA [14,20], RA [3,16], and RCC5 and RCC8 [25].

2.1   Point Algebra and Interval Algebra

Definition 1 (Point Algebra (PA) [29]**).** *Let $\mathcal{U}$ be the set of real numbers. The Point Algebra is the Boolean subalgebra generated by the JEPD set of relations $\{<, >, =\}$, where $<, >, =$ are defined as usual.*

PA contains eight relations, viz. the three basic relations $<, >, =$, the empty relation, the universal relation $\star$, and three non-basic relations $\leq, \geq, \neq$.

Definition 2 (Interval Algebra (IA) [1]**).** *Let $\mathcal{U}$ be the set of closed intervals on the real line. Thirteen binary relations between two intervals $x = [x^-, x^+]$ and $y = [y^-, y^+]$ are defined by the order of the four endpoints of x and y, see Table 1. The Interval Algebra is generated by these JEPD relations.*

We write

$$\mathcal{B}_{\mathrm{IA}} = \{\mathsf{b}, \mathsf{m}, \mathsf{o}, \mathsf{s}, \mathsf{d}, \mathsf{f}, \mathsf{eq}, \mathsf{fi}, \mathsf{di}, \mathsf{si}, \mathsf{oi}, \mathsf{mi}, \mathsf{bi}\} \qquad (1)$$

for the set of basic IA relations. Ligozat [19] defines the *dimension* of a basic interval relation as 2 minus the number of equalities appearing in the definition of the relation (see Table 1). That is, for basic relations we have

$$\dim(\mathsf{eq}) = 0, \dim(\mathsf{m}) = \dim(\mathsf{s}) = \dim(\mathsf{f}) = 1, \dim(\mathsf{b}) = \dim(\mathsf{o}) = \dim(\mathsf{d}) = 2.$$

For a non-basic relation R we define

$$\dim(R) = \max\{\dim(\theta) : \theta \text{ is a basic relation in } R\}. \qquad (2)$$

Using the conceptual neighbourhood graph (CNG) of IA [15], Ligozat [19] gives a geometrical characterisation for ORD-Horn relations. Consider the CNG of IA (shown in Table 1 (ii)) as a partially ordered set $(\mathcal{B}_{int}, \preceq)$ (by interpreting any relation to be smaller than its right or upper neighbours). For $\theta_1, \theta_2 \in \mathcal{B}_{int}$ with $\theta_1 \preceq \theta_2$, we write $[\theta_1, \theta_2]$ as the set of basic interval relations θ such that $\theta_1 \preceq \theta \preceq \theta_2$, and call such a relation a *convex* interval relation. An IA relation R is called *pre-convex* if it can be obtained from a convex relation by removing one or more basic relations with dimension lower than R. For example, $[\mathsf{o}, \mathsf{eq}] = \{\mathsf{o}, \mathsf{s}, \mathsf{fi}, \mathsf{eq}\}$ is a convex relation and $\{\mathsf{o}, \mathsf{eq}\}$ is a pre-convex relation. Ligozat has shown that ORD-Horn relations are precisely pre-convex relations. Every path consistent network over $\mathcal{H}$ is consistent [24]. In addition, every path consistent network over $\mathcal{C}_{\mathrm{IA}}$ is globally consistent and minimal [19].

Table 1. IA basic relations (i) definitions and (ii) conceptual neighbourhood graph, where $x = [x^-, x^+], y = [y^-, y^+]$ are two intervals.

Relation	Symb.	Conv.	Dim.	Definition
before	b	bi	2	$x^+ < y^-$
meets	m	mi	1	$x^+ = y^-$
overlaps	o	oi	2	$x^- < y^- < x^+ < y^+$
starts	s	si	1	$x^- = y^- < x^+ < y^+$
during	d	di	2	$y^- < x^- < x^+ < y^+$
finishes	f	fi	1	$y^- < x^- < x^+ = y^+$
equals	eq	eq	0	$x^- = y^- < x^+ = y^+$

(i)

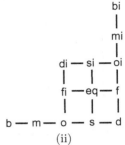

(ii)

2.2   RCC5 and RCC8

The RCC5/8 constraint language is a fragment of the Region Connection Calculus (RCC) [25]. The RCC is a first order theory based on a binary connectedness relation and has canonical models defined over connected topological spaces [18,28]. Since applications in GIS and many other spatial reasoning tasks mainly consider objects represented in the real plane, in this paper, we interpret regions as non-empty regular closed sets in the plane, and say two regions are *connected* if they have non-empty intersection.

Definition 3 (RCC5 and RCC8 Algebras). *Let $\mathcal{U}$ be the set of non-empty regular closed sets, or* regions, *in the real plane. The RCC8 algebra is generated by the eight topological relations*

$$\mathbf{DC, EC, PO, EQ, TPP, NTPP, TPP^{-1}, NTPP^{-1}},$$

where $\mathbf{DC, EC, PO, TPP}$ *and* $\mathbf{NTPP}$ *are defined in Table 2,* $\mathbf{EQ}$ *is the identity relation, and* $\mathbf{TPP^{-1}}$ *and* $\mathbf{NTPP^{-1}}$ *are the converses of* $\mathbf{TPP}$ *and* $\mathbf{NTPP}$ *respectively (see Fig. 1 for illustration). RCC5 is the sub-algebra of RCC8 generated by the five part-whole relations*

$$\mathbf{DR, PO, EQ, PP, PP^{-1}},$$

where $\mathbf{DR = DC \cup EC, PP = TPP \cup NTPP,}$ *and* $\mathbf{PP^{-1} = TPP^{-1} \cup NTPP^{-1}}$.

2.3   Cardinal Relation Algebra and Rectangle Algebra

Definition 4 (Cardinal Relation Algebra (CRA) [14,20]). *Let $\mathcal{U}$ be the real plane. Define binary relations $NW, N, NE, W, EQ, E, SW, S$ and SE as in Fig. 2. The Cardinal Relation Algebra is generated by these nine JEPD relations.*

Table 2. Topological interpretation of basic RCC8 relations in the plane, where a, b are regions, and a°, b° are the interiors of a, b, respectively.

Relation	Definition	Relation	Definition
DC	$a \cap b = \varnothing$	**TPP**	$a \subset b, a \not\subset b^\circ$
EC	$a \cap b \neq \varnothing, a^\circ \cap b^\circ = \varnothing$	**NTPP**	$a \subset b^\circ$
PO	$a \not\subset b, b \not\subset a, a^\circ \cap b^\circ \neq \varnothing$	**EQ**	$a = b$

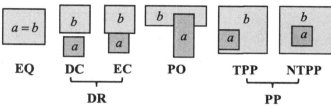

EQ   DC   EC   PO   TPP   NTPP

DR   PP

Fig. 1. Illustration for basic relations in RCC5/RCC8

CRA can be viewed as an extension of PA to the plane. Similarly, IA can also be extended to regions in the plane. We assume an orthogonal basis in the Euclidean plane. For a bounded region a, its *minimum bounding rectangle* (MBR), denoted by $\mathcal{M}(a)$, is the smallest rectangle which contains a and whose sides are parallel to the axes of the basis. We write $I_x(a)$ and $I_y(a)$ as, respectively, the x- and y-projections of $\mathcal{M}(a)$. The basic rectangle relation between two bounded regions a, b is $\alpha \otimes \beta$ iff $(I_x(a), I_x(b)) \in \alpha$ and $(I_y(a), I_y(b)) \in \beta$, where α, β are two basic IA relations (see Fig. 4 for illustration). We write $\mathcal{B}_{\text{RA}}$ for the set of basic rectangle relations, i.e.,

$$\mathcal{B}_{\text{RA}} = \{\alpha \otimes \beta : \alpha, \beta \in \mathcal{B}_{\text{IA}}\}. \tag{3}$$

There are 169 different basic rectangle relations in $\mathcal{B}_{\text{RA}}$. The Rectangle Algebra (RA) is the algebra generated by relations in $\mathcal{B}_{\text{RA}}$ [3].

Relation	Definition
NW	$x < x', y > y'$
N	$x = x', y > y'$
NE	$x > x', y > y'$
W	$x < x', y = y'$
EQ	$x = x', y = y'$
E	$x > x', y = y'$
SW	$x < x', y < y'$
S	$x = x', y < y'$
SE	$x > x', y < y'$

Fig. 2. Basic relations of CRA.

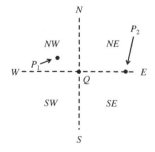

Fig. 3. Examples: $P_1 \ NW \ Q$ and $P_2 \ E \ Q$

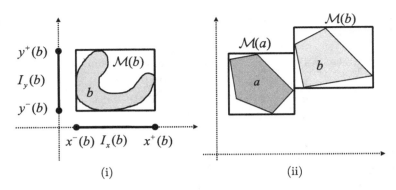

Fig. 4. (i) The minimum bounding rectangle $\mathcal{M}(b)$ of a region b; (ii) the RA relation of a to b is $\mathsf{m} \otimes \mathsf{o}$.

Henceforth, for two IA relations R, S, we will write $R \otimes S$ for the (non-basic) relation $\{\alpha \otimes \beta : \alpha \in R, \beta \in S, \alpha, \beta \in \mathcal{B}\}$; analogously, for two subclasses of IA relations $\mathcal{R}$ and $\mathcal{S}$, we will write $\mathcal{R} \otimes \mathcal{S}$ for the set of RA relations $\{R \otimes S : R \in \mathcal{R}, S \in \mathcal{S}\}$. The following lemma is straightforward.

Lemma 1. *Let* $\Delta = \{v_i(R_{ij} \otimes S_{ij})v_j\}_{i,j=1}^n$ *be an RA network, where* R_{ij} *and* S_{ij} *are arbitrary IA relations. Then* Δ *is satisfiable iff its projections* $\Delta^x = \{x_i R_{ij} x_j\}_{i,j=1}^n$ *and* $\Delta^y = \{y_i S_{ij} y_j\}_{i,j=1}^n$ *are satisfiable IA networks.*

As a consequence, we know $\mathcal{H} \otimes \mathcal{H}$ is a tractable subclass of RA. No maximal tractable subclass has been identified for RA, but a larger tractable subclass of RA has been identified in [3].

2.4 Properties of Qualitative Calculi

While PA, IA, CRA and RA are all closed under composition, the composition of two basic RCC5/8 relations is not necessarily a relation in RCC5/8 [13,18].

For two RCC5/8 relations R and S, recall that we write $R \diamond S$ for the weak composition of R and S. Suppose α, β, γ are three basic RCC5/8 relations. Then we have

$$\gamma \in \alpha \diamond \beta \Leftrightarrow \gamma \cap (\alpha \circ \beta) \neq \varnothing. \tag{4}$$

The weak composition of two (non-basic) relations R and S is computed as follows:

$$R \diamond S = \bigcup \{\alpha \diamond \beta : \alpha \in R, \beta \in S\}.$$

Because PA, IA, CRA and RA are closed under composition, we have

Proposition 1. *For* $\mathcal{M}$ *being PA, IA, CRA or RA, weak composition is the same as composition, i.e. for any* $R, S \in \mathcal{M}$, *we have* $R \circ S = R \diamond S$.

Proposition 2 (See [12]). *With the weak composition operation $\diamond$, the converse operation $^{-1}$, and the identity relation, PA, IA, RCC5/8, CRA, and RA are relation algebras. In particular, the weak composition operation $\diamond$ is associative. Moreover, for PA, IA, RCC5/8, CRA, and RA relations R, S, T, we have the following cycle law*

$$(R \diamond S) \cap T \neq \varnothing \Leftrightarrow (R^{-1} \diamond T) \cap S \neq \varnothing \Leftrightarrow (T \diamond S^{-1}) \cap R \neq \varnothing. \qquad (5)$$

Figure 5 gives an illustration of the cycle law.

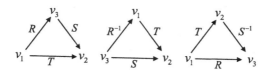

Fig. 5. Illustration of the cycle law (from [17]).

In the following, we assume that $\diamond$ takes precedence over $\cap$.

We say a network $\mathcal{N} = \{v_i R_{ij} v_j : 1 \leq i, j \leq n\}$ is *path consistent* if for every $1 \leq i, j, k \leq n$, we have

$$\varnothing \neq R_{ij} \subseteq R_{ik} \diamond R_{kj}.$$

In general, path consistency can be enforced by calling the following rule until an empty constraint occurs (then $\mathcal{N}$ is inconsistent) or the network becomes stable

$$R_{ij} \leftarrow (R_{ik} \diamond R_{kj}) \cap R_{ij},$$

where $1 \leq i, j, k \leq n$ are arbitrary. A cubic time algorithm, henceforth called the *path consistency algorithm* or PCA, has been devised to enforce path consistency. For any qualitative constraint network $\mathcal{N}$, the PCA either detects inconsistency of $\mathcal{N}$ or returns a path consistent network, written $\mathcal{N}_p$, which is equivalent to $\mathcal{N}$ and also known as the *algebraic closure* or *a-closure* of $\mathcal{N}$ [21]. It is easy to see that in this case $\mathcal{N}_p$ refines $\mathcal{N}$, i.e., we have $S_{ij} \subseteq R_{ij}$ for each constraint $(v_i S_{ij} v_j)$ in $\mathcal{N}_p$.

Definition 5. *Let $\mathcal{M}$ be a qualitative calculus with universe $\mathcal{U}$. Suppose $\mathcal{N} = \{v_i T_{ij} v_j : 1 \leq i, j \leq n\}$ is a QCN over $\mathcal{M}$ and $V = \{v_1, ..., v_n\}$. For a pair of variables $v_i, v_j \in V$ $(i \neq j)$ and a basic relation α in T_{ij}, we say α is feasible if there exists a solution $(a_1, a_2, ..., a_n)$ in U of $\mathcal{N}$ such that (a_i, a_j) is an instance of α. We say $\mathcal{N}$ is minimal if α is feasible for every pair of variables v_i, v_j $(i \neq j)$ and every basic relation α in T_{ij}.*

A basic network is a network in which every relation basic. A scenario of $\mathcal{N}$ is a basic network with form $\Theta = \{v_i \theta_{ij} v_j : 1 \leq i, j \leq n\}$, where each θ_{ij} is a basic relation in T_{ij}. A scenario is consistent if it has a solution. We say $\mathcal{N}$ is weakly globally consistent (globally consistent, respectively) if any consistent *scenario (solution, respectively) of* $\mathcal{N}\!\downarrow_{V'}$ *can be extended to a consistent scenario (solution, respectively) of* $\mathcal{N}$, *where* V' *is any nonempty subset of* V *and* $\mathcal{N}\!\downarrow_{V'}$ *is the restriction of* $\mathcal{N}$ *to* V'.

It is clear that every (weakly) globally consistent network is consistent and minimal.

In the following, we assume the qualitative calculus $\mathcal{M}$ has the following properties:

$$\mathcal{M} \text{ is a relation algebra with operations } \diamond, id_{\mathcal{U}}, \text{ and } ^{-1}; \tag{6}$$

$$\text{Every path consistent basic network over } \mathcal{M} \text{ is consistent.} \tag{7}$$

Interestingly, (6) and (7) hold for every qualitative calculus $\mathcal{M}$ mentioned in this paper, i.e. PA, IA, RCC5/8, CRA and RA.

3   Distributive Subalgebras

Definition 6. *[17] Let* $\mathcal{M}$ *be a qualitative calculus. A subclass* $\mathcal{S}$ *of* $\mathcal{M}$ *is called a* subalgebra *if* $\mathcal{S}$ *contains all basic relations and is closed under converse, weak composition, and intersection. A subalgebra* $\mathcal{S}$ *is* distributive *if weak composition distributes over non-empty intersections of relations in* $\mathcal{S}$, *i.e.* $R \diamond (S \cap T) = (R \diamond S) \cap (R \diamond T)$ *and* $(S \cap T) \diamond R = (S \diamond R) \cap (T \diamond R)$ *for any* $R, S, T \in \mathcal{S}$ *with* $S \cap T \neq \varnothing$.

Suppose $\mathcal{X}$ is a subclass of $\mathcal{M}$. We write $\widehat{\mathcal{X}}$ for the subalgebra of $\mathcal{M}$ generated by $\mathcal{X}$, i.e. $\widehat{\mathcal{X}}$ is the closure of $\mathcal{X}$ in $\mathcal{M}$ under intersection, weak composition, and converse. In particular, $\widehat{\mathcal{B}}$ denotes the closure of $\mathcal{B}$ in $\mathcal{M}$.

Proposition 3. *Let* $\mathcal{M}$ *be one of the calculi PA, IA, RCC5/8, CRA, RA and* $\mathcal{B}$ *the set of basic relations of* $\mathcal{M}$. *Then* $\widehat{\mathcal{B}}$ *is a distributive subalgebra.*

This shows that the above definition of distributive subalgebra is well-defined for these calculi and every distributive subalgebra of $\mathcal{M}$ contains $\widehat{\mathcal{B}}$ as a subclass.

3.1   Distributive Subalgebra is Helly

Helly's theorem [9] is a very useful result in discrete geometry. For n convex subsets of $\mathbb{R}$, it says if the intersection of any two of them is non-empty, then the intersection of the whole collection is also non-empty. Interestingly, relations in a distributive subalgebra have a similar property as convex sets in the real line and, moreover, relations having such property are exactly those in a distributive subalgebra.

Definition 7. *A subclass* $\mathcal{S}$ *of a qualitative calculus is called* Helly *if, for every* $R, S, T \in \mathcal{S}$, *we have*

$$R \cap S \cap T \neq \varnothing \quad \text{iff} \quad R \cap S \neq \varnothing, \ R \cap T \neq \varnothing, \ S \cap T \neq \varnothing. \tag{8}$$

If S is a subalgebra, then it is straightforward to prove that S is Helly if and only if, for any n relations $R_1, ..., R_n$ in S, we have

$$\bigcap_{i=1}^{n} R_i \neq \varnothing \quad \text{iff} \quad (\forall 1 \leq i \neq j \leq n) \; R_i \cap R_j \neq \varnothing \tag{9}$$

The following result is first proven for RCC5/8 in [17]. Following a similar proof, it is straightforward to show this holds in general.

Lemma 2 ([17]). *Suppose $\mathcal{M}$ is a qualitative calculus that satisfies (6), i.e. $\mathcal{M}$, with the weak composition, the converse operation, and the identity relation, is a relation algebra. Then every distributive subalgebra of $\mathcal{M}$ is Helly.*

Surprisingly, the above condition is also sufficient.

Theorem 1. *Suppose $\mathcal{M}$ is a qualitative calculus that satisfies (6). Let S be a subalgebra of $\mathcal{M}$. Then S is distributive if and only if it is Helly.*

Proof. Since Lemma 2 already shows the "only if" part, we only need to show the "if" part. Suppose R, S, T are three relations in S. We first note $R \diamond (S \cap T) \subseteq R \diamond S \cap R \diamond T$. Furthermore, for any basic relation γ, by using the cycle law twice, we have

$$\gamma \notin R \diamond (S \cap T) \Leftrightarrow \{\gamma\} \cap R \diamond (S \cap T) = \varnothing$$
$$\Leftrightarrow R^{-1} \diamond \gamma \cap S \cap T = \varnothing$$
$$\Leftrightarrow R^{-1} \diamond \gamma \cap S = \varnothing \text{ or } R^{-1} \diamond \gamma \cap T = \varnothing$$
$$\Leftrightarrow \{\gamma\} \cap R \diamond S = \varnothing \text{ or } \{\gamma\} \cap R \diamond T = \varnothing$$
$$\Leftrightarrow \gamma \notin R \diamond S \text{ or } \gamma \notin R \diamond T.$$

This shows $R \diamond (S \cap T) = R \diamond S \cap R \diamond T$. That is, S is Helly only if it is distributive. $\square$

3.2 Path Consistency Implies Weakly Global Consistency

We have the following very important result for distributive subalgebras.

Theorem 2. *Let $\mathcal{M}$ be a qualitative calculus that satisfies (6) and (7). Suppose S is a distributive subalgebra of $\mathcal{M}$. Then every path consistent network over S is weakly globally consistent and minimal.*

This result was first proved for RCC5/8 in [17]. If every path consistent network over S is *consistent*, then, following the proof in [17, Theorem 18], we can show every path consistent network over S is also weakly globally consistent and minimal. From the analysis in the following section, we can easily see that this is the case for PA, IA, CRA, RA, and RCC5/8. A detailed proof of the general case can be found in the appendix of [22].

4   Maximal Distributive Subalgebras

A distributive subalgebra $\mathcal{S}$ is *maximal* if there is no other distributive subalgebra that properly contains $\mathcal{S}$. In this section, we compute and list all maximal distributive subalgebras for RA, IA, CRA, and RA. For RCC5/8 we refer to [17, Appendix B].

4.1   Maximal Distributive Subalgebras of PA, IA, RCC5, and RCC8

Let $\mathcal{M}$ be PA, IA, RCC5, or RCC8 and $\mathcal{X}$ a subclass of $\mathcal{M}$. Recall we write $\widehat{\mathcal{X}}$ for the subalgebra of $\mathcal{M}$ generated by $\mathcal{X}$ and write $\mathcal{B}$ for the set of basic relations in $\mathcal{M}$. To compute the maximal distributive subalgebras of $\mathcal{M}$, we first compute $\widehat{\mathcal{B}}$, and then check by a program if $\widehat{\mathcal{B}} \cup \mathcal{Z}$ satisfies distributivity for some subset $\mathcal{Z}$ of $\mathcal{M}$.

Write $\mathcal{D}$ for the set of relations R in $\mathcal{M}$ such that $\widehat{\mathcal{B} \cup \{R\}}$ satisfies distributivity. We then check for every pair of relations R, S in $\mathcal{D}$ if $\widehat{\mathcal{B} \cup \{R, S\}}$ satisfies distributivity. If this is the case, then we say R has d-relation to S. Fortunately, the result shows that there are precisely two disjoint subsets $\mathcal{X}$ and $\mathcal{Y}$ (which form a partition of $\mathcal{D}$) such that each relation R in $\mathcal{X}$ ($\mathcal{Y}$, respectively) has d-relation to every other relation in $\mathcal{X}$ ($\mathcal{Y}$, respectively), but has no d-relation to any relation in $\mathcal{Y}$ ($\mathcal{X}$, respectively). Moreover, $\widehat{\mathcal{B} \cup \mathcal{X}}$ and $\widehat{\mathcal{B} \cup \mathcal{Y}}$ are both distributive subalgebras of $\mathcal{M}$. It is clear that these are the only maximal distributive subalgebras of $\mathcal{M}$.

In the following, we list the maximal distributive subalgebras of PA and IA and refer to [17, Appendix B] for those of RCC5 and RCC8.

PA. The closure of basic relations of PA contains 4 non-empty relations

$$\widehat{\mathcal{B}}_{\text{PA}} = \{<, >, =, \star\}. \tag{10}$$

One of the maximal distributive subalgebras contains 6 non-empty relations

$$<, >, =, \star, \leq, \geq, \tag{11}$$

which is exactly the subclass $\mathcal{C}_{\text{PA}}$ of convex PA relations; the other contains 5 non-empty relations

$$<, >, =, \star, \neq, \tag{12}$$

which is exactly the subclass $\mathcal{S}_{\text{PA}}$ identified in [2].

IA. The closure of basic IA relations, $\widehat{\mathcal{B}}_{\text{IA}}$, contains 29 non-empty relations (see Table 3). Our computation shows that IA has two maximal distributive subalgebra, one contains additional 53 non-empty relations, shown in Table 4, which is exactly the subclass $\mathcal{C}_{\text{IA}}$ of convex IA relations; the other contains additional 52 non-empty relations, shown in Table 5, which is exactly the subclass $\mathcal{S}_{\text{IA}}$ identified in [2].

Table 3. The closure of basic IA relations, $\mathcal{B}_{IA}$, contains 29 non-empty relations.

eq	d oi f
fi	d o s
f	d di o oi s si f fi eq
f fi eq	bi
si	bi oi mi
s	bi di oi mi si
s si eq	bi d oi mi f
mi	bi d di o oi mi s si f fi eq
m	b
oi	b o m
o	b di o m fi
di	b d o m s
di oi si	b d di o oi m s si f fi eq
di o fi	b bi d di o oi m mi s si f fi eq
d	

Table 4. Additional relations contained in $\mathcal{C}_{IA}$.

fi eq	di fi	d di o oi mi s si f fi eq
f eq	di si	d di o oi m s si f fi eq
si eq	di si fi eq	d di o oi m mi s si f fi eq
s eq	di oi si f f fi eq	bi mi
oi f	di oi mi si	bi oi mi f
oi si	di oi mi si f f fi eq	bi oi mi si
oi si f eq	di o s si fi eq	bi oi mi si f eq
oi mi	di o m fi	bi di oi mi si f f fi eq
oi mi f	di o m s si fi eq	bi d oi mi s si f eq
oi mi si	d f	bi d di o oi m mi s si f fi eq
oi mi si f eq	d s	b m
o fi	d s f eq	b o m fi
o s	d oi s si f eq	b o m s
o s fi eq	d oi mi f	b o m s fi eq
o m	d oi mi s si f eq	b di o m s si fi eq
o m fi	d o s f fi eq	b d o m s f fi eq
o m s	d o m s	b d di o oi m mi s si f fi eq
o m s fi eq	d o m s f fi eq	

4.2   Maximal Distributive Subalgebras of CRA

The procedure to compute the maximal distributive subalgebras of CRA is similar to the procedure for PA, IA, RCC5 and RCC8, but with some differences.

First, we compute $\widehat{\mathcal{B}}$, and then check by a program if $\widehat{\mathcal{B} \cup \mathcal{Z}}$ satisfies distributivity for some subset $\mathcal{Z}$ of CRA.

Write $\mathcal{D}$ for the set of relations R in CRA such that $\widehat{\mathcal{B} \cup \{R\}}$ satisfies distributivity. There are 8 different subalgebras in the set of subalgebras $\{\widehat{\mathcal{B} \cup \{R\}}\}$:

Table 5. Additional relations contained in $\mathcal{S}_{IA}$.

f fi	bi d di o oi s si	b d di o oi m s si
s si	bi d di o oi s si f fi eq	b bi d di o oi
di oi	bi d di o oi mi	b bi d di o oi f fi
di o	bi d di o oi mi f fi	b bi d di o oi s si
d oi	bi d di o oi mi s si	b bi d di o oi s si f fi eq
d o	b o	b bi d di o oi mi
d di o oi	b di o	b bi d di o oi mi f fi
d di o oi f fi	b di o fi	b bi d di o oi mi s si
d di o oi s si	b di o m	b bi d di o oi mi s si f fi eq
bi oi	b d o	b bi d di o oi m
bi di oi	b d o s	b bi d di o oi m f fi
bi di oi si	b d o m	b bi d di o oi m s si
bi di oi mi	b d di o oi	b bi d di o oi m s si f fi eq
bi d oi	b d di o oi f fi	b bi d di o oi m mi
bi d oi f	b d di o oi s si	b bi d di o oi m mi f fi
bi d oi mi	b d di o oi s si f fi eq	b bi d di o oi m mi s si
bi d di o oi	b d di o oi m	
bi d di o oi f fi	b d di o oi m f fi	

$R \in \mathcal{D}\}$. We call these 8 distributive subalgebras the *seed* subalgebras. Among these, only 4 are not contained in any other ones. We call these the *candidate* subalgebras. We then verify the following three facts:

1. For any pair of different candidate subalgebras $\mathcal{S}_i$ and $\mathcal{S}_j$, we have $\widehat{\mathcal{S}_i \cup \mathcal{S}_j}$ is not distributive.
2. For any pair of non-candidate subalgebras $\mathcal{S}_i$ and $\mathcal{S}_j$, we have $\widehat{\mathcal{S}_i \cup \mathcal{S}_j}$ is either a candidate subalgebra or not distributive.
3. For any pair of subalgebras $\mathcal{S}_i$ and $\mathcal{S}_j$ s.t. $\mathcal{S}_i$ is a candidate subalgebra, $\mathcal{S}_j$ is a non-candidate subalgebra, and $\mathcal{S}_j \not\subseteq \mathcal{S}_i$, we have $\widehat{\mathcal{S}_i \cup \mathcal{S}_j}$ is not distributive.

Based upon the above facts, we show that the four candidate subalgebras are the only maximal distributive subalgebras of CRA.

To prove the maximality, suppose $\mathcal{S}$ is one of the four candidate subalgebras. Let R be a relation in CRA which is not in $\mathcal{S}$. Then $\widehat{\mathcal{S} \cup \{R\}}$ is not distributive. This is because, by the above facts either $\widehat{\mathcal{B} \cup \{R\}}$ is not distributive or $\widehat{\mathcal{B} \cup \{R\}}$ is one of the 8 subalgebras and $\widehat{\mathcal{B} \cup \{R\} \cup \mathcal{S}}$ is not distributive.

To prove that there are no other maximal distributive subalgebras, suppose $\mathcal{S}'$ is a distributive subalgebra that is not a subset of any of the four candidate subalgebras. $\mathcal{S}'$ must contain at least two relations in $\mathcal{D}$, say R_1 and R_2. By the above facts, we know the closure of the union of $\widehat{\mathcal{B} \cup \{R_1\}}$ and $\widehat{\mathcal{B} \cup \{R_2\}}$ is either not distributive or one of the four maximal distributive subalgebra. If it is the latter case, then $\mathcal{S}'$ would be either not distributive or a superset of one of the four maximal distributive subalgebras. Note that the latter situation cannot happen as it contradicts the maximality of the four maximal distributive subalgebras.

Interestingly, these four maximal distributive subalgebras of CRA correspond exactly to the Cartesian products of the maximal distributive subalgebras of PA, viz. $\mathcal{C}_{PA} \otimes \mathcal{C}_{PA}, \mathcal{C}_{PA} \otimes \mathcal{S}_{PA}, \mathcal{S}_{PA} \otimes \mathcal{C}_{PA}, \mathcal{S}_{PA} \otimes \mathcal{S}_{PA}$, where we interpret in a natural way a CRA relation e.g. $\{NW, N\}$ as $\{<, =\} \otimes \{>\}$.

4.3   Maximal Distributive Subalgebras of RA

Unlike the other small calculi we have discussed, RA has a large number (169) of basic relations, resulting a total of 2^{169} relations in it. It becomes infeasible to exploit the former brute-force procedure to compute the maximal distributive subalgebras of RA. However, noting that the maximal distributive subalgebras of CRA are exactly the Cartesian products of the two maximal distributive subalgebras of PA, we conjecture that a similar situation happens to RA. This is indeed true.

Theorem 3. *RA has exactly four maximal distributive subalgebras, which are the Cartesian products of the two maximal distributive subalgebras of IA.*

Proof. For convenience, we write $\mathcal{D}_1$ and $\mathcal{D}_2$ for the maximal distributive subalgebras $\mathcal{C}_{IA}$ and $\mathcal{S}_{IA}$. It is straightforward to show that their Cartesian products $\mathcal{D}_i \otimes \mathcal{D}_j$ $(1 \leq i, j \leq 2)$ are all distributive subalgebras of RA.

In order to show the maximality of $\mathcal{D}_i \otimes \mathcal{D}_j$, suppose $R \notin \mathcal{D}_i \otimes \mathcal{D}_j$. We show that the subalgebra $\widehat{\{R\} \cup \mathcal{D}_i \otimes \mathcal{D}_j}$ is not distributive. Let $R_x = \{\alpha \in \mathcal{B}_{IA} \mid \exists \beta \in \mathcal{B}_{IA}$ s.t. $(\alpha, \beta) \in R\}$ and define R_y similarly. Note that R is always contained in $R_x \otimes R_y$. There are two cases.

Case 1. $R \subsetneq R_x \otimes R_y$. Then there exist $\alpha_0 \in R_x$ and $\beta_0 \in R_y$ s.t. $\alpha_0 \otimes \beta_0 \notin R$. Let $S = \{\alpha_0\} \otimes \star$ and $T = \star \otimes \{\beta_0\}$. Note that $\widehat{\mathcal{B}}_{RA}$ is strictly contained in $\mathcal{D}_i \otimes \mathcal{D}_j$. Thus $S, T \in \widehat{\{R\} \cup \mathcal{D}_i \otimes \mathcal{D}_j}$. It is easy to see that $R \cap S \neq \varnothing$, $R \cap T \neq \varnothing$, and $S \cap T \neq \varnothing$, but $R \cap S \cap T = \varnothing$. By Theorem 1, this implies that $\widehat{\{R\} \cup \mathcal{D}_i \otimes \mathcal{D}_j}$ is not distributive.

Case 2. $R = R_x \otimes R_y$. Then we have either $R_x \notin \mathcal{D}_i$ or $R_y \notin \mathcal{D}_j$. Take $R_x \notin \mathcal{D}_i$ as an example. Then $\widehat{\{R_x\} \cup \mathcal{D}_i}$ is not distributive. This implies that there exist $R_0, S_0, T_0 \in \widehat{\{R_x\} \cup \mathcal{D}_i}$ which do not satisfy Helly's condition (8). Note that $R_0 \otimes \star$, $S_0 \otimes \star$, and $T_0 \otimes \star$ are all in $\widehat{\{R\} \cup \mathcal{D}_i \otimes \mathcal{D}_j}$. However, the three relations $R_0 \otimes \star$, $S_0 \otimes \star$, and $T_0 \otimes \star$ do not satisfy (8), which means that $\widehat{\{R\} \cup \mathcal{D}_i \otimes \mathcal{D}_j}$ is not distributive.

The above proves the maximality of $\mathcal{D}_i \otimes \mathcal{D}_j$. To show the uniqueness, suppose $\mathcal{S}$ is a distributive subalgebra. We show $\mathcal{S}$ is a subset of $\mathcal{D}_i \otimes \mathcal{D}_j$ for some i, j.

First, we show for every $R \in \mathcal{S}$ we have $R = R_x \otimes R_y$. Suppose not. Then there exist $\alpha \in R_x$ and $\beta \in R_y$ s.t. $\alpha \otimes \beta \notin R$. Similar to the proof of the maximality, we know both $\{\alpha\} \otimes \star$ and $\star \otimes \{\beta\}$ are in $\widehat{\mathcal{B}}$ and, hence, in $\mathcal{S}$. The three relations $R, \{\alpha\} \otimes \star, \star \otimes \{\beta\}$, however, do not satisfy Helly's condition (8).

Next, we show that $\mathcal{S}$ is a subset of $\mathcal{D}_i \otimes \mathcal{D}_j$ for some i, j. Write $\mathcal{S}_x = \{R_x : R \in \mathcal{S}\}$ and $\mathcal{S}_y = \{R_y : R \in \mathcal{S}\}$. We assert that $\mathcal{S}_x$ and $\mathcal{S}_y$ are both distributive subalgebras of IA. We first note that if $R = R_x \otimes R_y \in \mathcal{S}$, then both $R_x \otimes \star$ and $\star \otimes R_y$ are in $\mathcal{S}$. This is because, for instance, $\{\mathsf{eq}\} \otimes \star$ is a relation in $\widehat{\mathcal{B}}_{\mathrm{RA}} \subseteq \mathcal{S}$ and $(R_x \otimes R_y) \diamond (\{\mathsf{eq}\} \otimes \star) = R_x \otimes \star$. It is easy to check that $\{R_x \otimes \star : R_x \otimes R_y \in \mathcal{S}\}$ is a distributive subalgebra which is contained in $\mathcal{S}$. Now, it is clear that $\mathcal{S}_x$ is a distributive subalgebra of IA and, hence, contained in either $\mathcal{D}_1$ or $\mathcal{D}_2$. The same conclusion applies to $\mathcal{S}_y$. Therefore, $\mathcal{S}$ is a subset of $\mathcal{D}_i \otimes \mathcal{D}_j$ for some i, j. □

The above proof also applies to CRA.

5 Partial Path Consistency and Variable Elimination

In this section, we present two nice properties of distributive subalgebras, which will play an important role in reasoning with large sparse constraint networks.

5.1 Variable Elimination

In [31], Zhang and Marisetti proposed a novel variable elimination method for solving (classical and finite) connected row convex (CRC) constraints [10]. The idea is to eliminate the variables one by one until a trivial problem is reached. Although very simple, the algorithm is able to make use of the sparsity of the problem instances and performs very well. One key property of CRC constraints is that any strong path consistent CRC constraint network is globally consistent. Recall that a similar property has been identified in our Theorem 2 for constraint networks over a distributive subalgebra. The following lemma and theorem show that the same variable elimination method also applies to constraint networks over a distributive subalgebra,

Lemma 3. *Let $\mathcal{M}$ be a qualitative calculus that satisfies (6) and (7). Suppose $\mathcal{N} = \{v_i R_{ij} v_j \mid 1 \leq i, j \leq n\}$ is a network over a distributive subalgebra $\mathcal{S}$ of $\mathcal{M}$ and $V = \{v_1, ..., v_n\}$. If $R_{ij} \subseteq R_{in} \diamond R_{nj}$ for every $1 \leq i, j < n$, then $\mathcal{N}_{-n}$ is consistent only if $\mathcal{N}$ is consistent, where $\mathcal{N}_{-n} = \{v_i R_{ij} v_j \mid 1 \leq i, j \leq n - 1\}$ is the restriction of $\mathcal{N}$ to $\{v_1, ..., v_{n-1}\}$.*

Proof. Suppose $\{\delta_{ij} : 1 \leq i, j < n\}$ is a consistent scenario of $\mathcal{N}_{-n}$. First, write T_i for $R_{n,i}$ and let $\widehat{T}_i = \bigcap_{j=1}^{n-1} T_j \diamond \delta_{ji}$. We only need to show $\widehat{T}_j \subseteq \widehat{T}_i \diamond \delta_{ij}$. Note

$$\widehat{T}_i \diamond \delta_{ij} = (\bigcap_{j=1}^{n-1} T_{j'} \diamond \delta_{j'i}) \diamond \delta_{ij} = \bigcap_{j'=1}^{n-1} (T_{j'} \diamond \delta_{j'i} \diamond \delta_{ji}) \supseteq \bigcap_{j'=1}^{n-1} T_{j'} \diamond \delta_{j'j} = \widehat{T}_j.$$

Second, we show $\widehat{T}_i$ is not empty. To this end, by Helly's condition (8), we only need to show $T_j \diamond \delta_{ji} \cap T_{j'} \diamond \delta_{j'i} \neq \varnothing$ for any $j \neq j'$. Using the cycle law twice, we have

$$T_j \diamond \delta_{ji} \cap T_{j'} \diamond \delta_{j'i} \neq \varnothing \quad \text{iff} \quad T_{j'} \diamond \delta_{j'i} \diamond \delta_{ij} \cap T_j \neq \varnothing$$
$$\text{iff} \quad T_{j'}{}^{-1} \diamond T_j \cap \delta_{j'i} \diamond \delta_{ij} \neq \varnothing$$
$$\text{iff} \quad R_{j'n} \diamond R_{nj} \cap \delta_{j'i} \diamond \delta_{ij} \neq \varnothing.$$

Because $\delta_{j'j} \subseteq R_{j'n} \diamond R_{nj}$ and $\delta_{j'j} \subseteq \delta_{j'i} \diamond \delta_{ij}$, we have $R_{j'n} \diamond R_{nj} \cap \delta_{j'i} \diamond \delta_{ij} \neq \varnothing$, hence $T_j \diamond \delta_{ji} \cap T_{j'} \diamond \delta_{j'i} \neq \varnothing$. $\qquad\square$

From the above lemma, it will be easy to prove the following theorem.

Theorem 4. *Let $\mathcal{M}$ be a qualitative calculus that satisfies* (6) *and* (7). *Suppose $\mathcal{N} = \{v_i R_{ij} v_j \mid 1 \leq i, j \leq n\}$ is a network over a distributive subalgebra $\mathcal{S}$ of $\mathcal{M}$ and $V = \{v_1, ..., v_n\}$. Denote $\mathcal{N}^*_{-n} = \{v_i \widehat{R}_{ij} v_j \mid \widehat{R}_{ij} = R_{ij} \cap R_{in} \diamond R_{nj}, 1 \leq i, j \leq n - 1\}$, i.e. $\mathcal{N}^*_{-n}$ is the network after eliminating v_n and updating the corresponding constraints. Then $\mathcal{N}^*_{-n}$ is consistent if and only if $\mathcal{N}$ is consistent.*

By the previous theorem, we can directly devise an efficient variable elimination algorithm for constraint networks over a distributive subalgebra. At each step, we choose the node which has the smallest degree for deleting. In particular, we can simply remove all nodes with degree 1 from the constraint network without affecting its consistency. This is especially useful for efficient reasoning with large sparse constraint networks. In fact we have done a small experiment of variable elimination on the NUTS RDF dataset[1], which has 2236 variables but contains only 3176 constraints (without their converses). The result is very intriguing that consistency is decided within 15 ms and only 5569 constraints are visited (the number of compositions is only 2075), compared to the $\mathcal{O}(n^3)$ time path consistency algorithm.

5.2   Partial Path Consistency

Another efficient method for solving sparse constraint networks is the partial path consistency (PPC) algorithm proposed by Bliek and Sam-Haroud [6]. The idea is to enforce path consistency (PC) on sparse graphs by triangulating instead of completing them. The authors demonstrated that, as far as CRC constraints are concerned, the pruning capacity of PC on triangulated graphs and their completion are identical on the common edges. Recently, PPC has also been extended to qualitative spatial and temporal constraint solving [8,26], where the authors proved that any PPC constraint network over a maximal tractable subclass of IA or RCC8 is always consistent. However, for constraint networks over these subclasses, the pruning capacity of PC on triangulated graphs and their completion may not be identical on the common edges. In this section, we show that the result is affirmative for constraint networks over distributive subalgebras.

We first recall several basic notions related to PPC introduced in [6].

An undirected graph $G = (V, E)$ is *triangulated* or *chordal* if every cycle of length greater than 3 has a chord, i.e. an edge connecting two non-consecutive

---
[1] http://nuts.geovocab.org/.

vertices of the cycle. For each $v \in V$, the adjacency set $Adj(v)$, is defined as $\{w \in V : \{v,w\} \in E\}$. A vertex v is *simplicial* if $Adj(v)$ is complete. Every chordal graph has a simplicial vertex. Moreover, after removing a simplicial vertex and its incident edges from the graph, a chordal graph remains chordal. The order in which simplicial vertices are successively removed is called a *perfect elimination order*.

Lemma 4 ([6]). *If $G = (V,E)$ is an incomplete chordal graph, then one can add a missing edge (u,w) with $u, w \in V$ such that*

- *the graph $G' = (V, E \cup \{\{u,w\}\})$ is chordal graph; and*
- *the graph induced by $X = \{x | \{u,x\}, \{x,w\} \in E\}$ is complete.*

For a constraint network $\mathcal{N} = \{v_i R_{ij} v_j : 1 \leq i, j \leq n\}$ over $V = \{v_1, ..., v_n\}$, the *constraint graph* of $\mathcal{N}$ is the undirected graph $G(\mathcal{N}) = (V, E(\mathcal{N}))$, for which we have $\{v_i, v_j\} \in E(\mathcal{N})$ iff $R_{ij} \neq \star$. Given a constraint network $\mathcal{N}$ and a graph $G = (V, E)$, we say $\mathcal{N}$ is *partial path consistent w.r.t. G* iff for any $1 \leq i, j, k \leq n$ with $\{v_i, v_j\}, \{v_j, v_k\}, \{v_i, v_k\} \in E$ we have $R_{ik} \subseteq R_{ij} \diamond R_{jk}$ [8].

The following result was first proved for RCC8 in [27]. The proof given there is also applicable to other calculi. We here give a slightly different proof which does not use the weakly global consistency result.

Theorem 5. *Let $\mathcal{M}$ be a qualitative calculus that satisfies (6) and (7). Suppose $\mathcal{N} = \{v_i R_{ij} v_j \mid 1 \leq i, j \leq n\}$ is a network over a distributive subalgebra $\mathcal{S}$ of $\mathcal{M}$ and $V = \{v_1, ..., v_n\}$. Assume in addition that $G = (V, E)$ is a chordal graph such that $E(\mathcal{N}) \subseteq E$. Then enforcing partial path consistency on G is equivalent to enforcing path consistency on the completion of G, in the sense that the relations computed for the constraints in G are identical.*

Proof. The proof is similar to the one given for CRC constraints [6, Theorem 3]. Suppose we have a chordal graph $G = (V, E)$ such that $G(\mathcal{N}) \subseteq G$ and $\mathcal{N}$ is PPC w.r.t. G. We will add to G the missing edges one by one until the graph is complete. To prove the theorem, we show that the relations of the constraints can be computed from the existing ones so that each intermediate graph, including the complete graph, is path consistent.

In the following we assume the order $v_1, \ldots, v_n$ is a perfect elimination order of chordal graph G. Denote $S_i = \{v_{n-i+1}, \ldots, v_n\}$, $G_i = G(S_i)$ (the induced subgraph of G by S_i), and $F_i = \{v_k \in N(v_{n-i}) : k > n - i\}$, where $N(v_{n-i}) = \{v_j : \{v_j, v_{n-i}\} \in E\}$.

We add the missing edges one by one to G in the following manner:

1. choose the largest i such that G_i is complete;
2. choose vertices v_{n-i}, v_j in G;
3. label the edge $\{v_{n-i}, v_j\}$ (and resp. its reverse) with

$$R_{n-i,j} = \bigcap_{v_k \in F_i} R_{n-i,k} \diamond R_{k,j}.$$

After adding one edge, we prove G', the resulting graph, is still path consistent.

First, we show the added label is non-empty. To show this, by Theorem 1, we need only show $R_{n-i,k} \diamond R_{k,j} \cap R_{n-i,k'} \diamond R_{k',j} \neq \varnothing$ for any $v_k \neq v_{k'} \in F_i$. Such a pairwise intersection is not empty because, by the cycle law of relation algebra, we have

$$R_{n-i,k} \diamond R_{k,j} \cap R_{n-i,k'} \diamond R_{k',j} \neq \varnothing \quad \text{iff} \quad R_{k,n-i} \diamond R_{n-i,k'} \cap R_{k,j} \diamond R_{j,k'} \neq \varnothing.$$

Since $G(F_i \cup \{v_{n-i}\})$ and G_i are complete and path consistent, we have $R_{k,k'} \subseteq R_{k,n-i} \diamond R_{n-i,k'}$ and $R_{k,k'} \subseteq R_{k,j} \diamond R_{j,k'}$. This shows $R_{k,n-i} \diamond R_{n-i,k'} \cap R_{k,j} \diamond R_{j,k'} \neq \varnothing$ and, hence, $R_{n-i,k} \diamond R_{k,j} \cap R_{n-i,k'} \diamond R_{k',j} \neq \varnothing$.

We then need to show the constraint network is path consistent for the three paths $\langle n-i, j, k' \rangle$, $\langle n-i, k', j \rangle$, and $\langle k', n-i, j \rangle$.

For $\langle n-i, j, k' \rangle$, note that, for any $k \in F_i$, we have $R_{n-i,k'} \subseteq R_{n-i,k} \diamond R_{k,k'} \subseteq R_{n-i,k} \diamond R_{k,j} \diamond R_{j,k'}$. Therefore, we have $R_{n-i,k'} \subseteq \bigcap_{k \in F_i} R_{n-i,k} \diamond R_{k,j} \diamond R_{j,k'}$. By distributivity, we know $R_{n-i,k'} \subseteq (\bigcap_{k \in F_i} R_{n-i,k} \diamond R_{k,j}) \diamond R_{j,k'} = R_{n-i,j} \diamond R_{k,j}$.

For $\langle n-i, k', j \rangle$, by the construction of $R_{n-i,j}$, we have $R_{n-i,j} \subseteq R_{n-i,k'} \diamond R_{k',j}$.

For $\langle k', n-i, j \rangle$, we need to show $R_{k',j} \subseteq R_{k',n-i} \diamond R_{n-i,j}$. Note $R_{n-i,j} = \bigcap_{v_k \in F_i} R_{n-i,k} \diamond R_{k,j}$. By distributivity, it is sufficient to show, for each $k \in F_i$, $R_{k',j} \subseteq R_{k',n-i} \diamond R_{n-i,k} \diamond R_{k,j}$. Because $G(F_i \cup \{v_{n-i}\})$ is complete and PC, $R_{k',k} \subseteq R_{k',n-i} \diamond R_{n-i,k}$. Moreover, because $G(F_i \cup \{v_j\})$ is complete and PC by construction and induction, $R_{k',j} \subseteq R_{k',k} \diamond R_{k,j} \subseteq R_{k',n-i} \diamond R_{n-i,k} \diamond R_{k,j}$.

Thus, after adding a missing edge, the resulting graph remains path consistent. At last we will get the complete graph, which is equivalent to the completion of G. Note that the label of every edge in G is not changed. This finishes the proof.   □

6   Further Discussion

In this section we discuss the relation of distributive subalgebras with conceptual neighbourhood graphs (CNGs) [15] and star distributivity [23] of classical CSPs.

6.1   Distributive Subalgebras and Conceptual Neighbourhood Graph

As we have seen, the classes of convex IA and RCC8 relations are maximal distributive subalgebras of IA and RCC8 respectively. For IA, Ligozat [19] characterises the convex relations by using the CNG of IA [15] (shown in Table 1 (ii)). An IA relation is convex if it is an "interval" $[\alpha, \beta]$ containing all the relations between its two endpoint relations α, β in the CNG. The subclass of convex IA relations is exactly the maximal distributive subalgebra $\mathcal{C}_{\mathrm{IA}}$.

Similar idea applies to PA and RCC5 directly. For PA, the CNG is shown in the left of Fig. 6. From the CNG of PA, we observe the "convex" relations correspond to relations in $\mathcal{C}_{\mathrm{PA}} = \{<, =, >, \leq, \geq\}$, one of the maximal distributive subalgebras of PA. For RCC5, the CNG is shown in the middle of Fig. 6.

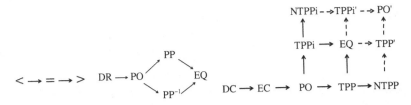

Fig. 6. CNGs of PA, RCC5, and RCC8

The subclass of convex RCC5 relations is precisely the maximal distributive subalgebra $\mathcal{D}_{14}^5$ specified in [17].

The CNG of CRA is constructed by using the CNG of PA. For example, note that < and = are conceptual neighbours in the CNG of PA, and NW is defined as $x < x'$ and $y > y'$ and N is defined as $x = x'$ and $y > y'$. Then N and NW should be conceptual neighbours in the CNG of CRA. The complete CNG of CRA is given in [20] and the subclass of convex CRA relations corresponds to the maximal distributive subalgebra $\mathcal{C}_{PA} \otimes \mathcal{C}_{PA}$. Like CRA, the CNG of RA is constructed by using the CNG of IA. The subclass of convex RA relations [3] is the maximal distributive subalgebra $\mathcal{C}_{IA} \otimes \mathcal{C}_{IA}$.

For RCC8, the situation is a little different. We need to revise the CNG by introducing three imaginary relations **TPP′**, **TPP^{-1}′** and **PO′** (see Fig. 6, right). After this modification, Chandra and Pujari [7] identified the class of convex RCC8 relations, which is precisely the maximal distributive subalgebra $\mathcal{D}_{41}^8$ specified in [17].

A natural question arises as, "Can we obtain each maximal distributive subalgebra by designing an appropriate CNG?" The answer seems negative as the maximal distributive subalgebra $\mathcal{S}_{PA}$ contains $\neq$ but does not contain either $\leq$ or $\geq$.

6.2 Relation with Classical CSPs

For finite domain CSPs, Montanari observed properties similar to the distributivity in this paper. In [23], Montanari defined two different concept related to distributivity. One is a *distributive set of relations w.r.t. set X_k* and the other is *star-distributive constraint network*. The second concept is very similar to our notion of distributivity, except that it only requires the relations to form a closure w.r.t. the network. A constraint network over a distributive subalgebra is always star-distributive, but it is not clear whether a star-distributive network is always over a distributive subalgebra.

As we have seen, relations in a distributive subalgebra exhibit convexity in Helly's sense. In finite CSP, row convex constraints [5] and (the more general) tree convex constraints [30] enjoy a similar property, which is specified w.r.t. the "rows" or "images" of the constraints rather than the constraints themselves. The relations R, S, T in Table 6 are all CRC constraints. Moreover, we have $R \diamond (S \cap T) \neq R \diamond S \cap R \diamond T$ and $R \cap S \neq \varnothing$, $R \cap T \neq \varnothing$, $S \cap T \neq \varnothing$ but

$R \cap S \cap T = \varnothing$. This shows that CRC constraints are not always distributive and do not always satisfy Helly's condition (8).

Table 6. Example showing that CRC constraints are not always distributive.

$\begin{pmatrix} 1\,0\,0 \\ 1\,1\,0 \\ 0\,0\,1 \end{pmatrix}$	$\begin{pmatrix} 1\,1\,1 \\ 0\,0\,1 \\ 0\,0\,1 \end{pmatrix}$	$\begin{pmatrix} 0\,0\,1 \\ 1\,1\,1 \\ 0\,1\,0 \end{pmatrix}$	$\begin{pmatrix} 0\,0\,1 \\ 0\,0\,1 \\ 0\,0\,0 \end{pmatrix}$	$\begin{pmatrix} 0\,0\,1 \\ 1\,1\,1 \\ 0\,0\,0 \end{pmatrix}$
R	S	T	$R \diamond (S \cap T)$	$R \diamond S \cap R \diamond T$

7 Conclusion

In this paper, we gave a detailed discussion of the important concept of distributive subalgebra proposed in a recent work [17]. We proved that distributive subalgebras are exactly subalgebras which are Helly in our sense and showed path consistent networks over a distributive subalgebra are weakly globally consistent. We also found all maximal distributive subalgebras for PA, IA, CRA, and RA. Finally, we proposed two nice properties of distributive subalgebras which will be used for efficient reasoning of large sparse constraint networks. Future work will implement and empirically evaluate and compare these two methods by using real datasets.

References

1. Allen, J.F.: Maintaining knowledge about temporal intervals. Commun. ACM **26**(11), 832–843 (1983)
2. Amaneddine, N., Condotta, J.-F.: From path-consistency to global consistency in temporal qualitative constraint networks. In: Ramsay, A., Agre, G. (eds.) AIMSA 2012. LNCS, vol. 7557, pp. 152–161. Springer, Heidelberg (2012)
3. Balbiani, P., Condotta, J.F., Fariñas del Cerro, L.: A new tractable subclass of the rectangle algebra. In: IJCAI-99, pp. 442–447 (1999)
4. van Beek, P.: Approximation algorithms for temporal reasoning. In: IJCAI, pp. 1291–1296 (1989)
5. van Beek, P., Dechter, R.: On the minimality and global consistency of row-convex constraint networks. J. ACM **42**, 543–561 (1995)
6. Bliek, C., Sam-Haroud, D.: Path consistency on triangulated constraint graphs. In: IJCAI-99, pp. 456–461 (1999)
7. Chandra, P., Pujari, A.K.: Minimality and convexity properties in spatial CSPs. In: ICTAI, pp. 589–593 (2005)
8. Chmeiss, A., Condotta, J.: Consistency of triangulated temporal qualitative constraint networks. In: ICTAI, pp. 799–802 (2011)
9. Danzer, L., Grünbaum, B., Klee, V.: Helly's theorem and its relatives. In: Proceedings of the Seventh Symposium in Pure Mathematics of the American Mathematical Society (Convexity), pp. 101–179. American Mathematical Society Providence, RI (1963)

10. Deville, Y., Barette, O., Van Hentenryck, P.: Constraint satisfaction over connected row-convex constraints. Artif. Intell. **109**(1), 243–271 (1999)
11. Duckham, M., Li, S., Liu, W., Long, Z.: On redundant topological constraints. In: KR-2014 (2014)
12. Düntsch, I.: Relation algebras and their application in temporal and spatial reasoning. Artif. Intell. Rev. **23**(4), 315–357 (2005)
13. Düntsch, I., Wang, H., McCloskey, S.: A relation-algebraic approach to the region connection calculus. Theor. Comput. Sci. **255**(1–2), 63–83 (2001)
14. Frank, A.U.: Qualitative spatial reasoning with cardinal directions. In: ÖGAI-91, pp. 157–167. Springer (1991)
15. Freksa, C.: Temporal reasoning based on semi-intervals. Artif. Intell. **54**(1), 199–227 (1992)
16. Guesgen, H.W.: Spatial Reasoning Based on Allen's Temporal Logic. International Computer Science Institute, Berkeley (1989)
17. Li, S., Long, Z., Liu, W., Duckham, M., Both, A.: On redundant topological constraints. Artif. Intell. **225**, 51–78 (2015)
18. Li, S., Ying, M.: Region connection calculus: its models and composition table. Artif. Intell. **145**(1–2), 121–146 (2003)
19. Ligozat, G.: Tractable relations in temporal reasoning: pre-convex relations. In: Proceedings of Workshop on Spatial and Temporal Reasoning, ECAI-94, pp. 99–108 (1994)
20. Ligozat, G.: Reasoning about cardinal directions. J. Vis. Lang. Comput. **9**(1), 23–44 (1998)
21. Ligozat, G., Renz, J.: What is a qualitative calculus? a general framework. In: Zhang, C., W. Guesgen, H., Yeap, W.-K. (eds.) PRICAI 2004. LNCS (LNAI), vol. 3157, pp. 53–64. Springer, Heidelberg (2004)
22. Long, Z., Li, S.: On distributive subalgebras of qualitative spatial and temporal calculi. arXiv : 1506.00337 [cs.AI] (2015). http://arxiv.org/abs/1506.00337
23. Montanari, U.: Networks of constraints: fundamental properties and applications to picture processing. Inf. Sci. **7**, 95–132 (1974)
24. Nebel, B., Bürckert, H.J.: Reasoning about temporal relations: a maximal tractable subclass of Allen's interval algebra. J. ACM **42**(1), 43–66 (1995)
25. Randell, D.A., Cui, Z., Cohn, A.G.: A spatial logic based on regions and connection. In: KR-92, pp. 165–176 (1992)
26. Sioutis, M., Koubarakis, M.: Consistency of chordal RCC-8 networks. In: ICTAI, pp. 436–443 (2012)
27. Sioutis, M., Li, S., Condotta, J.F.: Efficiently characterizing non-redundant constraints in large real world qualitative spatial networks. In: IJCAI (2015, to appear)
28. Stell, J.G.: Boolean connection algebras: a new approach to the region-connection calculus. Artif. Intell. **122**(1), 111–136 (2000)
29. Vilain, M.B., Kautz, H.A.: Constraint propagation algorithms for temporal reasoning. In: Proceedings of AAAI, pp. 377–382 (1986)
30. Zhang, Y., Freuder, E.C.: Properties of tree convex constraints. Artif. Intell. **172**(12–13), 1605–1612 (2008)
31. Zhang, Y., Marisetti, S.: Solving connected row convex constraints by variable elimination. Artif. Intell. **173**(12–13), 1204–1219 (2009)

What is in a Contour Map?
A Region-Based Logical Formalization of Contour Semantics

Torsten Hahmann[1]([✉]) and E. Lynn Usery[2]

[1] National Center for Geographic Information and Analysis,
School of Computing and Information Science, University of Maine,
Orono, ME 04469, USA
torsten@spatial.maine.edu
[2] Center of Excellence for Geospatial Information Science, U.S. Geological Survey,
Rolla, MO 65401, USA

Abstract. Contours maps (such as topographic maps) compress the information of a function over a two-dimensional area into a discrete set of closed lines that connect points of equal value (isolines), striking a fine balance between expressiveness and cognitive simplicity. They allow humans to perform many common sense reasoning tasks about the underlying function (e.g. elevation).

This paper analyses and formalizes contour semantics in a first-order logic ontology that forms the basis for enabling computational common sense reasoning about contour information. The elicited contour semantics comprises four key concepts – contour regions, contour lines, contour values, and contour sets – and their subclasses and associated relations, which are grounded in an existing qualitative spatial ontology. All concepts and relations are illustrated and motivated by physical-geographic features identifiable on topographic contour maps. The encoding of the semantics of contour concepts in first-order logic and a derived conceptual model as basis for an OWL ontology lay the foundation for fully automated, semantically-aware qualitative and quantitative reasoning about contours.

Keywords: Contour maps · Isolines · Knowledge representation · First-order logic · Spatial ontology · Region-based space · Naive geography · Physical reasoning

1 Introduction

Contour maps effectively convey information about measures taken at spatial locations within a bounded or unbounded region of space, such as elevation information on a topographical surface, bathymetric information about the depths of lakes and seas, or meteorological information, e.g., barometric pressure or annual rainfall.[1] A contour map compactly represents such a function of space

---

[1] Such measures form a *field* when the space is unbounded. We use the term field more loosely, including both bounded and unbounded variants.

© Springer International Publishing Switzerland 2015
S.I. Fabrikant et al. (Eds.): COSIT 2015, LNCS 9368, pp. 375–399, 2015.
DOI: 10.1007/978-3-319-23374-1_18

by reducing it to a small set of contours (also called *contour lines* or *isolines*), each of which has an associated *contour value*. Each contour separates points in the underlying space based on whether their function value is below or above the contour value. For example, contours on topographic maps (see Fig. 1) separate the areas where the elevation is higher than the contour value from those where the elevation is lower than the contour value. Thereby, contours reduce the complexity of the underlying field by extracting a representation that is conceptually simple enough to effectively communicate the encoded information to humans, yet sufficiently expressive and detailed so that humans can use it to answer many common sense questions about the underlying field.

The idea of contours dates, at least, back to the abstractions that Charles Hutton [10] employed in the 18th century to estimate the mean density of the Earth; even earlier uses are discussed in [17]. The earliest USGS topographic maps from the 1880s already employ contours to visualize terrain elevation [24]. Despite the widespread use of contours by humans for geospatial data visualization and analysis, representations for contours are missing from all currently available geospatial data standards, including the OGC Reference Model [15][2], and have not been utilized for machine-based naive geographical (in the sense of [4]) or physical-geographical processing and reasoning. Many common sense questions, such as identifying high points in a terrain, finding paths that minimize elevation differences, or locating where water flows and collects, could be directly answered by a automated deduction system, e.g. a theorem prover or similar inference mechanism, from a declarative specification of a field's contours as facts about their qualitative spatial relationships on top of the proposed ontology. It could completely forego the process of having to produce a (printed or digital) map image and does not need to employ the full dataset of the (possibly unknown) underlying field nor full-fledged three-dimensional spatial algorithms. Encoding the knowledge represented by a field in a computer-interpretable format that captures the essence of contour maps opens up many possibilities for qualitative reasoning [3,9] that goes beyond the most simplistic qualitative formalisms concerned only with abstract regions of space.

1.1   Objective

The objective of this paper is to lay the semantic foundations that will enable computers to perform such common sense reasoning with contour information without human interaction. More specifically, we investigate the use of a region-based formalism of contour semantics, building on existing ontologies of qualitative space [7] and measured quantities [20], and formalizing the identified key concepts and relations explicitly in a first-order logical ontology. The resulting ontology can facilitate machine processing and reasoning of contour information, either directly by utilizing automated theorem provers, or indirectly by deriving

---

[2] OGC's reference model and more specific standards such as GeoSPARQL and GML include *coverage* data types to represent fields, but offer no way of representing fields using contours.

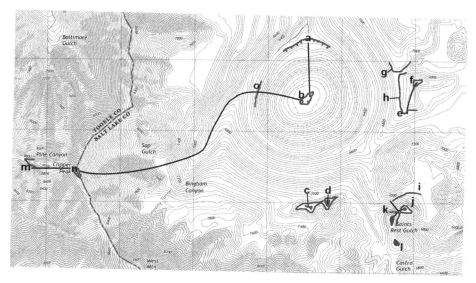

Fig. 1. A topographic map (Bingham Canyon area in Utah) with contour lines spaced in a contour interval of 50 ft apart. The contours *e–o* denote ascending contours, while the hachured contours *a–d* denote descending contours. All named contours are completely or partially highlighted by bold green lines. *b* is at the bottom of the Bingham Canyon Mine, an open-pit copper mine in operation since the beginning of the 20th century. To the west (left-hand side), the terrain rises steeply (along the superimposed line that runs perpendicular to contours), with Clipper Peak (*n*) reaching above 9200 feet. See the text for examples and more explanations.

lightweight versions of the ontology. Ultimately, we want an automated reasoner to be able to utilize contour information encoded in the ontology to answer simple qualitative or Boolean queries such as:

- Is the path along a given two-dimensional vector (in the underlying, implicit field) relatively flat, ascending, descending, or a combination thereof?
- Is there a path between two points that does not involve a change in elevation?
- Which of two areas has a greater maximal elevation?
- Which of two areas has a greater roughness (turtourocity)?

It turns out that basic integer operations and spatial sums suffice to capture the quantitative aspects of contour semantics (to constrain contour values and points values). Including these operations in the formalization of contour semantics greatly increases the utility and expressiveness of the otherwise purely qualitative representation.

Our specific contributions are: (1) analyzing whether a *contour region* is an adequate primitive notion for formalizing contour semantics; (2) formalizing contour semantics as a region-based first-order ontology using a small set of contour-related primitives (contour set, ascending and descending contour regions, and a relation between contour regions and their boundary values);

(3) relating the contour values associated with contour regions to measures at individual points; (4) identifying the additional quantitative operations that are required to support reasoning about contours; (5) mathematically characterizing important classes of contour sets and maps; and (6) deriving a light-weight conceptual model from the first-order formalization.

1.2   Background

While much work has been done to automatically derive contour maps from three-dimensional data, e.g. terrain data [18] as represented in Digital Elevation Models (DEM), less focus has been put on making such derived contour maps available for automated reasoning. Early work on digital representations of contour maps [1,6,11,13] has focused on utilizing tree structures to identify pits, peaks, saddles, ridges and similar topographic features, but has not attempted to formalize contour semantics. Moreover, the developed data structures, so-called *contour trees*, emphasize the nesting and adjacency relationship between contour lines and have been effective for surface and feature extraction, but are conceptually difficult to reason with. Our objective is to devise a conceptually and mathematically simpler representation of contours by using a region-based approach that emphasizes the regions enclosed by contours over the contours themselves. This will enable the development of an ontology of contour semantics that extends qualitative, mereotopology-based representations of space, e.g. [2,5,9,19], while also being able to define contour lines and the relationships among contour lines as well as between contour lines and the regions they bound. This would not be possible in the simpler formalisms from [2,5,19], that do not capture relationships between spatial entities of different dimensions, such as between regions and curves.

As basis for such a region-based formalization of contour semantics, we focus on the ontological concepts that underlie *contour map conceptualizations* (the idea of a set of nested contours, wherein all contours are closed curves, i.e. curves without endpoints) and not *contour map depictions* (the visualization of a typically rectangular portion of a contour map conceptualization wherein contours may end at the edge of a map). Each contour conceptualization is treated as a set of holeless *contour regions* (each bounded by a contour) on a two-dimensional plane, related by *spatial containment* (one contour region being a subregion of another one). We are agnostic about *how* the contour representation is obtained in the first place: it may be derived from a primary data source such as a DEM or it may be the only available source, as is the case for many historic maps or maps sketched by humans. But even when the primary data are available for querying, a contour representation may provide a cognitively simpler, yet semantically rich model for human-computer interaction about the underlying field. We are at no point concerned with recovering the original field through approximations/interpolations from a contour conceptualization.

Table 1. Summary of the spatial concepts and relations from [7] and their information definitions that are used in the formalization of contour semantics.

Spatial predicate	Informal definition
$Cont(x,y)$	x is spatially contained in y (all points in x are also in y) independent of the dimensions of x and y
$PO(x,y)$	x and y spatially overlap, that is, their intersection is non-empty. $Cont(x,y)$ and $Cont(y,x)$ are specializations of $PO(x,y)$
$BCont(x,y)$	x is boundary-contained in y; that is, $Cont(x, \mathrm{bd}(y))$ holds
$TCont(x,y)$	x is spatially contained in y and part (either a point or a segment) of the boundary of x is contained in the boundary of y
$Con(x)$	x is a self-connected spatial region, that is, every part of x is connected to its complement
$Point(x)$	x is a point, i.e., an indivisible zero-dimensional spatial region
$Curve(x)$	x is a curve, i.e., a one-dimensional spatial region (which includes straight lines as well as straight and curved line segments)
$SimpleLoopCurve(x)$	x is a curve that represents a closed manifold, i.e., it is self-connected, not self-intersecting and has no branching points
$ArealRegion(x)$	x is a two-dimensional spatial region
$ClosedArealRegion(x)$	x is a two-dimensional spatial region that represents a closed manifold or a manifold with boundary bounded by a simple loop curve

1.3 The Underlying Multidimensional Qualitative Spatial Ontology

The presented formalization of contour semantics reuses the theory $CODIB$ from [7,8] that formalizes basic qualitative spatial relations between *abstract spatial regions* (subsequently we drop the implicit qualifier "abstract") of different dimensions, including points, linear features (curve and line segments and complex curve/line configurations), areal regions and voluminous regions. $CODIB$ is briefly reviewed here; Table 1 summarizes all relevant predicates and their intended interpretations with the full details available in [7]. It is formalized using the basic spatial relation of containment, $Cont(x,y)$, meaning that the spatial region x is spatially contained (a subregion of) in the spatial region y – independent of the dimensions of x and y (x can, of course, never have a greater dimension than y), and $BCont(x,y)$ meaning that x is contained in the boundary $\mathrm{bd}(y)$ of y and thus necessarily of a lower dimension than y.

Independent of their dimension, all spatial regions are topologically closed, that is, every point on their boundary is contained in the set. Additionally, a predicate $\leq_{\dim}(x,y)$ relatively compares the dimensions of two arbitrary spatial regions. The following kinds of spatial regions can be defined and distinguished based on the spatial relations they partake in and their relative dimensions: $Point(x)$, $Curve(x)$, and $ArealRegion(x)$. We particularly require *simple loop curves*, curves that form a loop, and *closed areal regions*, which are either closed manifolds (like a sphere) or regions bounded by a simple loop curve (manifolds with boundaries). Because all spatial regions are topologically closed, curves that have endpoints always include their endpoints (a loop curve or infinite curve without endpoints as well as rays with a single endpoint are still topologically closed) and areal regions always include their boundary curve(s) if one exists (a plane or the surface of a sphere without boundaries as well as half-planes with a partial boundary are still topologically closed).

The formalization in *CODIB* makes no assumption about the specific digital representations of spatial regions. For example, arbitrary curves may be represented as polylines, as sets of Bezier curves, or in other ways. The actual representation does not impact the semantic constraints that they must adhere to. This also holds true for the proposed formalization of contour semantics: it is not tied to whether the underlying, often implicit, field is encoded using a DEM, TIN, a simple raster representation, a vector representation, or any other suitable format.

2   The Basic Contour Concepts and Their Formalization

We are interested in the most general, yet essential semantics for representing a field as a contour map independent of what kind of measure the field represents. However, we will use elevation contours as found on topographic maps to motivate and illustrate the presented axioms and definitions. Nevertheless, we intend our ontology to apply equally to all kinds of contour maps with other kinds of measures such as bathymetric elevation, barometric pressure, temperature, or precipitation (rainfall/snow). The ontology also applies to interpolated measures such as population density, household income, or crime rate, which cannot be directly measured at individual points and whose contour lines are referred to as *isopleths* for that reason. The only constraint we impose on the underlying field is that its values form a *continuous surface*, ruling out, for example, the existence of overhanging rock faces that could yield multiple contours. The underlying field is not required to be *functional* (*single-valued*), that is, we allow "vertical cliffs" that yield more than a single elevation value at a point, though these must be continuous in the previous sense in order to clearly indicate an instantaneous rise or drop in value.

For our formalization, we will treat contour regions as abstract spatial regions that are distinct from physical entities, such as an actual mountain or a depression, and from other nonphysical entities, such as a conceptual or real map of population density, that it may represent. Contour regions are abstract spatial regions that describe the spatial extent (or parts thereof) of the represented mountain or depression as a (simplified) mathematical abstraction. Contour regions differ from arbitrary closed areal region (the subclass of abstract spatial regions they specialize) in that they must be associated to a contour value that describes a physical property of the represented physical object or field within that region. So in order to distinguish them from other closed areal regions, we introduce a new primitive notion. Analogously, we must distinguish contour lines from arbitrary simple loop curves, the kind of abstract spatial regions they specialize.

Multiple concepts would qualify as primitive notions, including the notions of a "contour line" (the lines usually drawn on a contour map which indicate the contour value at the points on the line), a "region of equal contour value" (the area between two "adjacent" contour lines), and of a "contour region" (the entire area enclosed by a contour line). We chose the last option as primitive for the following reasons. First, contour regions and lines are interdefinable, that is, contour lines are simply the boundaries of contour regions and contour regions are spatial regions bounded by contour lines. But by relying on contour lines as primitive makes it logically more difficult to define contour regions because contour lines separate two half-planes, whose formal distinction is nontrivial. Moreover, contour regions are spatially more well-behaved than contour lines in the sense that they have a natural spatial containment ("nesting") relation. For the latter reason, the contour regions are also preferable over the "regions of equal contour value", which are not spatially nested but spatially encircle one another without overlap or intersection. The spatial relationship among "regions of equal contour value" requires distinguishing the notion of "topologically inside" from "topologically outside". Moreover, the nesting of contour regions is robust when the contour interval is changed: a finer contour interval defines regions that are spatially contained in the regions obtained using a coarser contour interval. This is not the case for the "regions of equal contour value". Finally, contour regions are always holeless regardless of the resolution (the contour interval) of the contour map.

2.1 Contour Regions

Our new primitive spatial concept is that of a contour region, in addition to the purely spatial primitives introduced in [7,8]). A contour region is a special kind of holeless *ClosedArealRegion*, a topologically closed two-dimensional ("areal") region bounded by a single closed curve. We call the bounding curve of such a region a *contour*, formalized in more detail in Sect. 2.6. Intuitively, the contour that bounds a contour region is a curve that is both simple, i.e. without branching points, and a closed manifold, i.e. without endpoints (CR-A0)[3]. While in map drawings contours may end at the edge of the map, conceptually they continue beyond the visible portion of the contour map, forming a closed curve. Since the presented ontology is about the conceptualization of a contour map, not about its actual drawing, the encoding of a specific field would always result in a set of closed contours such that the outermost contours define the extent

---

[3] All presented axioms, definitions and theorems are first-order sentences which are implicitly universally quantified over any variables that are not explicitly quantified.

of the, possibly irregular-shaped and disconnected, conceptualized contour map. CR-A0 is included for completeness and to fully integrate the presented contour ontology with the spatial ontology $CODIB$ [7]. However, the contour ontology can be reused without importing $CODIB$ and all its dependencies, as long as all spatial predicates are interpreted as described informally in Table 1 and formally in [7].

(CR-A0) $CR(x) \rightarrow ClosedArealRegion(x) \wedge Con(x') \wedge SimpleLoopCurve(\mathrm{bd}(x))$
(all contour regions are closed areal regions without holes, that is, whose complement is self-connected, bounded by a simple curve with no end- or branch points).

2.2 Contour Values

Contour regions differ from other abstract spatial regions in that they are associated with observations or measures (for the purpose of this paper, we do not distinguish between those) of a specific field's quality, such as elevation, which could also be an interpolated, derived, estimated, or simulated quality. We call the actual measure of this quality the contour region's *contour boundary value*, denoted by the relation $ctrBdV(x, v_x)$. It expresses that v_x is the measure, which we call the *measured quantity*, in the underlying field at every point on the boundary of contour region x. $ctrBdV$ is a functional relation for contour regions, meaning that every contour region has exactly one contour boundary value associated with it (CV-A2, CV-A3). A contour inherits the contour boundary value from the contour region it bounds (CtrV-D) (Fig. 2).

We reuse some of the core ideas from the Observation and Measurement Ontology [20] in dealing with contour values. Specifically, contour values are special kinds of *measured quantities* (also called *observed quantities*), with each such contour value being associated with some *measured quality* and *measurement scale* (CV-A4). Ontologically, measured quantities and thus contour values are *qualities*, which are distinct from the categories of endurants and perdurants (compare [12]). Most importantly, CV-A6 captures the fundamental property that a contour's boundary value is equivalent to the measured quantity of all point measures along its bounding contour line, which in turn allows inferring that the measured quantities along the boundary of a single contour region are constant (CV-T1)[4].

(CV-A1) $CR(x) \rightarrow \exists v_x[ctrBdV(x, v_x)]$ (all contour regions have a contour boundary value)

---

[4] We assume that any two measured quantities x and y with $MQuantity(x)$ and $MQuantity(y)$ can be directly compared using standard (in)equality so that the result is not a mere comparison of their numeric values (denoted by mValue(x) and mValue(y)) but takes their associated units mUnit(x) and mUnit(y) into account. E.g., if $x = 1\,\mathrm{km}$ and $y = 100\,\mathrm{m}$, then $x > y$ is true. All comparisons of measured quantities, even between quantities in the same unit, require a common measured qualities (mQuality(x) = mQuality(y)), e.g., both are elevations.

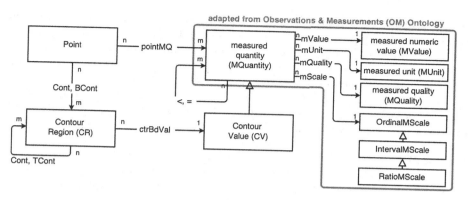

Fig. 2. The concepts of contour region and contour value in relationship to the concepts and relations concerning measures. Each contour region has a unique contour value associated to via the *ctrBdV* relation. Each contour value is a measured quantity or observation, which has a numerical value, a unit, a quality that is measures (such as elevation or annual precipitation), and an associated scale. Measures can also be taken at points, which are spatially related to the contour regions via the relations *Cont* and *BCont*.

(CV-A2) $ctrBdV(x, v_x) \rightarrow CR(x) \wedge CV(v_x)$ (*ctrBdV* is a relation between contour regions and contour values)

(CV-A3) $ctrBdV(x, v_1) \wedge ctrBdV(x, v_2) \rightarrow v_1 = v_2$ (each contour region has a unique contour boundary value, that is, *ctrBdV* is a function on the domain CR)

(CV-A4) $CV(x) \rightarrow MQuantity(x) \wedge \exists y[mScale(x, y) \wedge OrdinalMScale(y)]$ (each contour value is a measured quantity that has a ordinal scale associated to it)

(CV-A5) $pointMQ(p, v_p) \rightarrow Point(p) \wedge MQuantity(v_p)$ ($pointMQ(p, v_p)$ relates a measured quantity to the point it has been measured at)

(CV-A6) $ctrBdV(x, v_x) \wedge pointMQ(p, v_p) \wedge \text{mQuality}(v_x) = \text{mQuality}(v_p) \wedge BCont(p, x) \rightarrow v_p = v_x$ (point measures on a contour region's boundary that measure the same quality as the contour boundary value have the same quantity as the contour boundary value)

(CV-T1) $ctrBdV(x, v_x) \wedge pointMQ(p, v_p) \wedge pointMQ(q, v_q) \wedge BCont(p, x) \wedge BCont(q, x) \wedge \text{mQuality}(v_x) = \text{mQuality}(v_p) = \text{mQuality}(v_q) \rightarrow v_p = v_q$ (any two points on the boundary of a contour region that measure the same quality as the contour region must have the same measured value; follows from CV-A6)

Contours only make sense for measured qualities associated with a scale that has a linear order defined over all possible measured values. Otherwise the nesting of contour regions bears no useful information. For example, for any two elevation measures a and b either $a > b$, $a = b$ or $a < b$ is true. Thus, contour values must have at least an ordinal scale associated with it (CV-A4). The scale is either a discrete or a continuous ordered set of values, possibly bounded by some lowest and/or highest value (e.g. no point has a precipitation less than 0 liters per m^2 per year) or be unbounded. Our restriction to ordinal scales disallows representing fields of measured qualities that use nominal scales ("categorizations", see [23]) in our ontology, because they lack clear semantics of what it entails when two contour regions are spatially nested. The semantics of special sets of contour regions defined later in the paper relies on the scale's ordering of the associated contour values. More powerful scales, such as interval or ratio scales, specialize ordinal scale (CV-A7, CV-A8) and are thus permissible.

(CV-A7) $IntervalMScale(x) \rightarrow OrdinalMScale(x)$ (qualities measured using interval scales are a subclass of those using ordinal scales)

(CV-A8) $RatioMScale(x) \rightarrow IntervalMScale(x)$ (qualities measured using ratio scales are a subclass of those using interval scales)

Alternatively to aligning our formalization with the observation and measurement pattern from [20], it could also be aligned with the joint OGC-ISO standard on geographic observations and measurements [16]. The only drawback of such an alignment is that the OGC standard does not explicitly distinguish the different measurement scales on which our formalization relies. In other respects, the OGC standard offers additional modeling capabilities, such as more fine-grained classifications of observations, ways to represent dynamic (temporally changing) observations, and representation of sampling, that are unnecessary for modeling static contour maps.

2.3   Contour Sets

A *contour set* captures the idea of a *contour map* as a set of nested contour regions in a very flexible way. It can capture arbitrary subsets of the contour regions from a single source field or combine contours from multiple source fields of the same physical quality into one. Here, we formalize their basic semantics.

A generic contour set, denoted as $GenCS(x)$, is a nonempty, but potentially infinite, set of contour regions whose contour values are about the same measured quality (CS-A1). For example, a generic contour set may include various ascending and descending contour regions (properly defined further down) whose boundary values are all elevation measures. Such a set may not contain contour region about, e.g., precipitation measures. This ensures that comparisons between contour values within a contour set are always meaningful to humans. The relation $inCS(x, s)$ indicates that the contour region x is in the contour set s (CS-A2). Contour regions in the same contour set cannot properly overlap, that is, they are either in a containment relation or do not overlap at all (CS-A3). They can, however, be in tangential containment (*TCont*) by sharing a portion of their border as happens at a vertical cliff. The proper nesting defined by the containment relation *Cont* constructs the nice spatial structure between contour regions seen in Fig. 1.

(CS-A1) $inCS(x, s) \rightarrow CR(x) \wedge GenCS(s)$ ($inCS$ is a relation between a contour set and a contour region)

(CS-A2) $inCS(x, s) \wedge inCS(y, s) \wedge ctrBdV(x, v_x) \wedge ctrBdV(y, v_y) \rightarrow$ mQuality(v_x) = mQuality(v_y) (the boundary values of two contour regions x and y in the same contour set s have the same measured quantities)

(CS-A3) $inCS(x, s) \wedge inCS(y, s) \wedge PO(x, y) \rightarrow Cont(x, y) \vee Cont(y, x)$ (Any two overlapping contour regions in the same set are in a containment relation)

(CS-A4) $inCS(x, s) \wedge inCS(y, s) \wedge Cont(x, y) \wedge Cont(y, x) \rightarrow x = y$ (contour regions in the same contour set that contain each other are identical, that is, distinct contour regions in the same set have distinct spatial extents)

(CS-A5) $GenCS(s) \rightarrow \exists x[inCS(x, s)]$ (every contour set contains a contour region)

(CS-A6) $GenCS(s) \wedge GenCS(t) \wedge s \neq t \rightarrow \exists x[CR(x) \wedge (inCS(x, s) \vee inCS(x, t)) \wedge (\neg inCS(x, s) \vee \neg inCS(x, t))]$ (two distinct $GenCS$ differ in at least one CR)

(CSS-D) $CSubSet(s, t) \equiv GenCS(s) \wedge GenCS(t) \wedge \forall x[inCS(x, s) \rightarrow inCS(x, t)]$ (a contour set s is a subset of contour set t iff every contour region is s is also in t)

(CSArea-D) $GenCS(s) \rightarrow csArea(s) = \Sigma_{(x|inCS(x,s))} x$ (the two-dimensional footprint of any generic contour set s as the sum of its contained contour regions)

In Sect. 3 we will look at more specialized classes of contour sets that have a fixed (constant) interval or a single contour region that contains all others and which more accurately describe contours typically used in topographic maps.

2.4 Parent-Child Relations Between Contour Regions in a Contour Set

Parent-child relations between contour regions are central to the idea of contour maps. We specify them using the ternary relation $ChildCR(x, y, s)$, which expresses that among all contour regions in s, x (the *child*) is spatially contained in y (the *parent*) and no other contour region in s contains x but not y. In other words, x is the next immediate contour region nested in y^5. For example, in Fig. 3, $a2$ is a child of $a1$ and $b3$ a child of $b2$. A parent may have multiple children, for example all of $a1$, $b1$ and $c1$ are children of p, but a child has exactly one parent (CR-T2). A special case of the parent-child relation occurs when the child x is *tangentially contained* in y ($TCont(x, y)$), for example at a cliff. This requires special handling in line-based contour formalisms [6,13], but poses no problem for our region-based formalism. In fact, cliffs are thus easily definable in terms of the $TCont$ relation between parents and their children.

---

[5] Our parent-child relations are based on spatial containment among regions and are similar to the parent-child relation in the *enclosure trees* from [1]. The resulting structure is closely related to the graphs known as *contour trees* [11] that essentially uses a dual version of our representation by representing regions as arcs and contours as nodes.

The children of a common parent are called *siblings* of one another, denoted by $SiblingCR(x, y, s)$[6]. In Fig. 1, c and d are siblings, as are g and h. More than two regions can be siblings of one another. Note that parent and child and siblings do not need to be of the same type (ascending or descending), as demonstrated by the siblings $b1$ and $c1$ in Fig. 3. Topographically, we know that the parent region of two or more siblings contains a saddle, called a "pass" [6] when the siblings are both ascending and a "bar" when they are both descending [6].

(ChildCR-D) $ChildCR(x, y, s) \equiv inCS(x, s) \land inCS(y, s) \land Cont(x, y) \land \forall z[inCS(z, s) \land x \neq z \land Cont(x, z) \rightarrow Cont(y, z)]$ (x is a child contour region of y in contour set s iff y spatially contains x and all other contour regions in s that spatially contain x also spatially contain y)

(SiblingCR-D) $SiblingCR(x, y, s) \equiv inCS(x, s) \land inCS(y, s) \land x \neq y \land \exists z[ChildCR(x, z, s) \land ChildCR(y, z, s)]$ (contour regions x and y in s are siblings iff they share a parent z in s)

(CR-T1) $ChildCR(x, y, s) \rightarrow \neg ChildCR(y, x, s)$ (the parent-child relation is asymmetric)

(CR-T2) $ChildCR(x, y, s) \land ChildCR(x, z, s) \rightarrow y = z$ (each child can have only a single parent)

(CR-T3) $SiblingCR(x, y, s) \rightarrow SiblingCR(y, x, s)$ (the sibling relation is symmetric in the first two parameters)

(CR-T4) $SiblingCR(x, y, s) \rightarrow \neg PO(x, y)$ (sibling contour regions cannot spatially overlap; follows from CS-A4, ChildCR-D, and SiblingCR-D).

2.5   Ascending and Descending Contour Regions

Contour regions come in two variants: either the contour boundary value (and thus the enclosing contour line) denotes the *minimal* or the *maximal contour value* for the bounded contour region. This means either all points inside the contour region and not inside any nested contour region have a measured quantity (of the same measured quality as the contour boundary value) that is *above* (or equal) or *below* (or equal) to the contour boundary value. As extreme cases, the measured quality could be constant and equivalent to the contour boundary value at all points in the contour region.

By convention, contour lines in topographic maps that denote a minimal contour value (indicating a hill) are displayed as regular (non-hachured) lines, while those that denote a maximal contour value (indicating a depression) have dashes (hachures) pointing inwards as shown for $b3$–$b5$ and $c1$–$c3$ in Fig. 3[7]. Two disjoint and jointly exhaustive subclasses of $CR(x)$ are introduced to formalize this distinction: ascending contour regions, $AscCR(x)$, denoting a contour region

---

[6] In order to capture the contour set that forms the context for the parent-child and sibling relations, we chose to model them as ternary predicates. In the derived conceptual model in Fig. 5, the parent-child and sibling relations are expressed using a new helper class each, together with new relations between the helper classes and the parents/children/siblings.

[7] Other conventions about label direction and positioning are also commonly used.

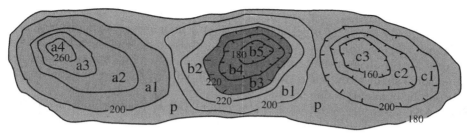

Fig. 3. Examples of a contour set that consists of 13 contour regions: the ascending contour regions p, $a1$–$a4$, $b1$, and $b2$ and the descending contour regions (with hachured borders, the hachures pointing in the direction of the descent) $b3$–$b5$ and $c1$–$c3$. Some contour lines are labeled with their boundary values while others are implied because this map is an example of a fixed interval contour set. For example, $b1$ has a *ctrBdV* of 200 ft, meaning that any point on the boundary of or inside $b1$, but not inside $b2$ is guaranteed to have an elevation of at least 200 ft and less than 220 ft. Equally, points inside $b3$ but not inside $b4$ have elevation values not greater than 220 ft but greater than 200 ft (the implied contour boundary value of $b4$). This makes clear that not all points in $b1$ have an elevation above 200 ft: the points inside $b4$ are in fact all below 200 ft (this happens, e.g., in a volcanic crater).

This 13-element contour set forms a single parent contour set (*SPCS*) with p being the root contour region that spatially contains all others. This contour set has many subsets, of which the sets $a1$–$a4$, $b1$–$b5$, and $c1$–$c3$ are all totally ordered contour sets (*TOCS*). Among them, $a1$–$a4$ is an ascending set (*AscTOCS*) and $c1$–$c3$ a descending set (*DescTOCS*), while $b1$–$b5$ is neither. However, $b3$–$b5$ is again descending (*DescTOCS*).

with an ascent towards the inside (CR-A1), and descending contour regions, $DescCR(x)$, denoting a contour region with a descent towards the inside (CR-A2). Note that we can only make claims above the points inside the contour region that are not also within any smaller nested contour region because the next contained contour region may be of opposite nature. For example, all points in region $b2 - b3$ are above 220 feet in elevation, but some points in $b2$, namely those in $b3$, are below 220 feet in elevation. CR-A1 and CR-A2 merely impose necessary conditions for contour regions being called ascending or descending, they do not suffice to infer that $AscCR$ and $DescCR$ are exhaustive and disjoint subclasses of CR. Rather, this is imposed by our knowledge about contour semantics, which we must explicitly postulate (CR-A3). Furthermore, because $AscCR$ and $DescCR$ theoretically both permit that all contained points have measured quantities equivalent to their boundary values, we also have to explicitly postulate their disjointness (CR-A4).

(CR-A1) $AscCR(x) \rightarrow \forall s, v_x, p, v_p [inCS(x,s) \wedge ctrBdV(x,v_x) \wedge Point(p) \wedge$
    $Cont(p,x) \wedge pointMQ(p,v_p) \wedge \mathrm{mQuality}(v_p) = \mathrm{mQuality}(v_x) \wedge \forall y [inCS(y,s) \wedge$
    $Cont(p,y) \rightarrow Cont(x,y)] \rightarrow v_p \geq v_x]$ (an ascending contour region x is a
    contour region such that every point that is contained in x but not contained
    in some smaller contour region has a measured quantity greater or equal to
    the contour boundary value of x)

(CR-A2) $DescCR(x) \rightarrow \forall s, v_x, p, v_p [inCS(x, s) \wedge ctrBdV(x, v_x) \wedge Point(p) \wedge$
$Cont(p, x) \wedge pointMQ(p, v_p) \wedge \text{mQuality}(v_p) = \text{mQuality}(v_x) \wedge \forall y[inCS(y, s) \wedge$
$Cont(p, y) \rightarrow Cont(x, y)] \rightarrow v_p \leq v_x]$ (a descending contour region x is a
contour region such that every point that is contained in x but not contained
in some smaller contour region has a measured quantity less or equal to the
contour boundary value of x)

(CR-A3) $CR(x) \rightarrow AscCR(x) \vee DescCR(x)$ (all contour regions are either
ascending or descending)

(CR-A4) $\neg AscCR(x) \vee \neg DescCR(x)$ (ascending and descending contour regions
are disjoint)

The axioms CR-A5–CR-A8 formalize the relationships between the contour
values of a child and a parent in all four combinations of ascending and descending contour regions. This restricts the contour values of the involved contour
regions, but not the measured quantities at points inside the contour regions.
While contour regions of the same type must have strictly increasing/decreasing
contour boundary values (CR-A5, CR-A6), this is not necessarily true for contour regions of opposite types. For example, in Fig. 3 the ascending region $b2$
and its descending child $b3$ have the same contour value. This is permissible
because the points in between can vary in between 220 ft and 240 ft (assuming a
fixed interval of 20 ft), while the descending contour value of 220 ft indicates that
further inside all point measures dip below 220 ft. This is reflected in CR-A7 (for
a descending child in an ascending parent) and CR-A8 (for an ascending child
in a descending parent). Moreover, any two siblings of the same kind must have
identical contour values (CR-A9). By CR-A5 and CR-A6, a child with the same
$ctrBdV$ as its parent must be of the opposite type as the parent (CR-T5).

(CR-A5) $AscCR(x) \wedge AscCR(y) \wedge ChildCR(x, y, s) \wedge ctrBdV(x, v_x) \wedge$
$ctrBdV(y, v_y) \rightarrow v_x > v_y$ (an ascending contour region x that is a child of
another ascending contour region y in s has a higher contour value than y)

(CR-A6) $DescCR(x) \wedge DescCR(y) \wedge ChildCR(x, y, s) \wedge ctrBdV(x, v_x) \wedge$
$ctrBdV(y, v_y) \rightarrow v_x < v_y$ (a descending contour region x that is a child of
another descending contour region y in s has a higher contour value than y)

(CR-A7) $DescCR(x) \wedge AscCR(y) \wedge ChildCR(x, y, s) \wedge ctrBdV(x, v_x) \wedge$
$ctrBdV(y, v_y) \rightarrow v_x \geq v_y$ (a descending contour region x that is a child of a
ascending contour region y in s has a higher or equal contour value as y)

(CR-A8) $AscCR(x) \wedge DescCR(y) \wedge ChildCR(x, y, s) \wedge ctrBdV(x, v_x) \wedge$
$ctrBdV(y, v_y) \rightarrow v_x \leq v_y$ (an ascending contour region x that is a child of a
descending contour region y in s has a lower or equal contour value as y)

(CR-A9) $SiblingCR(x, y, s) \wedge ctrBdV(x, v_x) \wedge ctrBdV(y, v_y) \wedge [[AscCR(x) \wedge$
$AscCR(y)] \vee [DescCR(x) \wedge DescCR(y)]] \rightarrow v_x = v_y$ (sibling contour regions
of the same kind must have identical contour values)

(CR-T5) $ChildCR(x, y, s) \wedge ctrBdV(x, v_x) \wedge ctrBdV(y, v_y) \wedge v_x = v_y \rightarrow$
$[AscCR(x) \wedge DescCR(y)] \vee [DescCR(x) \wedge AscCR(y)]$ (a child contour region
with the same measured quantity as its parent is of the opposite type as its
parent)

While CV-A6 already postulated that each point on the boundary of a contour region must have a point measure equivalent to the contour boundary value, we can now be more precise: any childless ascending (descending) contour region contains only measured point values higher (lower) than the contour boundary value (CV-A9). For nested contour regions of the same type, point measure inside a parent region but not inside any of its children have a measured value in between the contour boundary values of the parent and the children (CV-A10). Use Fig. 3 as example: all points in p that are not in either of $a1$, $b1$, or $c1$ have a value between 180 ft and 200 ft. This would be the case even if $c1$ had a boundary value of 180 ft.

(CV-A9) $inCS(x, s) \land \forall y[\neg ChildCR(y, x, s)] \land ctrBdV(x, v_x) \land Point(p) \land$ $Cont(p, x) \land pointMQ(p, v_p) \land mQuality(v_x) = mQuality(v_p)$ $\rightarrow [[AscCR(x) \rightarrow v_p \geq v_x] \land [DescCR(x) \rightarrow v_p \leq v_x]]$ (a childless ascending (descending) CR x contains only points whose values of the same measured quality are higher (lower) or equal to x's contour boundary value)

(CV-A10) $ChildCR(x, y, s) \land ctrBdV(x, v_x) \land ctrBdV(y, v_y) \land Point(p) \land$ $Cont(p, y) \land pointMQ(p, v_p) \land \forall z[ChildCR(z, y, s) \rightarrow \neg Cont(p, z)] \land$ $mQuality(v_p) = mQuality(v_z) \rightarrow [[AscCR(x) \land AscCR(y) \rightarrow v_y \leq v_p < v_x]$ $\land [DescCR(x) \land DescCR(y) \rightarrow v_x < v_p \leq v_y]]$ (any point measure v_p of the same quality as the contour boundary value v_y that is within the ascending (descending) CR y and not within any of it children z must have value in between v_y and the boundary contour value v_x of all its ascending (descending) child CR).

2.6   Contour Lines

Next we define a contour line as the boundary of a contour region (CL-D) with its contour value being the contour boundary value of the contour region it encloses (CtrV-D). CtrV-D is included for cognitive clarity only and is not needed later on. Just as we distinguish ascending and descending contour regions, we distinguish the contour lines that bound them. Alluding to geographic intuition, we call the boundary of an ascending contour region a *hill contour* (HillC-D) and the boundary of a descending contour region a *depression contour* (DeprC-D). Because ascending and descending contour regions are disjoint and exhaustive subclasses of contour regions, hill and depression contours are also disjoint and exhaustive (CL-T1, CL-T2).

(CL-D) $Contour(x) \equiv \exists y[CR(y) \land x = bd(y)]$ (a contour line is the boundary of a contour region)

(CtrV-D) $ctrV(bd(x)) = v \equiv ctrBdV(x, v)$ (the contour value of a contour line is the contour boundary value of the contour region it encloses)

(HillC-D) $HillContour(x) \equiv \exists y[AscCR(y) \land x = bd(y)]$ (a hill contour is the boundary of an ascending contour region)

(DeprC-D) $DepressionContour(x) \equiv \exists y[DescCR(y) \land x = bd(y)]$ (a depression contour is the boundary of a descending contour region)

(CL-T1) $Contour(x) \rightarrow HillContour(x) \vee DepressionContour(x)$ (every contour is either a hill or a depression contour)

(CL-T2) $\neg HillContour(x) \vee \neg DepressionContour(x)$ (hill and depression contours are disjoint)

We can now also formally define *adjacency* between contours within a contour set. Contours that bound regions that are in a parent-child or in a sibling relation are adjacent, as are contours of two parent-less contour regions – a case that is only possible in contour sets without a single parent. This definition captures the fundamental adjacency relation between contour lines used in the contour data structures developed in [6].

(AdjCL-D) $AdjacentContours(m, n) \equiv \exists s, x, y \Big[m = \mathrm{bd}(x) \wedge n = \mathrm{bd}(y) \wedge inCS$
$(x, s) \wedge inCS(y, s) \wedge \big[ChildCR(x, y, s) \vee SiblingCR(x, y, s) \vee \forall z [inCS(z, s) \rightarrow$
$(x = z \vee \neg Cont(x, z)) \wedge (y = z \vee \neg Cont(y, z))] \big] \Big]$ (two contour lines are adjacent iff they bound contour regions x and y in a common contour set s such that they are either in a parent-child relation, in a sibling relation, or any other contour region in s properly contains neither x nor y).

3   More Specialized Contour Sets

So far, the class *GenCS* denotes arbitrary sets of contour regions whose contour values are about a common measured quality without further restrictions on acceptable contour values, their spatial configuration, or the presence/absence of ascending or descending contour regions. Next we will refine this generic class to more interesting and more narrowly defined subclass as shown in the hierarchy of contour sets in Fig. 4.

3.1   Fixed (Constant) Interval Contour Sets (*FICS*)

Most contour maps that display terrain information have a constant interval between adjacent contours. For example, the National Map commonly uses 20 feet contour intervals, but also intervals of 5, 10, 40, or 50 feet. That means all contour regions within a single contour set have only contour boundary values that are a multiple of 20 (or 5, 10, 40 or 50) feet. We capture this idea in the class of *fixed interval contour sets*, FICS. Each fixed interval contour set has a unique contour interval defined (FICS-D, FICS-A1) as indicated by the relation $ctrInterval(s, i)$ between a contour set s and its contour interval i. Conversely, the existence of a contour interval suffices to ensure that a contour set s is a fixed interval contour set – a property that helps express subsequent axioms and theorems more succinctly.

FICS-A2 captures the essential condition of what it means for a contour set to have a fixed interval: the difference between any two contour boundary values in the set is equivalent to $baseValue + k \cdot contourInterval$ for some integer k and

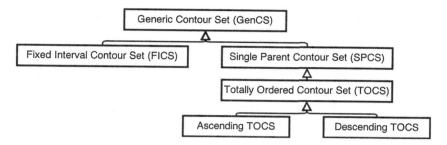

Fig. 4. The hierarchy of classes of contour sets. The class *FICS* uses the existence of a constant interval between contours as distinguishing criterion, which is independent of the criteria used for distinguishing the classes *SPCS*, *TOCS*, *AscTOCS*, and *DescTOCS* from *GenCS*. Thus, the class *FICS* is not necessarily disjoint with any of the other classes of contour sets.

constant values *baseValue* and *contourInterval*. This applies to sets of contours that are all multiples of a contour interval, but also accommodates the more general case of sets of contours with constant intervals yet unusual base values (e.g. contours at 5, 25, 45, and 65 feet). Within our formalization there is no need to fix the base value – in fact, it is never explicitly mentioned. Finally, fixed interval contour sets require that the measured property uses at least an interval scale (FICS-A3), otherwise the contour interval lacks clear semantics.

(FICS-D) $FICS(s) \rightarrow GenCS(s) \land \exists i[ctrInterval(s,i)]$ (each fixed interval contour set is a contour set that has a contour interval)

(FICS-A1) $ctrInterval(s,i) \land ctrInterval(s,j) \rightarrow i = j$ (each contour set has at most one contour interval; it is a functional attribute)

(FICS-A2) $ctrInterval(s,i) \land inCS(x,s) \land inCS(y,s) \land ctrBdV(x,v_x) \land ctrBdV(y,v_y) \rightarrow \exists k[Integer(k) \land v_x = v_y + k \cdot i]$ (the difference between the contour boundary values of any two contour regions in a fixed interval contour set is a multiple of the set's contour interval)

(FICS-A3) $ctrInterval(s,i) \rightarrow FICS(s) \land MQuantity(i) \land \exists x[mScale(i,x) \land IntervalMScale(x)]$ ($ctrInterval(s,i)$ denotes that the contour set s has a fixed measured quantity i as its contour interval with an interval scale associated to it)

(FICS-A4) $ctrInterval(s,i) \land ChildCR(x,y,s) \land ctrBdV(x,v_x) \land ctrBdV(y,v_y) \rightarrow (x - y = i \lor y - x = i \lor x - y = 0)$ (the absolute difference between a parent and a child contour region in a fixed interval contour set is either the contour interval or 0)

Now the additional properties FICS-T1–FICS-T5 about the contour boundary values of parents, children, and siblings in a fixed interval contour set become provable.

(FICS-T1) $ctrInterval(s,i) \land ChildCR(x,y,s) \land ctrBdV(x,v_x) \land ctrBdV(y,v_y) \land AscCR(x) \land AscCR(y) \rightarrow v_x - v_y = i$ (in a fixed interval contour set, the difference between the boundary values of an ascending child contour region and its ascending parent is the contour interval)

(FICS-T2) $ctrInterval(s, i) \wedge ChildCR(x, y, s) \wedge ctrBdV(x, v_x) \wedge ctrBdV(y, v_y) \wedge DescCR(x) \wedge DescCR(y) \rightarrow v_y - v_x = i$ (in a fixed interval contour set, the difference between the boundary values of a descending parent contour region and its descending child is the contour interval)

(FICS-T3) $FICS(s) \wedge SiblingCR(x, y, s) \wedge ctrBdV(x, v_x) \wedge ctrBdV(y, v_y) \wedge AscCR(x) \wedge AscCR(y) \rightarrow v_x = v_y$ (ascending sibling contour regions in a fixed interval contour set have equivalent contour boundary values)

(FICS-T4) $FICS(s) \wedge SiblingCR(x, y, s) \wedge ctrBdV(x, v_x) \wedge ctrBdV(y, v_y) \wedge DescCR(x) \wedge DescCR(y) \wedge \rightarrow v_x = v_y$ (descending sibling contour regions in a fixed interval contour have equivalent contour boundary values)

(FICS-T5) $ctrInterval(s, i) \wedge inCS(x, s) \wedge inCS(y, s) \wedge ctrBdV(x, v_x) \wedge ctrBdV(y, v_y) \wedge Integer(k) \wedge v_x < ki < v_y \rightarrow \exists z[\wedge inCS(z, s) \wedge ctrBdV(z, ki)]$ (for any multiple k of a contour set's contour interval i, if ki is between the boundary contour values v_x and v_y of contour regions x and y in s, then some contour region z in s has ki as contour boundary value)

Mathematical Characterization of *GenCS and FICS*. The contour regions within a generic contour set and their spatial containment relation exhibit mathematically well-defined structures in all models of the ontology of generic contour maps **GCtrMap** $= \{$CR-A1–CR-A9, CV-A1–CV-A10, CS-A1–CS-A6, CRChild-D, CRSibling-D$\}^8$. Each model forms a forest – a set of trees – as formalized by the following theorem. The superscript notation $P^{\mathcal{M}}$ denotes the interpretation of predicate P under the model $\mathcal{M}$.

Theorem 1. *Let $\mathcal{M}$ be an arbitrary model of the ontology **GCtrMap**. For each $\mathcal{S} \in GenCS^{\mathcal{M}}$, let $(x, y) \in Cont \Leftrightarrow (x, \mathcal{S}), (y, \mathcal{S}) \in inCS^{\mathcal{M}}$ and $(x, y) \in Cont^{\mathcal{M}}$. Then $(\mathcal{S}, Cont)$ is a forest.*

Proof (Sketch). Recall that any two contour regions in a common contour set are either in a spatial containment relation or do not overlap at all (CS-A3). If two contour regions are in a spatial containment relation, they form a parent-child relation in the tree. By CR-T2, each child has at most one parent and *Cont* is antisymmetric (see [7]), hence no cycles exist. Thus, each parent-less contour region in $\mathcal{S}$ forms a tree. Because $(\mathcal{S}, Cont)$ can contain multiple parent-less contour regions, it is a forest. □

A consequence of this theorem is that the underlying field of a contour set is covered by the spatial extent of all its parent-less (root) contour regions. Each spatial location of the underlying field is thus in exactly one of the root contour regions.

This theorem extends to the structures $(\mathcal{S}, Cont)$ defined over instances of *FICS*, which are also forests. The only difference is that in fixed interval contour

---

[8] For brevity, this ontology excludes all definitions that are unnecessary for the characterization.

sets the interval between the contour boundary values of a parent and its children in the tree is constrained. In the next section, we will look at other types of contour sets that further constrain the underlying mathematical structure over the spatial containment relation.

3.2   Totally Ordered Contour Sets ($TOCS$)

Generic and fixed interval contour sets do not have a single *root contour region* as just shown by the mathematical characterization. Now we specifically define a class of contour sets – the *single parent contour sets* denoted as $SPCS(s)$ – that have a unique contour region that spatially contains all other contour regions in the set (SPCS-D). This region serves as the root of the contour set. Note that the two subclasses of generic contour sets, the single parent contour sets $SPCS$ and the fixed interval contour sets $FICS$, are not necessarily disjoint and neither is a subclass of the other.

As further specializations of single parent contour sets, we call a single parent contour set s a *totally ordered contour set*, $TOCS(s)$, if, and only if, no contour region therein has a sibling. We distinguish two subclasses of totally ordered contour sets: *ascending totally ordered contour sets*, $AscTOCS(s)$, and *descending totally ordered contour sets*, $DescTOCS(s)$. These two classes capture specific geographic phenomenon – hills and depressions – of particular interest to elevation contours as displayed on topographic maps. For example, in Fig. 3 the contour set consisting of $a1$–$a4$ is an ascending totally ordered contour set that describes a hill, while the two contour sets consisting of $b3$–$b5$ and $c1$–$c3$, respectively, form descending totally ordered contour sets that describe depressions (mine holes in this case).

(SPCS-D) $SPCS(s) \equiv GenCS(s) \wedge \exists x[inCS(x, s) \wedge \forall y[inCS(y, s) \rightarrow Cont(y, x)]]$
    (A single parent contour set has a unique largest contour region that spatially contains all other contour regions in the set)

(TOCS-D) $TOCS(s) \equiv SPCS(s) \wedge \neg\exists x, y[SiblingCR(x, y, s)]$ (a totally ordered contour set is a $SPCS$ that does not contain any siblings)

(AscTOCS-D) $AscTOCS(s) \equiv TOCS(s) \wedge \forall x[inCS(x, s) \rightarrow AscCR(x)]$ (an ascending $TOCS$ is a $TOCS$ that consists only of ascending contour regions)

(DescTOCS-D) $DescTOCS(s) \equiv TOCS(s) \wedge \forall x[inCS(x, s) \rightarrow DescCR(x)]$ (an descending $TOCS$ is a $TOCS$ that consists only of descending contour regions)

(TOCS-T1) $TOCS(s) \wedge ChildCR(x, z, s) \wedge ChildCR(y, z, s) \rightarrow x = y$ (each contour region z in a $TOCS$ contains no more than one child)

(TOCS-T2) $AscTOCS(s) \wedge ChildCR(x, y, s) \wedge ctrBdV(x, v_x) \wedge ctrBdV(y, v_y) \rightarrow v_x > v_y$ (boundary values in an ascending $TOCS$ increase towards the inside)

(TOCS-T3) $DescTOCS(s) \wedge ChildCR(x, y, s) \wedge ctrBdV(x, v_x) \wedge ctrBdV(y, v_y) \rightarrow v_x < v_y$ (boundary values in a descending $TOCS$ decrease towards the inside)

Mathematical Characterization of $TOCS$, $AscTOCS$, and $DescTOCS$. Now, we can extend Theorem 1 to characterize the substructures $(\mathcal{S}, Cont)$ formed by contour sets that are (ascending/descending) totally ordered contour sets in models of the ontology **TOCtrMap** = **GCtrMap** $\cup$ {SPCS-D, TOCS-D, AscTOCS-D, DescTocs-D}. This new ontology **TOCtrMap** is a definitional extension of **GCtrMap**.

Theorem 2. *Let $\mathcal{M}$ be an arbitrary model of the ontology **TOCtrMap**. For each $\mathcal{S} \in GenCS^{\mathcal{M}}$, let $(x, y) \in Cont \Leftrightarrow (x, \mathcal{S}), (y, \mathcal{S}) \in inCS^{\mathcal{M}}$ and $(x, y) \in Cont^{\mathcal{M}}$. If $\mathcal{S} \in SPCS^{\mathcal{M}}$, then $(\mathcal{S}, Cont)$ is a tree. If $\mathcal{S} \in TOCS^{\mathcal{M}}$, then $(\mathcal{S}, Cont)$ is a chain.*

Proof (Sketch). Each so-defined structure $(\mathcal{S}, Cont)$ is a forest by Theorem 1. Now suppose $\mathcal{S} \in SPCS^{\mathcal{M}}$. Then by SPCS-D, $\mathcal{S}$ has a unique root and thus is single-rooted and, thereby, connected. But every connected forest is a tree. Now suppose $\mathcal{S} \in TOCS^{\mathcal{M}}$. Then by TOCS-D each parent has at most one child. Because each contour region has at most one parent (CR-T2), it follows that spatial containment totally orders the tree's contour regions such that the tree is a chain. □

This characterization of every instance of $SPCS$ forming a tree over the spatial containment relation matches the description of each contour line forming an *enclosure tree* over its nested contours lines given by [1].

4   Definable Areas of Physical-Geographical Significance

For many purposes, spatial regions that are not contour regions themselves, but that can be defined in terms of them, are of practical interest. These include *regions of equal contour value*, each of which is a contour region minus the region of all its children. For example, in Fig. 3, the area of p minus the areas of $a1$, $b1$, and $c3$ forms a region of equal contour value with elevations between 180 and 200 feet. Within that area, we cannot distinguish the elevation measure between any two points – it hits the resolution threshold inherent in the chosen contour interval. Such areas are possibly holed and/or disconnected[9]. We define this area using the sum and difference operations on regions (CA$_=$-D). By CV-A10, all points in such an ascending (descending) region x have measured quantities between x's contour boundary value and the minimum (maximum) contour boundary value of all its ascending (descending) children.

The definition of $cArea_=(x, s)$ for arbitrary regions enables us to define the entire, possibly disconnected subarea of a contour region that is above (below) the contour boundary value of x. To define such area, all contained contour areas with boundary values above (below) that of x are summed up (CA$_>$-D, CA$_<$-D). To make these definitions more palatable, we first define a subset of

---

[9] The region of equal contour value is only disconnected when separated by two or more cliffs, which are points/segments of its containing contour region where it shares a portion of its boundary with one or multiple child contour regions.

all contour regions in s that are contained in x and have at least (at most) the contour boundary value of x (GCS$_>$-D and GCS$_<$-D). It immediately follows that within an ascending (descending) totally ordered contour set, the spatial extent of any contour region is equivalent to its contour area $cArea_>$ ($cArea_<$) (CA$_>$-T1, CA$_<$-T1).

(CA$_=$-D) $inCS(x,s) \rightarrow cArea_=(x,s) = x - \Sigma_{(y|ChildCR(y,x,s))}y$ (the *contour area of equal value*, $cArea_=(x,s)$, of contour region x in s is the area of x minus the area of all its children)

(GCS$_>$-D) $GenCS_>(t,s,x) \equiv \forall y[inCS(y,t) \leftrightarrow inCS(y,s) \wedge Cont(y,x) \wedge ctrBdV(x,v_x) \wedge ctrBdV(y,v_y) \wedge v_y \geq v_x]$ (t is the subset of s's contour regions that are spatially contained in x and have a boundary value no smaller than x)

(CA$_>$-D) $cArea_>(x,s) = \Sigma_{(y|inCS(y,t) \wedge GenCS_>(t,s,x))} cArea_=(y,s)$ (the area within x with measured values above the contour boundary value of x)

(GCS$_<$-D) $GenCS_<(t,s,x) \equiv \forall y[inCS(y,t) \leftrightarrow inCS(y,s) \wedge Cont(y,x) \wedge ctrBdV(x,v_x) \wedge ctrBdV(y,v_y) \wedge v_y \leq v_x]$ (t is the subset of s's contour regions that are spatially contained in x and have a boundary value no bigger than x)

(CA$_<$-D) $cArea_<(x,s) = \Sigma_{(y|inCS(y,t) \wedge GenCS_<(t,s,x))} cArea_=(y,s)$ (the area within x with measured values above the contour boundary value of x)

(CA$_>$-T1) $AscTOCS(s) \wedge inCS(x,s) \rightarrow cArea_>(x,s) = x$ (each contour region in a ascending $TOCS$ has itself as ascending contour area)

(CA$_<$-T1) $DescTOCS(s) \wedge inCS(x,s) \rightarrow cArea_<(x,s) = x$ (each contour region in a descending $TOCS$ has itself as descending contour area)

The areas $cArea_>(x,s)$ and $cArea_<(x,y)$ can be visualized as the areas that are included (or excluded) when the surface defined by the underlying field is intersected with a horizontal plane at the elevation of x's boundary. These areas are geographically significant in many applications. For example, on elevation contour maps the area $cArea_<(x,s)$ indicates the portion of the landscape that is submerged in a lake with a certain water level as indicated by x as the lake boundary. Equally, a certain elevation threshold may be used to identify areas prone to flooding because of their elevation below a certain threshold expressed as a contour region with suitable contour boundary value. The areas above the threshold that are completely surrounded by lower-lying areas will become "islands" in the case of flooding. The areas $cArea_>(x,s)$ and $cArea_<(x,y)$ can also be used to identify vegetation or habitat zones defined by a certain elevation threshold.

5 Discussions and Conclusions

In this paper, we formally analyzed the inherent, yet implicit semantics of contour maps and axiomatized it as a rigorous first-order ontology amendable to automated reasoning as necessary for achieving computational spatial intelligence. While humans adhere to the implicit semantics buried in contour maps without difficulties, the explicit axiomatization is a necessary first step towards employing contour information directly for fully automated reasoning by general purpose reasoners (such as theorem provers), which can answer broader ranges of queries as compared to the highly customized spatial algorithms that are currently used in geographic information systems.

The resulting ontology is largely based on the purely qualitative representation of space from [7] that relies exclusively on mereotopological (of contact and parthood) relations among abstract spatial regions of different dimensions such as points, curves (here specifically simple loop curves), areal regions, and associated spatial relations defined in terms of spatial containment (the subregion relation). The ontology adapts the ontology pattern from [20] about measured quantities to specify how contours and contour regions within a contour map are associated with a shared measured quality, such as elevation, and how the values on contour lines are related to point measures inside the regions they bound. Besides the imported spatial concepts and relations, the formalization requires very few additional primitive concepts and relations: contour regions CR, its subclasses $AscCR$ and $DescCR$ and its associated relation $ctrBdV$ as well as primitives to specify sets of contour regions (contour sets $GenCS$ and the related membership relation $inCS$).

Foundation for Quantitatively Enhanced Qualitative Spatial Reasoning. Purely qualitative representations of space [2,3,7,9] can answer only the simplest common sense questions about space. In many cases, such representations must be supplemented by simple quantitative reasoning, as exemplified by the presented formalization, which through the use of simple quantitative operations greatly increases the utility of a predominantly qualitative representation of space. In this spirit, the presented formalization lays the foundation for automated *contour-based spatial reasoning* by not only providing a set of axioms that explicitly captures the semantics of contour maps, but also through the identification of the quantitative operations necessary to support reasoning about them. This contour-based spatial reasoning takes conventional qualitative, region-based reasoning formalisms such as the RCC [19] or *CODIB* [7,8] to the next level by integrating it with quantitative information – the measures/observations – at locations within such spatial regions. Our ontology show that simple integer operations (sum, difference, multiplication, and (in)equalities) for contour values and a spatial sum operation (Σ) over sets of spatial regions suffice to support qualitative-quantitative reasoning about contours. That means a computational framework for reasoning about contours can rely primarily on non-quantitative mereotopological reasoning (similar to reasoning with the Region Connection Calculus [19]) if complemented by a quantitative reasoner that can efficiently perform basic integer operations and calculate sums of spatial regions. We anticipate that such hybrid qualitative-quantitative reasoning can answer simple, common sense geographical queries as exemplified in Sect. 1.1, such as identifying the elevation profile of a path, comparing regions with respect to their max/min elevation or roughness, estimating volumes of depressions, waterbodies and reservoirs, or identifying the extent of areas prone to flooding, without having to employ a full quantitative model of a field as currently done in geographic information systems build on data types and algorithms from computational geometry.

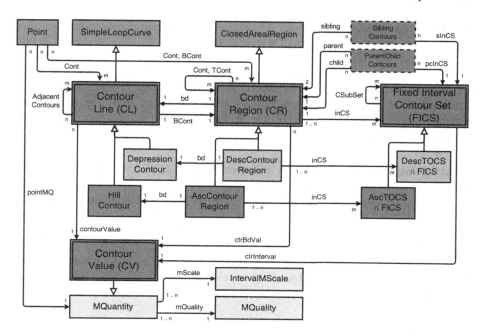

Fig. 5. The manually derived conceptual model of *conceptual contour maps* in UML notation based on the first-order ontology. This simplified model excludes concepts and relations pertaining to the various defined areas from Sect. 4 and is limited to fixed interval contour sets. It introduces new concepts *ParentChildContours* and *SiblingContours* (displayed with dashed borders) that result from encoding the ternary relations *ChildCR* and *SiblingCR* using binary ones. While the conceptual model provides a visual summary of the ontology, it is no adequate replacement. Many of the intricate relationships in contour maps, such as between contour regions and their contour values within a contour set, are not at all visible in the conceptual model.

Derived Conceptual Model. While the intricate interplay between qualitative and quantitative constraints in contour maps justifies the use of first-order logic for our detailed ontological analysis, the developed ontology is not necessarily ideal for storing and querying information of *particular contour maps*. Thus, a translation of the ontology into a less expressive, yet still computer-interpretable language (e.g., from the OWL family) will foster broader adoption and reuse. While such extraction could be largely automated, certain first-order logical constructs must be remodeled to adapt to the restrictions that less expressive languages impose. For example, the ternary relations *ChildCR*, *SiblingCR*, $GenCS_>$, and $GenCS_<$ need to be split into new concepts and associated sets of binary relations in order to represent them in OWL. The portion of the semantics that can be expressed in, e.g., OWL Full and the constraints that would be lost in such a representation remain to be investigated. As a step towards fostering broader use, we manually derived the simplified conceptual model shown in Fig. 5 that relates the key contour concepts: contour regions, contour lines, contour values, and contour sets. It further distinguishes between ascending (shown in brown) and descending (in blue) contour regions, lines, and sets. While *AscCR* and *DescCR*, and *HillContour* and *DepressionContour* exhaustively classify contour regions and lines, respectively, *AscTOCS* and *DescTOCS* are non-exhaustive special cases of contour sets. To improve readability, the full classification of contour sets from Fig. 4 has been omitted from Fig. 5, focusing instead on contours with fixed contour intervals, which are prevalent in many applications, including topographic maps.

Relationship to Surface Water Networks. We have justified many contour-related concepts using simple physical-geographical features that we would like to locate on topographic contour maps, such as hills, peaks, depressions, pits, saddles, and cliffs. However, many more geographical features, in particular those that play a central role in surface networks [21] and surface water networks [14, 22], can be more accurately spatially grounded in elevation contour maps. This requires a fuller investigation of how surface water features, including channels, pour points, and watershed basins, are related to the identified concepts on contour maps.

Acknowledgments. Part of this paper is based on discussions about key concepts in contour maps at the joint SOCoP (Spatial Ontology Community of Practice) and GeoVoCamp workshop in Madison, WI in June 2014, continued at a GeoVoCamp meeting at USGS in Reston, VA in December 2014. We gratefully acknowledge and thank those who contributed to these preliminary discussions: Carl Sack (at Madison), Joshua Lieberman, Dave Kolas and John (Ebo) David (at Reston), and in particular Dalia Varanka for her valuable contributions at both workshops. We further thank the workshop organizers Nancy Wiegand and Gary Berg-Cross, whose efforts to organize these events enabled those fruitful discussions. Finally, we appreciate the constructive comments of Philip Thiem and four anonymous reviewers, which helped improve the final paper.

References

1. Boyell, R., Ruston, H.: Hybrid techniques for real-time radar simulation. In: IEEE Fall Joint Computer Conference (IEEE 1963), pp. 445–458 (1963)
2. Casati, R., Varzi, A.C.: Parts and Places. MIT Press, Cambridge (1999)
3. Cohn, A.G., Renz, J.: Qualitative spatial representation and reasoning. In: van Harmelen, F., Lifschitz, V., Porter, B. (eds.) Handbook of Knowledge Representation. Elsevier, Amsterdam (2008)
4. Egenhofer, M.J., Mark, D.M.: Naive geography. In: Kuhn, W., Frank, A.U. (eds.) COSIT 1995. LNCS, vol. 988, pp. 1–15. Springer, Heidelberg (1995)
5. Egenhofer, M.J., Sharma, J.: Topological relations between regions in R^2 and Z^2. In: Abel, D.J., Ooi, B.-C. (eds.) SSD 1993. LNCS, vol. 692, pp. 316–336. Springer, Heidelberg (1993)
6. Freeman, H., Morse, S.: On searching a contour map for a given terrain elevation profile. J. Franklin Institute **284**(1), 1–25 (1967)

7. Hahmann, T.: A reconciliation of logical representations of space: from multidimensional mereotopology to geometry. Ph.D. thesis, University of Toronto, Department of Computer Science (2013)
8. Hahmann, T., Grüninger, M.: A naïve theory of dimension for qualitative spatial relations. In: Symposium on Logical Formalizations of Commonsense Reasoning (CommonSense 2011). AAAI Press (2011)
9. Hahmann, T., Grüninger, M.: Region-based theories of space: mereotopology and beyond. In: Hazarika, S.M. (ed.) Qualitative Spatio-Temporal Representation and Reasoning: Trends and Future Directions, pp. 1–62. IGI, USA (2012)
10. Hutton, C.: An account of the calculations made from the survey and measures taken at Schehallien. Philos. Trans. R. Soc. Lond. **68**, 689–788 (1778)
11. Kweon, I., Kanade, T.: Extracting topographic terrain features from elevation maps. CVGIP: Image Underst. **59**, 171–182 (1994)
12. Masolo, C., Borgo, S., Gangemi, A., Guarino, N., Oltramari, A.: Wonderweb deliverable D18 - ontology library (final report). National Research Council - Institute of Cognitive Science and Technology, Trento, Technical report (2003)
13. Morse, S.: Concepts of use in contour map processing. Commun. ACM **12**, 147–152 (1969)
14. O'Callaghan, J., Mark, D.M.: The extraction of drainage networks from digital elevation data. Comput. Vis. Graph. Image Process. **28**(3), 323–344 (1984)
15. Open Geospatial Consortium (OGC): OGC reference model. OGC 08–062r7, December 2011. http://www.opengeospatial.org/standards/orm
16. Open Geospatial Consortium (OGC): ISO 19156: geographic information - observations and measurements. OGC 10–004r3, September 2013. http://www.opengeospatial.org/standards/om
17. Pike, R., Evans, I., Hengle, T.: Geomorphometry: a brief guide. In: Hengl, T., Reuter, H. (eds.) Geomorphometry: Concepts, Software Applications. Elsevier, Amsterdam (2009)
18. Rana, S. (ed.): Topological Data Structures for Surfaces. Wiley, New York (2004)
19. Randell, D.A., Cui, Z., Cohn, A.G.: A spatial logic based on regions and connection. In: KR 1992: Principles of Knowledge Representation and Reasoning, pp. 165–176 (1992)
20. Rijgersberg, H., van Assem, M., Top, J.: Ontology of units of measure and related concepts. Semant. Web J. **4**(1), 3–13 (2013)
21. Sinha, G., Kolas, D., Mark, D., Romero, B.E., Usery, L.E., Berg-Cross, G., Padmanabhan, A.: Surface network ontology design patterns for linked topographic data, May 2014
22. Sinha, G., Mark, D., Kolas, D., Varanka, D., Romero, B.E., Feng, C.-C., Usery, E.L., Liebermann, J., Sorokine, A.: An ontology design pattern for surface water features. In: Duckham, M., Pebesma, E., Stewart, K., Frank, A.U. (eds.) GIScience 2014. LNCS, vol. 8728, pp. 187–203. Springer, Heidelberg (2014)
23. Stevens, S.S.: On the theory of scales of measurement. Science **103**(2684), 677–680 (1946)
24. Usery, E.L., Varanka, D., Finn, M.P.: A 125 year history of topographic mapping and GIS in the U.S. Geological Survey 1884–2009, Part 1, 1884–1980, March 2015

Navigation by Humans and Machines

Learning Spatial Models for Navigation

Susan L. Epstein[1]([⊠]), Anoop Aroor[1], Matthew Evanusa[1],
Elizabeth I. Sklar[2,3], and Simon Parsons[2]

[1] Hunter College and The Graduate Center of The City University of New York,
New York, NY 10065, USA
susan.epstein@hunter.cuny.edu, aaroor@gc.cuny.edu,
matthew.evanusa@gmail.com
[2] University of Liverpool, Liverpool, UK
s.d.parsons@liverpool.ac.uk
[3] Kings College, London, UK
elizabeth.sklar@kcl.ac.uk

Abstract. Typically, autonomous robot navigation relies on a detailed, accurate map. The associated representations, however, do not readily support human-friendly interaction. The approach reported here offers an alternative: navigation with a spatial model and commonsense qualitative spatial reasoning. Both are based on research about how people experience and represent space. The spatial model quickly develops as the result of incremental learning while the robot moves through its environment. In extensive empirical testing, qualitative spatial reasoning principles that reference this model support increasingly effective navigation in a variety of built spaces.

Keywords: Navigation · Learning · Spatial model · Heuristics · Qualitative spatial reasoning

1 Introduction

A person who travels without a map to multiple locations (*targets*) relies on local perception to build a mental model that supports her goals. That model is replete with *spatial affordances*, abstractions that remove perceived but irrelevant details [11] and support spatial reasoning. The thesis of our work is that, despite sensor noise and actuator uncertainty, an autonomous robot can quickly learn to travel effectively when it too relies on commonsense qualitative spatial reasoning and models spatial affordances. This paper reports on *SemaFORR*, a hierarchical architecture for autonomous robot navigation. SemaFORR both makes navigation decisions and identifies spatial affordances. The principal results reported here are SemaFORR's ability to learn a serviceable spatial mental model from spatial affordances quickly, and to navigate with increasing effectiveness to a sequence of targets, guided by that model.

The robot begins from a corner of its environment and then tries to reach a set of targets in a prespecified order as quickly as possible. Instead of a map,

© Springer International Publishing Switzerland 2015
S.I. Fabrikant et al. (Eds.): COSIT 2015, LNCS 9368, pp. 403–425, 2015.
DOI: 10.1007/978-3-319-23374-1_19

the robot has limited local perception, that is, it senses obstructions in a few directions and only in its immediate vicinity. Each time the robot finishes with a target, SemaFORR analyzes the perceptual history from that trip to learn and refine a spatial mental model of the environment. The revised model then serves as input for navigation to the next target.

Navigation is viewed here as a sequence of actions selected one at a time. To select the robot's next action, SemaFORR pragmatically capitalizes on the synergy among commonsense spatial rationales. Each rationale is a reactive procedure whose input is the robot's current percepts and spatial model. Some rationales (e.g., "move to the target you perceive directly before you") are applicable to any environment. Other rationales (e.g., "move through this exit") exploit affordances present in the model. The resultant system is transparent, human-friendly, and could advance human-robot collaboration.

The environments investigated here are three small, built spaces with different topologies, and a real-world indoor space of considerably greater complexity. By construction, SemaFORR can operate either physical robots in our laboratory or simulated ones in a virtual world. The thorough and extensive empirical work reported here, however, would have dramatically taxed our robot hardware. It also would have required considerably more elapsed time to recharge and recalibrate each robot periodically. The results reported here, therefore, are in simulation, but with realistically noisy actuators that may reposition the robot somewhat more or less than intended, as they do the physical robots.

The next section of this paper summarizes related work in intelligent architectures and robot navigation. Subsequent sections describe how SemaFORR decides and learns, and provide the experimental design and results. The paper closes with a discussion that considers the ramifications of our system decisions, and outlines our current work.

2   Related Work

FORR (FOr the Right Reasons) is a general architecture for learning and problem solving [7]. A FORR-based program is built to learn quickly, adapt rapidly, and restructure its own decision-making process. These properties provide robustness in complex, unpredictable situations. FORR was confirmed as cognitively plausible on human game players [29], and has since learned successfully in a variety of application domains, including game playing, constraint satisfaction, and human-machine dialogue.

Ariadne was an early FORR-based application for a simulated robot in a grid world [8,9]. Ariadne's task was idealized, however. The robot operated alone in a static environment. Its sensors had no range limit; its sensors and actuators were both noise-free. The robot moved perfectly and only orthogonally. It had no physical footprint; instead it occupied an entire grid cell. Moreover, what Ariadne learned, while intuitively appealing, was not based on what we now know about how people represent and experience space. As a result, Ariadne fared best in random environments without organizational principles, or in environments with extensive, centralized open space. Environments built for people (e.g., a set of offices) proved considerably more difficult for Ariadne.

In contrast, SemaFORR is intended for dynamic, partially observable environments including complex office buildings, warehouses, factories, and search-and-rescue settings. In such environments, maps may be unreliable for path planning, and landmarks may be obscured or obliterated. Communication may be sporadic and slow, sensors and actuators noisy, and barriers or passageways unanticipated. Moreover, a SemaFORR robot, while autonomous, is intended to work with others.

SemaFORR is part of the *HRTeam* (Human-Robot Team) project, where a person collaborates with a set of heterogeneous, autonomous, low-end physical robots. HRTeam's long-term goal is to support the person as the team investigates environments presumed unsafe for people. The person contributes to decision making but does not control it. This motivates our approach, in which each robot uses commonsense qualitative spatial reasoning and a spatial mental model to determine its actions. The HRTeam framework includes software to assign targets, a central server for communication, a shared knowledge store for the components of the spatial model, and a controller for each robot. Because HRTeam's software framework is built on Player/Stage [13], physical and simulated robots use the same decision maker (which selects actions in the controller), and the same driver (which sends commands from the controller to the robot's motors) (See [31] for further details on HRTeam.)

SemaFORR draws from research on how people experience, remember, and move through space (e.g., [15,16]). Its spatial model is inspired by what researchers now know about human spatial perception and navigation. Instead of an image-like metric map, people rely on what appears to be a gradually acquired collage of different kinds of knowledge [36]. Because metrically or topologically impossible environments do not deter people [41], neurophysiologists have suggested that human mental models remove perceived but irrelevant details [11]. For clarity, additional related cognitive work is cited in the next section.

SemaFORR's cognitive underpinnings have led to key differences from traditional approaches to robot navigation. The state-of-the-art approach to navigation in mobile robotics is to construct a detailed metric map using probabilistic *SLAM* (simultaneous localization and mapping) [2,6]. Once a map has been constructed, the robot can *localize* (find its position there) and plan a path between any two points on the map. While the robot constructs the map, it also localizes within the map segment it has constructed (hence "simultaneous") and can plan paths within that segment. Plans can also be constructed to points outside the map segment, although unknown features may interfere with their execution. This can be somewhat mitigated if path planning and execution are combined with obstacle avoidance. SemaFORR, however, has no map.

A purely reactive robot navigation architecture can support modular software design and flexibility in an environment not specifically structured for it. Such an architecture uses "if ⟨*sensor - value*⟩ then ⟨*action*⟩" rules to select actions. To cover a variety of reactive behaviors, early approaches relied, for example, on subsumption architectures [5] or potential fields [1]. Subsumption architectures, however, require careful engineering to sequence all their applicable rules, and neither subsumption nor potential fields learns spatial features in the environment. Instead, SemaFORR's decision process integrates obvious correct reactions (e.g., "don't move into a wall") with commonsense qualitative spatial reasoning principles (e.g., "move in the direction of the target"). Many of these responses reference SemaFORR's mental model. Layered robot architectures typically partition control based on functionality (e.g., with layers for reactive feedback control, planning, and low-level action selection [12]). In contrast, SemaFORR makes only low-level action decisions.

Deliberative robot navigation architectures rely on a *plan*, a sequence of *way-points* the robot should go to on its way to its target (e.g., [24]). The A* algorithm [18] produces optimal paths, but it explores many alternatives and assumes full knowledge of a static environment. Despite a reliable map, however, a realistic environment may include noisy actuators, dynamic map changes, and other moving agents, and thereby may necessitate plan repair or replanning. Rather than cache all pairs of shortest paths [4] or plan from previous searches [22], SemaFORR reacts to its local perceptions and its spatial model. In other work, HRTeam relies on a skeletal version of SemaFORR with an A* path planner that uses a global (i.e., full) map of the environment [10]. Here, however, we test the bounds of what a single robot can achieve alone, without a map and without a human or robot partner.

Finally, *semantic mapping* seeks to abstract spatial representations constructed for robots so that they can also support communication with people [23]. Most semantic mapping commits first to SLAM and then tries to explain its results in more human-friendly terms, often by augmentation of metric maps with objects (e.g., desks) or labels (e.g., "office"). Some work in semantic mapping deliberately steers the robot (e.g., [27,38]); SemaFORR's robot is autonomous. Other work in semantic mapping is restricted to extremely simple environments with labeled training examples (e.g., [40]); SemaFORR's environments can, as in Fig. 6, be quite complex. In summary, semantic mapping performs inference on metric maps derived from sensor data, while SemaFORR derives affordances directly from sensor data. Thus, instead of recording obstructions, SemaFORR learns ways to facilitate navigation. SLAM addresses "where am I?" while SemaFORR addresses "why should I chose this action?" This approach supports more transparent reasoning and more natural communication with people.

3   SemaFORR

A SemaFORR robot's task is to *visit* (come within ϵ of) each of a pre-sequenced set of targets. To support this goal, SemaFORR learns a spatial model of its environment that emerges as it explores and reasons about space. This section explains SemaFORR's decision context and how it learns a spatial model. Then it describes how SemaFORR chooses an action and explains the individual components of that reasoning mechanism with a unifying example.

3.1   The Decision Context

At a *decision point*, SemaFORR selects the robot's next action. The robot's *action repertoire* is its set of possible actions: forward linear moves (henceforth, simply *moves*), clockwise and counterclockwise rotations (*turns*), and a pause (a no-op). Although the robot could theoretically make a move or turn of any size, SemaFORR restricts that choice to a discrete set of possibilities. The *intensity* of an action is a qualitative representation of how far the robot is intended to travel or turn. Intensities are ordinal labels calibrated to correspond to a particular physical robot and its environment. A move has intensity only between 1 and 5; a turn has intensity between 1 and 4, either clockwise or counterclockwise. Thus there are 14 possible actions in the robot's action repertoire.

The outcome of an action, however, is realistically non-deterministic. This paper focuses on the Surveyor SRV-1 Blackfin, a small platform in our laboratory with a webcam and 802.11 g wireless. We have extensively observed and measured the actuator noise on a set of Blackfins there, and model it probabilistically here. As a result, when SemaFORR decides to act with intensity i, the robot acts with intensity $i \pm \delta$, where δ is an increasing function of i.

For localization, the robot relies on a system of overhead cameras, simulated here. The *position* of a robot is its location coordinates (x, y) and its orientation θ on the true map (henceforth, the *world*). The location of a target is specified with respect to the same coordinate system.

The robot's knowledge store is a set of descriptives that capture its experience, goals, and behavior. A *descriptive* is a shared data object with functions that determine how and when to update it. HRTeam's *DM* (Descriptive Manager) provides a shared knowledge store of descriptives for all team members. The DM receives messages from HRTeam's central server, extracts relevant data from them, and provides the current value of any descriptive to the robot on demand. Basic descriptives include the robot's position, its *agenda* (list of target locations to visit), its current target, and the history of its decisions made thus far on its way to that target.

The robot's percepts are represented as a descriptive called the *wall register*. It simulates a set of limited-range measurements for the distance from the robot to the nearest wall in 10 directions. From the robot's heading of 0°, these measurements are taken on either side at 8.87°, 17.5°, 37.2°, 74.5°, and 195°. In the example in Fig. 1(a), not every ray touches an obstacle; some halt at their maximum range. Note that wall-register values are egocentric, while the positions of the robot and its targets are allocentric. (The walls also have a buffer that thickens them slightly for these measurements, to prevent unintended collisions from noisy actuators.) SemaFORR builds and refines its spatial model incrementally, from a history of its percepts and positions.

3.2   The Spatial Model

The components of SemaFORR's spatial model are spatial affordances, ways the environment provides opportunities to address a goal [14]. SemaFORR's affordances support its reasoning and its explicit representation about two-dimensional space (in the spirit, but without the finer granularity, of [20]). An affordance is calculated from sensor data and the robot's decision points; it describes spatial knowledge that supports effective navigation.

Instead of a map or a formal logic, SemaFORR's affordances summarize its experience in its environment as locations, lines, and areas. Each category of affordance is represented as a separate descriptive. SemaFORR has three kinds of affordances: trails with markers, regions with exits, and conveyors. Trails and conveyors are learned only from *successful travel* (i.e., immediately after the robot reaches its target); regions are learned whether or not travel succeeds.

A *trail* affords the ability to travel along a familiar, ordered sequence of locations. As the robot travels, the DM records its *path* as a sequence of *decision states* $t_1, t_2, \ldots, t_k$, where each state t_i is the robot's current position and wall-register values. The trail-learning algorithm is analogous to the way people compute return paths [17], but with locations rather than landmarks or viewpoints. An example of a path and its trail appears in Fig. 1(b). When the robot reaches its target, the algorithm begins with a trail that is merely a copy of the decision states that formed the path. Then, from its last decision state t_k, the algorithm looks for the smallest i where the wall register at t_i perceived t_k. If it finds such a t_i the algorithm reduces the trail to $t_1, t_2, \ldots, t_i, m_k$, where the *trail marker* m_k is the location of t_k and the wall register values at t_k. This process repeats for each decision state along the trail, moving from the target backwards to the starting point. The resultant trail is a (typically shorter) ordered set of trail markers with line segments between consecutive trail markers. In the worst case, learning time is quadratic in the path length. A trail reduces the computational and physical effort required to travel between the target and the robot's starting point. Although a trail is likely to be suboptimal, it is more direct and has fewer digressions than the path from which it originates.

A *region* is an obstruction-free local area. Wall register vectors are limited only by obstacles (e.g., walls) and the sensors' range. The robot only senses (i.e., produces wall register values) from a position where it is about to make a

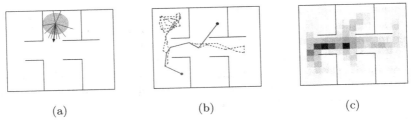

(a)   (b)   (c)

Fig. 1. (a) The robot's position produces values for its wall register, which provides its local view. The arrow represents the robot's heading; the subtended circle is the detected region. (b) A (dotted) path begun in the lower left room and the (solid) trail derived from it, with dots at the trail markers. (c) A conveyor grid; darker cells are more often crossed by trails.

decision. At each such position, it computes a region as a circle whose center is its current location and whose radius is the length of its shortest wall register value, as in Fig. 1(a). Regions are reminiscent of some human mental models [19, 30] and of areas in online mapping [35], but do not require that the robot map all walls first. As the robot travels, regions may gradually grow or shrink, but they never overlap. (This was a pragmatic design decision; there is a tradeoff between the number of regions and the cost to maintain and use them.) An *exit* from a region is a point on its circumference that intersects with a path.

Finally, a *conveyor* affords visitation to a small area that has often facilitated travel (similar to [25]). The *conveyor grid* covers the footprint of the world with cells about 1.5 times the size of the robot's footprint. The algorithm that learns conveyors tallies the frequency with which all trails pass through its grid cells. High-count cells are conveyors; they appear darker in the example in Fig. 1(c). Together the descriptives form a knowledge store over which SemaFORR reasons.

3.3   The Reasoning Mechanism

SemaFORR selects one action at a time, that is, it does not plan. To select an action, it executes a *decision cycle*. At the beginning of a decision cycle, SemaFORR retrieves the current descriptive values from the DM and caches them in the robot's memory store. These include its position and the current target and spatial model. Then SemaFORR reasons about which action to choose from its 14-action repertoire. The output of a decision cycle is the selected action.

In SemaFORR, a *rationale* is a plausible reason to select an action. An *Advisor* is a boundedly rational (i.e., resource-limited) procedure that applies a rationale to evaluate actions. SemaFORR's use of multiple rationales is consistent with the recent result that multiple wayfinding strategies best predict human route selection [33]. The input to an Advisor is a set of possible actions and the descriptives' values. The output of an Advisor is a (possibly empty) set of *comments*, each of which expresses an opinion about the appropriateness of a single action from the perspective of the Advisor's rationale.

SemaFORR partitions its Advisors into tiers that correspond to Montello's distinction between locomotion (*tier 1*) and wayfinding (*tier 3*) [26]. As shown in Fig. 2, a decision cycle first invokes the tier-1 Advisors in a predetermined order. One at a time, they have the opportunity to comment. In tier 1, Advisors' rationales are quick to compute and assumed to be correct. Each tier-1 rationale gives rise to a single Advisor that mandates or vetoes obvious reactions. If any tier-1 Advisor mandates an action, that becomes the decision and no further Advisors are consulted. If a tier-1 Advisor vetoes an action, it is eliminated from the set of possible actions passed to the next tier-1 Advisor.

Despite possible vetoes, tier-1 processing always retains some action, so that decision making always returns a value. If at any point in tier 1 only "pause" remains, it becomes the decision, and no other Advisors are consulted, that is, the robot does nothing until its next decision cycle. Otherwise, the unvetoed actions are forwarded to tier 3, which chooses among them. (Tier 2, which supports the

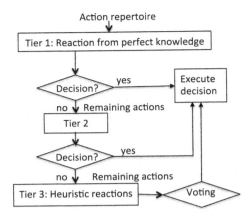

Fig. 2. A schematic for SemaFORR's control structure.

selection of action sequences, is the focus of current development, and not used in the work reported here.)

In tier 3, SemaFORR alternately chooses a pause or a move on one decision cycle, and then a turn on the next. Thus, if in Fig. 2 the previous decision cycle were for turns, only the unvetoed moves and "pause" would be forwarded to tier 3. Note that a pause does not halt movement or sensing; it merely defers a decision to the next cycle, and thereby permits longer consecutive moves. Turns of intensity one serve the same purpose as the pause, that is, they permit longer turns in the same direction.

In tier 3, all Advisors comment before any decision is made. SemaFORR's tier-3 Advisors have deliberately disparate spatial rationales. To resolve their differences of opinion and to capitalize on the synergy among them, *voting* tallies the comment strengths from all tier-3 Advisors. When Advisor i comments on action j with strength s_{ij}, voting returns an action with maximum total strength:

$$argmax_j \sum_i s_{ij} \qquad (1)$$

Ties in voting are broken at random. The remainder of this section explains the commonsense rationales and how they rely on local perception and the spatial model, and provides a unifying example of a SemaFORR decision.

3.4 Tier-1 Advisors

Table 1 lists SemaFORR's 13 Advisor rationales by tier. There are three tier-1 Advisors. The first, VICTORY, comments if it does not sense an obstruction in its "line of sight" to the target. VICTORY mandates the move that will bring the robot closest to the target or the turn that will head the robot most directly toward it. The second, AVOIDWALLS opposes actions likely to bring the robot too

Table 1. Rationales that underlie SemaFORR's 22 Advisors. There is one Advisor for each tier-1 rationale. Each tier-3 rationale is implemented by a move Advisor and a turn Advisor (with the exception of †, applicable to turns only).

Tier 1, in order	
VICTORY	Go to an unobstructed target
AVOIDWALLS	Do not go within ϵ of a wall
NOTOPPOSITE	Do not return to last heading
Tier 3 heuristics vote	
	Based on commonsense reasoning
BIGSTEP	Take a long step or turn in the direction of a long step
ELBOWROOM	Go where there is room to move
EXPLORER	Go to unfamiliar locations
GOAROUND†	Turn to avoid obstacles directly in front of you
GREEDY	Go closer to the target
	Based on learned spatial affordances
CONVEY	Go to frequent, distant conveyors
ENTER	Go into a region via an exit
EXIT	Leave a region via an exit
TRAILER	Pursue a useful trail segment
UNLIKELY	Do not enter a leaf region unless it contains the target

close to a wall and thereby risk collision due to its noisy actuators. ("Likely" is within γ of a wall, given the wall register values.) Finally, NOTOPPOSITE vetoes turns that would simply restore the robot's immediately preceding heading.

Thus, by design, tier 1 selects an action toward to an unobstructed target, and otherwise forwards to tier 3 only actions that avoid collisions with walls and do not oscillate in place. Recall that tier-1 Advisors are expected to be correct. Given the uncertainty inherent in the robot's actuators and its partial view of the world, little else can be safely asserted. The remainder of SemaFORR's reasoning is necessarily heuristic.

3.5 Tier-3 Advisors

SemaFORR's tier-3 Advisors have heuristic rationales. With the exception of GOAROUND, each tier-3 rationale gives rise to two Advisors: one for moves (name ends in M) and one for turns (name ends in T). A turn Advisor considers how its associated move Advisor would comment after each possible turn. A turn decision is not a classical robotics plan, however, because it makes no commitment to a subsequent move; it only anticipates one. For example, GREEDYM comments on moves with strengths that are inversely proportional to the distance they are expected to place the robot from the target. In tandem, GREEDYT calculates its comment strengths from how close the robot could come to the target if it were to turn and then make GREEDYM's most preferred move.

Each tier-3 Advisor has a metric that assigns a real value to each possible action. To comment on a given set of n actions, an Advisor applies its metric, ranks the actions in descending order by their metric values, and then assigns corresponding *comment strengths* from n down to 1. The larger the strength, the more the Advisor prefers the action.

In addition to GREEDY, four more tier-3 rationales also represent common-sense and rely only on local perception. BIGSTEP supports large actions, with comment strengths proportional to the action's size. ELBOWROOM supports actions that keep the robot further from walls. When the robot is facing a wall, GOAROUNDT supports turns that veer away from it, and prefers larger turns more strongly when the wall is closer. Finally, EXPLORER advocates exploration to reduce uncertainty, which people also do in noisy, dynamic environments [32]. It supports actions toward locations that are relatively novel with respect to the current target (i.e., minimize the total Euclidean distance to previous decision points).

The remaining tier-3 rationales exploit learned spatial affordances. CONVEY supports actions to high-count conveyors, with preference for those further from the robot. When high-count conveyors are near one another, CONVEY thereby advocates travel through those locations rather than merely to them.

TRAILER is a case-based reasoning mechanism for trails. A trail is *accessible* if and only if a ray from the robot's current wall register intersects some line segment of the trail. Unless it has already done so, TRAILER identifies an accessible trail that has a marker m within sensory range of the target (as indicated by the wall register at m). If there is such a trail, TRAILER's comments greedily support actions toward trail markers further along the trail segment that leads to the target. TRAILERM's comment strengths reflect the ability of each move to get the robot farther along that trail on its way to the vicinity of the target. There is no plan-like commitment to a trail, however, because the other Advisors may draw the robot elsewhere. (Indeed, once the robot arrives in the immediate vicinity of the target, VICTORY will take control.) Furthermore, if TRAILER cannot sense any marker on its selected trail for four consecutive decision cycles, it is "lost," does not comment, and looks for a new trail on the next decision cycle.

Three Advisor rationales reference regions. A *leaf region* is defined as one whose exits all lie within an arc of no more than 90°. (With perfect knowledge, a leaf region would be a dead-end.) When the target lies in region T and the robot in region R adjacent to T, ENTER supports actions into T; UNLIKELY opposes actions into a leaf region other than T; and EXIT supports actions toward any exit from R if the target is not in R, in the spirit of [3].

Figure 3 provides a unifying example. It superimposes on the true map the robot, its current target, and the current spatial model, with leaf regions shown somewhat lighter. At this point in the decision cycle, AVOIDWALLS has already vetoed the move of intensity 5 because actuator error could drive the robot into

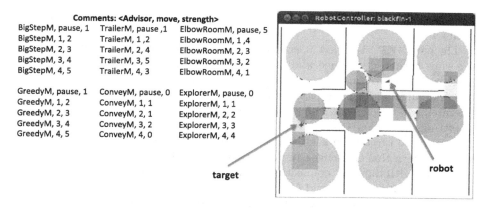

Comments: <Advisor, move, strength>		
BigStepM, pause, 1	TrailerM, pause ,1	ElbowRoomM, pause, 5
BigStepM, 1, 2	TrailerM, 1 ,2	ElbowRoomM, 1 ,4
BigStepM, 2, 3	TrailerM, 2, 4	ElbowRoomM, 2, 3
BigStepM, 3, 4	TrailerM, 3, 5	ElbowRoomM, 3, 2
BigStepM, 4, 5	TrailerM, 4, 3	ElbowRoomM, 4, 1
GreedyM, pause, 1	ConveyM, pause, 0	ExplorerM, pause, 0
GreedyM, 1, 2	ConveyM, 1, 1	ExplorerM, 1, 1
GreedyM, 2, 3	ConveyM, 2, 1	ExplorerM, 2, 2
GreedyM, 3, 4	ConveyM, 3, 2	ExplorerM, 3, 3
GreedyM, 4, 5	ConveyM, 4, 0	ExplorerM, 4, 4

target

robot

Fig. 3. A spatial model and the comments generated by the tier-3 Advisors when the robot is at the indicated decision point on its way to a target. For clarity, the walls (unknown to the robot) and only the trail selected by the Advisors are shown.

a wall. Because SemaFORR turned on its previous decision, only pause and moves with intensity 1, 2, 3, and 4 along the robot's current heading have been forwarded to the tier-3 Advisors.

The comments in Fig. 3 clearly reflect the rationales and the current decision point. GoAroundT does not comment because no turns are currently available. BigStepM's comment strengths reflect the size of each move. Since the target is far from the robot's location, GreedyM comments on 4 with strength 5, and on the other moves with decreasing strengths. For clarity, Fig. 3 shows only the selected trail that Trailer wants to pursue. ConveyM's comment strengths direct the robot to the darker grid cell that lies in front of it. ElbowRoomM's comments encourage moves to locations further from walls. By this time the robot had visited much of the room it was in, so ExplorerM's comment strengths seek to drive it out into the hallway, to less familiar territory. EnterM and UnlikelyM do not comment because the target is not within any region; ExitM does not comment because the robot is not in any region. The vote in Fig. 3 selects the move of intensity 3, which has the highest support.

4 Empirical Design

To examine how SemaFORR's knowledge and skill evolve in world W, the robot is given an agenda, a sequence of targets T to visit. In a *run*, the robot begins at the same position (in the lower left of W) and attempts to visit each target in the fixed but randomized order of T. Once the robot reaches a target, it addresses the next one on its agenda from its current position. If the robot does not reach a target after 250 decisions, it is deemed to have *failed* on that target, and addresses the next target in T from its current position. The evaluation criteria for a run are the *success rate* (percentage of reached targets in T), the total elapsed wall-clock time in seconds, and the total distance traveled in centimeters.

The pair $\langle W, T \rangle$ is called a *setting*. There are two sources of non-determinism in a run: actuator variance and random tie-breaking during voting. Thus, to gauge performance consistency within a given setting, results are averaged over 5 runs. To gauge performance consistency within a given W, there are 5 randomly generated, 40-target sets T for each world W, and results are averaged over all settings. Thus our data describes navigation performance on 1000 targets (25 runs of 40 targets) in each world. To gauge performance consistency across different navigation challenges, our experiments investigate three worlds built for people but with different connectivities [28]. World A simulates an office space, world B a rotunda, and world C a warehouse or library stacks.

The experiments reported here compare SemaFORR with *SemaFORR-A**, a gold standard for robot planning. From a map of the world, SemaFORR-A* plans a shortest (A*) path to each new target τ. This path is represented as a sequence of waypoints from the robot's initial location to τ and avoids walls on the map. To make a decision, SemaFORR-A* selects the action intended to bring the robot closest to its next waypoint in the plan. Such navigation would be perfect were it not for actuator errors, which may move the robot to a position where a waypoint is obstructed or too far away. In that case, SemaFORR-A* must replan. To reduce the impact of actuator error and thereby help adhere to the plan, SemaFORR-A* selects only small (intensity-1) moves and turns. Comparison to SemaFORR-A* evaluates the impact of reactivity and a local, rather than a global, view.

To tease apart SemaFORR's navigation skills, we also test five ablated versions of SemaFORR. *SemaFORR-B* navigates with only commonsense qualitative spatial reasoning, as represented by tier 1 plus four qualitative commonsense rationales in tier 3: BigStep, ElbowRoom, GoAround, and Greedy. SemaFORR-B has no spatial affordances; comparison to it evaluates the impact of commonsense spatial reasoning. To gauge the impact of exploration, *SemaFORR-E* augments SeamFORR-B with Explorer. To evaluate the impact of the individual spatial affordances, *SemaFORR-C, SemaFORR-R,* and *SemaFORR-T*, each add a single spatial affordance (conveyors, regions, or trails, respectively) to SemaFORR-E, along with their associated Advisors.

5   Results

5.1   Learned Spatial Models

A qualitative evaluation metric is the appropriateness of SemaFORR's learned spatial models. Figure 4(a), (b), (c) and (d) show how a spatial model for world A evolved during a single run. Note how the model develops quickly, with few changes after the first 10 targets. Videos of the robot's travel show how a model evolves after each target despite actuator error: https://www.youtube.com/watch?v=3C_675H6-xk, https://www.youtube.com/watch?v=4WF8unQlSm8.

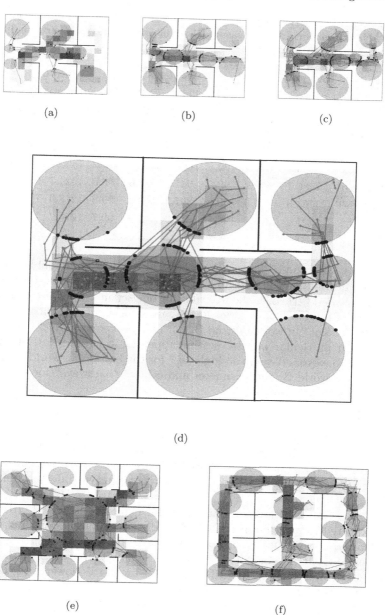

(a) (b) (c)

(d)

(e) (f)

Fig. 4. Spatial models in SemaFORR, overlaid on the corresponding true map, which the robot does not have. Conveyors are shown as grid cells (darker is more salient), and regions as circles with dotted exits. Trails are line segments with dots at the trail markers. The spatial model for world A as it evolves after (a) 10 (b) 20 and (c) 30 targets. After 40 targets, SemaFORR's learned spatial models of (d) world A, (e) world B, and (f) world C.

Table 2. Navigation performance of SemaFORR-A* (an ideal planner), SemaFORR, and five ablated versions of it, measured by time in seconds, distance, and success (percentage of reached targets per run).

Navigator	World A			World B			World C		
	Time	Dist.	Succ.	Time	Dist.	Succ.	Time	Dist.	Succ.
SemaFORR-A*	1035.9	400.1	100.0	884.6	335.1	100.0	1119.9	437.9	100.0
SemaFORR-B	1823.7	974.5	94.1	1124.6	565.0	97.8	2473.5	887.5	82.2
SemaFORR-E	1415.1	1018.8	99.4	1090.2	730.9	99.2	1497.7	983.2	98.5
SemaFORR-C	1323.8	977.4	99.8	1009.8	698.7	99.2	1303.2	933.2	99.7
SemaFORR-R	1280.1	892.2	99.6	941.5	612.6	99.6	1524.7	919.7	98.2
SemaFORR-T	1163.8	813.0	99.5	867.2	553.3	99.8	1278.0	775.7	99.6
SemaFORR	1221.2	854.5	99.5	835.3	554.9	99.7	1275.7	798.2	99.8

Figure 4(d), (e), and (f) show, for each world, the spatial model learned after a single run, overlaid on the walls of its true map. Inspection indicates that these final models varied little over the five runs for one setting. They also varied little from one set of targets T to another in the same world. Observe how the regions capture the "rooms" in worlds A and B, but only two of the cubicles in world C; targets in that particular world-C setting appeared in cubicles less often. Note, too, how the conveyors develop a "highway" for the hallway in world A, diagonal "highways" for world B, and perimeter and central "highways" for world C.

5.2 Performance

Performance results appear in Table 2. For all our navigators, the target sets in world B are clearly the easiest, and those in world C the most difficult. In the following discussion, high variance caused both by actuator error and randomized target sets T makes some apparent differences inconclusive; differences cited here are at $p < 0.05$.

Without a map and given its penchant for exploration, SemaFORR should not be expected to match SemaFORR-A*'s distance along its optimal paths in a complete map. Nonetheless, in world B SemaFORR reaches its targets just as fast as SemaFORR-A*. In worlds A and C, SemaFORR travels further but is only slightly slower than SemaFORR-A* (18 % and 14 % slower, respectively).

Both SemaFORR-A* and SemaFORR spend most of their time in travel rather than decision making. SemaFORR-A* devotes about 19 % of its time to decisions in all 3 worlds. SemaFORR devotes 17 % of its time to decisions in worlds A and B, and 18 % in C. Moreover, in every world, SemaFORR's learning requires less than 0.01 % of the elapsed time.

Compared to SemaFORR-B, SemaFORR improves navigation: time is substantially reduced, distance decreases in worlds A and C, and reliability (as measured by success rate) rises. SemaFORR-E demonstrates the improvement exploration brings to the commonsense reasoning of SemaFORR-B, and the

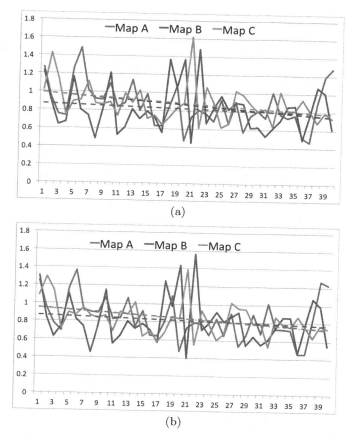

(a)

(b)

Fig. 5. For each world, plotted across 40 targets and averaged over all runs, the ratio of SemaFORR's (a) time and (b) distance to those of SemaFORR-E on each target. All (dashed) regression lines have negative slopes. This indicates that performance improved across each target set. Regression lines for worlds A and C are nearly identical.

price the robot pays for it. Travel time is reduced by 22 %, 3 %, and 39 %, in worlds A, B, and C, respectively, and the success rate rises, particularly in world C. Exploration also increases travel distance, however, by 5 %, 29 %, and 11 %, respectively. Moreover, as one would expect, SemaFORR-E does not demonstrate any improvement across targets; it fails as late as on the 32nd target in world A, the 36th in world B, and the 29th in world C. Compared to exploration plus commonsense, each spatial affordance alone reduces travel time (except for regions in world C) and distance, with improvements in every success rate for all but SemaFORR-C in world B. The trails (SemaFORR-T) are particularly effective; they can reach the targets as quickly in world B as SemaFORR-A* does, presumably because they allow longer, highly effective steps.

SemaFORR's superiority to SemaFORR-B demonstrates that learning spatial affordances is an important component in SemaFORR's performance. There are two ways to gauge if that learning is online, that is, if SemaFORR's performance improves across a sequence of targets. First, with more experience, SemaFORR should fail on targets less often. While SemaFORR-B has runs in which it fails as late as on the 40th (final) target, SemaFORR's last failure on any setting and run in map A was on the 19th and in map C on the 23rd. (There was one failure in world B on the 35th, probably brought on by actuator error because higher-intensity moves are more often possible in B.) Second, SemaFORR should reach its targets faster. Because random generation makes some targets intrinsically more difficult to reach, we normalize performance by how hard it is to reach a target, estimated here by the performance of SemaFORR-E on the same target.) Fig. 5 shows that the regression trend lines for this ratio descend across 40 tasks in all three worlds for both time and distance. In other words, with experience SemaFORR learns to reach its targets more quickly and in a shorter distance.

6   Discussion

SemaFORR is not envisioned as a replacement for SLAM, but as a companion to it, one that facilitates robot-human interaction. A cognitively plausible mental model like SemaFORR's can be shared with a person at a level of abstraction that is both meaningful and parsimonious. For example, "I'm at the center of the biggest region" is considerably more informative for people than "My pose is $< 10, 20, 39° >$."

When, as in HRTeam, a person collaborates with an agent, the ability to explain the agent's reasoning in a human-friendly manner is a first step toward transparent, more natural communication and the establishment of trust. Instead of a metric map, SemaFORR relies on commonsense reasons that support low-level actions. As a result, a SemaFORR robot can explain any of its reasons. A comment from EXPLORER, for example, can be paraphrased as "Let's go this way because we haven't explored it much yet," and one from ENTER as "I want to go into this dead-end [leaf region] because it contains the target."

SemaFORR has a mechanism to translate a decision made in tier 3 into natural language that reflects both the rationales of the Advisors whose comments supported it and the degree of their preference (i.e., comment strength). For example, a vote where the strength of ELBOWROOM's comment is considerably higher than that of any other comment becomes "I really want to move into open space." In contrast, a vote in which the comments from BIGSTEP, EXPLORER, and GREEDY dominate but with somewhat lower strengths than in the previous example becomes "It seems reasonable to move as far as possible, into a new area, and toward our target." Furthermore, because of its modularity, SemaFORR can readily incorporate new explanations as a spatial model becomes more elaborate. Recent work on the construction of a depiction from a verbal description (e.g., [37]) and the negotiation of a route between a robot and person (e.g., [39]) could also be applicable here. The remainder of this section examines the components that drive SemaFORR, its real-world applicability, and our current work.

6.1   What Makes SemaFORR Work

SemaFORR uses its sensors' values to compute and store a simplistic model of its spatial environment. This model is dynamic; it consists of affordances (regions, trails, and conveyors) that proliferate and change as the robot experiences new parts of the environment. Because the model is based on heuristic learning algorithms that analyze only the robot's local views and actual travel, it is necessarily a set of approximations. The interdependence among affordances (e.g., the presence of some high-scoring conveyors along a trail) is deliberate; it both mediates disagreement among possibly disparate approximations and enriches their usefulness. SemaFORR's spatial representation provides flexibility and efficiency, and provides a human-like basis for strategy formation [34].

SemaFORR's architecture exploits the synergy among naïve commonsense reasons. For example, in a mix of heuristics that argue with various intensities for or against a set of moves, GREEDY is just one among many reasons to move down a hallway toward the right room. Qualitative reasoning also allows the robot to correct for, and even anticipate, the inaccuracies of its actuators. For example, when a large move incurs a large actuator error that draws the robot off its intended trail and into an irrelevant room in world A, EXPLORER and EXIT soon pull it back out again.

SemaFORR's model of the robot is realistic. As the floor surface changes or as batteries drain, it is reasonable to assume that a real-world robot's position after it executes a decision may not be precisely what it anticipated. This motivated the replication of laboratory-observed actuator error during simulation. Advisors' comment strengths also deliberately smooth data already recognized as approximate. SemaFORR discretizes movement in continuous space into a sequence of frequent decisions. It chooses an action at least once per second and as often as 3 times per second, depending upon the length of the intervening moves.

SemaFORR's learning is pragmatic; it only infers conveyors, trails, and exits from regions when it manages to reach its target. (Regions, however, are simple local observations, learned either when the robot arrives at its current target, or it reaches its decision limit.) Moreover, SemaFORR's learning algorithms are heuristically honed for fast computation and retrieval. Thus, the resultant spatial model is necessarily an approximation and not a map. The robot represents only what it experiences. If a setting does not take the robot to an area where it can capture a local view, the model will not include that area. In the sensor placement described here, there is also a bias toward the robot's heading, which collects more information about what is in front of the robot. Nonetheless, learning during navigation supports flexibility and gradual improvement. Reinforcement learning that relies on an abstract map [21] is somewhat similar, but SemaFORR extracts and labels its own training examples heuristically, from its experience.

SemaFORR is similar to robotics work both in subsumption and in potential fields. Tier 1 is analogous to a subsumption architecture, where rules are carefully engineered and ordered. (The robot vacuum cleaner Roomba^{TM}, for example, has a subsumption architecture.) Tier 1, however, has only 3 Advisors, and they make only 24.13 %, 37.09 %, and 22.75 % of all decisions in worlds A, B, and C, respectively; the heuristic Advisors in tier 3 make all the others. SemaFORR's tier 3 is analogous to a potential field where forces attract or repel the robot with vectors analogous to Advisors' comment strengths. Potential fields, however, are vulnerable to local minima and lack the wider range of behaviors (action durations) provided by SemaFORR's move intensities. SemaFORR avoids local minima through two mechanisms: exploration and randomization. Exploration draws it to locations less recently visited; ties in voting broken at random provide enough non-determinism to extricate it from repetitive behavior.

Robotics has traditionally relied on precision planning from a global view; if the robot has no map it immediately tries to construct one. People in a complex environment, however, lack the working memory to construct an A* plan. Instead, they satisfice with a spatial model, commonsense qualitative reasoning, and the ability to learn. SemaFORR tests the extent to which such behavior supports navigation. People can also concatenate previously successful routes to construct a new one. While SemaFORR could similarly piece together trail segments, computation over extensive stored experience would soon become costly. Instead, SemaFORR's tier-3 Advisors foreshadow some of the approaches anticipated for tier 2, which is currently under development. A planner could, for example, support a turn decision followed immediately by the move that made it attractive, or it could follow a trail more closely.

6.2   Transition to the Real World

Although the results reported here are for simulation, Player/Stage simplifies the transition to physical-world execution. SemaFORR-A* controls physical Blackfins in a laboratory whose walls are replicated in world A. Indeed, the values used here for how close the robot can come to a wall and how close it must come to the target were gleaned from the metrics already used in the lab. (The same values were also applied, unchanged, to worlds B and C.) An earlier study demonstrated that some performance metrics gathered in HRTeam's simulation are good predictors of behavior in the physical world [31]. Current work includes on-the-floor experiments to confirm that transfer from simulation to the physical world remains effective. Meanwhile, we continue to hone SemaFORR in simulation, where we can run online experiments quickly.

Both the descriptives and the Advisors were developed in world A, but are sufficiently generic to apply to worlds B and C as well. To test the extent of that generality, and to see how SemaFORR's approach scales, we have reproduced for simulation a considerably more elaborate environment, a wing from the floor of a large building that includes one of our labs. This is a realistic built space, about 40 times the area of world A, with 3 hallways and about 7 times as many rooms. It is also considerably more complex; there are rooms with multiple doors and rooms that are accessible only through a sequence of other rooms (i.e., not directly off a hallway). The complexity of this world necessitates more targets and a larger decision limit before failure. Here we used 50 randomly-chosen targets and a 400 decision-cycle limit; all other settings and algorithms remained the same.

The ease with which SemaFORR scales is visible in Fig. 6, shown after a single run. Figure 6 makes clear that the model only includes areas in which the robot travels; more points would likely be necessary to cover this world and to refine the regions that appear to cross a wall. (Regions typically reduce with more travel through them.) It also shows some opportunities for further improvements in SemaFORR. It is clear that for less square rooms it may be worthwhile to merge regions in some way. Moreover, as the complexity of the space increases, the likelihood that any trail matches a new target declines significantly. One way to address this issue would be to combine subsequences of trails at runtime.

We expect SemaFORR's reactive approach to support a variety of other behaviors, some of which go unaddressed by modern, plan-based navigation. In particular, an agent with an agenda need not consider it in a prescribed order. A brief detour to address a target on the agenda but not the current focus of attention is an obvious extension to SemaFORR. Moreover, it should be possible to transfer knowledge between similar environments, such as floors in some office buildings. In addition, when an individual robot needs repair or recharging, a clone given the acquired knowledge is a near-seamless replacement (subject to its idiosyncratic actuator and sensor noise). Finally, how often a robot must expend energy to sense is an open question. Because SemaFORR senses only between actions, a variety of tested intensities could provide a preliminary answer.

Current work capitalizes on SemaFORR's modularity to support its gradual development. One current research avenue is the use of a team of robots that addresses a setting simultaneously, with targets assigned to particular robots. Each robot remains autonomous, with its own copy of SemaFORR, but all robots share in the construction and use of the same spatial model, stored on the DM. We are now testing rationales (analogous to AVOIDWALLS, ELBOWROOM, and BIGSTEP) to avoid robot-robot collisions and crowding.

SemaFORR's modularity includes the ability to support robots on different platforms (e.g., a Blackfin and a Turtlebot^{TM}) with different maneuverability and different footprints. Features that may appear platform-specific here are actually modular and readily replaced without hand tuning, subject to calibration. For example, the number of discrete commands was intended for the Blackfin, but a slightly larger set of intensities should pose no difficulty for the architecture. (The maximum-intensity command actually reflects the furthest one would want the robot to move or turn without sensing again; the minimum intensity reflects the shortest move or turn a particular robot could travel in response to a brief motor command. Both are learnable as a function of the world and the robot platform.) How close the robot must come to the target to be successful is a function of the robot's footprint. Different sensors (e.g., a Kinect or 20 equally-spaced infrared units along the robot's perimeter) could be

Fig. 6. What SemaFORR has learned after 50 targets in a challenging real-world environment. The robot enters at the stairwell, marked with an S.

readily accommodated. Indeed, better sensors should provide considerably more accurate local views that could further improve performance. Thus, we believe that a SemaFORR-supported heterogeneous multi-robot team is a tractable next step.

Meanwhile, SemaFORR quickly learns features of an environment that facilitate effective autonomous navigation without costly mapping or planning. That knowledge transfers from one task to another. When a SemaFORR robot travels, it moves around obstacles and toward its target, with big steps where its world permits. It also anticipates access within regions, uses markers from old trails, turns around obstacles, explores new locations, and recovers from its own errors. Remarkably, that suffices to reach targets in these environments, and quickly builds a simple spatial model of the world that facilitates explainable, human-friendly, effective navigation.

Acknowledgments. This work was supported in part by the National Science Foundation under IIS 11-17000, IIS 11-16843 and CNS 08-51901, and by The City University of New York, Collaborative Incentive Research Grant 1642. We thank Tuna Ozgelen, Eric Schneider, and our anonymous reviewers for their insights and guidance. Tereza Shterenberg produced the map in Fig. 6.

References

1. Arkin, R.C.: Integrating behavioral, perceptual, and world knowledge in reactive navigation. Robot. Auton. Syst. **6**, 105–122 (1990)
2. Bailey, T., Durrant-Whyte, H.: Simultaneous localization and mapping: part II. IEEE Robot. Autom. Mag. **13**(3), 108–117 (2006)
3. Björnsson, Y., Halldórsson, K.: Improved heuristics for optimal path-finding on game maps. In: Laird, J., Schaeffer, J. (eds.) AIIDE 2006, pp. 9–14. AAAI, Marina del Rey (2006)
4. Botea, A., Harabor, D.: Path planning with compressed all-pairs shortest paths data. In: Borrajo, D., Fratini, S., Kambhampati, S., Oddi, A. (eds.) ICAPS-2013, pp. 288–292. AAAI, Rome (2013)
5. Brooks, R.A.: Intelligence without representation. Artif. Intell. **47**(1–3), 139–160 (1991)
6. Durrant-Whyte, H., Bailey, T.: Simultaneous localization and mapping: part I. IEEE Robot. Autom. Mag. **13**(2), 99–110 (2006)
7. Epstein, S.L.: For the right reasons: the FORR architecture for learning in a skill domain. Cogn. Sci. **18**(3), 479–511 (1994)
8. Epstein, S.L.: Spatial representation for pragmatic navigation. In: Hirtle, S.C., Frank, A.U. (eds.) COSIT 1997. LNCS, vol. 1329, pp. 373–388. Springer, Heidelberg (1997)
9. Epstein, S.L.: Pragmatic navigation: reactivity, heuristics, and search. Artif. Intell. **100**(1–2), 275–322 (1998)
10. Epstein, S., Schneider, E., Ozgelen, A.T., Munoz, J.P., Costantino, M., Sklar, E.I., Parsons, S.: Applying FORR to human/multi-robot teams. In: Human-Agent-Robot Teamwork Workshop at 7th ACM/IEEE International Conference on Human-Robot Interaction (2012)
11. Frommberger, L., Wolter, D.: Spatial abstraction: aspectualization, coarsening, and conceptual classification. In: Freksa, C., Newcombe, N.S., Gärdenfors, P., Wölfl, S. (eds.) Spatial Cognition VI. LNCS (LNAI), vol. 5248, pp. 311–327. Springer, Heidelberg (2008)
12. Gat, E.: On three-layer architectures. In: Kortenkamp, D., Bonnasso, R.P., Murphy, R. (eds.) Artificial Intelligence and Mobile Robots. AAAI Press, Cambridge (1998)
13. Gerkey, B., Vaughan, R., Howard, A.: The player/stage project: tools for multi-robot and distributed sensor systems. In: The International Conference on Advanced Robotics, pp. 317–323 (2003)
14. Gibson, J.J.: The theory of affordances. In: Shaw, R., Bransford, J. (eds.) Perceiving, Acting, and Knowing: Toward an Ecological Psychology, pp. 67–82. Lawrence Erlbaum, Mahwah (1977)
15. Golledge, R.G.: Path selection and route preference in human navigation: a progress report. In: Kuhn, W., Frank, A.U. (eds.) COSIT 1995. LNCS, vol. 988. Springer, Heidelberg (1995)
16. Golledge, R.G.: Human wayfinding and cognitive maps. In: Golledge, R.G. (ed.) Wayfinding Behavior, pp. 5–45. Hopkins University Press, Baltimore (1999)
17. Hamburger, K., Dienelt, L.E., Strickrodt, M., Röser, F.: Spatial cognition: the return path. In: Knauff, M., Pauen, M., Sebanz, N., Wachsmuth, I. (eds.) Cognitive Science 2013, pp. 537–542. Cognitive Science Society, Austin (2013)
18. Hart, P.E., Nilsson, N.J., Raphael, B.: A formal basis for the heuristic determination of minimum cost paths. IEEE Trans. Syst. Sci. Cybern. SSC4 **4**(2), 100–107 (1968)

19. Hölscher, C., Tenbrink, T., Wiener, J.M.: Would you follow your own route description? cognitive strategies in urban route planning. Cognition **121**, 228–247 (2011)

20. Jonietz, D., Timpf, S.: An affordance-based simulation framework for assessing spatial suitability. In: Tenbrink, T., Stell, J., Galton, A., Wood, Z. (eds.) COSIT 2013. LNCS, vol. 8116, pp. 169–184. Springer, Heidelberg (2013)

21. Joshi, S., Schermerhorn, P., Khardon, R., Scheutz, M.: Abstract planning for reactive robots. In: Parker, L.E. (ed.) ICRA 2012, pp. 4379–4384. IEEE, Saint Paul (2012)

22. Koenig, S., Likhachev, M., Furcy, D.: Lifelong planning A*. Artif. Intell. **155**(1), 93–146 (2004)

23. Kostavelis, I., Gasteratos, A.: Semantic mapping for mobile robotics tasks: a survey. Robot. Auton. Syst. **66**, 86–103 (2015)

24. Mataric, M.: Integration of representation into goal-driven behavior-based robots. IEEE Trans. Robot. Autom. **8**(3), 304–312 (1992)

25. Meilinger, T.: The network of reference frames theory: a synthesis of graphs and cognitive maps. In: Freksa, C., Newcombe, N.S., Gärdenfors, P., Wölfl, S. (eds.) Spatial Cognition VI. LNCS (LNAI), vol. 5248, pp. 344–360. Springer, Heidelberg (2008)

26. Montello, D.R., Sas, C.: Human factors of wayfinding in navigation. In: Karwowski, W. (ed.) International Encyclopedia of Ergonomics and Human Factors, 2nd edn, pp. 2003–2008. CRC Press, London (2006)

27. Mozos, Ó.M., Triebel, R., Jensfelt, P., Rottmann, A., Burgard, W.: Supervised semantic labeling of places using information extracted from sensor data. Robot. Auton. Syst. **55**, 391–402 (2007)

28. Ozgelen, A.T., Sklar, E.I.: Toward a human-centric task complexity model for interaction with multi-robot teams. In: Workshop on Human-Agent Interaction Design and Models (HAIDM) at AAMAS-2014 (2014)

29. Ratterman, M.J., Epstein, S.L.: Skilled like a person: a comparison of human and computer game playing. In: Moore, J.D., Lehman, J.F. (eds.) Cognitive Science 95, pp. 709–714. Lawrence Erlbaum Associates, Pittsburgh (1995)

30. Reineking, T., Kohlhagen, C., Zetzsche, C.: Efficient wayfinding in hierarchically regionalized spatial environments. In: Freksa, C., Newcombe, N.S., Gärdenfors, P., Wölfl, S. (eds.) Spatial Cognition VI. LNCS (LNAI), vol. 5248, pp. 56–70. Springer, Heidelberg (2008)

31. Sklar, E.I., Ozgelen, A.T., Munoz, J.P., Gonzalez, J., Manashirov, M., Epstein, S.L., Parsons, S.: Designing the HRTeam framework: lessons learned from a rough-and-ready human/multi-robot team. In: Dechesne, F., Hattori, H., ter Mors, A., Such, J.M., Weyns, D., Dignum, F. (eds.) AAMAS 2011 Workshops. LNCS, vol. 7068, pp. 232–251. Springer, Heidelberg (2012)

32. Speekenbrink, M., Konstantinidis, E.: Uncertainty and exploration in a restless bandit task. In: Bello, P., Gaurini, M., McShane, M., Scassellati, B. (eds.) Cognitive Science 2014, pp. 1491–1496. Cognitive Science Society, Austin (2014)

33. Takemiya, M., Ishikawa, T.: Strategy-based dynamic real-time route prediction. In: Tenbrink, T., Stell, J., Galton, A., Wood, Z. (eds.) COSIT 2013. LNCS, vol. 8116, pp. 149–168. Springer, Heidelberg (2013)

34. Tenbrink, T., Bergmann, E., Konieczny L.: Wayfinding and description strategies in an unfamiliar complex building. In: 33rd Annual Conference of the Cognitive Science Society, pp. 1262–1267 (2011)

35. Thrun, S., Bücken, A., Burgard, W., Fox, D., Fröhlinghaus, T., Hennig, D., Hofman, T., Krell, M., Schmidt, T.: Map learning and high-speed navigation in RHINO. In: Kortenkamp, D., Bonasso, R.P., Murphy, R. (eds.) AI-based Mobile Robots: Case Studies of Successful Robot Systems, pp. 21–52. MIT Press, Cambridge (1998)

36. Tversky, B.: Cognitive maps, cognitive collages, and spatial mental models. In: Frank, A.U., Campari, I. (eds.) Spatial Information Theory: A Theoretical Basis for GIS. LNCS, vol. 716, pp. 14–24. Springer, Berlin (1993)

37. Vasardani, M., Timpf, S., Winter, S., Tomko, M.: From descriptions to depictions: a conceptual framework. In: Tenbrink, T., Stell, J., Galton, A., Wood, Z. (eds.) COSIT 2013. LNCS, vol. 8116, pp. 299–319. Springer, Heidelberg (2013)

38. Vasudevan, S., Gächter, S., Nguyen, V., Siegwart, R.: Cognitive maps for mobile robots —an object based approach. Robot. Auton. Syst. 55(5), 359–371 (2007)

39. Weiser, P., Frank, A.U.: Cognitive transactions – a communication model. In: Tenbrink, T., Stell, J., Galton, A., Wood, Z. (eds.) COSIT 2013. LNCS, vol. 8116, pp. 129–148. Springer, Heidelberg (2013)

40. Wolf, D.F., Sukhatme, G.S.: Semantic mapping using mobile robots. IEEE Trans. Robot. 24(2), 245–258 (2008)

41. Zetzsche, C., Galbraith, C., Wolter, J., Schill, K.: Representation of space: image-like or sensorimotor? Spat. Vis. 22(5), 409–424 (2009)

Defensive Wayfinding: Incongruent Information in Route Following

Martin Tomko[✉] and Kai-Florian Richter

Department of Geography, University of Zürich, Zürich, Switzerland
{martin.tomko,kai-florian.richter}@geo.uzh.ch

Abstract. Extensive research has focused on what constitutes *good* route directions, identifying qualities such as the logical sequential ordering, the inclusion of landmarks, and ergonomic ways of referring to turns as critical to delivering cognitively adequate instructions. In many cases, however, people are not actually provided with route directions adhering to these qualities. Yet, often people are still able to successfully navigate to the planned destinations, despite poor or even erroneous direction giving. In this paper, we introduce the concept of *defensive wayfinding* as the particular type of problem solving people undertake when presented with route directions incongruent with their experience of the environment. We present a systematic investigation of the incompatibilities that may occur between route descriptions and the environment. We note that the content of route directions is produced by the direction giver based on observations of the environment. We develop a classification of the impacts of uncertainty in these observations based on the theory of measurement scales of Stevens [33]. We then relate uncertainty to its impact on route following and the ability of the wayfinder to detect problems during wayfinding. We conclude with a discussion of the impacts of common-sense expectations on the need to engage in defensive wayfinding.

1 Introduction

It is not uncommon to receive route directions from a friend, someone on the street, or a spatial assistance system that do not match conventions or immediate experience of the environment during wayfinding. Whether the names of the streets do not match, the hotel room is not found at an expected floor, or a landmark is not on the expected side of the road, people have to resort to common sense to cope with such mismatches between the expected spatial configurations and the experienced environment.

In this paper, we approach this resilience of people in face of uncertainty and conflicting information and analyze in a formal manner how it is possible to deal with mismatches between the information provided and the perceived environment. While wayfinders exhibit a high level of confidence in automated route directions [21], erroneous route directions are not uncommon, due to uncertainty inherent in the underlying spatial data, the algorithms used, or in the case of human route directions due to distortions in spatial memory.

© Springer International Publishing Switzerland 2015
S.I. Fabrikant et al. (Eds.): COSIT 2015, LNCS 9368, pp. 426–446, 2015.
DOI: 10.1007/978-3-319-23374-1_20

We propose a typology of route direction uncertainty that can occur in diverse types of route directions, such as turn-based directions [29] and destination descriptions [38]. We have chosen route directions as a prototypical representative of a pragmatic type of spatial communication that must uniquely and unambiguously describe the destination to the wayfinder [38]. We do not consider characteristic descriptions [37] or illustrative and artistic types of spatial communication [40] in this paper.

To our knowledge, such a typology of uncertainty impacting on the quality of route directions has not been proposed elsewhere, in contrast to the in-depth analysis of the errors that may be committed by wayfinders. The proposed typology is, in our opinion, a necessary step to the understanding of the resilience demonstrated by human wayfinders in following erroneous or ambiguous route directions. If we can understand how humans apply common sense and combine other information available to them during wayfinding to resolve uncertainty, we may also start building automata that will be able to follow directions provided by humans—such as the self driving intelligent cars coming to our streets in the next few years.

The next section further sets the context of our work by presenting relevant research from different areas. Then, we first introduce the concept of *defensive wayfinding* and other key concepts (in Sect. 3). We provide a theoretical analysis and typology of uncertainty which may occur in route directions in Sect. 4. Section 5 discusses mismatches between instructions and/or information perceived from the environment with a wayfinder's common-sense understanding of the world. The paper ends with a discussion of this typology in light of current state of wayfinding research in Sect. 6.

2 Background

In this section, we review relevant research on wayfinding, wayfinding errors and cognitively ergonomic route directions. Most importantly, we link verbal route directions and their elementary route instructions to linguistic work on referring expressions, a fundamental concept in our discussion throughout this paper. Finally, we briefly introduce the classification of measurement levels from applied statistics. This classification forms the basis of the typology developed in this paper. As we shall argue, uncertainty in route directions is the consequence of problems in the observation of the world by the direction giver, or of differing conceptualizations of the world by the direction giver and the wayfinder.

2.1 Aided Goal-Oriented Wayfinding

Wayfinding, as defined by Golledge is the dominantly cognitive activity related to *"the process of determining and following a path or route between an origin and destination"* ([14], p.6). As such, this is an inherently goal-oriented activity. In this paper, we focus on aided wayfinding [42], where the route is communicated verbally or depicted graphically. Both descriptions and depictions can take the form of a procedural discourse [36] as either turn-based route directions [28,44], or that of destination descriptions [38] (a type of descriptive discourse), or their combination [29]. Route directions as a form of spatial communication have been widely researched by a range of different disciplines.

Properties of good route directions have been discussed extensively (e.g., [1, 7,19]. Allen [1] distinguishes the following verbal devices used in communication of route directions: (1) Environmental features called *nominals* (e.g., 'church tower') (2) Delimiters such as distance and direction designations, relational terms specifying relative positions and relative movement, and modifiers providing additional detail to more precisely specify referents or actions (e.g., '... right in 500 m'); (3) Verbs of movement (e.g., 'turn'); and (4) State of being verbs (e.g., 'You are...'). Allen [1] also highlights as a principal characteristics of good turn-based route directions that instructions in their sequential ordering need to be aligned with the order in which wayfinders experience the environment. Furthermore, descriptives, which relate instructions to the decision points along the route, have been identified as being important [1]. This is supported by Denis and colleagues [7,25], who among others have identified landmarks at decision points to significantly contribute to high quality route directions.

2.2   Wayfinding Errors and Route Directions

When following a set of directions during wayfinding, the content of the directions is matched with the perceived information stimuli [14]. The more complex an intersection, the more complex the directions provided by instruction givers must be [15]. Haque et al. [15] also introduce the concept of *instruction equivalent choices*, or, in other words, the actions one can make to satisfy an instruction. The more equivalent choices there are for a given instruction, the more detailed instructions are offered by human direction-givers, with more nuanced descriptions of decision points [17]. In a similar line of research, Hirtle et al. [16] have explored the way complex parts of a route result in more detailed route directions.

The process of interpretation of route directions as proposed in [3] has been applied to study the errors committed by wayfinders when following route directions. Brunye et al. [2] discussed the preferential choices, i.e., different kinds of heuristics, to deal with situations where the wayfinder encounters detectable incongruences between the environment and the instructions. They discuss landmark-based and direction-based preferential heuristics, with a stronger trust in turns provided by GPS navigation systems and a higher trust in landmarks provided by human direction givers.

In this paper, we explicitly focus on the classification of disparities between the information *about* the route acquired from route directions and the perceived stimuli from the physical environment, including signage [26]. We analyze the types of uncertainty that may influence the quality of the provided directions. This is the fundamental distinction from [2,43] and other research on human decision-making, which focuses on errors committed by wayfinders themselves (e.g., mistakes, slips and lapses), while the information content of the directions is taken as unambiguous and factually correct. We also focus exclusively on the explicit information content of the directions and do not consider other uncertainty markers that may provide additional clues about the quality of the provided directions (e.g., prosody) [35], or the clarity of the instructions themselves [31].

2.3 Route Directions and Pragmatic Communication

Route directions transfer highly pragmatic information content communicated in a succinct manner [13,38]. Frank [13] proposed the measurement of the content of route directions in terms of their *pragmatic* value, following previous work on the concept of correctness of maps and spatial communication with maps using multi-agent simulation [12].

Expressions aimed at uniquely identifying a referent from a set of potential referents are called *referring expressions* [5,8,34], and route directions can be considered a particular case of such expressions. Route descriptions either uniquely identify the locations and types of actions to be taken (turn-based route directions), or in the case of destination descriptions uniquely identify the destination of the route [38].

Each instruction contained within a turn-based route description should also uniquely and succinctly identify the decision point and the action to be taken [18]. In this sense, an instruction is ideally a referring expression in its own right. In this paper, we argue that ambiguities in interpretations of instructions particularly arise when these are not referring expressions, but they are either underspecified or overspecified—i.e., their information content is either insufficient or excessive [22,32].

2.4 Classifications of Measurements and Uncertainty

Stevens [33] defines a measurement as the assignment of values to an observation based on consistent rules. As we will demonstrate, this equally applies to the observations of the world that translate into the acquisition of spatial knowledge, either stored in a computer or held by a person. In case of collecting data in a mapping workflow, the applied rules are formalized and made explicit; in cognitive mapping they are subjective and usually not available for conscious reflection. Still, based on the spatial knowledge acquired through observations, the speaker or a computer can provide route directions.

Stevens' typology of measurement consists of the following four distinct scales, in order of decreasing statistical transformations that are permissible: *ratio, interval, ordinal and nominal*. Beyond the statistical operations, Stevens also notes the empirical transformations relating to these measurement scales, as they are likely to be experienced during our common-sense interaction with the geographic space. These are the determination of respective ratios, the determination of equality of intervals or differences, determination of relative magnitude, and determination of equality between observations [33]. Our observations of the geographic space are biased and prone to errors [10], and it is exactly the impact of this uncertainty in the observations of different measurement scales that we study in this paper.

We are well aware of the criticism of Stevens' typology from the measurement sciences as well as from statisticians. This criticism focused on the definitions of the measurement scales, the types of levels of measurement themselves, as well as on the rigidity and prescriptive nature of the classification disliked by statisticians. Nevertheless, the practical applicability of this systematization (as proven by its widespread application in the behavioral and social sciences) shows the pragmatics of its application. In GIScience, an extension to the classification was proposed by Chrisman [4]. As the original classification of Stevens is sufficient for our purpose, we do not further discuss this extended classification.

Each measurement is influenced to a certain extent by uncertainty, but so far, most research has focused on route directions that were supposed to be *correct*, or certain. Fisher [11] identifies two major types of uncertainty: (1) the probabilistic type occurring when measuring well defined objects or their classes (often called *error* in measurement science and relating to the difference between the observed and the true values), and (2) the type of uncertainty originating from a vague or ambiguous definition of the object or its class. This latter type of ambiguity includes sub-types relating to poor definition of the object or class themselves (e.g., the boundary of a forest), or to the discrepancies between multiple classification systems (e.g., when two observers cannot agree whether an angular deviation is best referred to as *straight* or it is already a *turn*). We will build on this classification of uncertainty in our systematization.

3   Defensive Wayfinding

This section will introduce the concept of *defensive wayfinding*. We will start by defining a few key terms, which will then be used to disentangle the types of uncertainty that people (or machines) may include in their instructions. The typology of uncertainty, presented in the next section, will be applied to study the impact on observations belonging to different types of measurements [33]. It distinguishes between descriptions of actions and descriptions of location specifications.

3.1   Defensive Wayfinding and the Interpretation of Route Directions

We call the need to deal with instructions that are incongruent with the expected scene due to uncertainty or a breach of common-sense expectations *defensive wayfinding*, inspired by the term used for defensive driving—driving under adverse conditions. We define it as:

Definition 1. *Defensive wayfinding is a form of goal-oriented wayfinding occurring under adverse conditions, where the wayfinder must exert excessive mental effort to align, correct, supplement or find alternative interpretations of the information acquired from route instructions or signage because of a mismatch with the perceived structure of the environment or with their own expectations about the environment.*

Route directions generated by both human direction givers and computer systems can include uncertainty originating from multiple sources: (1) the data (or mental knowledge) from which they are generated are incomplete, inaccurate, or based on an incompatible conceptualization of the world and its properties; (2) the reference selection algorithms and heuristics applied are wrong or do not consider some parameters of the data; or (3) the process translating the computed path and spatial object references into verbal directions or graphical depictions is inaccurate or ambiguous. In the remainder of this paper, we do not focus on the causes of these issues anymore, but rather study their manifestation.

Expanding on Frank [12], the process of interpreting route directions by the wayfinder includes the reconciliation of four main sources of information:

1. the existing spatial knowledge of the wayfinder (if available);
2. information from the perceived physical environment;
3. the information contained in the instructions, usually communicated through two main forms: textual or verbal route directions, including destination descriptions. and visualizations as maps, including sketches;
4. common sense heuristics based on learned patterns of behaving in a given environment (e.g., that people drive on the left side of the street in the UK, that highway exits are numbered consecutively, or that the first digit in a hotel room number relates to the floor on which it is found).

The misalignment between the environment and the received wayfinding instructions closely relates to the concept of correctness of maps as introduced by Frank [12]. Uncertainty impacts on route following when a route instruction is insufficiently detailed (e.g., incomplete), or when some of the referents or their attributes are not selected correctly. This is the case of ambiguous directions, including situations when erroneous information is provided as part of excessive detail and the instruction therefore contains contradictions.

When common-sense assumptions of the wayfinder about the world are not satisfied during wayfinding without prior notice in the route descriptions or some additional source of information, the wayfinder may also commit wayfinding errors or have to exert excessive mental effort to follow the route. In this paper, we specifically address the content provided to wayfinders in route descriptions, along with the compatibility of these descriptions with their common-sense assumptions (discussed in Sect. 5). We leave the discussion of the role of prior spatial knowledge and the accuracy of their own perception of the environment to future work.

3.2   Structure and Content of Route Instructions

Keeping in mind the research on good route directions (e.g., [1,19]), we take an abstract perspective on route directions in this paper. That is, we are not concerned with concrete verbal or graphical instantiations of instructions. We consider them generally consisting of two parts: an action part (i.e., a turn instruction), and a location specification, which may be a place description or a street name. We will use the following terms throughout the definition and discussion of our typology of route direction uncertainty.

Instruction. The main element holding information for the wayfinder to follow at a decision point. It consists of a combination of an action description and of a location specification, where one of these parts may or may not be present, and each of them may or may not be incomplete, incorrect, or not matching the common-sense expectations of the wayfinder;

Action Description. The specification of the actions to take at a given decision point (e.g., left, veer right).

Location Specification is a description aiming at facilitating the identification of the location where a turn ought to happen. It may consist of references to diverse elements of the city form [20] and usually includes a reference to the most salient properties of the elements referred to, selected from all possible referents available in the intersection scene [30]. These properties may be, for example, colour, name, shape, or egocentric position as it should be perceived by the wayfinder at the decision point. Note that location specifications can range widely in granularity (e.g., references to landmarks vs. references to entire districts and city parts, such as the CBD).

Route is defined as a sequence of following segments and turns at intersections that allow the wayfinder to follow a specific path. Routes are assumed to be correct and complete, i.e., actually leading from origin to destination. Further, it should also be possible to generate a correct and complete set of instructions for a route that would guide a wayfinder to the destination, i.e., not only does the route exist, but it is also describable.

Route Directions are a sequence of instructions communicated to the wayfinder and intended to allow the wayfinder to reach the destination. In the case of turn-based instructions, route directions prescribe the *how* of following the path, while in the case of destination descriptions, they describe the *where* of the destination. Route directions may be impacted by uncertainty—for example, be incomplete and incorrect.

Intersection Scene provides the hypothetical full description of the environment at a potential decision point. It contains references to all the possible turns and landmarks that can be included in an instruction, as well as signage. The concept of intersection scene is important to be able to identify distractors in location specifications or action descriptions.

3.3   Uncertainty in Route Directions

If individual instructions describing the route do not represent minimum length referring expressions, i.e., exact, shortest possible specifications [5] of the action and location, the instruction may not be unambiguously decoded and the route may not be followed successfully. Wayfinders may apply heuristics and combine the instructions with their own environmental knowledge to follow such ambiguous instructions. When wayfinders are able to detect one or multiple discrepancies with expected observations of the environment, they may reduce their trust in either the provided directions and/or the information they encounter in the environment (e.g., signage), and approach subsequent information cautiously.

Instructions that are not minimum-length referring expressions are sometimes issued to provide wayfinders with a heightened sense of security about the provided directions (e.g., 'turn left at the red building, it is next to the church'). Contradictions in route instructions may occur when a redundant amount of information is provided and some part of this redundant information is incorrect (e.g., there is no church next to the red building). If the information content is contradictory, this strategy of providing extra information is actually counterproductive, as the excessive information leads to insecurity about which part of the instruction is correct [2].

Instructions may also be underspecified, i.e., not provide enough information to unambiguously specify what to do. This may happen because the instruction giver has incomplete spatial knowledge, does not remember the spatial configuration correctly, or does not realise that more information is required for some other reasons. Accordingly, we define under- and overspecification of instructions as:

Definition 2. *Underspecification in a route instruction is the provision of an insufficient amount of information about the action to be taken or the location where it should be taken. It occurs when the instruction is not enough to decide which referent or action is to be selected from the set of confounding elements of the intersection scene.*

Definition 3. *Overspecification in a route instruction is the provision of too much information about the action to be taken or the location where it should be taken. It occurs when the instruction contains excess information that is contradictory and, thus, does not allow the correct referent or action to be selected from the set of confounding elements of the intersection scene.*

We apply elements of Fisher's [11] classification of uncertainty as follows:

Definition 4. *Error in a route instruction is the specification of an inaccurate, albeit possibly precise observation value, and relates to the probabilistic component of uncertainty.*

Definition 5. *Ambiguity in a route instruction is an imprecise (to an extent that it may also be missing) specification of an action or referent or its properties that allows for multiple valid interpretations of the part of the instruction by the wayfinder.*

Note that we do not explicitly distinguish vagueness from ambiguity here. Spatial concepts referred to in route directions often have a vague nature (e.g., a slight turn referred to as *veer right*). As we focus on the interpretation of the information by the wayfinder, the problem is manifested as a *discord* between the conceptualizations of the direction provider and the recipient [11]. We believe that a further distinction is not necessary at this stage.

3.4   Detectability of Uncertainty in Route Instructions

In a range of cases, the uncertainty in the provided route instructions may not be immediately evident to the wayfinder and may not be detected while taking an action at a decision point. We call such uncertainty *not detectable*. However, the errors and ambiguities may become manifest later along the route. If sufficient signals from the environment allow wayfinders to realize that they are lost, they may be able to retrace their steps and choose an alternative interpretation of the instructions and alter their actions at a decision point where they suspect the instructions were not correct. The identification of this place may not be always possible (this is when the wayfinder is truly lost). In particular, if an observation can be matched to a non-intended referent observable in a scene at a specific decision point, it means that the value of a property referred to is inaccurate but precise. This is (in isolation) a non-detectable error that may lead to a wrong action by the wayfinder.

Accordingly, our final definition captures this issue of misleading instructions that may (literally) throw the wayfinder off track.

Definition 6. *Detectability* *is the ability of a wayfinder to identify* in-situ *the possibility that there is a problem with the given route instruction. This allows the wayfinder to proceed cautiously (if possible, e.g., using heuristics) and to consciously backtrack and take an alternative action at this point if the first heuristic proves incorrect.*

4   Route Following with Uncertain Instructions

We now analyze uncertainty in view of its detectability *in-situ* during route following, by exploring the nature of the uncertainty, which again depends on the nature of the observation (location, action, or their properties) expressed in the instructions. We argue that only if wayfinders can identify a problematic instruction can they also resiliently resolve the situation during route following. We will introduce observations in function of the applicable measurement levels that captured them, using Stevens' classification scheme [33]. In our examples, we use a simplified model of route instructions, considering only turn specifications for the action descriptions. These represent the most common type of action descriptions and sufficiently illustrate the concept.

Tables 1 and 2 present the proposed typology of possible uncertainty types in route directions, organized along two dimensions. First, we separate by component of an instruction (the *action description* and the *location specification*). This separation is reflected by having two separate tables; one for each component. Second, we use the scales of measurements [33] to classify the way in which uncertainty impacts on different types of observations. This uncertainty is then further studied with respect to (1) underspecification; (2) the selection of an erroneous referring expression; or (3) overspecification. We then evaluate whether this uncertainty is detectable *in situ*.

Figure 1 illustrates this typology with an example wayfinding situation. The correct behavior in this fictitious example would be to turn left into the second street (Second Ave). The situation in Fig. 1(a) is underspecified, as it is not clear which street to the left to take. The situation in (b) depicts a referring expression, which unambiguously describes an action description/location specification

Table 1. Typology of uncertainty in the action (turn) description part of route instructions. Values in the *D?* column [Y,N] specify whether the uncertainty is potentially detectable by the wayfinder immediately at the decision point (as opposed to later along the route); a question mark indicates that detectability is dependent on the concrete spatial situation.

Expression	Underspecified	D?	Full referring expression	D?	Overspecified	D?
Observation						
1. Nominal	Unspecified or ambiguous turn reference: • turn reference with multiple matching possibilities 'turn LEFT at ...' where there are two options that could be interpreted as left.	Y?	Erroneously specified turn reference: • 'left' instead of 'right'.	N	Excessive specification of the turn reference that may confuse the wayfinder: • 'veer left', which is perceived as just 'left' by the wayfinder, although both interpretations may be considered correct.	Y
2. Ordinal	Missing or vague specification of the ordinal number of the turn: • 'BLANK right at the roundabout'; • 'take SOME exit from the freeway'	Y	Erroneous specification of the ordinal number of the turn: • 'take the first exit from the roundabout', where it should be the second.	N	Conflicting specification of the ordinal number of the turn: • 'turn at the second left, at the red house', but the red house is at the first.	Y
3. Interval	Ambiguous or vague specification of the relative turn magnitude: • 'turn to 11 BLANK UNITS...'	Y	Erroneous specification of the relative magnitude of the turn: • 'turn to 3 o'clock', where it should be 9 o'clock.	N	Conflicting specification of the magnitude of the turn: • 'turn right to 9 o'clock'.	Y
4. Ratio	Ambiguous or vague specification of the heading: • 'turn to azimuth approx. 30°'	Y	Erroneous specification of the turn heading • 'turn to 30°', where it should be −30°).	N	Conflicting specification of the headings • 'turn right to −90°'.	Y

combination, but it is the wrong one for this particular route. The instruction in Fig. 1(c) finally is overspecified and, thus, provides conflicting information, as the red house is in fact located at the first left turn. In these examples, the ambiguity in examples (a) and (c) is detectable, i.e., a wayfinder has a chance to realize that something is not quite clear when reaching the first street (First Ave). The situation in (b) is not detectable as it seems to be a perfectly valid instruction that is clearly executable. Here, a wayfinder may only realize that

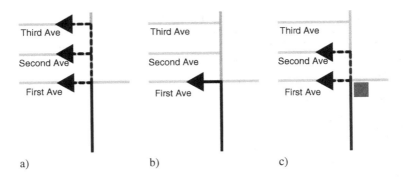

Fig. 1. Example illustrations for underspecified instructions, incorrect referring expression, and overspecified instructions. The correct action would be to turn left at Second Ave: (a) 'turn left into some street'; (b) 'turn left into the first street'; (c) 'turn left at the second street, where the red house is' (Color figure online).

something is wrong when trying to execute one of the next instructions, in case these cannot be matched to the environment anymore.

Two further types of uncertainty have been identified that are not further included in the tables below, due to their specific nature:

Instruction Omissions: this is the situation where a sequence of route instructions is incomplete and does not fully specify the route. The instruction giver omits in its entirety the reference to a decision point and the action to take there. Such a set of instructions does not represent a full route description. The wayfinder essentially has no means to recover from such a situation, and depending on the nature of the environment and prior knowledge, may or may not be able to detect the omission and resolve the problem. In measurement terms, this would be the case of a missing measurement.

Out-of-Sequence Instruction: this situation occurs when route instructions are not provided in the same sequence as decision points are encountered along the route. Again, the resulting route description is not a referring expression. The wayfinder is likely to be able to detect the situation, and may apply heuristics to solve this situation. An instruction commission (the inclusion of an excessive instruction) is in principle a special case of an out-of-sequence instruction. Depending on the nature of the environment, the wayfinder may be able to recover from this type of uncertainty, but an excessive mental effort is certainly needed to match the directions to the environment.

4.1   Measurement Scales and Route Instructions

On the nominal scale, elements of an intersection scene are simply differentiated by attaching different labels to them. For example, an instruction may differentiate different turn options by using 'left' or 'right' as a label. However, depending on the spatial configuration, turns may not be easily named in this way. Then some form of counting may be used, corresponding to an ordinal scale (e.g., 'take

Table 2. Typology of uncertainty in the location specification part of route instructions. Values in the *D?* column [Y,N] specify whether the uncertainty is potentially detectable by the wayfinder immediately at the decision point (as opposed to later along the route); a question mark indicates that detectability is dependent on the concrete spatial situation.

Expression	Underspecified	D?	Full referring expression	D?	Overspecified	D?
Observation						
1. Nominal	Unspecified referent or its property: • missing or incomplete street or landmark name or type 'turn left into BLANK Ave' where you can see First, Second, and Third Ave; • missing or ambiguous street or landmark property (e.g., physical characteristic) 'turn at the yellow house', but there are several yellow houses visible	Y?	Erroneously specified referent or its property: • Incorrect street or landmark name and/or type 'turn left into First Ave' where Second Ave would be correct; • incorrect street or landmark property reference uniquely identifying a non-intended (possibly non-existent) referent 'turn left at the yellow house' where the turn should be at the red house	N	Reference to an excessive street or landmark or their property that may confuse the wayfinder: • 'turn left at the big yellow-brick church', however the big church is red-brick	Y
2. Ordinal	Missing or vague specification of the ordinal number of the referent or of the type of the referent: • 'turn left at the BLANK roundabout'; • 'turn left at the fifth BLANK'	Y	Erroneous specification of the ordinal number of the referent: • 'turn after the third petrol station', but the correct one would be the second.	N	Conflicting specification of the ordinal number of the referent: • 'take the third roundabout, where the red house is', but the house is at the first roundabout).	Y
3. Interval	Unspecified or vague specification of the interval, or missing or vague specification of the referent described by the interval: • 'in five BLANK.UNITS'; 'after SOME minutes'.	Y	Erroneous specification of the interval or its type for the referent: • 'turn left in 20m', but correct turn would be in 200m	N	Conflicting specification of the interval or it's type for the referent: • 'turn left in 200m, at the red house', but the red house is only 50m away	Y
4. Ratio	Missing or ambiguous specification of distance (quantity and/or units, metric or temporal) from start of the route (only applicable ratio scale): • 'in 10 BLANK from start turn left'.	Y	Erroneous specification of distance (quantity and/or units, metric or temporal) from start of the route: • 'turn left in 5min from start', but correct turn would be in 10min.	N	Conflicting specification of the distance (quantity and /or units, metric or temporal) from start of the route: • 'turn left in 10min from start, at the red house', but the red house is 2min from start	Y

the third exit from the roundabout'). These may be the two most commonly used scales in route directions.

As actions are by their nature local, they are rarely anchored in a reference frame that would allow the action descriptions to be related to an absolute origin. Accordingly, interval and ratio scale are very rarely applicable in action descriptions. In the case of turn specifications, interval scale requires the indication of a turn magnitude relative to the wayfinder. One such case is the clock position angular specification. It is a system used in aviation and other mostly natural environments. We believe that azimuth (heading) based, (and usually device assisted) navigation is the only case where a (cyclic) ratio measurement scale is applicable [4]. These turns are anchored in an absolute reference frame[1].

Interval and ratio scale may be more common for the location specification component. Still, employing the ratio scale for location specifications, while possible in principle, is rarely used. It is only applicable to instrument-assisted wayfinding (e.g., using an odometer or GPS coordinates), and requires an absolute, continuous measurement from the start of the route or some other absolute reference system. This measurement scale is used in rally navigation[2], where ordinal and interval observations are still more common.

An instruction employs an interval scale any time it mentions some kind of explicit distance information (e.g., 'in 2 km turn right', 'after 5 mins you'll have to go straight'). In human-generated route directions, such distance information usually only indicates estimations (rather than precise measurements). Often, the quantitative information may be complemented by statements about uncertainty, such as 'about' (e.g., 'in *about* 5 km'). They still, however, allow for a comparison of distance intervals, for example, 2 km (meaning 'a while') vs. 200 m (meaning 'soon'). Ordinal and nominal scale capture references to spatial objects, be they landmarks or streets. If there are multiple similar referents that get distinguished through counting ('the third roundabout'), the description is on an ordinal scale, otherwise it is nominal ('the roundabout').

4.2   Detectability of Uncertainty

For the action component of a route instruction (Table 1), errors and ambiguity are possibly detectable in case the turn component is under- or overspecified. Since overspecification leads to uncertainty due to confusion—i.e., there is excessive information which is internally in conflict and, thus, does not match the encountered situation—this will always be detectable. Underspecified turn instructions are detectable if a wayfinder can realise that there are multiple options and the correct one has not been unambiguously specified. This will be the case for instructions on the ordinal scale; here the ordinal number is missing or only vaguely defined. On the nominal scale, underspecification is only

---

[1] e.g.,   http://www.toujourspret.com/techniques/orientation/exploration/releve_gil well.gif.

[2] e.g., http://www.ladakar.com/wp-content/uploads/2011/12/Dakar-Assistance-Road book-1.jpg.

detectable if any other applicable option is available that could be referred to with the same referents. This is usually called a *distractor*.

Imagine that in the example in Fig. 1(a) the instruction would be 'turn left'. This would be technically underspecified as there are multiple streets leading to the left, however, a valid common-sense assumption would be to turn left at the first available option. Thus, this kind of underspecification may be non-detectable (see also the discussion in Sect. 5). Conversely, uncertainty in over-specification is often detectable. For instance, in the case 'turn left at the big yellow-brick church', the church in question is indeed larger than usual, but it is made of red bricks. The wayfinder would most likely be able to detect the error in the instruction and act accordingly.

Arguments about the detectability of uncertainty in the location specification component run along similar lines as those for the action component (Table 2). Incorrect referring expressions are essentially always not detectable in-situ as they unambiguously describe spatial configurations that are actually there and can be identified. Only in cases where an instruction erroneously refers to an element that is actually not existing (e.g., 'Smith St.' instead of 'Miller St.'), this may be ultimately detectable. However, arguably, such an instruction would not be a full referring expression anymore. Overspecification again will lead to confusion and conflicts, which are always detectable. Likewise, underspecification is detectable if it is evident from the instruction/spatial configuration combination that the instruction is vague or ambiguous.

5   Defensive Wayfinding with Incongruent Expectations

Beyond the information inherent in the route descriptions, other expectations of the wayfinder significantly impact on the interpretation of the references in route directions. In this section, we briefly discuss some of these expectations that facilitate wayfinding. While we believe that an exhaustive catalogization of such expectations may not be possible, we identify affordance expectations and common-sense beliefs among the primary expectations that a wayfinder makes when relating the route directions to the environment. If these expectations are not satisfied, the wayfinder resorts to defensive wayfinding. We note that the wayfinder may only realize that they resorted to conscious defensive wayfinding if a certain (likely subjective) threshold of incongruity is exceeded.

Affordance Mismatch: this ambiguity occurs when an instruction that is otherwise correct cannot be followed by a wayfinder due to the mismatch of the environmental affordances with the wayfinder's accessibility constraints [39]. For instance, a city wayfinder may expect all of the streets referred to in the route directions to be accessible—either legally (no private roads) or physically (e.g., no stairs on a path for a person with physical disability). Affordance mismatches often occur when roads are closed during roadworks or accidents. Depending on the nature of the route and the extent of the mismatch, the wayfinder may be able to solve this problem by applying heuristics (e.g., make a detour). This situation is certainly detectable by the wayfinder and has a good likelihood of being solved successfully.

Common-Sense Beliefs: common-sense beliefs fundamentally enable our daily interaction with the environment, and are grounded in the assumption that others respect them as well. They may originate from codified frameworks and laws (e.g., traffic code stating the side of the road where one should drive), or be incrementally enforced by the frequency with which they are experienced (e.g., the experience that the first digit of a hotel room number relates to the floor). If a source of information *about* the route or perceivable *in* the environment violates such common sense assumptions, the route giver should—if possible—emphasize this to the wayfinder during direction giving.

Signage interpretation is significantly grounded in common sense, with the signage expected to satisfy a number of expectations: (1) the pictograms to have a standardized meaning; (2) an expectation about the units used to indicate magnitudes in a given cultural environment (e.g., kilometers for distances in Europe); (3) numbered items (e.g., freeway exits, hotel room numbering) to be sequentially ordered; and others.

Hotel Wayfinding Scenario. We illustrate how the violation of some of these assumptions can result in the confusion of a wayfinder in an indoor scenario. The environment will be familiar to many participants of the COSIT 2013 conference—the Royal Hotel Scarborough, UK. Our traveler–the first author of this paper–has been given the keys to room 229. The following common sense beliefs are usually associated with such information: (1) the room will be on the second floor; (2) the room numbers will be organized sequentially; (3) room 229 will be located in the vicinity of room 228 and possibly room 230; (4) signage will be placed at decision points throughout the environment, especially where the circulation system violates these assumptions.

At first, the wayfinder's common sense beliefs hold. He takes the stairs to the second floor, where his common sense beliefs are reinforced, as directions to rooms in the 200 range are indicated (Fig. 2(b)). The ranges include number 229, to the right (Fig. 2(c)). The wayfinder circumnavigates the atrium (*void* in floorplan on Fig. 2(a)), but to no avail. Back to the first decision point he notices the signs on the opposite wall (d). The ranges indicated, however, do not contain room 229! The wayfinder is now actively in a defensive wayfinding mode. He assumes there is an error in the signage (room 229 not found in the signed ranges) as well as an incompatibility with previous assumptions (Room 601 in the range of the 200s, violated sequential ordering).

At this stage, a typical automaton would interrupt its action or require inputs from an operator, while our wayfinder resorts to heuristics and tries his luck. The wayfinder continues through a door, towards location (e). Again, 229 is not present, but 230 is, and he expects 229 to be nearby. The trust towards the signage is now low and the wayfinder is truly exploring. A room 218 is found along the corridor, matching the information on the signs. At decision point (f), a door signed *fire exit* is in the way and the wayfinder is unsure whether one is allowed to use it (possible affordance mismatch). The daring wayfinder passes

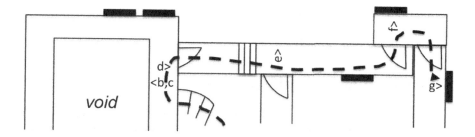

(a) Floorplan of the second floor of the hotel

(b) First decision point

(c) Signage at first decision point

(d) Second decision point

(e) Fourth decision point

(f) Fifth decision point

(g) Destination

Fig. 2. Sequence of information at decision points when looking for room 229. The route (dashed line in floorplan) to room 229 (g). Arrows $>$ with letters b, c, d, e, f and g indicate placement of signs depicted in Figure b, c, d, e, f and g.

through anyway and is confronted with room 601. Following a nondescript door behind him, the wayfinder finds the sign indicating the rooms 230–233, with room 229 there as well. Finally!

6   Discussion and Conclusions

In this paper, we have introduced the concept of defensive wayfinding as a type of wayfinding requiring excessive mental effort to align, correct and supplement assistive route information. Further, we have developed a typology of potential uncertainty in route directions based on Stevens' measurement scales [33] and Fisher's typology of uncertainty [11]. We have used this typology to explore whether the wayfinder will be able to detect whether the route directions provided contain a degree of uncertainty that would let them engage in defensive wayfinding. If not, the wayfinder has no ability to seek complementary route information or apply heuristic strategies to increase the likelihood of success during wayfinding. Finally,we have explored the role some common-sense assumptions play in wayfinding and how they combine with route directions. If these assumptions are incongruous with the received information or the perceived environment, the wayfinder needs to engage with defensive wayfinding.

In the following, we will discuss our findings in the context of the current state of wayfinding research. However, while we have focused on route directions in this paper, we believe that the general approach can be applied to systematizing errors in other forms of pragmatic communication as well, for example, in the description of object locations in table-top or vista scenarios (e.g., [24,41]).

Conciseness as a Way to Control Confusion: As shown by [6], people favor conciseness in the production of route directions, in particular avoiding excessive details in the descriptions of landmarks. Based on our formal exploration, we concur that excessive but vague or erroneous information included in route directions may lead to confusion of wayfinders and ultimately to erroneous heuristics applied by the wayfinders (cf. [32]). A separate stream of research worth of investigation is the coupling between verbal and pictorial, or verbal and gestural route instructions. How would uncertainty in overspecified instructions be handled by wayfinders? Similarly, how would statements of uncertainty or confirmation (e.g., "I think it's a left turn", "You can't miss it!" [31]) be interpreted and do they aid the wayfinder at all?

Route Knowledge as Interpretation of Survey Knowledge: Verbal route directions provide sequentially presented, limited spatial information about the route. While maps—a medium to capture survey spatial knowledge—have been extensively studied in terms of spatial data quality, to our knowledge such frameworks do not exist for route spatial knowledge. We are hopeful that this paper will provide a starting point for the evaluation of diverse representations of route spatial knowledge. Local systematic errors in the spatial data used for the generation of these route directions are hard to identify by the wayfinders due to the selective nature of the information communicated. While a map reader may be able to identify systematic errors in a map (e.g., all shops situated on the wrong side of the street, or erroneous labels on all streets in a suburb), a wayfinder receiving a limited description of a decision point where only the *interpretation* of the original measurement impacted by the systematic error is communicated (e.g., 'turn right *after* the landmark, as opposed to *before*'; '...you will see a shop on your *left*', as opposed to right) usually has no such opportunity. This interpretation is often linked to the transformation of the observation from one measurement scale to another, which may further impede on the detectability of the error.

Defensive Wayfinding and Resilient Heuristics in Wayfinding: We have classified the different types of measurements expressed in route directions. We believe that such a systematization provides a starting point for the evaluation and benchmarking of the outputs of route direction services. The question is whether uncertainty in route directions and incongruence with common sense could be modeled formally, in order to establish a computational agent able to evaluate the quality of the provided information, as well as diverse defensive wayfinding heuristics and strategies. Individual propensity to challenge the received information is likely to play a large role in the strategies employed by individuals. The above systematization may be modeled following the observation/knowledge model from Raubal and Worboys [27], and the extent to which the wayfinder has lost trust in the provided directions may be operationalized using a slot approach (cf. [9,23]).

As noted, Brunyé et al. [2] studied heuristics to deal with situations where a wayfinder detects differences between the environment and the instructions. It remains to be explored to what extent would the accumulation of errors in a description lead to a change in heuristics or a decrease in trust to a set of directions. In other words, our typology could be used to better formulate hypotheses about the heuristics applicable in different direction following situations [2,43], leading to a systematic study of the mechanisms with which wayfinders deal with defensive wayfinding situations. Similarly, our systematization will serve as a means to estimate whether route directions from different systems may confuse the wayfinder and ultimately, could be used to implement strategies in systems that need to understand human instructions in a resilient manner (such as mobile robots, or indeed self-driving cars).

References

1. Allen, G.L.: From knowledge to words to wayfinding: issues in the production and comprehension of route directions. In: Hirtle, S., Frank, A. (eds.) Spatial Information Theory. Lecture Notes in Computer Science, vol. 1329, pp. 363–372. Springer, Berlin (1997)
2. Brunyé, T.T., Gagnon, S.A., Gardony, A.L., Gopal, N., Holmes, A., Taylor, H.A., Tenbrink, T.: Where did it come from, where do you go? Direction sources influence navigation decisions during spatial uncertainty. The Q. J. Exp. Psychol. **68**(3), 585–607 (2015)
3. Burns, P.C.: Wayfinding errors while driving. J. Environ. Psychol. **18**(2), 209–217 (1998)

4. Chrisman, N.R.: Rethinking levels of measurement for cartography. Cartography Geogr. Inf. Syst. **25**(4), 231–242 (1998)
5. Dale, R., Reiter, E.: Computational interpretations of the Gricean maxims in the generation of referring expressions. Cogn. Sci. **18**, 233–263 (1995)
6. Daniel, M.P., Denis, M.: The production of route directions: investigating conditions that favour conciseness in spatial discourse. Appl. Cogn. Psychol. **18**(1), 57–75 (2004)
7. Denis, M., Pazzaglia, F., Cornoldi, C., Bertolo, L.: Spatial discourse and navigation: an analysis of route directions in the city of venice. Appl. Cogn. Psychol. **13**, 145–174 (1999)
8. Donnellan, K.: Reference and definite descriptions. Philos. Rev. **75**(3), 281–304 (1966)
9. Duckham, M., Kulik, L.: "Simplest" paths: automated route selection for navigation. In: Kuhn, W., Worboys, M.F., Timpf, S. (eds.) COSIT 2003. LNCS, vol. 2825, pp. 169–185. Springer, Heidelberg (2003)
10. Egenhofer, M.J., Mark, D.: Naive geography. In: Frank, A., Kuhn, W. (eds.) Spatial Information Theory. Lecture Notes in Computer Science, vol. 988, pp. 1–15. Springer, Berlin (1995)
11. Fisher, P.F.: Models of uncertainty in spatial data. In: Longley, P.A., Goodchild, M., Maguire, D.J., Rhind, D.W. (eds.) Geographical Information Systems, vol. 1, 2nd edn., pp. 191–205. Longman, Essex (1999)
12. Frank, A.U.: Spatial communication with maps: defining the correctness of maps using a multi-agent simulation. In: Habel, C., Brauer, W., Freksa, C., Wender, K.F. (eds.) Spatial Cognition 2000. LNCS (LNAI), vol. 1849, pp. 80–99. Springer, Heidelberg (2000)
13. Frank, A.: Pragmatic information content: how to measure the information in a route description. In: Duckham, M., Goodchild, M., Worboys, M. (eds.) Foundations of Geographic Information Science, pp. 47–68. Taylor and Francis, London (2003)
14. Golledge, R.G.: Human wayfinding and cognitive maps. In: Golledge, R.G. (ed.) Wayfinding Behavior: Cognitive Mapping and other Spatial Processes, pp. 5–45. Johns Hopkins University Press, Baltimore (1999)
15. Haque, S., Kulik, L., Klippel, A.: Algorithms for reliable navigation and wayfinding. In: Barkowsky, T., Knauff, M., Ligozat, G., Montello, D.R. (eds.) Spatial Cognition 2007. LNCS (LNAI), vol. 4387, pp. 308–326. Springer, Heidelberg (2007)
16. Hirtle, S., Richter, K.F., Srinivas, S.: This is the tricky part: when directions become difficult. J. Spat. Inf. Sci. **1**, 53–73 (2010)
17. Klippel, A., Tenbrink, T., Montello, D.R.: The role of structure and function in the conceptualization of directions. In: van der Zee, E., Vulchanova, M. (eds.) Motion Encoding in Language and Space. Oxford University Press, Oxford (2010)
18. Klippel, A., Richter, K.F., Hansen, S.: Cognitively ergonomic route directions. In: Karimi, H. (ed.) Handbook of Research on Geoinformatics, Chap. XXIX, pp. 230–237. IGI: Information Science Reference, Hershey (2009)
19. Lovelace, K.L., Hegarty, M., Montello, D.R.: Elements of good route directions in familiar and unfamiliar environments. In: Freksa, C., Mark, D.M. (eds.) COSIT 1999. LNCS, vol. 1661, pp. 65–82. Springer, Heidelberg (1999)
20. Lynch, K.: The Image of the City. The MIT Press, Cambridge (1960)
21. Ma, R., Kaber, D.B.: Effects of in-vehicle navigation assistance and performance on driver trust and vehicle control. Int. J. Indus. Ergon. **37**(8), 665–673 (2007)

22. Mackaness, W., Bartie, P., Espeso, C.S.-R.: Understanding information requirements in "Text Only" pedestrian wayfinding systems. In: Duckham, M., Pebesma, E., Stewart, K., Frank, A.U. (eds.) GIScience 2014. LNCS, vol. 8728, pp. 235–252. Springer, Heidelberg (2014)
23. Mark, D.M.: Automated route selection for navigation. IEEE Aerosp. Electron. Syst. Mag. 1(9), 2–5 (1986)
24. Mast, V., Wolter, D.: A probabilistic framework for object descriptions in indoor route instructions. In: Tenbrink, T., Stell, J., Galton, A., Wood, Z. (eds.) COSIT 2013. LNCS, vol. 8116, pp. 185–204. Springer, Heidelberg (2013)
25. Michon, P.-E., Denis, M.: When and Why are visual landmarks used in giving directions? In: Montello, D.R. (ed.) COSIT 2001. LNCS, vol. 2205, pp. 292–305. Springer, Heidelberg (2001)
26. Mollerup, P.: Wayshowing: A Guide to Environmental Signage – Principles and Practices. Lars Müller Publishers, Baden (2005)
27. Raubal, M., Worboys, M.F.: A formal model of the process of wayfinding in built environments. In: Freksa, C., Mark, D.M. (eds.) COSIT 1999. LNCS, vol. 1661, pp. 381–399. Springer, Heidelberg (1999)
28. Richter, K.-F., Klippel, A.: A model for context-specific route directions. In: Freksa, C., Knauff, M., Krieg-Brückner, B., Nebel, B., Barkowsky, T. (eds.) Spatial Cognition IV. LNCS (LNAI), vol. 3343, pp. 58–78. Springer, Heidelberg (2005)
29. Richter, K.F., Tomko, M., Winter, S.: A dialog-driven process of generating route directions. Comput. Environ. Urban Syst. 32(3), 233–245 (2008)
30. Richter, K.F., Winter, S.: Landmarks – GIScience for Intelligent Services. Springer, Cham (2014)
31. Riesbeck, C.K.: "You can't miss it!": Judging the clarity of directions. Cogn. Sci. 4(3), 285–303 (1980)
32. Schneider, L.F., Taylor, H.A.: How to get there from here? Mental representations of route descriptions. Appl. Cogn. Psychol. 13, 415–441 (1999)
33. Stevens, S.S.: On the theory and scales of measurement. Science 103(2684), 677–680 (1946)
34. Strawson, P.F.: On referring. Mind 59(235), 320–344 (1950)
35. Tenbrink, T., Bergmann, E., Konieczny, L.: Wayfinding and description strategies in an unfamiliar complex building. In: Proceedings of the 33rd Annual Conference of the Cognitive Science Society. Cognitive Science Society, Boston (2011)
36. Tom, A., Denis, M.: Referring to landmark or street information in route directions: what difference does it make? In: Kuhn, W., Worboys, M.F., Timpf, S. (eds.) COSIT 2003. LNCS, vol. 2825, pp. 362–374. Springer, Heidelberg (2003)
37. Tomko, M., Purves, R.S.: Venice, city of canals: characterizing regions through content classification. Trans. GIS 13(3), 295–314 (2009)
38. Tomko, M., Winter, S.: Pragmatic construction of destination descriptions for urban environments. Spat. Cogn. Comput. 9(1), 1–29 (2009)
39. Tomko, M., Winter, S.: Describing the functional spatial structure of urban environments. Comput. Environ. Urban Syst. 41, 177–187 (2013)
40. Ullmer-Ehrich, V.: The structure of living space descriptions. In: Jarvella, R.J., Klein, W. (eds.) Speech, Place, Action, pp. 219–249. John Wiley and Sons, Chichester (1982)
41. Vasardani, M., Timpf, S., Winter, S., Tomko, M.: From descriptions to depictions: a conceptual framework. In: Tenbrink, T., Stell, J., Galton, A., Wood, Z. (eds.) COSIT 2013. LNCS, vol. 8116, pp. 299–319. Springer, Heidelberg (2013)
42. Wiener, J.M., Büchner, S.J., Hölscher, C.: Taxonomy of human wayfinding tasks: a knowledge-based approach. Spat. Cogn. Comput. 9(2), 152–165 (2009)

43. Williamson, J., Barrow, C.: Errors in everyday routefinding: a classification of types and possible causes. Appl. Cogn. Psychol. **8**(5), 513–524 (1994)
44. Wuersch, M., Caduff, D.: Refined route instructions using topological stages of closeness. In: Li, K.-J., Vangenot, C. (eds.) W2GIS 2005. LNCS, vol. 3833, pp. 31–41. Springer, Heidelberg (2005)

A Wayfinding Grammar Based on Reference System Transformations

Peter Kiefer[✉], Simon Scheider, Ioannis Giannopoulos, and Paul Weiser

Institute of Cartography and Geoinformation, ETH Zürich,
Stefano-Franscini-Platz 5, 8093 Zürich, Switzerland
{pekiefer,sscheider,igiannopoulos,pweiser}@ethz.ch

Abstract. Wayfinding models can be helpful in describing, understanding, and technologically supporting the processes involved in navigation. However, current models either lack a high degree of formalization, or they are not holistic and perceptually grounded, which impedes their use for cognitive engineering. In this paper, we propose a novel formalism that covers the core wayfinding processes, yet is modular in nature by allowing for open slots for those spatial cognitive processes that are modifiable, or not yet well understood. Our model is based on a *formal grammar grounded in spatial reference systems* and is both interpretable in terms of observable behavior and executable to allow for empirical testing as well as the simulation of wayfinding.

Keywords: Wayfinding · Navigation · Spatial cognitive processes · Formal grammar · Reference systems

1 Introduction and Related Work

Navigation, i.e., the combined endeavor of both locomoting and wayfinding, is an activity most people carry out on a daily basis. While locomotion can be defined as the coordinated movement in the nearby environment in order to avoid obstacles, wayfinding refers to "the planning and decision-making necessary to reach a destination" [32]. Successful wayfinding consists of a wide variety of cognitive processes that can be distributed through time and among individuals (cf. [10,13,43,44]), as well as involve the coordination of internalized and externalized spatial knowledge (cf. [30]). In fact, human cognition goes beyond "what is inside our heads alone" by encompassing "the cognitive roles of the social and material world" [12].

Research in spatial cognition has long wondered about the nature of the cognitive processes that make up navigation and wayfinding and attempted to model them. For example, Downs and Stea proposed that wayfinding consists of orientation (Where am I?), route choice (Where should I go?), monitoring (Am I still on track?), and goal recognition (Am I there yet?) [3]. Golledge identified various sub-processes that involve "to determine turn angles, to identify segment lengths and directions of movement, to recognize en route and distant

© Springer International Publishing Switzerland 2015
S.I. Fabrikant et al. (Eds.): COSIT 2015, LNCS 9368, pp. 447–467, 2015.
DOI: 10.1007/978-3-319-23374-1_21

landmarks, and to embed the route to be taken in some larger reference frame" [8, p. 7]. Arthur and Passini suggested that wayfinding consists of information processing (perception and interpretation of the environment), decision making (constructing a hierarchical action plan) and decision execution (transformation of a plan into behavior) [2].

In contrast to the aforementioned models that remain descriptive, there also exist formal approaches that are capable of simulating wayfinding based on rule sets that denote condition-action pairs [1,9,24,25,41]. However, Golledge [7] criticized that such decision models prescribe a particular way of how decision making takes place that does not match established theories on human cognition. Haken and Portugali [11] suggested to model the interaction of internal and externalized spatial knowledge with neural synergetic networks which involve a variety of feedback loops. Raubal and Worboys [35], in turn, proposed a graph model of possible knowledge and location transitions in an environment that allows representing navigation as a path. However, the model does not address how this graph can be built. A way to summarize route knowledge based on a formal grammar was proposed by Klippel [21]. Yet, while these models are formally specified they only represent some aspects of the wayfinding process. Formal grammars in which rule applications can be spatially constrained were proposed by Schlieder [38] and Kiefer [15,16]. However, these formalisms do not aim specifically at modeling wayfinding, they rather model general intentional behavior.

To conclude this brief review, there is still a lack of formal and operational models that are cognitively plausible and capture the processes of wayfinding from a holistic point of view, without prescribing questionable assumptions regarding decision and search procedures.

In this paper, we propose a novel wayfinding process model based on a formal grammar which can be termed a *simulation meta-model*. It covers the core wayfinding processes, yet is modular in nature by allowing for *open slots* to account for those spatial cognitive processes that are modifiable, not well understood, or for which there is no reliable theory yet. These open slots can then later be filled with (ad-hoc) process models, and tested. Our model is interpretable in terms of observable behavior (including perceptual processes and actions) and at the same time executable, such that the wayfinding processes can be simulated. Furthermore, since a major part of the relevant cognitive and perceptual processing consists of *interactions with spatial reference systems*[1] our model is grounded in reference systems.

The remainder of this paper is structured as follows. We informally explain our model in Sect. 2 and introduce the grammar in Sect. 3, before we illustrate how different kinds of wayfinding scenarios can be simulated (Sect. 4). We discuss the limitations of our approach and conclude with an outlook in Sect. 5.

---

[1] These include: cognitive reference frames, mental survey representations as well as geographic reference systems (cf. [36]).

2 Wayfinding in Terms of Reference System Transformations

In this section we describe the theoretical principles and components of our model. We illustrate how the different wayfinding processes can interact with each other. Note, we do not make any claim on how the processes are actually "implemented" in the human cognitive system. However, we suggest that a central role in generating interaction constraints is played by reference systems and their transformations (cf. [5]).

2.1 Spatial Reference Systems in Wayfinding

Spatial reference systems are used to refer to locations across individuals and across time [23]. The way how spatial reference is determined and with respect to which ground phenomena is characteristic to the particular reference system. It affects how and to what extend a location can be transformed from one system into another [37]. Cognitive (internalized) reference frames play a fundamental role in learning and remembering space [39], while spatial coordinate (externalized) systems establish the semantics of maps and other forms of spatial data [23]. Each reference system comes with particular kinds of operations that play an important role in the wayfinding process:

1. *Egocentric Reference Systems.* An egocentric frame is centered and aligned with the body of a perceiving ego. It can be aligned with the direction into which the eyes look (*retinal*), or it can be *aligned with the head* or the *body front* of a person [14]. In any case, an egocentric frame captures a momentary *perceptual array*[2] of the ego, with *objects* and *locations* perceived in a certain angle and distance from the self (self-to-object). Objects can be both *places* and *bodies* with surfaces [14]. Egocentric frames roughly correspond to *Vista* space [33]. They are closely connected to the perceptual array and thus to direct experience, and are kept primarily in short-term or working memory. They take an important role in motor-control, as well as in projecting locomotion into the perceived environment, and are probably located in the brain's parietal cortex [14]. Their role in wayfinding is that they provide input for self-localization (where am I?) and are output of path-localization (where do I need to go?), both of which form major parts of the required attentional effort.

2. *Allocentric Reference Systems and Route Knowledge.* Allocentric systems encode locations relative to other objects (object-to-object). Humans can easily transform egocentric locations into allocentric ones (and vice versa) by taking egocentric locations with respect to perceivable ground objects and orientations [5,14,39]. For this reason, allocentric systems are able to render locations inter-subjective (i.e., they can be shared among others) and

---

[2] Note that we use this term in an intermodal sense, i.e., not restricted to vision and thus integrating different modalities of perception.

independent of a point of view or movement. They give a particular fixed meaning to *qualitative spatial locations* [5] (e.g., "in front of"), which are taken with respect to ground objects and some perceivable orientation, and are termed *relation templates* (cf. [27]). Allocentric reference systems constitute the meaning of large parts of human spatial language [26], as well as spatial memory [39]. They may constitute what Montello calls *environmental space*, i.e., space that needs to be apprehended through locomotion [33]. We assume that wayfinding knowledge is largely encoded in an allocentric form, more particularly in terms of *short term and long-term route memory* consisting of sequences of actions and allocentric locations with *landmarks* as ground objects [31]. For example, we may remember our way to work in terms of the sequence: turn right in front of the church, turn right at the bank, then enter the parking lot. Their role in wayfinding is that they represent route instructions which implement plans based on spatial memory.

3. *Survey Reference Systems.* While allocentric (cognitive) reference systems already constitute a kind of inter-subjective knowledge, spatial reference remains uncertain when ground objects are not in view or have never been experienced. This renders them unsuitable for survey planning. In the wayfinding process, survey knowledge is indispensable whenever the way extends beyond any location that is describable relative to known ground objects. For this reason, people have learned to use reference systems that represent the geometric configuration of unknown objects and locations [31]. Their role in wayfinding is therefore to support the construction of *possible ways to go*, i.e., the *planning* of wayfinding and its *simulation* in case route memory fails. Survey reference systems are grounded relative to the *earth's surface* or other ground phenomena that remain in view. They roughly correspond to Montello's *geographic* space which needs to be "learned via symbolic representations" [33]. One example is a *geographic reference system* on which geographic maps are based. Their cognitive counterparts are mental representations of survey knowledge ("cognitive maps") which are bird-eye views kept in long-term memory, constructed by cumulative spatial experience or by memorizing geographic maps, and which allow for perspective taking and making spatial inferences (cf. [31,42]). Transformations of survey locations to allocentric or egocentric systems are only possible when ground objects and orientations can be mapped. To what extent cognitive maps resemble cartographic maps and whether this analogy is rather a metaphor is debatable [20]. However, it seems fair to assume that some kind of survey knowledge (either internalized or externalized) is necessary for the purpose of wayfinding.

2.2   Wayfinding Processes and Their Dependencies

In this section we informally discuss reference system transformations and other cognitive processes that occur during wayfinding. Figure 1 illustrates our model as a transition graph on different kinds of spatial knowledge and the processes connecting them.

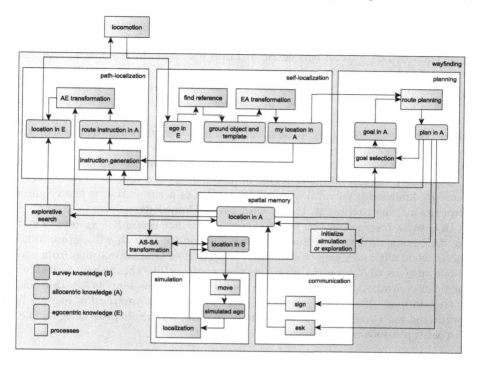

Fig. 1. Modeling the wayfinding process based on reference system transformations.

We describe our model using the example of Susi who is trying to find her way to some destination. Susi starts with her egocentric field of view ("ego in E", Fig. 1). Her first task is to figure out where she is, i.e., she needs to *localize herself* w.r.t. some allocentric system (cf. "orientation" [3]; Fig. 1: box "self-localization"). That is, she needs to find a spatial referent – a suitable ground object and relation template – and transform her position into an allocentric location. Then Susi starts *planning* a route by selecting a goal (Fig. 1: box "planning") and by initializing a plan – being simply an intention to go there – which can recursively be divided into sub-plans by selecting subgoals from memory (cf. "route choice" [3] and "decision making" [2]).

Plans can be implemented in several ways. If Susi already knows from past experience how to turn the plan into an *instruction*, she will be able to retrieve this instruction from memory in terms of route knowledge, e.g., "turn left in front of the church" (Fig. 1: box "path-localization"). If, however, spatial memory does not allow to generate an instruction, she will have several possibilities: For example, she can start an *explorative search*. In this case, Susi will just "follow her nose" in her egocentric system until she reaches a location that is familiar to her (from where she can again start planning), or until she recognizes the goal location. In addition, Susi can *ask* other people to acquire route knowledge, consult some form of signage (Fig. 1: box "communication"), or search for a route on a map. For instance, a sign pointing into a certain direction describes an allocentric location relative to that sign ('location in A') which then needs to be transformed into an egocentric location ('location in E') through an 'AE transformation'.

Searching for a route on a map is specifically interesting in terms of reference system transformations because then survey knowledge becomes relevant. In our model, we describe map-based wayfinding as *navigation simulation* on a survey reference system (Fig. 1: box "simulation"). Susi performs a search on the survey map, moving her attention (and/or possibly her finger) to the goal and memorizes this as a sequence of imagined actions and allocentric locations. We assume that a similar simulation can be performed on a mental representation of survey knowledge (mental map). The output of a simulation is the required (sequence of) instruction(s), thus yielding route knowledge.

In case an instruction can be *transformed into her egocentric system* (Fig. 1: box "AE transformation"), Susi can start locomoting, which will populate her field of view with new objects that may trigger new transformations from egocentric systems to allocentric ones. Susi may continue like this as long as instructions can continuously be generated from memory and as long as locations can be transformed. If available memory runs out, or if an instruction cannot be interpreted w.r.t. perception, she will need to start the whole process again (i.e., start with self-localization based on "ego in E").

3   Wayfinding Grammar

We model wayfinding as the process of planning and constructing wayfinding instructions, and translating them from survey or allocentric locations into egocentric reference systems (i.e., locations in the field of view of the ego). Instructions can be generated by experience, survey simulation, or communication (see also Fig. 1).

For this purpose, we use a formal *grammar*, i.e., a set of rewriting rules that transform a non-terminal start symbol into a string of terminal symbols. Using a formal grammar, wayfinding can be represented as a sequence of rule applications generating terminals which stand for behaviors, such as locomotion or visual search. Thus, a wayfinding process can be *simulated* in terms of the rewriting process. Furthermore, similar to the syntax of a language, the grammar defines allowable sequences of behaviors ("well-formed formulae"), so that a parser can check whether a given string of measured behaviors can denote a wayfinding process.

We first define the grammar in terms of symbol sets and rules, before explaining how the elements of the grammar interact. The scenarios in Sect. 4 illustrate how the grammar can be used.

Definition 1 (LOCATIONS). *Let* $\mathbf{S}$ *denote a finite set of objects in a survey reference system,* $\mathbf{A}$ *a finite set of allocentric systems, and* $\mathbf{E}$ *a finite set of objects in an ego-centric reference system. We define* $\mathbf{L^S} \subseteq \mathbf{S} \times \mathbf{Rel}$, $\mathbf{L^A} \subseteq \mathbf{A} \times \mathbf{Rel}$, and $\mathbf{L^E} \subseteq \mathbf{E} \times \mathbf{Rel}$ *as the sets of survey, allocentric, and egocentric locations w.r.t a (finite) set of spatial relation templates* $\mathbf{Rel}$. *The set of all locations is then defined as* $\mathbf{L} = \mathbf{L^S} \cup \mathbf{L^A} \cup \mathbf{L^E}$.

Spatial relation templates $rel_i \in \mathbf{Rel}$ can be used to describe a location relative to a reference object, such as $L^A_{frontOfChurch} = (A_{church}, rel_{inFrontOf})$, with $A_{church} \in \mathbf{A}$, $L^A_{frontOfChurch} \in \mathbf{L^A}$.

Definition 2 (WAYFINDING GRAMMAR). *A wayfinding grammar is a context-free production system* $(\mathbf{B}, \mathbf{N}, \mathbf{R}, E_{ego}, L^A_g, Acc)$ *with*

- $\mathbf{B}$ *denoting a set of wayfinding behaviors (the terminals, see Definition 3)*
- $\mathbf{N} = \{P_{[L^A_i][L^A_j]}, I_{[L^A_k][L^A_i]}\} \cup \mathbf{L} \cup \mathbf{E} \cup \mathbf{S}$ *defining the non-terminals. The plan non-terminal* $P_{[L^A_i][L^A_j]}$ *and the instruction non-terminal* $I_{[L^A_k][L^A_i]}$ *are attributed with locations from* $\mathbf{L^A}$ *(the 'from' and the 'to' location).*
- $\mathbf{R}$, *a set of production rules from* $\mathbf{N}$ *to* $(\mathbf{N} \cup \mathbf{B})^+$ *(see Table 1)*
- $E_{ego} \in \mathbf{E}$ *is the start symbol denoting the perceiving ego*
- $L^A_g \in \mathbf{L^A}$ *denoting the allocentric goal location of the wayfinding process.*
- $Acc \subseteq \mathbf{L^A} \times \mathbf{L^A} \times \mathbf{H}$ *represents a wayfinder's route knowledge as a graph over* $\mathbf{L^A}$ *with edge labels from the set of headings* $\mathbf{H}$, *which denote turn directions such as "turn left" or "turn north".* $(L^A_i, L^A_j, h_k) \in Acc$ *means "the wayfinder knows that* L^A_j *is directly accessible from* L^A_i *in direction* h_k*".*

Definition 3 (WAYFINDING BEHAVIORS). *The set of wayfinding behaviors* $\mathbf{B}$ *contains the following elements*

$goto[L^E_i]$	*The wayfinder locomotes to* L^E_i.
$\mathbf{H}$	*The wayfinder changes heading to* $h_i \in \mathbf{H}$.
$s_{AE}[L^A_i]$	*Search for allocentric location* L^A_i *in egocentric system.*
$s_{EA}[L^E_i]$	*Search for egocentric location* L^E_i *in allocentric system.*
$s_{AS}[L^A_i]$	*Search for allocentric location* L^A_i *in survey system.*
$s_{SA}[L^S_i]$	*Search for survey location* L^S_i *in allocentric system.*
s_{exp}	*Search during explorative wayfinding.*
s_{sign}	*Search on signage.*
s_{refE}	*Search for referent object in egocentric system.*
s_{refS}	*Search for referent object in survey system.*
sim	*The wayfinder simulates navigation in a survey system.*
sim_{end}	*The wayfinder stops survey simulation.*
$ask[L^A_s][L^A_d]$	*The wayfinder requests instructions to go from* L^A_s *to* L^A_d.

We use an *attributed context-free grammar* [22] (see Definition 2). Non-terminals for plan and instruction are attributed with allocentric locations denoting the "from" and "to" locations. This is used to constrain the application of rules and it effectively turns grammar rules into *meta-rules*: allowable rule sets can be derived by substituting variables with elements from $\mathbf{L}$ (cf. Table 1).

Rules are successively applied to non-expanded non-terminals in order to expand them, generating a *production tree*, similar to normal context-free grammars (cf. Fig. 2b). A left-to-right and depth-first traversal on these trees determines the temporal order of the resulting terminal sequence of wayfinding behaviors.

Table 1. Meta rules of the wayfinding grammar.

Production meta rule	Triggers	Description
PLANNING		
$L_i^A \rightarrow P_{[L_i^A][L_g^A]}$ with L_g^A = goal from grammar definition, $L_i^A \neq L_g^A$; rule applied at most once	$f_{Sub\text{-}plan}$, f_{Impl}	(1) Plan initialization
$P_{[L_s^A][L_d^A]} \rightarrow P_{[L_s^A][L_i^A]} \, P_{[L_i^A][L_d^A]}$ with $L_s^A \neq L_d^A$, $L_s^A \neq L_i^A$, $L_i^A \neq L_d^A$	$f_{Sub\text{-}plan}$, f_{Impl}	(2) Sub-planning
$P_{[L_s^A][L_d^A]} \rightarrow I_{[L_s^A][L_d^A]}$ with $L_s^A \neq L_d^A$	f_{Instr}, $f_{Re\text{-}plan}$	(3) Instruction substitution from memory
$I_{[L_s^A][L_d^A]} \rightarrow P_{[L_s^A][L_d^A]}$ with $L_s^A \neq L_d^A$	$f_{Sub\text{-}plan}$, f_{Impl}	(4) Re-planning
$I_{[L_s^A][L_d^A]} \rightarrow h_i \, L_d^A$ with $(L_s^A, L_d^A, h_i) \in Acc$, $L_s^A \neq L_d^A$		(5.1) Instruction generation
$I_{[L_s^A][L_d^A]} \rightarrow h_i \, L_z^A \, I_{[L_z^A][L_d^A]}$ with $(L_s^A, L_z^A, h_i) \in Acc$, $L_s^A \neq L_d^A$, $L_s^A \neq L_z^A$, $L_z^A \neq L_d^A$	f_{Instr}, $f_{Re\text{-}plan}$	(5.2) Instruction generation
TRANSFORMATION		
$L_i^A \rightarrow s_{AE}[L_i^A] \, L_j^E$		(6) Egocentric matching
$L_i^E \rightarrow s_{EA}[L_i^E] \, L_j^A$		(7) Reverse egoc. matching
$L_i^A \rightarrow s_{AS}[L_i^A] \, L_j^S$		(8) Survey-matching
$L_i^S \rightarrow s_{SA}[L_i^S] \, L_j^A$		(9) Reverse survey-matching
$E_i \rightarrow s_{refE} \, L_j^E$		(10) Egocentric localization
$S_i \rightarrow s_{refS} \, L_j^S$	f_{RemS}	(11) Survey localization
LOCOMOTION		
$L_i^E \rightarrow goto[L_i^E]$	f_{Perc}, f_{RemE}	(12) Locomotion of the ego to the new location L_i^E
$P_{[L_s^A][L_d^A]} \rightarrow L_d^A$	f_{Exp}	(13) Initialize exploration
$L_i^A \rightarrow h_i \, s_{exp} \, L_j^E \, L_i^A$ with $(L_i^A \rightarrow s_{AE}[L_i^A L_j^E]) \notin \mathbf{R}$	f_{Exp}	(14) Explorative search for L_i^A
COMMUNICATION		
$P_{[L_s^A][L_d^A]} \rightarrow s_{sign} \, I_{[L_s^A][L_d^A]}$	f_{RemC}	(15) Use signage
$P_{[L_s^A][L_d^A]} \rightarrow ask[L_s^A][L_d^A] \, I_{[L_s^A][L_d^A]}$	f_{RemC}	(16) Ask for instructions
SIMULATION		
$P_{[L_s^A][L_d^A]} \rightarrow L_s^A \, I_{[L_s^A][L_d^A]}$ with $L_s^A \neq L_d^A$	f_{Sim}	(17) Instruction substitution with simulation
$L_i^S \rightarrow sim \, h_i \, S_{egosim}$	Replace: (11)	(18) Simulated locomotion
$L_i^A \rightarrow sim_{end}$	Remove: (19)	(19) End of simulation

The wayfinder can execute a terminal (e.g., start locomoting) as soon as there is no unexpanded non-terminal or non-executed terminal left of it. Example production trees for our grammar are presented in Sect. 4.

Table 2. Kinds of procedures for wayfinding production. Numbers in brackets refer to rules in Table 1

Procedure	Purpose	Result	Triggered on
PRODUCTION PROCEDURES			
f_{Init}	Generate start symbol, static rules and select goal	Generate E_{ego} (1) (8) (9) (11), set L_g^A	Start
PLAN UPDATING			
$f_{Sub\text{-}plan}$	Select a sub-goal for a given plan from internal memory	Generate (2)	(1) and (4)
f_{Impl}	Generate rules that implement a plan	Generate (3) (13) (15) (16) (17)	(1) (2) (4)
f_{Instr}	Select edge from route knowledge (Acc) to build instruction	Activate (6) deactivate (8) generate (5.1) (5.2)	(3) (5.2) (15) (16) (17)
$f_{Re\text{-}plan}$	Offer a re-planning possibility	Generate (4)	(3) (5.2) (15) (16) (17)
EGOCENTRIC UPDATING			
f_{Perc}	Replace egocentric locations and their transformations	Update (6) (7) (10) (12)	(12) and start
f_{Exp}	Select egocentric location for explorative search	Generate (14)	(13)
SURVEY UPDATING			
f_{Sim}	Switch to simulation mode and simulate	Activate (8) deactivate (6) generate (19)(18)	(17)
MEMORY UPDATING			
f_{RemE}	Store locomotion experience in memory	Update edges in Acc	(12)
f_{RemS}	Store simulation experience in memory	Update edges in Acc	(11)
f_{RemC}	Store communicated knowledge in memory	Update edges in Acc	(15) (16)

Note that the modeled wayfinding behavior (the terminals) may or may not be observable in a specific case, depending on the sensor technology available. If certain terminals are not measurable in a given scenario they can simply be removed from the grammar. For instance, terminals $goto[L_i^E]$ and h_i require a position and directional sensor respectively, while the search behaviors s_x can be derived from measuring eye movements with eye tracking [19]. Terminals denoting search for specific objects are attributed with the object for which is searched. It is known that eye movements can be aggregated and classified to certain types of visual search [18], and we assume that the listed types of search

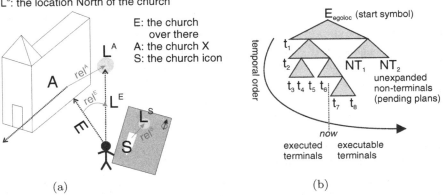

L^E: the location right of the church
L^A: the location behind the church
L^S: the location North of the church

E: the church over there
A: the church X
S: the church icon

Fig. 2. (a) Reference systems in wayfinding. The same location may be egocentric "right of", allocentric "behind", and "North of" the church on the survey level. (b) Production tree of a wayfinding grammar. Non-terminals are substituted (topdown direction) and terminals are executed (left-right direction).

can be distinguished accordingly. During visual search on a map-based survey system the eye tracking data need to be matched with features on the map [17] to detect which objects exactly are looked at.

Our *grammar is not static* since the sets of egocentric non-terminals **E** and locations **L^E**, the rule set **R**, and the accessibility graph *Acc* representing route knowledge can be changed during grammar execution. This is to account for the fact that wayfinding knowledge and the perceptual array change as ego finds its way through an environment. We model the dynamic aspects by using updating procedures (See Table 2), which are triggered as indicated in column 2 of Table 1. Updating procedures generate new rules and activate or deactivate others depending on the state of the wayfinding process. They remain *external to our grammar* and represent open slots in our model, thus need to be filled with realistic sub-models, perceptual processes or measurement procedures.

Furthermore, since the wayfinding process can fail, *grammar execution can fail*, too. For example, if an instruction is not up-to-date, it may not be transformable into egocentric locations because its description cannot be matched with perception. In terms of our grammar, this simply means that the instruction cannot be fully expanded to terminals. In this case, the old production tree is discarded and the production process simply starts anew with self-localization.

In the following, we explain the semantic interpretation of the grammar and the updating processes. Numbers in parentheses refer to the corresponding rules in Table 1.

3.1   Self Localization

The start symbol of our grammar is the ego E_{ego}. The first thing ego needs to do is to self-localize itself w.r.t to an object in its field of view. That is, an egocentric ground object from **E** (objects in ego ref. system) and a relation template from **Rel** must be selected which specify ego's location. For instance, $L^E_{egoloc} = (E_{pub}, rel_{outside})$ would describe an egocentric location "outside of the pub from the perspective of ego". Rule (10) models this self-localization. It involves a search for a referent object and a relational template in the egocentric system (terminal s_{refE}). There is usually more than one way of describing the egocentric location, so the wayfinder can choose from several rules of type (10). These rules are created for E_{ego} by procedures f_{Init} and f_{Perc} (See Table 2). Once L^E_{egoloc} has been determined it can be used to transform the location of ego to an allocentric location which provides the input for planning. Rule (7) expresses that an egocentric location from $\mathbf{L^E}$ can be transformed into an allocentric representation from $\mathbf{L^A}$ with a search (terminal $s_{EA}[L^E_i]$), taking into account a perceivable orientation and the geometry of the ground object. For instance, the location L^E_{egoloc} from the example above could be transformed into the allocentric location "in front of the pub" (In a reference system grounded in the pub and looking into the direction of its entrance): $L^A_{egoloc} = (A_{pub}, rel_{infront})$.

3.2   Planning

The wayfinding process often involves planning on several levels of abstraction [34]. We start the process with rule (1) which initializes the plan to go from the current location to the goal location L^A_g. Sub-planning with intermediate locations is achieved by (potentially multiple) applications of rule (2). The choice and order of applying planning rules determines the wayfinder's *planning strategy* which can be modeled by procedures $f_{Sub\text{-}plan}$ and f_{Impl} (See Table 2) but are out of the scope of this paper.

In our grammar, there are four ways of continuing with a given plan. First, one can implement the plan by using an *instruction* from route memory (rule (3)). Second, one can *simulate* wayfinding in a survey reference system (rules (17) to (19)). Third, one can *ask* some other agent (rule (16)) or consult signage (rule (15)) in order to obtain an instruction. Finally, one can do *explorative search* for the goal just by following one's nose (rules (13) and (14)). We will explain the latter three possibilities in the following subsections.

Using rule (3), plans can be substituted by instructions which are detailed sequences of actions and accessible allocentric locations constructed based on route knowledge stored in the accessibility graph Acc. The process of selecting and adding information to an instruction can be modeled by f_{Instr} (See Table 2). An instruction is given by a heading information h_i and an allocentric location L^A_i (rules (5.1) and (5.2)): the wayfinder turns into a certain direction where L^A_i is supposed to be found. If L^A_i is not yet the destination of the instruction sequence (rule (5.2)) further instructions are necessary.

Only if the left-most branch of the resulting production tree has been expanded with an instruction, the ego can start transforming L_i^A from this instruction into an egocentric location which enables locomotion. This transformation is described by rule (6) which requires a search (terminal $s_{AE}[L_i^A]$). That is, the allocentric location L_i^A (e.g., "behind the church") is turned into an egocentric location accessible from the wayfinder's current position (e.g., "right of the church") (See Fig. 2a). Note, if L_i^A could not be successfully interpreted (i.e., matched onto the environment), one can either do an explorative search for L_i^A by rule (14), or abort and start anew[3].

3.3 Locomotion

Locomotion moves the ego through space and, in this way, continuously generates new fields of view. In our formalism, this means updating the set of egocentric (perceived) objects $\mathbf{E}$ by procedure f_{Perc} (See Table 2), i.e., new egocentric objects enter and old objects leave the field of view, similar to the flow in a perceptual array. Corresponding egocentric locations $\mathbf{L^E}$ and transformation rules need to be updated as well. That is, transformation rules with egocentric locations as rule head or body that are out of view will be deleted and transformation rules with novel egocentric objects enter the rule set. f_{Perc} is triggered every time a new physical locomotion to a given egocentric location is performed by generating $goto[L_i^E]$ using rule (12).

Locomotion can be triggered in *wayfinding by instruction* as well as in *explorative search* (following one's nose). Rule (13) models the intialization of explorative search from a plan that cannot be resolved to sub-plans (with rule (2)) or substituted with an instruction (with rule (3)). The wayfinder switches from a structured 'planning mode' to an 'explorative mode' (triggering f_{Exp} in Table 2). Exploration is then continued by applying rule (14) and (12) in an iterative fashion in order to move ego in her field of view without instruction (see example in Sect. 4.2), until the plan destination is reached by chance, i.e., until the allocentric destination on the right hand side of (14) can be transformed to the egocentric system with rule (6).

The rules described so far (rules (1–7), (10), (12–14)) are sufficient for modeling the wayfinding process as a sequence of multi-level planning down to the instruction level or as explorative search, followed by transformations finally leading to changes in heading and locomotion. We have not described, however, the manipulation of (external and internal) spatial memory, but simply assumed that route knowledge is always given. In the following, we describe how new knowledge can be added.

3.4 Generating Route Knowledge

In finding our way in an environment, we essentially need to know that we can access a location from another one "directly". If we know that being at L_a^A

---

[3] Note that survey simulation was deactivated with rule (8) by f_{Instr} in Table 2.

implies that we can view and access another location L_b^A by heading in direction h_{dir}, then we know how to get to L_b^A once we are at L_a^A (and so on), even without planning and without physically being at L_a^A.

We model this knowledge with the accessibility graph Acc (see Definition 2). The graph may contain edges at the beginning of wayfinding process if the wayfinder has previous knowledge about an area. The graph can be updated during wayfinding in two ways: either based on egocentric experience, or based on movement simulation in a survey reference system[4].

In egocentric experience, if we have successfully moved from where we are now to an egocentric location we may add an edge between corresponding allocentric locations to the accessibility graph. That is, the procedure f_{RemE} in Table 2, which is triggered after application of rule (12), takes the current and the previous $goto[L_i^E]$ terminals (based on an inverse traversal of the production tree), retraces them to back to allocentric locations $L_{current}^A$ and $L_{previous}^A$, determines the last direction h_i, and creates an according edge in Acc.

3.5   Movement Simulation in Survey Reference System

The second way of adding knowledge to Acc is provided by *movement simulation in a survey system* (see rules (17)-(19) in Table 1). Note that we make no further assumptions about the nature of these reference systems, e.g., as to whether they are externalized maps or cognitive systems. In fact, they can denote one or the other.

For example, if the wayfinder cannot directly substitute a plan to go from L_s^A to L_d^A with an instruction because route knowledge is missing, she may decide to simulate a path by moving her finger on a physical paper map. This process is started by rule (17), which creates an L_s^A and triggers f_{Sim} in Table 2 switching the grammar to simulation mode, meaning that allocentric locations are now mappable only to survey locations (rule (8) activated), and not to egocentric ones (rule (6) deactivated), and the planning destination is set as simulation goal by adding a corresponding rule (19). L_s^A can then be mapped to a survey location L_s^S by rule (8) and becomes the start location for simulation. Next, the simulated ego is moved by rule (18), yielding a new survey position S_{egosim} (the "finger" is moved on the map). By iterative applications of (18) and (11) a route sequence is recorded until the simulation reaches the destination, leading to an application of rule (19). The simulation terminals (*sim* and *sim_{end}*) appearing during movement simulation could, for instance, be measured by eye or hand movements. An example for simulation in a survey reference system is presented in Sect. 4.3.

The main purpose of the simulation process consists in creating new edges in Acc, similar to the edge updating by ego-centric movement: if we are able to successfully simulate a path of the simulated ego to a location L_j^S on a geographic map, and if there exist allocentric transformation rules for these locations, then

---

[4] In principle, route knowledge may also be removed (forgotten) or overwritten by new experience or simulations.

we can add a corresponding edge to the accessibility graph (procedure f_{RemS} in Table 2). That is, after application of rule (11), the right-hand location L_j^S and its grandparent node L_i^S (see production tree in Fig. 6) are determined. By looking up reverse survey-matching rules (9) for L_i^S and L_j^S, according allocentric locations L_i^A and L_j^A are determined. These are combined with h_i (the child node of L_i^S) and added to Acc.

3.6   Obtaining Instructions from Signage or by Asking

As an alternative to simulation, wayfinders frequently use signage or other communication acts to obtain instructions. In our grammar we represent this with two rules:

Rule (15) models a search on a sign (terminal s_{sign}) which triggers f_{RemC} in Table 2 leading to an updated accessibility graph and a new instruction. This instruction could now be mapped with rule (5.2) to an allocentric location in direction h_i in the reference system of the sign L_{sign}^A (the place the sign is pointing to). An example is presented in Sect. 4.4.

Asking for an instruction from L_s^A to L_d^A is modeled in an equivalent way by the act of asking for directions (terminal $ask_{[L_s^A][L_d^A]}$, rule (16)). The wayfinder may either ask another agent or a pedestrian navigation system. All communication processes lead back to the same instruction $I_{[L_s^A][L_d^A]}$ which may now be partially processable using the new route knowledge. We illustrate this with an example in Subsection 4.5.

4   Wayfinding Scenarios

In this section, we illustrate with examples how different types of wayfinding behavior can be modeled using our wayfinding grammar.

Suppose the following scenario: Susi is visiting a foreign city of which she has so far acquired only limited spatial knowledge. She has just left the pub and is now trying to find her way back to the hotel (See Fig. 3).

4.1   Following a Path from Internal Memory

In the first example, we assume Susi remembers the path from previous experience, i.e., her accessibility graph is initialized as follows

$$Acc = (L_{pub1}^A, L_{church1}^A, h_{NW}), (L_{church1}^A, L_{m1}^A, h_N)$$
$$(L_{m1}^A, L_{road1}^A, h_W), \quad (L_{road1}^A, L_g^A, h_W)$$

with $L_{pub1}^A = (A_{pub}, rel_{infront})$, $L_{church1}^A = (A_{church}, rel_{left})$,
$\quad L_{m1}^A \; = (A_{market}, rel_{on})$, $L_{road1}^A \; = (A_{mainroad}, rel_{on})$,
$\quad L_g^A \; = (A_{hotel}, rel_{infront})$

Susi particularly remembers the market as a central place in this city. Thus, even though the accessibility graph is complete, she applies a planning strategy

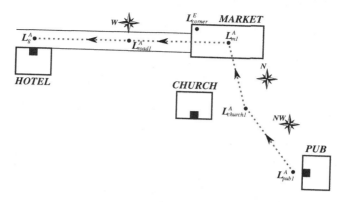

Fig. 3. Example wayfinding scenario.

by structuring her plan into two sub-plans (pub to market, market to goal). The sub-plans are converted into instruction sequences which are sequentially turned into allocentric locations. Since her route memory is reliable, these allocentric locations can successfully be transformed into egocentric locations towards which she locomotes.

Figure 4 illustrates the production tree after applying rules in the following order: (10), (7), (1), (2), (3), (5.2), (6), (12), (5.1), (6), (12). The resulting terminal sequence is $s_{refE} \ s_{EA}[L^E_{pub}] \ h_{NW} \ s_{AE}[L^A_{church1}] \ goto[L^E_{church}] \ h_N \ s_{AE}[L^A_{m1}]$ $goto[L^E_{market}]$. The exactly same procedure could now be applied to the second sub-plan $P_{[L^A_{m1}][L^A_g]}$ which would lead Susi to her goal.

4.2 Explorative Path Search

Let us now look at Susi's remaining path from the market to the hotel. In the second example, Susi again has complete knowledge on accessibility, but a construction fence obscures the transition from the market to the main road. Thus, the rule transforming the allocentric location L^A_{road1} into an egocentric location is not available in the rule set (the rule set was updated in f_{Perc} when Susi entered the market with $goto[L^E_{market}]$, see Table 2).

Susi decides to explore her environment to find a way around the construction fence. She stays in direction h_W and, with a visual search (s_{exp}), identifies one promising egocentric location at the corner of the market (L^E_{corner}) from which she thinks the road might be accessible. She locomotes to L^E_{corner} and, indeed, from this new location she is able to find an egocentric location in her field of view that corresponds to L^A_{road1}. Figure 5 illustrates how explorative path search appears in the production tree. Note that rule (14) can be applied multiple times to yield a longer search process.

4.3 Map-Based Wayfinding

In this example, Susi only knows the way from the pub to the market, but not how to get from the market to the hotel, i.e., $(L^A_{m1}, L^A_{road1}, h_W)$ and $(L^A_{road1}, L^A_g, h_W)$ are missing in Acc.

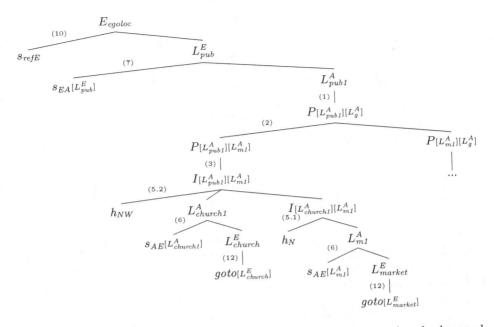

Fig. 4. An example production tree for wayfinding based on complete background knowledge with sub-planning (example in Sect. 4.1).

Fortunately, Susi finds a public you-are-here map at the market which will help her updating her accessibility graph. First, she transforms her current allocentric location to a location in the map reference system (i.e., she localizes herself on the map). Then she successively moves her visual attention on the map towards West, detecting two survey locations L_{road}^S and L_{hotel}^S which are acessible from her location. She notices that the second, L_{hotel}^S, corresponds to her allocentric goal location L_g^A. Thus, her accessibility graph is completed, and she stops using the map. Figure 6 illustrates the according production tree.

4.4   Use of Signage

As in Sect. 4.3, Susi does not know the way from the market to the hotel. This time she uses a sign with the name of her hotel pointing towards West. Reading the sign (s_{sign}) enables her to update the accessibility graph Acc (with function f_{RemC}, See Table 2). She now knows that she can go from the market to the road ((L_{m1}^A, L_{road1}^A, h_W) ∈ Acc), and that she can create an instruction from this (function f_{Instr})[5]. Thus, she now heads towards West, searches for the allocentric location L_{road1}^A, matches it to an egocentric location, and starts locomoting. Figure 7 (left) illustrates this example.

---

[5] A complex sign can add more than one edge to Acc.

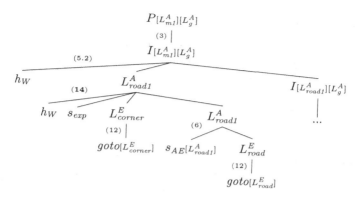

Fig. 5. Partial production tree for wayfinding with explorative path search (example in Sect. 4.2).

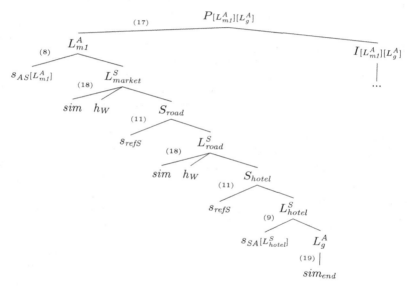

Fig. 6. Partial production tree for wayfinding with a map (example in Sect. 4.3).

Fig. 7. Partial production trees for wayfinding with signage (left, see Sect. 4.4) and with asking for directions (right, see Sect. 4.5).

4.5   Asking for Directions

Again, Susi has no previous knowledge on the second part of her journey. She asks a passersby for directions from the market to her hotel ($ask[L_{m1}^A][L_g^A]$). The answers update her accessibility graph, and she is now able to follow these instructions to the goal (see Fig. 7, right).

5   Discussion and Outlook

We argued that, in contrast to wayfinding decision models based on condition-action rules (cf. Sect. 1), our approach does not make claims on how people actually make wayfinding decisions and how they search for locations and objects. The open slots in our model correspond precisely to these knowledge gaps and are simply the *procedures* necessary to generate the language defined by our grammar (cf. Table 2). The flexibility we gain is that we now have a precise model that separates "what we know" from "what we do not yet know".

In addition, we can now test our model together with particular (ad-hoc) procedures which might fill these slots based on embedding them in wayfinding simulations and comparing results with empirical observations. Testing could be achieved in three steps. First, the wayfinding behaviors defined in Definition 3 need to be tracked, e.g., using mobile eye tracking technology, as well as sensors such as GPS, accelerometer, and compass. Second, the collected data need to be processed and analyzed in order to match the actions to wayfinding behaviors. For visual search behaviors we propose to learn classifiers trained with machine learning from empirically collected data [18]. Third, the production trees need to be generated and compared against the empirical data. A parser (e.g., a modified version of an Earley Parser [4]) could be developed which produces all trees that are possible given the recognized behaviors and finally, the most likely parse tree could be selected using a selection algorithm and a probabilistic model. The probabilities of certain productions will depend on the complexity of a particular wayfinding situation [6], i.e., a model of the environment, the user, and the instruction complexity. Another challenge here, as in most approaches based on formal grammars, consists in ambiguities: sometimes there will be more than one possible parse tree that explains a given behavior/terminal sequence. Quantifying the amount of this ambiguity in real wayfinding situations is one opportunity for future research.

Even though our proposed model is amodal, processes as terminals (See Definition 3) might obviously be translated into *visual* behavior. However, this ignores that people also use auditory and haptic senses during wayfinding. For instance, a ringing church bell can be used to localize a church in an ego-centric reference system, while haptic maps enable non-visual search in a survey system [28]. Loomis et al. [29] argue that, instead of using a visual image, humans convert all types of sensory input to a *spatial image*, i.e., an amodal representation kept in working memory that abstracts from the sensor(s) from which it has been derived. A possible way to model different modalities in our grammar could be to replace the 'search' terminals (s_x) by according $Sense_x$ non-terminals. Further production rules (called 'convert sensory input to spatial image') would map these $Sense_x$ non-terminals to either one or a sequence of terminals for auditory, visual, and haptic sensing. Finding sensoric interpretations for the wayfinding actions in our model and integrating them over different perceptual modalities is a topic for future research.

Using a computational formalism to simulate the wayfinding process can be useful, keeping in mind that a cognitive process is organized very differently from an algorithm in a computing machine. This leads to some general open methodical questions. First, grammars are not the only formalism that can be used for modeling wayfinding. One specific advantage of grammars compared to other rule-based systems and workflow engines, is that their expressiveness as well as computational complexity of parsing are well-understood. Second, cognitive and brain processes are typically highly parallel, i.e., different spatial representations are computed simultaneously [40], while our grammar execution is modeled as a single process. For example, we decided to represent route knowledge primarily in terms of allocentric reference systems, while in cognition, it may be stored in both egocentric and allocentric form [31]. Second, while our grammar in principle covers different kinds of externalized and internalized knowledge sources, it currently does not distinguish different kinds of systems of the same reference type, for example, different maps for indoor and outdoor that involve different sensed behaviors.

Furthermore, our model should be enriched by further aspects of the wayfinding process. This includes, i.e., modeling the continuous modification of the perceptual array during movement, getting lost, aspects of aided and unaided wayfinding [45], the knowledge exchange between people (cf. [10,43,44]), and the interaction of people with artifacts.

References

1. Anderson, J.R., Matessa, M., Lebiere, C.: ACT-R: a theory of higher level cognition and its relation to visual attention. Hum.-Comput. Interact. **12**(4), 439–462 (1997)
2. Arthur, P., Passini, R.: How the wayfinding process works. In: Wayfinding: People, Signs, and Architecture, pp. 26–39. McGraw-Hill, New York (1992)
3. Downs, R.M., Stea, D.: The world in the head. In: Maps in Minds: Reflections on Cognitive Mapping, chap. 4, pp. 125–135. Harper & Row Series in Geography, Harper & Row (1977)
4. Earley, J.: An efficient context-free parsing algorithm. Commun. ACM **13**, 94–102 (1970)
5. Frank, A.U.: Formal models for cognition - taxonomy of spatial location description and frames of reference. In: Freksa, C., Habel, C., Wender, K.F. (eds.) Spatial Cognition 1998. LNCS (LNAI), vol. 1404, pp. 293–312. Springer, Heidelberg (1998)
6. Giannopoulos, I., Kiefer, P., Raubal, M., Richter, K.-F., Thrash, T.: Wayfinding decision situations: a conceptual model and evaluation. In: Duckham, M., Pebesma, E., Stewart, K., Frank, A.U. (eds.) GIScience 2014. LNCS, vol. 8728, pp. 221–234. Springer, Heidelberg (2014)
7. Golledge, R.G.: Place recognition and wayfinding: making sense of space. Geoforum **23**, 199–214 (1992)

8. Golledge, R.G.: Human wayfinding and cognitive maps. In: Wayfinding Behavior: Cognitive Mapping and Other Spatial Processes, chap. 1, pp. 5–45. The Johns Hopkins University Press (1999)

9. Gopal, S., Klatzky, R.L., Smith, T.R.: Navigator: a psychologically based model of environmental learning through navigation. J. Environ. Psychol. **9**(4), 309–331 (1989)

10. Hahn, J., Weiser, P.: A quantum formalization for communication coordination problems. In: Atmanspacher, H., Bergomi, C., Filk, T., Kitto, K. (eds.) QI 2014. LNCS, vol. 8951, pp. 177–188. Springer, Heidelberg (2015)

11. Haken, H., Portugali, J.: Synergetics, inter-representation networks and cognitive maps. In: The Construction of Cognitive Maps, pp. 45–67. Springer (1996)

12. Hutchins, E.: Distributed cognition. In: International Encyclopedia of the Social and Behavioral Sciences. Elsevier Science (2000)

13. Hutchins, E.L.: Cognition in the Wild, 2nd edn. MIT press, Cambridge (1996)

14. Kesner, R.P., Creem-Regehr, S.H.: Parietal contributions to spatial cognition. In: Handbook of Spatial Cognition, pp. 35–63. American Psychological Association (2013)

15. Kiefer, P.: Spatially constrained grammars for mobile intention recognition. In: Freksa, C., Newcombe, N.S., Gärdenfors, P., Wölfl, S. (eds.) Spatial Cognition VI. LNCS (LNAI), vol. 5248, pp. 361–377. Springer, Heidelberg (2008)

16. Kiefer, P.: Mobile Intention Recognition. Springer, New York (2011)

17. Kiefer, P., Giannopoulos, I.: Gaze map matching: mapping eye tracking data to geographic vector features. In: Proceedings of the 20th SIGSPATIAL International Conference on Advances in Geographic Information Systems, pp. 359–368. ACM, New York (2012)

18. Kiefer, P., Giannopoulos, I., Raubal, M.: Using eye movements to recognize activities on cartographic maps. In: Proceedings of the 21st SIGSPATIAL International Conference on Advances in Geographic Information Systems, pp. 498–501. ACM, New York (2013)

19. Kiefer, P., Giannopoulos, I., Raubal, M.: Where am I? investigating map matching during self-localization with mobile eye tracking in an urban environment. Trans. GIS **18**(5), 660–686 (2014)

20. Kitchin, R.M.: Cognitive maps: what are they and why study them? J. Environ. Psychol. **14**(1), 1–19 (1994)

21. Klippel, A., Tappe, H., Kulik, L., Lee, P.U.: Wayfinding choremes - a language for modeling conceptual route knowledge. J. Vis. Lang. Comput. **16**(4), 311–329 (2005)

22. Knuth, D.E.: Semantics of context-free languages. Math. Syst. Theor. **2**, 127–145 (1968)

23. Kuhn, W.: Semantic reference systems. Int. J. Geogr. Inf. Sci. **17**(5), 405–409 (2003)

24. Kuipers, B.: Modeling spatial knowledge. Cogn. Sci. **2**(2), 129–153 (1978)

25. Leiser, D., Zilbershatz, A.: The traveller: a computational model of spatial network learning. Environ. Behav. **21**(4), 435–463 (1989)

26. Levinson, S.C.: Space in Language and Cognition: Explorations in Cognitive Diversity, vol. 5. Cambridge University Press, Cambridge (2003)

27. Logan, G.D., Sadler, D.D.: A computational analysis of the apprehension of spatial relations. In: Language and Space. Language, Speech, and Communication, pp. 493–529. MIT Press (1996)

28. Lohmann, K., Eschenbach, C., Habel, C.: Linking spatial haptic perception to linguistic representations: assisting utterances for tactile-map explorations. In: Egenhofer, M., Giudice, N., Moratz, R., Worboys, M. (eds.) COSIT 2011. LNCS, vol. 6899, pp. 328–349. Springer, Heidelberg (2011)

29. Loomis, J.M., Klatzky, R.L., Giudice, N.A.: Representing 3D space in working memory: spatial images from vision, hearing, touch, and language. In: Multisensory Imagery, pp. 131–155. Springer (2013)

30. MacEachren, A.M.: How Maps Work: Representation, Visualization, and Design. Guilford Press, New York (1995)

31. McNamara, T.P.: Spatial memory: properties and organization. In: Handbook of Spatial Cognition, pp. 173–190. American Psychological Association (2013)

32. Montello, D.R.: Navigation. In: Cambridge Handbook of Visuospatial Thinking, pp. 257–294. Cambridge University Press (2005)

33. Montello, D.R.: Scale and multiple psychologies of space. In: Campari, I., Frank, A.U. (eds.) COSIT 1993. LNCS, vol. 716, pp. 312–321. Springer, Heidelberg (1993)

34. Passini, R.: Wayfinding: a conceptual framework. Urban Ecol. **5**(1), 17–31 (1981)

35. Raubal, M., Worboys, M.F.: A formal model of the process of wayfinding in built environments. In: Freksa, C., Mark, D.M. (eds.) COSIT 1999. LNCS, vol. 1661, pp. 381–399. Springer, Heidelberg (1999)

36. Richter, K.F., Winter, S.: Landmarks. Springer, Switzerland (2014)

37. Scheider, S.: Grounding Geographic Information in Perceptual Operations. Frontiers in Artifical Intelligence and Applications, vol. 244. IOS Press, Amsterdam (2012)

38. Schlieder, C.: Representing the meaning of spatial behavior by spatially grounded intentional systems. In: Rodríguez, M.A., Cruz, I., Levashkin, S., Egenhofer, M. (eds.) GeoS 2005. LNCS, vol. 3799, pp. 30–44. Springer, Heidelberg (2005)

39. Shelton, A.L., McNamara, T.P.: Systems of spatial reference in human memory. Cogn. Psychol. **43**(4), 274–310 (2001)

40. Simmering, V.R., Schutte, A.R., Spencer, J.P.: Generalizing the dynamic field theory of spatial cognition across real and developmental time scales. Brain Res. **1202**, 68–86 (2008)

41. Smith, T.R., Pellegrino, J.W., Golledge, R.G.: Computational process modeling of spatial cognition and behavior. Geogr. Anal. **14**(4), 305–325 (1982)

42. Tversky, B.: Cognitive maps, cognitive collages, and spatial mental models. In: Campari, I., Frank, A.U. (eds.) COSIT 1993. LNCS, vol. 716, pp. 14–24. Springer, Heidelberg (1993)

43. Weiser, P.: A Pragmatic Communication Model for Way-finding Instructions. Ph.D. thesis, Vienna University of Technology. Department of Geodesy and Geoinformation. Research Group Geoinformation (2014)

44. Weiser, P., Frank, A.U.: Cognitive transactions – a communication model. In: Tenbrink, T., Stell, J., Galton, A., Wood, Z. (eds.) COSIT 2013. LNCS, vol. 8116, pp. 129–148. Springer, Heidelberg (2013)

45. Wiener, J.M., Büchner, S.J., Hölscher, C.: Taxonomy of human wayfinding tasks: a knowledge-based approach. Spat. Cogn. Comput. **9**(2), 152–165 (2009)

Quantifying the Significance of Semantic Landmarks in Familiar and Unfamiliar Environments

Teriitutea Quesnot[✉] and Stéphane Roche

Center for Research in Geomatics, Pavillon Louis-Jacques Casault, Laval University,
1055 Avenue du Seminaire, Quebec City, QC G1V 0A6, Canada
teriitutea.quesnot.1@ulaval.ca, stephane.roche@scg.ulaval.ca

Abstract. During navigation, people tend to associate objects that have outstanding characteristics to useful landmarks. The landmarkness is usually divided into three categories of salience: the visual, the structural, and the semantic. Actually, the roles of visual and structural landmarks have been widely explored at the expense of the semantic salience. Thus, we investigated its significance compared to the two others through an exploratory experiment conducted on the Internet. Specifically, 63 participants were asked to select landmarks along 30 intersections located in Quebec City. Participants were split by gender and familiarity with the study area. Unsurprisingly, the results show that unlike strangers, locals tended to focus on highly semantic landmarks. In addition, we found that women were more influenced by the structural salience than men. Finally, our findings suggest that the side where travelers move compared to the road impacts on the landmark selection process.

Keywords: Gender difference · Landmarks · Semantic salience · Side of the road effect · Spatial knowledge · Wayfinding

1 Context and Research Goal

Landmarks are usually considered as an *organizing concept* of human spatial knowledge (cf. Siegel and White 1975), as *markers* used to find one's way (cf. Lynch 1960), but also as a *specific component of route directions* (cf. Tom and Denis 2003). Specifically, individuals who follow an itinerary with landmark-based instructions make less frequent breaks to ensure the route control than those who only rely on street names (Michon and Denis 2001 and Tom et al. 2003). In the same vein, wayfinding directions that contain a minimum set of landmarks are more easily memorized than those without any mention (Daniel et al. 2003). Such landmarks are usually described according to their *location* or their *intrinsic characteristics*.

Lovelace et al. (1999) proposed a categorization of landmarks based on their location, namely: (1) *off route landmarks*, which are considered as global orientation objects, (2) *on route landmarks*, i.e. objects located along the scheduled

© Springer International Publishing Switzerland 2015
S.I. Fabrikant et al. (Eds.): COSIT 2015, LNCS 9368, pp. 468–489, 2015.
DOI: 10.1007/978-3-319-23374-1_22

itinerary and used by travelers to ensure the route control, (3) *potential choice points*, which are entities located along an intersection where a turning decision is possible but not scheduled, and finally, (4), *choice points*, i.e. landmarks located at each intersection where a turn is planned. Sorrows and Hirtle (1999) proposed a typology of landmarks according to their individual attributes: (1) *visual* landmarks are objects easily distinctive by their visual characteristics (e.g. size, shape, color), (2) *structural* landmarks are remarkable because of their highly accessible position (e.g. a corner of a major street intersection), and (3), *cognitive* landmarks, which remain outstanding by their atypical meaning or function (e.g. a church surrounded by several shops). Concretely, a landmark rarely groups these three types of salience at the same time but it can be usually characterized by more than one of them.

Raubal and Winter reused this typology to formalize a model of landmark salience that was applied to the facades of buildings (2002). Similarly, Caduff and Timpf (2008) proposed an analogous model of landmarkness. According to them, three specific types of salience are involved in the process of landmark selection. The first one is the *perceptual salience*, which is limited to the visual attention in the context of wayfinding. The *cognitive salience* is the second one. It is linked to the observer's spatial knowledge. They distinguish the *degree of recognition* of objects (i.e. the more an object recurs, the more memorized and recognized it will be) from the *idiosyncratic relevance* of objects (i.e. the personal significance of objects according to the observer's experience). Finally, like Winter et al. (2005), the *contextual salience* is the last one mentioned by the authors. Indeed, tasks required during the navigation affect the observer's visual attention. For instance, a scene that contains a lot of objects requires more resources to extract relevant landmarks. In the same vein, a car driver obviously relies on a visual field more limited than a pedestrian. Thus, both might not necessarily choose the same landmark at the identical decision point.

Several research have been carried out according to those complementary approaches. For instance, Winter (2003) introduced the concept of *advance visibility* by assuming that buildings highly visible from a long distance catch a significant attention while navigating. He proposed to compute the advance visibility of objects by multiplying their *visibility coverage* and their *orientation*, in accordance with the direction of navigation. In parallel, Klippel and Winter (2005) enriched the concept of structural salience. According to them, the landmark candidate should ideally be located along the *same direction as the turn*, and *before the intersection*. This assumption was partly supported by Röser et al. (2012a; 2012b). However, research on the semantic salience of landmarks is currently rather limited. Since visual and structural saliencies rely on "easily" measurable criteria, or at least easily identifiable (e.g. size, color, location, etc.), research on the measure of landmarkness clearly focused on these two types of salience. In fact, the semantic salience is harder to estimate because it exclusively refers to the observer's spatial experience. Indeed, it relies on subjective indicators quite complex to assess. Actually, in the context of automatic landmark detection, giving route instructions that contain idiosyncratic landmarks is not feasible, unless the system accesses the traveler's spatial knowledge (through a history of check-ins for example). This functionality might be conceivable in the near future since the current version of Google Maps is able to design personal maps based on users' points of interest. But for now, the alternative consists of providing semantic landmarks collectively recognized at different scales (world, city, or district). Nonetheless, the following issue remains: *how to objectively establish that the Starbucks Coffee located at the corner of a given intersection is more recognized than the MacDonald's that faces it at the opposite corner?* Assigning a *precise* semantic salience score remains tricky, especially when one faces several famous venues. Before the advent of *Volunteered Geographic Information* (VGI) and Location-based Social Networks (e.g. Swarm), the detection of semantic landmarks was frequently limited to the most significant historical and cultural places and to few limited attributes (e.g. function, explicit signs, etc.).

VGI clearly changed the way of producing and consuming geographic data (Goodchild 2007). Citizens constantly connected to the Internet are now able to easily share - whether deliberately or not - their spatial position on social media. By this way, VGI gives an opportunity to capture the local geographic knowledge (Elwood et al. 2013 and Quesnot and Roche 2015b). In the specific context of landmark detection, Richter and Winter (2011) developed a tool that allows OpenStreetMap (OSM) users to tag landmarks on this platform. However, contributors dot not necessarily realize the usefulness of such approach since landmark-based instructions are not really publicized. An alternative consists in data mining the crowdsourced data that are produced without the will of contributing to any enrichment of spatial component. This kind of approach is in line with Tezuka and Tanaka's work on landmark extraction from web documents (2005). For instance, Crandall et al. (2009) proposed an interesting solution to detect landmarks by analyzing geotagged photos published on Flickr. However, their solution only extracts global landmarks (i.e. places from where lots of people publish photos) and neglects local landmarks. In addition, Quesnot and Roche (2015a) proposed to identify semantic landmarks by exploiting the check-ins shared on Facebook and Swarm. Unlike Crandall et al.'s approach, this alternative also takes local semantic landmarks into account. Indeed, their selection can be easily modulated according to the distribution of check-ins.

Except the focus on the *measure* of landmark semantic salience, only few research highlighted the *significance* of the semantic salience itself (e.g. Nothegger et al. 2004). To our best knowledge, Hamburger and Röser's research (2014) currently remains the only one that provides explicit empirical findings about the helpfulness of semantic landmarks in a wayfinding context. Unsurprisingly, the authors found that the memorization of the route was better when pictures of well-known landmarks were incorporated inside their virtual environment. Given this empirical finding, one can assume that an individual who travels inside an unfamiliar environment may not rely on the same landmarks as someone who knows the area by heart. By taking this assumption as a starting point, we ask the following research questions: *what is the significance of semantic salience compared to both visual and structural saliencies? Does this salience really represent a decisive parameter in the selection of landmarks for strangers? Moreover, is there a meaningful difference in the landmark selection between familiar and unfamiliar individuals?* The question of the semantic salience is crucial in the context of landmark selection. The current research attempts to provide additional *empirical* findings regarding Hamburger and Röser's work by also focusing on the visual and structural saliencies. Specifically, we first conducted a survey to estimate some structural salience scores according to the turning decision (Sect. 2). We reused those scores through an experiment conducted on the Internet where we asked participants to choose landmarks along 30 different intersections across Quebec City (Sect. 3).

2 Study 1

This study was the starting point of our research. Indeed, one of our goals was to compare the semantic salience with the visual and structural saliencies. Thus, we conducted a survey to establish some structural salience scores of reference. We then reused them in the main experiment that is detailed in the next section.

2.1 Participants

A total of 80 participants joined this survey (39 females and 41 males). We anonymously hired them *via* the general mailing list of Laval University. The participants' age varied from 18 to 55 with an average of 28.13 years.

2.2 Material

We used Google Form sheets to design this survey. We relied on an approach similar to the one proposed by Röser et al. (2012a; 2012b). In this way, we inserted two pictures of a cross intersection regarding two fields of view (*allocentric* and *egocentric*). We focused on this type of intersection since it remains one of the most commons in North America. We respectively included the allocentric and egocentric sketches (cf. Fig. 1) in the first and second sheets of the form.

2.3 Procedure

After having specified their gender and their age, the participants were invited to read a brief contextualization of the survey: *"Imagine that you are visiting an unfamiliar city. In theory, you would instinctively select landmarks along*

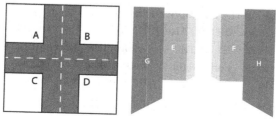

Fig. 1. Allocentric and egocentric sketches of the cross intersection.

your route to orient yourself and find your way (to go back to your hotel room for example). The main purpose of this survey is to identify the position of the landmark that you would select on a cross intersection according to a left turn and a right turn". We first presented the allocentric picture and we asked the participants two questions: *"Considering that you are arriving from the bottom of the figure, which place would you choose as a landmark if you were supposed to turn to the left?".* The same question was asked for the right turn. Participants were redirected to the second sheet of the form once they answered the questions about the allocentric picture. We then invited them to answer the same questions about the egocentric sketch.

2.4   Results and Discussion

Regardless of the perspective, buildings located along the same direction as the turn were clearly favored by both males and females (cf. Table 2). This finding is in line with the studies made in the same context (see Klippel and Winter 2005 and Röser et al. 2012a; 2012b). However, one can notice a slight difference between men and women by paying attention to the percentages on Fig. 2 and Table 2. Specifically, *women* tended to select more buildings located along the ssame direction as the turn for the *right* turn (92.31 % of women *vs* 85.37 % of men for the egocentric view) whereas *men* followed the same trend for the *left* one (90.24 % of men *vs* 87.18 % of women regarding the same view). In the same vein, men systematically favored the places located after the intersection when the egocentric view was proposed. Nonetheless, the p values calculated from the Fisher's exact test do not indicate any association between the gender and the position of the landmark candidate (cf. Table 1). Having said that, the most important finding to notice is the difference between the left turn and the right one. Indeed, the Fisher's test *suggests* a weak association between the turn and the position of the landmark according to the intersection (cf. Table 1). More precisely, 56.41 % of the women tended to select the buildings located after the intersection for the left turn whereas only 38.46 % of them chose those buildings for the right one (from an egocentric perspective). The pattern is identical for the men and for both allocentric and egocentric views. More precisely, building

Table 1. Test of independence between the position of the landmark and both the gender and the turn (p values of the Fisher's exact test).

	Position of the landmark candidate							
	Intersection (Before/After)				Direction of the turn (Same/Opposite)			
Gender	Allocentric		Egocentric		Allocentric		Egocentric	
	Left	Right	Left	Right	Left	Right	Left	Right
	0.37	1	0.82	0.27	0.75	0.67	0.73	0.48
Turn	0.26		0.11		0.18		1	

E seems to enjoy a greater attention from the participants for the left turn than building F for the right one (egocentric average: 52 % *vs* 37 %).

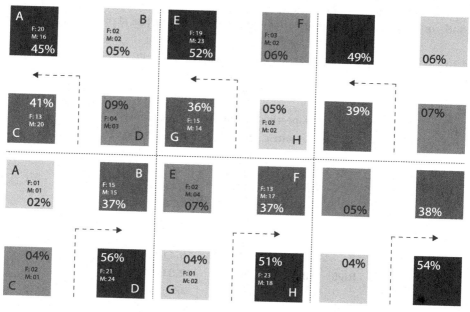

Fig. 2. Detailed results of the survey: allocentric view (left), egocentric view (center), and average scores (right). The top half refers to the left turn whereas the other half deals with the right one.

Actually, the visibility of buildings A and B (E and F in the egocentric sketch) is somewhat higher when the egocentric perspective is added (cf. Fig. 1). In fact, that is why Röser et al. (2012b) proposed a weight of 0.74 for the places located after the intersection and 0.24 for those located before the intersection. However, building A seems to be more visible than building B in participants' mind. On the one hand, we argue that they mentally positioned themselves on the *right side of the road*. Indeed, we do believe that the lines marked on the road in the allocentric sketch induced participants to reason as if they were driving a car. Yet, the side of the road where people drive here in Quebec City is the right one. Since the allocentric picture appeared first, we do think that participants relied on the same pattern for the egocentric one. This distinction was not highlighted by Röser et al. (2012a; 2012b) since the virtual traveler constantly moved in the middle of the road. However, one cannot ignore that individuals usually travel either on the left or on the right side of the road; and that is especially true for pedestrians. On the other hand, the way we drawn the buildings in the egocentric sketch might also had an impact in the participants' choice of landmarks. Specifically, the results could have been different if E and F were taller than G and H. To sum up, the findings of this survey suggest that the *direction of the turn* and the *side of the road* are two factors closely linked to the structural salience. The gender difference highlighted in Table 2 remains slight but is worth to be further investigated. In the end, we argue that the average between the allocentric and egocentric perspectives for both turns is a good compromise for a *quantitative* characterization of the structural salience.

Table 2. Summarized results of the survey (percentages). *St* corresponds to "Same direction as the Turn" and *Ot* to "Opposite direction of the Turn". *Bi* and *Ai* respectively correspond to "Before the Intersection" and "After the Intersection".

Turn	Perspective	Location	Female	Male
Left	Allocentric	**St**	**84.62**	**87.80**
		Ot	15.38	12.20
		Bi	43.59	**56.10**
		Ai	**56.41**	43.90
	Egocentric	**St**	**87.18**	**90.24**
		Ot	12.82	9.76
		Bi	43.59	39.02
		Ai	**56.41**	**60.98**
Right	Allocentric	**St**	**92.31**	**95.12**
		Ot	7.69	4.88
		Bi	**58.97**	**60.98**
		Ai	41.03	39.02
	Egocentric	**St**	**92.31**	**85.37**
		Ot	7.69	14.63
		Bi	**61.54**	48.78
		Ai	38.46	**51.22**

3   Study 2

We carried out the main experiment once the survey was completely finished. The second study was primarily done to bring empirical evidences about the significance of semantic landmarks in comparison with the visual and structural saliencies. We also wanted to further investigate the influence of both the gender and the side of road effect previously highlighted. In order to achieve these goals, we developed a web application based on the Google Maps API. By using the Google Street View service, the participants of this experiment were able to pick-up landmarks while navigating on the *right* side of the road.

3.1   Participants

A total of 63 individuals participated in this online experiment. We directly hired them by email. We systematically *excluded* individuals who were involved in the previous survey. Since our goal was to obtain a group divided by the *degree of familiarity* with the study area (i.e. Quebec City) and the *gender*, there were two hiring sessions. We first focused on familiar participants. We only selected individuals who lived at least *three years* in Quebec City. In the end, we hired 31 individuals (15 females and 16 males). This first sub-group was mainly composed of Laval University graduated students in geomatics, geography, urban planning, and neurosciences. The average age of this group was 26.6 years and the average time spent in Quebec City was 11.1 years. On the other hand, we selected 32 individuals (17 females and 15 males) who live in France and who have never been to Quebec City. This second sub-group was mainly composed of researchers, engineers, and technicians in geomatics, geography, and environment. Their age ranged from 24 to 45, with an average of 28 years. Each participant signed a consent form before participating in this experiment.

3.2 Material

We first determined the method of computation for the three saliencies and for the advance visibility. Afterward, we focused on the design of the experiment.

Visual Salience Score. We decided to compute the visual salience by making a combination of the attributes proposed by Raubal and Winter (2002) and Duckham et al. (2010) (cf. Table 3); namely: the *color* of the building, its *area*, its *proximity to road*, and the presence of *noticeable signs*. Indeed, these are standard attributes also used in similar research (e.g. Elias 2003).

Specifically, we assigned a score between 0 and 20 to the last two attributes. The *proximity to road* factor was computed on a GIS (ESRI ArcMap 10.2.2) using OSM Web Map Service as a base map. The score assigned to the color attribute was partially based on Raubal and Winter's approach for computing the visual salience (2002). However, instead of computing it on the basis of the

Table 3. Visual salience criteria scores.

Visual Attribute	Score Assignment					
	0	4	8	12	16	20
Explicit sign(s)						
Proximity to road (meters)	x>10	8<x<10	6<x<8	4<x<6	2<x<4	x<2
Color ([R,G,B])	Distance from the [R,G,B] score of the intersection scene.					
Area (square meters)	Distance from the average area of the four candidates.					

distance from the [R,G,B] mean score (computed itself from the visual salience score of each landmark located around a decision point), we based it on the distance from the [R,G,B] score of the *intersection scene*. More precisely, we define the [R,G,B] score of the intersection scene as the color that stands out the most from the *vista* that includes each potential landmark candidate. We used the *Color Thief* Javascript API to automatically extract the *dominant* color of the tiles associated with each intersection and each potential landmark. Afterward, we compared the [R,G,B] means of the dominant colors of both the intersection scene and the landmark candidate. The more the distance between the two means was significant, the higher the color score of the candidate was. In our opinion, computing a visual score through this approach remains sensible. Indeed, since participants virtually traveled inside a set of Google photo panoramas, the color of objects other than the potential landmark candidates (e.g. signboards, trees, parked cars, or walkers) might have interfered in the single visual salience score of the buildings. Similarly, we computed the *area score* by taking the distance from the *average area* of the candidates into account. We define the average area as the mean calculated from the areas of all landmark candidates. Note that there were systematically four landmark candidates per intersection (cf. Fig. 3). We also computed areas in square meters on the GIS. In addition, we transferred the color and area scores onto a [0–20] scale. In this way, we were able to compare these scores with the two others (i.e. *explicit signs* and *proximity to road*). The global visual salience score *GVis* was then computed with the advance visibility score (cf. Eq. 1).

$$GVis(p) = AV(p) \times \frac{(ES(p) + PR(p) + CO(p) + AR(p))}{80} \qquad (1)$$

where p = potential landmark, and the scores: AV = advance visibility, ES = explicit signs, PR = proximity to road, CO = color, and AR = area.

Advance Visibility Score. We computed the advance visibility score on ArcMap according to Winter's methodology (2003). Specifically, the *visibility coverage* was calculated by using the *Viewshed* command of the 3D Analysis

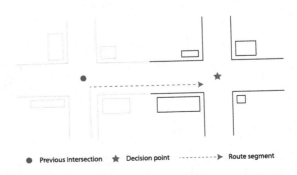

Fig. 3. Typical route segment.

Toolbox. In order to obtain a precise Digital Elevation Model, we combined a Digital Terrain Model at a scale of 1:20 000 with a polygon that contained locations, shapes, and elevations of the buildings of Quebec City. For each intersection, the lines of arriving were drawn according to the virtual route segments followed by the participants (cf. Figs. 3 and 6). Moreover, we took the orientation of the signboards into account in the computation of the *orientation scores* of the buildings. Indeed, venues perpendicularly oriented to the direction of arriving got an orientation score of 1 when they had an adequately oriented signboard (cf. Fig. 4). The *directions* were automatically computed by the GIS.

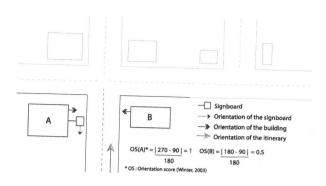

Fig. 4. Computation of the orientation score.

Semantic Salience Score. The computation of the semantic salience was done according to the authors' previous work (cf. Quesnot and Roche 2015a). However, we applied two specific changes. On the one hand, we improved the computation of the uniqueness score by applying a gradation. Thus, venues that belong to different supra-categories (e.g. religious monuments *vs* shops) got a uniqueness score higher than if they were belonging to the same infra-category (cf. Eqs. 2, 3, and 4).

$$Supra(p) = \frac{\sum_{i=1}^{n} P_i \in SC}{4} \qquad (2)$$

$$Infra(p) = \frac{\sum_{j=1}^{m} P_j \in IC}{4} \qquad (3)$$

where SC = supra-category of p, and IC its infra-category. Note that the denominator is set to 4 because the participants had to systematically choose one landmark among a set of four candidates.

Indeed, even though a clothing store and a coffee store are different *per se*, they remain *stores*. By this way, these two venues are not necessarily easily distinguishable during navigation. On the contrary, a church obviously stands out from the environment if it is surrounded by shops. On the other hand, we multiplied the uniqueness score and the geosocial activity score (i.e. the significance of a place on the social web) instead of summing them. Thus, we argue that the more function of an object stands out, the more its semantic (and visual) salience increases. We also decided to compute the global semantic score *GSem* according to the advance visibility score. Indeed, we consider the advance visibility as a decisive supra-factor that belongs to the *contextual* salience rather than the visual one. In the experiment, participants had to move along predetermined paths according to a specific field view. We would not have included the advance visibility score if the participants only had to choose the landmark candidates among a set of static panoramic images (e.g. Nothegger et al. 2004). In the end, we normalized GSem in order to fit on a [0–1] scale (cf. Eq. 5). To conclude, we want to precise that we harvested data from the Foursquare API v2.0 on February 10, 2015 to perform the geosocial activity scores.

$$GUnq(p) = \frac{1}{(Supra(p) + Infra(p))} \tag{4}$$

$$GSem(p) = AV(p) \times (Gunq(p) \times GSA(p)) \tag{5}$$

where $GUnq$ = graded uniqueness score, and GSA = geosocial activity score (cf. Quesnot and Roche 2015a).

Structural Salience Score. The structural score of each potential landmark candidate was computed according to the average scores obtained from the initial survey (cf. Fig. 2). Thus, each structural salience score only varied according to the turns (i.e. left or right). We did not combined the structural score with the advance visibility score since the configuration of the cross intersections was constantly the same at the end of each predetermined route segment.

Intersections, Potential Landmark Candidates, and Their Distribution.
We selected 120 potential landmarks across 30 cross intersections located in Quebec City. We determined a route segment for each decision point. Its starting point

Table 4. Top-ten semantic landmarks selected for the experiment.

Rank	Landmark candidate	Semantic salience score
1	Civilization Museum	1.00
2	Best Western Hotel	0.89
3	Le Hobbit restaurant	0.65
4	Yuzu Sushi restaurant	0.60
5	Brûlerie St-Jean	0.55
6	Les Fistons restaurant	0.55
7	Bistro B pub	0.50
8	Palace Royal Hotel	0.50
9	Brûlerie Vieux-Limoilou	0.47
10	Ralph et Laurie restaurant	0.39

was usually the intersection that preceded the decision point (cf. Fig. 3). A major concern was to select a strong enough diversity of venues. Our primary goal was to evaluate the significance of the semantic salience according to the familiarity with the area. Thus, we first chose highly recognized *local* landmarks *via* the analysis of Foursquare check-ins (cf. Table 4). We then added other places potentially recognizable by strangers such as Subway restaurants. Specifically, we included 22 places that belong to supra-categories memorable for strangers (cf. Table 5). These *global* landmarks covered 18 intersections. In the end, the visual salience varied from 0.012 to 0.986 with a mean of 0.531 and a standard deviation (SD) of 0.229. The distribution of the semantic salience was less scattered with a variation that went from 0 to 1 with a mean of 0.148 and a SD of 0.181. This low variance might be directly linked to the multiplication of the uniqueness and the geosocial activity scores.

Table 5. Categories of global landmarks and their occurrences in study 2.

Landmark category	Occurrence
Drugstore	8
Gas station (e.g. Shell)	5
Restaurant (e.g. Subway)	3
Hotel (e.g. Best Western)	2
Park	2
Church	1
Store (Crocs)	1

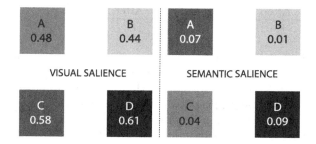

Fig. 5. Average salience scores according to their position.

As shown in the fifth figure, position D held the highest average score for both visual and semantic saliencies. On the contrary, position B was the less significant. Also, we can see that the most significant landmarks are located *before* the intersection. The distribution of the candidates according to their highest and lowest salience scores highlights a *link* between the semantic and visual saliencies. Indeed, among all candidates, 26 % of them held the highest visual salience score and the highest semantic salience score at the same time (cf. Table 6). In addition, almost half of them gathered the lowest visual and semantic salience scores together. Actually, this relationship seems quite expectable since the major semantic landmarks were essentially buildings that met almost all of the predetermined visual criteria (e.g. restaurants, pubs, hotel, etc.). Having said that, the structural salience was not at all disadvantaged. For instance, 28.3 % of the most significant structural landmarks held the highest visual salience score at the same time. The percentage reached 30 % with the semantic salience. Finally, as highlighted in the sixth table, around three-quarters of the landmarks that have the lowest visual salience score were located *after* the intersection. Although this asymmetry remains significant, one must keep in mind that 43.4 % of the highly visual landmarks were also located *after* the intersection. To conclude, position A hosted 30 % of the *major* semantic landmarks (*vs* 20 % of the minor ones) and 33.3 % of the *minor* visual landmarks (*vs* 16.6 of the major ones). One should take those statistics into account since we also investigate the significance of the *side of the road* factor.

Table 6. Distribution of the landmark candidates according to their highest and lowest salience scores, and their positions.

Sem / Vis	Highest	Lowest	Sem / Str	Highest	Lowest	Vis / Str	Highest	Lowest
Highest	26%	15%	Highest	30%	17.5%	Highest	28.3%	23.3%
Lowest	13%	47.5%	Lowest	18.3%	28.75%	Lowest	23.3%	23.3%

Sem / Pos	Highest	Lowest	Vis / Pos	Highest	Lowest
Before	53.3%	42.5%	Before	56.6%	23.3%
After	46.6%	57.5%	After	43.4%	76.7%

Online Experiment. Once the intersections and potential landmarks selected, we developed a web application that contained three components (cf. Fig. 6). The first one contained buttons located at the top of the webpage. By this way, users were able to manually move from an intersection to another. The second component was an HTML frame inserted at the extreme left of the webpage. This frame held a Google Form that was composed of four pages. The first page included basic information questions (age, familiarity with Quebec City, and time spent there in years). The three other pages contained questions about the landmark selection according to the turn (left or right) for each intersection (10 intersections per page). Finally, the last component was a viewport that contained a navigable Google Street View environment dynamically linked to a reducible window. We included a Google Maps base map inside this window to help participants to follow the route segment associated with each intersection. We used the Google Maps API V2 and the Street View Service to feed this viewport. We also added javascript listeners in order to ensure that participants keep the same field of view and only follow the predetermined route segments. In addition, dialog boxes appeared when users tried to move beyond those segments. We also added markers inside the environment to mark out their extremities.

3.3   Procedure

Walkthrough. After having electronically signed a consent form that explained the purpose of this experiment, participants were directed on a walkthrough page that guided them step by step in the understanding of the application.

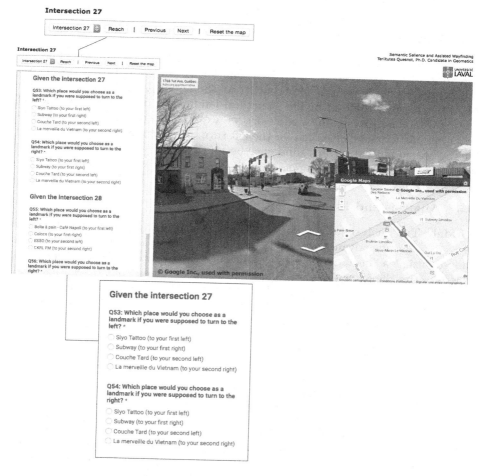

Fig. 6. Screenshot of the online experiment.

Basic Information. Participants had to specify basic information such as their *age*, their *gender*, their *familiarity with Quebec City*, and the *time spent there in years* for those who specified that they were familiars with the study area.

Landmark Selections. After having filled out the first sheet of the form, participants had to choose *two* landmarks between a set of four potential landmark candidates for each intersection. One for the left turn and another one for the right turn. Furthermore, they had to follow a predetermined route segment for each intersection. Additionally, they were also able to go back to the previous intersection and make a change if they had checked the wrong box by mistake.

3.4   Expected Results

First of all, we expect an association between the degree of familiarity and the semantic salience. Indeed, we do believe that locals will favor the main semantic landmarks we chose for this experiment. At the opposite, strangers might rely on either highly visible or highly structural landmarks. Secondly, if the side of the road factor proves to be significant, we expect that position A (see Fig. 5) will grab participants' attention for the left turn. Finally, given the results of the survey, we do not expect any difference between women and men regarding the selection of structural landmarks.

3.5   Analysis and Results

Model. The analysis of the data collected from the experiment required the usage of a specific advanced model known as the *random-parameters logit model.* Indeed, using a linear regression was inconceivable since the dependent variable (i.e. the choice of the participant) was not continuous but categorical and unordered. Consequently, we needed to perform a logistic regression; i.e. a regression based on the logit function (cf. Eq. 6). Note that in this case, one do not regress a measured variable but a *probability.*

$$Logit(P) = \ln \frac{P(1|X)}{1 - P(1|X)} = b_0 + b_1 x_1 + b_2 x_2 + \ldots + b_m x_m \qquad (6)$$

where $P(1{-}X) = $ the probability of choosing X, $b_0 = $ the intercept, $x_m = $ the regressor, and b_m its coefficient.

Since participants had to make a choice between four distinct places (i.e. alternatives = 4), the binary logit model (i.e. alternatives = 2) and ordered logit models (which suppose that the alternatives can be ordered together) were excluded. Moreover, we faced two types of regressor: *case-specific* (gender and familiarity) and *alternative-specific* (salience scores). In this context, a mixed logit model as defined by Cameron and Trivedi (2005, p. 500) - i.e. a combination of the *multinomial model* and the *pure conditional logit model* - is usually used (cf. Eq. 7).

$$P_{ij} = p(Y_i = j) = \frac{e^{X_{ij}\beta + W_i\gamma_j}}{\sum_{k=1}^{m} e^{X_{ik}\beta + W_i\gamma_k}} \tag{7}$$

where X_{ij} = the alternative-specific regressors,
and W_i = the case-specific regressors.

However, using this model implies that each observation is independent from the others. Yet, we worked on *longitudinal* data since each participant was invited to *successively* select a landmark across 30 intersections. For instance, the fact that the individual i tended to systematically choose the place located at his first right for the right turn needed to be taken into consideration. We attached importance to this parameter since the participants were also able to change their answers during the experiment. In the end, we used the *random-parameters logit model* because it includes random coefficients that induce correlations between choices (Cameron and Trivedi 2005). According to it, the probability that an individual i choose the alternative j is defined in the following equation:

$$P_{ij} = p(Y_i = j) = \frac{e^{X_{ij}\beta + W_i\gamma_j + X_{ij}v_i + W_i\sigma_{ji}}}{\sum_{k=1}^{m} e^{X_{ik}\beta + W_i\gamma_k + X_{ik}v_i + W_i\sigma_{kj}}} \tag{8}$$

Computation of the Probabilistic Model. The model was computed in SAS 9.4 using the *logistic* procedure: we took the characteristics of both the *alternatives* (i.e. the salience scores) and the *individuals* (i.e. the gender and the familiarity with Quebec City) into consideration in the first computation. We then focused on the locations of the places and the turns without taking the salience scores and the individual factors into account. This second computation allowed us to (1) analyze where participants tended to choose their landmarks according to the turns, and (2) compare the results of the experiment with those of the first study.

Outputs. The results of the first and second computations are respectively summarized in Tables 7 and 8. In order to facilitate the interpretation, we used the following variable coding for the second computation:

- **SB**: the place is located along the *same* direction as the turn, *before* the intersection;
- **SA**: the place is located along the *same* direction as the turn, *after* the intersection;
- **OB**: the place is located at the *opposite* direction of the turn, *before* the intersection;
- **OA**: the place is located at the *opposite* direction of the turn, *after* the intersection.

Interpretation of the Estimated Parameters. First of all, one can notice that the coefficients of the salience scores are statistically significant with a confidence interval of 99 % (cf. Table 7). In this case, the estimated parameters should be interpreted with extreme caution. Indeed, since the scores are not computed on the same unit basis, it does not make sense to interpret the parameters relatively together. However, by paying attention to the Wald Chi-Square measure, one can notice that the visual salience is more significant than the semantic salience. In the same vein, the structural salience is less significant than the two others (12.08 *vs* 20.47 and 26.21). Among the six potential interactions observed, only three of them are significant (cf. Table 7). Firstly, the interaction between the semantic salience score and the familiarity (p < 0.01) reveals that for equal visual and structural scores, locals are likely to choose the place with the highest semantic salience score. Secondly, the interaction between the structural salience and the gender (p < 0.02) indicates that for equal visual and semantic scores, women are likely to fit the average structural profile we chose (cf. Fig. 2). Thirdly, this trend is also observed with the locals (p = 0.0411). For instance, given the computed model, the probability that a *local woman* choose the venue A between the set of potential landmark candidates [A,B,C,D] located along a *cross intersection* is:

$$p(Y = A) = \frac{e^{2.52 \times Vis_A + (2.04+1.36+1.18) \times Str_A + (1.22+0.93) \times Sem_A}}{\sum_{k=A}^{D} e^{2.52 \times Vis_k + 4.58 \times Str_k + 2.15 \times Sem_k}} \tag{9}$$

where Vis_k = the visual salience of the landmark candidate k, Str_k its structural salience, and Sem_k its semantic salience.

Table 7. Analysis of conditional maximum likelihood estimates: *salience scores and individuals characteristics.*

Parameter	Ind.	DF	Coeff.	Std. Err.	Wald Chi2	Pr>Chi2
Structural		1	2.04373	0.58785	12.0869	**0.0005**
Visual		1	2.52575	0.49329	26.2163	**<.0001**
Semantic		1	1.22450	0.27062	20.4743	**<.0001**
Structural * Gender	Female	1	1.36616	0.58126	5.5241	**0.0188**
Structural * Gender	Male	0	0	.	.	.
Visual * Gender	Female	1	0.19949	0.49171	0.1646	0.6850
Visual * Gender	Male	0	0	.	.	.
Semantic * Gender	Female	1	-0.00258	0.28136	0.0001	0.9927
Semantic * Gender	Male	0	0	.	.	.
Structural * Familiarity	Yes	1	1.18518	0.58033	4.1708	**0.0411**
Structural * Familiarity	No	0	0	.	.	.
Visual * Familiarity	Yes	1	0.66235	0.48544	1.8617	0.1724
Visual * Familiarity	No	0	0	.	.	.
Semantic * Familiarity	Yes	1	0.93457	0.27522	11.5309	**0.0007**
Semantic * Familiarity	No	0	0	.	.	.

In addition, one can see that each coefficient of Table 8 is statistically signif-icant with the same confidence interval as the salience scores (i.e. 99 %). Specifi-cally, the parameters tell us that the participants were likely to choose the venue located along the same direction as the turn and before the intersection; regard-less of the turn. More precisely, the probability of choosing the place located before the intersection and along the same direction as the turn is 0.513 for the left turn and 0.579 for the right one. Also, the probability of selecting the place located after the intersection and still along the same direction as the turn is 0.304 for the left turn and 0.171 for the right one.

Table 8. Analysis of conditional maximum likelihood estimates: *locations and turns.*

Parameter	Code	Turn	DF	Coeff.	Std Error	Wald ChiSq	Pr>ChiSq
Location	SB		1	2.1952	0.1014	468.2858	<.0001
Location	OB		1	0.7817	0.1162	45.2750	<.0001
Location	SA		1	1.6740	0.1049	254.8525	<.0001
Location	OA		0	0	.	.	.
Location * Turn	SB	Right	1	-0.4636	0.1279	13.1395	0.0003
Location * Turn	SB	Left	0	0	.	.	.
Location * Turn	OB	Right	1	-0.4328	0.1493	8.4040	0.0037
Location * Turn	OB	Left	0	0	.	.	.
Location * Turn	SA	Right	1	-1.1580	0.1387	69.7452	<.0001
Location * Turn	SA	Left	0	0	.	.	.
Location * Turn	OA	Right	0	0	.	.	.
Location * Turn	OA	Left	0	0	.	.	.

To recap, each participant, regardless of the gender and the familiarity with the area, was influenced by the visual, semantic, and structural saliencies. Unlike men, women tended to select landmarks located along the same direction as the turn. We observe the same tendency for the local participants who additionally focused on semantic landmarks.

Model Validity. The 5:1 rule of thumb consists in having at least 5 observa-tions per explanatory variable; and ideally 10. Although this rule is sometimes called into question, following it implies to constitute an ideal sample size of 20 individuals and 30 landmark candidates. Since we worked with three times more individuals (more precisely 63) and four times more landmark candidates (120) than recommended, we estimate that the model and the parameters estimated are sufficiently reliable to draw strong conclusions.

3.6   Discussion

The findings around the semantic salience confirm our expectations. Except the fact that this salience was more significant than the structural one in the experiment (20.47 *vs* 12.08 Wald Chi-Square), one should notice a significant interaction with locals (p = 0.0007). In other words, locals favored significant semantic landmarks. This finding completes the recent research of Hamburger and Röser (2014). It *empirically* demonstrates that the semantic salience is clearly involved in the process of landmark selection. Just as the theory suggests (cf. Raubal and Winter 2002 and Caduff and Timpf 2008), this salience is also closely linked to the individual's knowledge of the environment. Given the distribution of the landmarks (cf. Table 6), the statistics computed from the model reveal that locals focused on semantic landmarks regardless of a low visual salience. Meanwhile, the strangers rather relied on highly visible landmarks. On the other hand, one must keep in mind that we included a higher amount of local landmarks for the purpose of this experiment (cf. Tables 4 and 5). The strength of this interaction would have been lower if we had incorporated more global landmarks. Consequently, we argue that the semantic salience should be computed according to a "*localness threshold*". Beyond that limit, the significance of the semantic salience for both locals and strangers might converge.

Unsurprisingly, the visual salience is the most significant score in the computed model. Indeed, wayfinding remains a process closely linked to the *visual attention*. Obviously, highly visible places easily grab travelers' attention (Lynch 1960 and Davies and Peebles 2010). On the contrary, the gender difference regarding the structural salience score was unexpected. This finding has not been discovered in previous similar studies (cf. Klippel and Winter 2005 and Röser et al. 2012a; 2012b). However, it does support the *global* gender difference in spatial ability already highlighted in several research (see Farr et al. 2012 for an overview). In addition, the fact that locals were likely to systematically select the candidates located along the same direction as the turn can be easily explained. In our opinion, they chose to *reduce* their visual attention to the side of the turn. Thanks to their meaningful environment knowledge, locals were able to select two different landmarks per intersection. At the opposite, strangers rather *extended* their visual attention to scan each candidate and pick-up the best one for both turns when necessary. In this context, the position of the landmark seems to be less important for strangers. Nevertheless, we acknowledge that the participants had a high degree of freedom during their virtual travel. Therefore, this assumption might be valid for *pedestrians* but not for car drivers since they cannot look away from the road more than few seconds.

Furthermore, one can see that the positions of the landmark candidates favored during the experiment slightly differ from those of the survey. In the experiment, participants were likely to choose the candidates located *before* the intersection for both turns. This observation is in line with the conclusions of Röser et al. (2012a; 2012b) and Klippel and Winter (2005). However, it is not really surprising since the most significant landmarks (visual and semantic) were located before the intersection (cf. Fig. 5). Having said that, it remains important to notice that the probability of selecting a landmark candidate located along the same direction as the turn and *after* the intersection is quite *higher* for the *left* turn (almost twice in comparison with the right turn). Once again, this evidence suggests that the side where participants moved compared to the road interfered in their choices. Participants favored position A for the left turn whereas it hosted one-third of the lowest visible landmarks and had an average visual salience score lower than position C (cf. Fig. 5).

As stated in the previous section, each salience score is statistically significant. Thus, the methods used for the computation of the salience scores were relevant. This observation supports the authors' proposition for the detection of local and global semantic landmarks using geosocial data (cf. Quesnot and Roche 2015a). Also, using the visual contrast between the scene and the places seems to be relevant for computing the visual salience. Since the visual and semantic salience scores were related to the advance visibility, the results also confirm its importance in the context of navigation. Having said that, one concern regarding the experiment needs to be highlighted. Indeed, participants selected landmarks by relying on both 2D (Google Maps base map) and 3D (Google Street View) media at the same time. Yet, individuals tend to adopt different route discourses according to the dimensionality of the supports (see Mast et al. 2010). Despite its optional display, the base map might have affected the participants' landmark selection.

4 Conclusion

The main purpose of this research was to evaluate the significance of semantic landmarks according to the observer's knowledge of the environment. The results of the experiment showed that the participants were all influenced by the visual, structural, and semantic saliencies. In addition, locals clearly favored local semantic landmarks. This empirical finding highlights the significance of the semantic salience in the landmark selection process. It also confirms that a semantic landmark should be included inside wayfinding instructions according to the traveler's spatial knowledge. Unless the place is famous, giving to strangers instructions based on semantic landmarks does not make sense. Surprisingly, the additional findings indicate that women were more influenced by the structural salience than men. In our opinion, this specific gender difference needs to be further investigated. In the context of automatic landmark detection, one should take those results into account for giving relevant wayfinding instructions and saving data consumption at the same time. In this way, geolocated data harvested from social media - especially *explicit platial data* (cf. Quesnot and Roche 2015b) - should be retrieved for computing meaningful *semantic* salience scores. At this stage, one of the major concerns is to pick-up either local or global semantic landmarks according to the traveler's familiarity with the area. Regarding the wayfinding instructions, it appears that the priority should be given to the visual salience for strangers; unless a well-known landmark is located at the decision point. On the other hand, the semantic salience remains more appropriate for locals. Structural salience scores should be applied if there are neither visual nor semantic outstanding landmarks. However, we believe that a *qualitative* characterization of the structural salience (e.g. "before the intersection") remains more relevant than a quantitative approach. Nonetheless, this characterization should absolutely take the side where travelers move compared to the road into account. Indeed, the results also suggest that this factor impacted on participants' selection of landmarks. In the end, it would be interesting to carry out similar studies for different types of intersection since this research only focused on cross intersections.

Acknowledgments. This research is funded by the *Social Sciences and Humanities Research Council of Canada* (SSHRC) through the second author's ordinary grant, and the *Geothink.ca* project. We would like to thank *Hélène Crépeau* from the *Mathematics and Statistics Department of Laval University* for her help in the computation of the logit model in SAS. We also sincerely thank the anonymous reviewers for their insightful comments. They helped us to significantly improve this paper. We finally thank the *Ministère de l'Energie et des Ressources Naturelles of the Quebec Province*, and the *Surveying and Cartography Department of Quebec City* for the geographic data used for the 3D analyses.

References

Caduff, D., Timpf, S.: On the assessment of landmark salience for human navigation. Cogn. Process. **9**(4), 249–267 (2008)

Cameron, A.C., Trivedi, P.K.: Microeconometrics: Methods and Applications. Cambridge University Press, New York (2005)

Crandall, D.J., Backstrom, L., Huttenlocher, D., Kleinberg, J.: Mapping the world's photos. In: Proceedings of the WWW 2009 Conference (2009)

Daniel, M.-P., Tom, A., Manghi, E., Denis, M.: Testing the value of route directions through navigational performance. Spat. Cogn. Comput. **3**, 269–289 (2003)

Davies, C., Peebles, D.: Spaces or scenes: map-based orientation in urban environments. Spat. Cogn. Comput. **10**(2–3), 135–156 (2010)

Duckham, M., Winter, S., Robinson, M.: Including landmarks in routing instructions. J. Location Based Serv. **4**(1), 28–52 (2010)

Elias, B.: Extracting landmarks with data mining methods. In: Kuhn, W., Worboys, M.F., Timpf, S. (eds.) COSIT 2003. LNCS, vol. 2825, pp. 375–389. Springer, Heidelberg (2003)

Elwood, S., Goodchild, M.F., Sui, D.S.: Prospects for VGI research and the emerging fourth paradigm. In: Sui, D.S., Elwood, S., Goodchild, M.F. (eds.) Crowdsourcing Geographic Knowledge, pp. 361–375. Springer, Netherlands (2013)

Farr, A.C., Kleinschmidt, T., Yarlagadda, P., Mengersen, K.: Wayfinding: a simple concept, a complex process. Transport Reviews **32**(6), 715–743 (2012)

Goodchild, M.F.: Citizens as sensors: the world of volunteered geography. GeoJournal **69**(4), 211–221 (2007)

Hamburger, K., Röser, F.: The role of landmark modality and familiarity in human wayfinding. Swiss J. Psychol. **73**(4), 205–213 (2014)

Klippel, A., Winter, S.: Structural salience of landmarks for route directions. In: Cohn, A.G., Mark, D.M. (eds.) COSIT 2005. LNCS, vol. 3693, pp. 347–362. Springer, Heidelberg (2005)

Lovelace, K.L., Hegarty, M., Montello, D.R.: Elements of good route directions in familiar and unfamiliar environments. In: Freksa, C., Mark, D.M. (eds.) COSIT 1999. LNCS, vol. 1661, pp. 65–82. Springer, Heidelberg (1999)

Lynch, K.: The Image of the City. MIT Press, Cambridge (1960)

Mast, V., Smeddinck, J., Strotseva, A., Tenbrink, T.: The impact of dimensionality on natural language route directions in unconstrained dialogue. In: Proceedings of the SIGDIAL 2010, pp. 99–102. Association for Computational Linguistics (2010)

Michon, P.E., Denis, M.: When and why are visual landmarks used in giving directions? In: Montello, D.R. (ed.) COSIT 2001. LNCS, vol. 2205, pp. 292–305. Springer, Heidelberg (2001)

Nothegger, C., Winter, S., Raubal, M.: Selection of salient features for route directions. Spat. Cogn. Comput. 4(2), 113–136 (2004)

Quesnot, T., Roche, S.: Measure of landmark semantic salience through geosocial data streams. ISPRS Int. J. Geo-Inf. 4(1), 1–31 (2015a)

Quesnot, T., Roche, S.: Platial or locational data? toward the characterization of social location sharing. In: Proceedings of the 48th Annual Hawaii International Conference on System Sciences, pp. 1973–1982. IEEE Computer Society Press, Washington, DC (2015b)

Raubal, M., Winter, S.: Enriching wayfinding instructions with local landmarks. In: Egenhofer, M., Mark, D.M. (eds.) GIScience 2002. LNCS, vol. 2478, pp. 243–259. Springer, Heidelberg (2002)

Richter, K.F., Winter, S.: Harvesting user-generated content for semantic spatial information: the case of landmarks in OpenStreetMap. In: Proceedings of the Surveying and Spatial Sciences Biennial Conference 2011, Wellington, NZ (2011)

Röser, F., Hamburger, K., Krumnack, A., Knauff, M.: The structural salience of landmarks: results from an on-line study and a virtual environment experiment. J. Spat. Sci. 57(1), 37–50 (2012a)

Röser, F., Krumnack, A., Hamburger, K., Knauff, M.: A four factor model of landmark salience - a new approach. In: Rußwinkel, N., Drewitz, U., Van Rijn, H., (eds.) Proceedings of the 11th International Conference on Cognitive Modeling (ICCM), Berlin, pp. 82–87 (2012b)

Siegel, A.W., White, S.H.: The development of spatial representations of large scale environments. In: Reese, W.H. (ed.) Advances in Child Development and Behavior, vol. 10. Academic Press, New York (1975)

Sorrows, M.E., Hirtle, S.C.: The nature of landmarks for real and electronic spaces. In: Freksa, C., Mark, D.M. (eds.) COSIT 1999. LNCS, vol. 1661, pp. 37–50. Springer, Heidelberg (1999)

Tezuka, T., Tanaka, K.: Landmark extraction: a web mining approach. In: Cohn, A.G., Mark, D.M. (eds.) COSIT 2005. LNCS, vol. 3693, pp. 379–396. Springer, Heidelberg (2005)

Tom, A., Denis, M.: Referring to landmark or street information in route directions: what difference does it make? In: Kuhn, W., Worboys, M.F., Timpf, S. (eds.) COSIT 2003. LNCS, vol. 2825, pp. 362–374. Springer, Heidelberg (2003)

Winter, S.: Route adaptive selection of salient features. In: Kuhn, W., Worboys, M.F., Timpf, S. (eds.) COSIT 2003. LNCS, vol. 2825, pp. 349–361. Springer, Heidelberg (2003)

Winter, S., Raubal, M., Nothegger, C.: Focalizing measures of salience for wayfinding. In: Meng, L., Reichenbacher, T., Zipf, A. (eds.) Map-based Mobile Services, pp. 125–139. Springer, Heidelberg (2005)

Author Index

Printed in the United States
By Bookmasters